Recent Tools for Computer- and Mobile-Assisted Foreign Language Learning

Alberto Andujar
University of Almeria, Spain

A volume in the Advances in Educational Technologies and Instructional Design (AETID) Book Series

IGI Global
DISSEMINATOR OF KNOWLEDGE

Published in the United States of America by
IGI Global
Information Science Reference (an imprint of IGI Global)
701 E. Chocolate Avenue
Hershey PA, USA 17033
Tel: 717-533-8845
Fax: 717-533-8661
E-mail: cust@igi-global.com
Web site: http://www.igi-global.com

Library of Congress Cataloging-in-Publication Data

Names: Andujar, Alberto, 1989- editor.
Title: Recent tools for computer- and mobile-assisted foreign language learning / Alberto Andujar, Editor.
Description: Hershey, PA : Information Science Reference, [2020] | Includes bibliographical references and index.
Identifiers: LCCN 2019024178 (print) | LCCN 2019024179 (ebook) | ISBN 9781799810995 (eISBN) | ISBN 9781799810971 (h/c) | ISBN 9781799810988 (s/c) | ISBN 9781799810971¬(h/c) | ISBN 9781799810988¬(s/c) | ISBN 9781799810995¬(eISBN)
Subjects: LCSH: Mobile communication systems in education. | Language and languages--Computer-assisted instruction. | Language and languages--Study and teaching--Technological innovations. | Educational technology.
Classification: LCC LB1044.84 (ebook) | LCC LB1044.84 .R43 2020 (print) | DDC 418.0078--dc23
LC record available at https://lccn.loc.gov/2019024178

This book is published in the IGI Global book series Advances in Educational Technologies and Instructional Design (AE-TID) (ISSN: 2326-8905; eISSN: 2326-8913)

British Cataloguing in Publication Data
A Cataloguing in Publication record for this book is available from the British Library.

Advances in Educational Technologies and Instructional Design (AETID) Book Series

Lawrence A. Tomei
Robert Morris University, USA

ISSN:2326-8905
EISSN:2326-8913

Mission

Education has undergone, and continues to undergo, immense changes in the way it is enacted and distributed to both child and adult learners. In modern education, the traditional classroom learning experience has evolved to include technological resources and to provide online classroom opportunities to students of all ages regardless of their geographical locations. From distance education, Massive-Open-Online-Courses (MOOCs), and electronic tablets in the classroom, technology is now an integral part of learning and is also affecting the way educators communicate information to students.

The **Advances in Educational Technologies & Instructional Design (AETID) Book Series** explores new research and theories for facilitating learning and improving educational performance utilizing technological processes and resources. The series examines technologies that can be integrated into K-12 classrooms to improve skills and learning abilities in all subjects including STEM education and language learning. Additionally, it studies the emergence of fully online classrooms for young and adult learners alike, and the communication and accountability challenges that can arise. Trending topics that are covered include adaptive learning, game-based learning, virtual school environments, and social media effects. School administrators, educators, academicians, researchers, and students will find this series to be an excellent resource for the effective design and implementation of learning technologies in their classes.

Coverage

- Web 2.0 and Education
- Instructional Design Models
- Adaptive Learning
- Virtual School Environments
- Bring-Your-Own-Device
- Digital Divide in Education
- E-Learning
- K-12 Educational Technologies
- Collaboration Tools
- Hybrid Learning

The Advances in Educational Technologies and Instructional Design (AETID) Book Series (ISSN 2326-8905) is published by IGI Global, 701 E. Chocolate Avenue, Hershey, PA 17033-1240, USA, www.igi-global.com. This series is composed of titles available for purchase individually; each title is edited to be contextually exclusive from any other title within the series. For pricing and ordering information please visit http://www.igi-global.com/book-series/advances-educational-technologies-instructional-design/73678. Postmaster: Send all address changes to above address.

Titles in this Series

For a list of additional titles in this series, please visit:
https://www.igi-global.com/book-series/advances-educational-technologies-instructional-design/73678

The Roles of Technology and Globalization in Educational Transformation
Blessing F. Adeoye (Walden University, USA) and Gladys Arome (Marian University, USA)
Information Science Reference • © 2020 • 259pp • H/C (ISBN: 9781522597469) • US $195.00

Utilizing Educational Data Mining Techniques for Improved Learning Emerging Research and Opportunities
Chintan Bhatt (Charotar University of Science and Technology, India) Priti Srinivas Sajja (Sardar Patel University, India) and Sidath Liyanage (University of Kelaniya, Sri Lanka)
Information Science Reference • © 2020 • 166pp • H/C (ISBN: 9781799800101) • US $165.00

Claiming Identity Through Redefined Teaching in Construction Programs
Sherif Mostafa (Griffith University, Australia) and Payam Rahnamayiezekavat (Western Sydney University, Australia)
Information Science Reference • © 2020 • 259pp • H/C (ISBN: 9781522584520) • US $185.00

Global Perspectives on Teaching and Learning Paths in Islamic Education
Miftachul Huda (Universiti Pendidikan Sultan Idris Malaysia, Malaysia) Jimaain Safar (Universiti Teknologi Malaysia, Malaysia) Ahmad Kilani Mohamed (Universiti Teknologi Malaysia, Malaysia) Kamarul Azmi Jasmi (Universiti Teknologi Malaysia, Malaysia) and Bushrah Basiron (Universiti Teknologi Malaysia, Malaysia)
Information Science Reference • © 2020 • 341pp • H/C (ISBN: 9781522585282) • US $195.00

Social Justice and Putting Theory Into Practice in Schools and Communities
Susan Trostle Brand (University of Rhode Island, USA) and Lori E. Ciccomascolo (University of Rhode Island, USA)
Information Science Reference • © 2020 • 359pp • H/C (ISBN: 9781522594345) • US $195.00

Handbook of Research on Diverse Teaching Strategies for the Technology-Rich Classroom
Lawrence A. Tomei (Robert Morris University, USA) and David D. Carbonara (Duquesne University, USA)
Information Science Reference • © 2020 • 426pp • H/C (ISBN: 9781799802389) • US $265.00

Emerging Technologies in Virtual Learning Environments
Kim Becnel (Appalachian State University, USA)
Information Science Reference • © 2019 • 348pp • H/C (ISBN: 9781522579878) • US $205.00

English Language Teaching in a Post-Method Paradigm
Paulette Joyce Feraria (The University of the West Indies, Jamaica)
Information Science Reference • © 2019 • 426pp • H/C (ISBN: 9781522592280) • US $175.00

IGI Global
DISSEMINATOR OF KNOWLEDGE

701 East Chocolate Avenue, Hershey, PA 17033, USA
Tel: 717-533-8845 x100 • Fax: 717-533-8661
E-Mail: cust@igi-global.com • www.igi-global.com

Editorial Advisory Board

List of Reviewers

Jennifer Dobberfuhl Quinlan, *Brigham Young University, USA*
Mathea Simons, *University of Antwerp, Belgium*
Julie Van de Vyver, *Université catholique de Louvain, Belgium*
Claudio Vanhees, *University of Antwerp, Belgium*
Jane Vinther, *University of Southern Denmark, Denmark*
Julia Weißenböck, *University of Salzburg, Austria*

Table of Contents

Detailed Table of Contents

This chapter explores students' engagement in a telecollaboration project between a Spanish and an American university. Students' cognitive, emotional, and behavioral engagement were evaluated throughout the project. A total of 53 students participated in an online exchange during one and a half months through two different applications, WhatsApp representing the text-based environment and Jitsi representing the synchronous videoconferencing platform. The engagement construct was explored using pre and post measures as well as tracking students' conversation in the platforms. Results yielded high levels of cognitive engagement as a result of the interaction. Values for emotional engagement were found to be higher in the instant messaging platform and behavioral engagement did not present significant values. Implications and recommendations for future research were drawn.

Reading from digital screens has become increasingly common practice in educational and recreational reading. The response to this digital shift has been twofold. Some suspect it will harm children's ability to perform deep reading; others highlight its potential to support reading among different groups. Digital reading tools, such as fiction with multimedia hyperlinks, could engage particularly reluctant readers or children from low-literate families. This chapter presents the results of an experimental, mixed-method study that identifies hyperlink type and frequency desirability in literary texts. A comparative analysis of respondent perspectives revealed that teachers mark on average more explanatory and enriching hyperlinks than pupils. Pupil and teacher hyperlink type desirability are significantly influenced by respectively literary genre and reading motivation, and importance of pupil reading motivation and media use. Pupil and teacher explanatory hyperlink frequency are significantly influenced by respectively literary genre, and importance of pupil reading motivation.

This chapter gives a brief overview on the use of social network sites (SNSs) for language learning and presents an empirical study of the use of Instagram (IG), one of the most popular SNSs, to assess learners' oral communication skills in the foreign language classroom. Several studies have mainly revealed the perceptions and preferences of learners in regards to using SNSs for learning grammar, vocabulary, L2 writing, reading, and speaking. In the current study, the performance scores of participants on an oral communication speaking task delivered both on IG and in class, as well as their scores on the Big Five personality traits as measured by the Quick Big Five Personality Test (QBFPT), were examined

trip to Scotland, teachers and two classes of 17-year-old students from a school in Salzburg, Austria, decided to present the outcome of their project work from their stay abroad in a more interactive and innovative way. Instead of inviting parents and friends to an evening of PowerPoint, the students designed AR posters and visitors were invited to an interactive gallery walk during which they had the chance to explore thematic posters related to Scotland. While at first glance the posters seem rather bland, digital content, which the students produced, opens up by scanning the pictures on the poster with a special AR app. This project successfully combines language and digital skills and has shown to motivate students to engage even deeper with their topic.

In this chapter, the author discusses the role of Massive Open Online Courses (MOOCs) in the field of foreign language education. It examines the relationship between MOOCs and self-regulated learning, and goes on to explore the repurposing of content-based MOOCs as complementary materials to foster students' self-regulated behaviour and practice of the target language inside and out the language classroom. A research project that has explored self-regulated language learning in a blended context with content-based MOOCs is also discussed here, with particular reference to the self-regulatory strategies employed by a group of language students at University level. A series of recommendations that supports the use of content-based MOOCs as part of face-to-face language courses are provided at the end of this chapter.

This chapter proposes to establish a starting point in the design of technology specialized in the development of listening comprehension skills from a theoretical perspective, when learning English as a second language. Therefore, an exploration about how technologies have evolved in the practice of learning English as a second language was required. Likewise, authors look at fundamental aspects of technology literacy and how this is rooted in users' contexts. From this, they conceptualize the symbolic competence through the ecological theory in order to design a computer-assisted language learning practice. Findings led to the conclusion that a guide on how to build specialized technology in English learning as a second language does not exist. What indeed exists is the application of recycled technology created for other purposes but used for English learning. Authors propose a practice where the meaning is explored through the understanding of what happens in the context by using 3D holograms as an optical illusion.

With more than 5 million college students taking at least one online or blended college course, (Seaman et al, 2018) and 5 percent of K-12 students taking an online or blended course (The Evergreen Education Group), research regarding student success in blended and online classes has begun to emerge and shed light on potential trouble spots. This chapter uses an action research model to further examine binge studying in two, beginning university Korean classes in relation to success on exams, overall course grades, language recall after a four-week break, and continued enrollment in Korean. Findings indicate a correlation between overall page views and course grade, between overall page views in 101 to retention in 102 and a correlation was also evident between binge studying and 102 assessment. A negative correlation was seen between binge studying and 101 final exam scores. The strongest correlation appeared to be between non-binge studiers, or those which dispersed their study throughout the semester, and language retention in Korean 102.

Gamification has emerged in recent years as a resource that incorporates game-related elements and mechanics into the classroom to foster students' motivation, engagement, and further competences. With the proliferation of English-medium instruction (EMI) and multilingual degree programs at university nowadays, new and innovative teaching tools are desirable to help students cope with the double-challenging task of acquiring new and complex disciplinary content through a foreign language. This chapter provides an overview of some recent computer-based gamification tools that may be applied in EMI and multilingual university settings. More specifically, the main features and advantages of these tools for these specific educational contexts are explored. This chapter may be useful for researchers and practitioners in the field of EMI and multilingual teaching in higher education.

The contribution of this chapter to the knowledge of motivation in relation to the learning of a foreign language lies in the extensive examination of student attitudes to a detailed field of involved factors. This chapter gives an account of affective factors in the digital classroom on the basis of the level of the digital literacy of the participating students. The digital learning of an L2 is correlated to the level of language learning anxiety, willingness to communicate in the L2, attitude to feedback, and preference for teaching methods. The results show that the level of digital literacy is so high that it should not in itself negatively impact attitude to other factors. The central finding of the investigation is that intrinsic motivation is the strongest motivational drive when compared to other motivational factors.

Preface

The constant growth of a globalized education model and the impact of technological tools on learning processes have involved changes at all educational levels. In this context, language learning processes have been no exception and during the last 30 years, disciplines such as Computer-Assisted Language Learning (CALL) or Mobile-Assisted Language Learning (MALL) have experienced an exponential growth as regards research and practice. The possibilities offered by technological devices to widen the opportunities for language learning and practice inside and outside the classroom environment have led to explore which tools are particularly beneficial for language development. Likewise, researchers have also investigated how an appropriate use of these tools may be conducive to the development of L2 skills such as reading, writing, listening or speaking among others.

This publication explores the current situation in the two aforementioned areas of study: CALL and MALL. The first area is understood as "the search for and study of the computer in language teaching and learning" (Levy, 1997). The second area of investigation explores mobile devices and applications as well as the ubiquitous characteristics of these devices to develop language learning processes. These two main areas encompass other specific areas such as: telecollaboration; mobile instant-messaging; text-to-speech systems; MOOCs; OER; social networks; flipped learning; apps; augmented reality; or gamification among others that will be also explored in this publication. Thus, the aim of the book is to explore recent developments in each of the areas under investigation as well as provide an insight into the technological tools in use for language learning processes. Likewise, this publication will serve as guideline for language teachers, higher institutions and researchers willing to explore and implement technological processes in the foreign language classroom. In this sense, experts in each of the areas addressed in this book will provide an insight into the current developments in their respective fields of study, as well as new improvements and potential language benefits observed. Furthermore, new research lines and technological tools to foster language development will be explored and an evaluation of the existing ones will be carried out.

The target audience of this book encompasses language teachers and practitioners, researchers in each of the fields under investigation, educational institutions and textbook publishing companies. The use of the information contained in this book may contribute to further understand and put into practice technological tools in second language teaching and learning. Furthermore, this publication will contribute to shaping the way technological tools are implemented within a classroom environment, and will help explore their potential in terms of language learning benefits. It will also serve as a guideline for language teachers and researchers on the aspects that need to be taken into consideration when using these technological devices. In addition, from a second language acquisition point of view, this publication will gather the main research in each of the fields under investigation, contributing to expanding language teaching skills in the foreign language classroom.

INSIDE THIS ISSUE

In this regard, the first chapter of this book offers insights into the benefits and drawbacks of adopting mobile learning in language education, both from a theoretical as well as a practical point of view. A survey was distributed to learners and teachers to explore pedagogically sound practices and provide a better understanding of the current and future role of MALL on language learning, specifically in higher education.

The second chapter offers a comprehensive overview of MALL studies, from 1994 through 2018, relating to the experimental implementation of mobile-based affordances and applications in the teaching and learning of languages, L1 as well as L2. It derives from the analysis of over 2000 MALL studies that have appeared in almost 500 distinct journals and conference presentations to nearly 300 different professional organizations.

The third chapter explores students' engagement in a telecollaboration project between a Spanish and an American university. Students' cognitive, emotional and behavioral engagement were evaluated throughout the project. A total of 53 students participated in an online exchange during one month and half through two different applications, WhatsApp representing the text-based environment and Jitsi representing the synchronous videoconferencing platform. The engagement construct was explored using pre and post measures as well as tracking students' conversation in the platforms.

The fourth chapter presents the results of an experimental mixed-method study on hyperlink desirability in literary texts. A comparative analysis of respondent perspectives revealed differences in terms of the type and frequency of explanatory and enriching hyperlinks that grade 7 pupils and their language teachers desired in five distinct literary genres, and explored which underlying characteristics significantly influenced hyperlink desirability.

The fifth chapter presents a detailed overview of the use of social network sites (SNSs) for language learning as well as an empirical study of the use of Instagram (IG) in the classroom. The study aims to demonstrate a solid classroom practice and its outcomes for researchers and practitioners. It also seeks to assess L2 learners' oral communication skills in a language classroom setting and on IG, and to explore how much personality traits can predict oral performance in the class and on IG.

The sixth chapter explores the use of an objective fluency measure, the CAF calculator. This tool and a workflow to compute fluency measures are introduced along with a study investigating utterance fluency of L2 learners of Japanese at two proficiency levels performing two tasks. The study found significant differences in speed, breakdown, and composite fluency measures between the two groups in both tasks. It also found that task type affects pause locations.

The seventh chapter describes the potential of text-to-speech synthesis (TTS) as a tool that can transform CALL platforms. Illustrating this point, a specific Irish language platform, *An Scéalaí* ('the Storyteller'), is presented. By incorporating TTS, this platform facilitates the training of literacy skills, writing and reading, with an emphasis at all times on the spoken language. The chapter has implications for language teaching in a broad sense, regardless of what language is involved.

The eighth chapter presents a project and study conducted with Austrian students using augmented reality (AR). Following the principles of the design-based learning approach, where students create digital content instead of merely consuming it, students had to create AR posters about a school trip to Scotland. An accompanying survey also provided an insight into the motivation and technology acceptance of the students.

The ninth chapter discusses the affordances of Massive Open Online Courses (MOOCs) and the use of content-based MOOCs in the language classroom to foster students' self-regulated learning and the practice of the target language, and goes on to provide some recommendations based on a case study that examined self-regulatory strategies of university students in an academic English course using content-based MOOCs.

The tenth chapter proposes to establish a starting point in the design of technology specialized in the development of listening comprehension skills from a theoretical perspective, when learning English as a second language. At the end of the contribution, authors propose a practice where the meaning is explored through the understanding of what happens in the context by using 3D holograms.

The eleventh chapter explores results of a study that examined binge studying in blended Korean 101 and 102 classes in relation to success on exams, overall course grades, language recall after a four-week break, and continued enrollment in Korean. Findings indicate correlation between overall page views and course grade as well as between overall page views in 101 to retention in 102. The strongest correlation appeared to be between non-binge studiers in 101 and language skills retention in Korean 102, thus suggesting a longer-term impact from not binge studying.

The twelfth chapter provides an overview of some recent computer-based gamification tools that may be applied in EMI and multilingual university settings. More specifically, the main features and advantages of these tools for these specific educational contexts will be explored. The contents of this chapter may be potentially useful for researchers and practitioners in the field of EMI and multilingual teaching in higher education.

The thirteenth explores the concept of motivation for language learning enhanced by digital/mobile means. The relevance or success of a given digital activity, be it self-directed or instructed, depends on the motivation of the learner to pursue the learning goal with energy and enthusiasm. The gist of the results of our empirical study is that learners perceive digital classrooms to be supportive of individual as well as group activities and that a new-found heightened language learning motivation encourages student creativity and willingness to communicate in the L2.

The fourteenth chapter discusses the importance of in-service teacher training (INSET) to promote the use of open natural language processing (NLP)-based technologies (NLPTs). This study investigates trainee feedback from the French-speaking version of the TELL-OP online training module using a mixed-methods approach and SWOT analysis. Finally, a perspective of the course design is provided, along with suggestions for improvement.

The fifteenth chapter explores, from a theoretical point of view, the literature investigating the use of WhatsApp and Instagram applications to foster language learning as well as the outcomes of the studies carried out in these two fields. Moreover, the possibilities of these two apps to develop blended learning environments are also regarded. Likewise, the theoretical frameworks in the existing literature for the possible implementation of these social networking tools in classroom contexts are considered.

CONCLUSION

The chapters presented in this book shed light on the possible implementation of some of the latest technological tools in foreign language learning contexts. In this sense, the possibilities offered by the tools presented in this book may need to be taken into consideration by language teachers and institutions willing to keep pace with the latest educational practices.

In certain countries, due to the absence of immersive language contexts, it becomes very difficult for students to develop L2 skills. Moreover, the practice time is often limited to in-class time which makes it even more difficult to receive a sufficient amount of input in the target language. This situation has led to the exploration of technological tools that can help foster students' development inside and outside the classroom. Likewise, the implementation of these tools has also contributed to the development of more autonomous and independent learners. Further factors such as the ubiquitous nature of some of the tools presented in this book have also contributed to break with the constraints of traditional classroom environments in which the learning process takes place at specific times and days.

Notwithstanding that the perceived benefits of these technological tools can contribute to improving language tuition and practice, their integration is still limited at an educational context where language policies and curriculums fail to incorporate the necessary elements for a successful implementation of these learning processes. Similarly, it also seems necessary to further understand the role of these technological elements within the classroom environment and which practices are more appropriate to foster successful language learning processes. Thus, this book does not exclusively focus on using certain tools to develop a foreign language, but on how to appropriately make use of these tools for language purposes and further understand the possibilities offered by each of them.

REFERENCES

Levy, M. (1997). *CALL: Context and conceptualisation.* Oxford, UK: Oxford University Press.

Acknowledgment

I would like to thank professor Maria Soledad Cruz Martínez and Professor Mary Spratt for the comments and help provided during the editing of this book. Likewise, I would like to thank all the reviewers who have kindly assisted and contributed to improving the quality of the chapters in this book with their insightful comments and expertise.

Chapter 1
Understanding Learner and Instructor Attitudes Toward and Use of Mobile-Assisted Language Learning

Ana Gimeno-Sanz
https://orcid.org/0000-0003-3366-0729
Universitat Politècnica de València, Spain

Valentina Morgana
https://orcid.org/0000-0001-6086-3228
Università Cattolica del Sacro Cuore, Italy

Julie Van de Vyver
https://orcid.org/0000-0001-8820-8380
Université catholique de Louvain, Belgium

ABSTRACT

This chapter offers insights into the benefits and drawbacks of adopting mobile learning in language education, both from a theoretical as well as a practical point of view. A survey was designed to explore pedagogically sound practices and provide a better understanding of the current and future role of MALL on language learning, specifically in higher education (HE). Comparison between the two target groups (learners and teachers) produced results to aid in aligning and narrowing distances between the learners' independent usage of MALL in informal learning and the instructors' perception of how mobile apps can or should be integrated into the language curriculum. The chapter proposes a number of research areas that require further exploration in MALL and with a set of recommendations in terms of embracing MALL practices in language learning and teaching.

DOI: 10.4018/978-1-7998-1097-1.ch001

INTRODUCTION

This chapter describes a multinational research project conducted by members of EuroCALL's[1] Mobile-Assisted Language Learning (MALL) Special Interest Group. Mobile learning involves the use of mobile technology, either alone or in combination with other information and communications technologies (ICT), to enable learning anytime and anywhere (UNESCO, 2012, p. 6). Mobile devices permeate our daily lives, providing unparalleled access to communication and information and, as the power, functionality and affordability of these devices increase, so does their potential to support learning in new ways and settings (UNESCO, 2013, p. 7). For the purpose of this study, mobile devices refer to smartphones or tablet computers that allow the use of mobile applications. Laptop computers are generally seen "as portable more than mobile" (Pegrum, 2014, p.54) therefore, in the context of this study, they are always mentioned in comparison with smartphones or tablets. The research study was thus designed to shed light on how mobile applications (apps) can have an impact on the way language learning is currently being approached by students and instructors. The multinational project targeted participants –both instructors and learners– based in Asia, North America and Europe and the data was collected by means of a large-scale online survey which investigated their practices and attitudes toward the use of MALL as a language learning tool in today's educational context, in an attempt to determine future trends. Data was analysed both qualitatively, based on Glaser & Strauss' Grounded Theory (1967) to conceptualize the data, and quantitatively through descriptive statistics. Comparison between the learners and teachers' data sets allowed us to draw a global picture of MALL usage in informal learning and attitudes toward the integration of mobile apps into formal learning. Consideration is also given in this chapter to the language skills and linguistic competences that learners and teachers perceive as the most appropriate and useful in MALL. The chapter finally suggests some MALL research areas to explore and provides a number of recommendations in terms of MALL practices.

BACKGROUND

Mobile-Assisted Language Learning in Higher Education

Mobile-Assisted Language Learning projects have been implemented globally at different education levels, from primary schools to higher education (HE) contexts (Bebell & O'Dwyer, 2010; Fleischer, 2012; Zucker & Light, 2009), although "statistically reliable measures of learning outcomes are few and far between" (Burston, 2015, p. 16). What has driven most researchers more recently is the idea that mobile learning involves not only the mobility of devices, but also the mobility of participants, experiences, and time (Pegrum, 2016; Lai et al., 2018). Teachers and learners use their personal mobile devices, such as tablets and mobile phones, in formal and informal settings, very often shifting the role of the device from content delivery to content creation thus allowing various and ubiquitous ways of learning (Kukulska-Hulme, 2016). In such a diverse and flexible situation, it has become crucial for researchers to investigate perspectives and attitudes toward the use of mobile devices both in formal (Kukulska-Hulme, 2016) and in informal language settings (Byrne & Diem, 2014; Ch'ng & Samsudin, 2013; Kukulska-Hulme, 2012) as the influence of self-directed use beyond the classroom can inform classroom practices and innovative methodologies (Lai et al., 2018).

In this respect, a number of MALL research studies have found evidence of the positive impact of the pedagogical use of mobile devices on the development of speaking and writing skills (Andujar, 2016; Lys, 2013; Tuttle, 2013), reading and listening comprehension skills (Hsu et al., 2013; Chen & Chang, 2011), and also on vocabulary learning (Li & Cummin, 2019; Stockwell & Liu, 2015) and on grammar reinforcement (Baleghizadeh & Oladrostam, 2010). In a recent study on the effects of texting on English vocabulary acquisition, Li and Cummin (2019) analysed the learning outcomes of 108 university students enrolled in an English for academic purposes (EAP) course. Using an experimental pre-test/post-test design, the researchers provided the intervention group with target academic vocabulary through text messages using the 'push mode' where information is sent directly to learners (Stockwell & Hubbard, 2013), while students in the control group used web-based vocabulary instructions or English dictionaries autonomously. Results showed important learning gains in terms of target academic vocabulary versus general academic vocabulary.

Various studies have also claimed that mobile learning fosters the development of important sub-skills such as collaborative learning (Roschelle et al., 2010), problem-solving and communication skills (Warschauer, 2007). The context of higher education seems particularly fit for such uses, as students will be required to have developed these skills, together with strong management abilities, when entering the labour market. University students and teachers seemed to be particularly attracted by the use and the integration of such devices into their language learning programmes. Various studies have focused on the perceptions, implementation and acceptance of mobile devices for language learning in HE (Gikas & Grant, 2018; Sung, Chang & Liu, 2015; Edmunds et al., 2012; TELL-OP, 2015) using both qualitative and quantitative data analysis. For instance, a study conducted by Edmunds et al. (2012) with 421 UK-based students at the Open University showed how learners' positive perceptions of 'usefulness at work' is one of strongest drivers of technology use.

Applications of MALL: Activities and Tasks

A study conducted by Steel (2015) with higher education students in Australia showed how mobile applications were the top five technologies used for language learning. In their recent review of MALL studies, Kukulska-Hulme and Viberg (2018) identified how teachers and learners use mobile devices for collaborative language learning and teaching. Interestingly, activities and tasks that provide flexibility, quick and personalized feedback, content creation, and personalization are the most often used and investigated. For example, Hazaea and Alzubi (2016) showed how the use of WhatsApp groups, online and offline dictionaries, camera and content-sharing apps in an extensive reading programme contributed to the improvement of learners' reading practices. Specifically, students used apps to collaborate and look up the meaning and pronunciation of specific words. Castrillo, Martín-Monje & Bárcena (2014) also found that the use of an instant messaging system (WhatsApp) to develop meaning negotiation strategies and improve foreign language skills has an impact on learners' attitudes and performance. Likewise, Gutiérrez-Colón et al.'s study (2015) concluded that an overwhelming majority of the students (90.9%) who were administered reinforcement reading comprehension activities via WhatsApp reported an increased disposition toward reading in English after receiving the exercises on their mobile phones. "Specifically, participants wrote that the experience had helped them to become more confident and less intimidated, thus leading to increased motivation overall to read in the L2." (p. 277). Some studies have explored the potentialities of language learning apps based on a specific MALL design framework. The study conducted by Read, Bárcena and Kukulska-Hulme (2016), for example, presents the prototype of an app designed to develop oral and written skills (ANT app) where users can also interact with other learners posting comments and opinions on social media. Findings reported positive reactions from users together with some concerns mainly related to technical issues (e.g. background noise for listening activities, cross-platform use, etc.).

In addition to the mainly qualitative research studies presented above, Sung, Chang & Liu (2015) conducted a systematic quantitative analysis of 110 studies in MALL in order to have a picture of the effects of MALL applications in various formal and informal settings. Results indicated positive effects of mobile learning implementation, but an enhancement of the quality of experimental design for MALL was advocated.

Positive Attitudes

Various studies in MALL have investigated how technology acceptance is often driven by the attitude learners and teachers have toward mobile devices (e.g. Edmunds et al., 2012; Botero et al., 2018). In particular, learners are generally driven by the perception of mobile learning as a channel that can help them reach their learning goals (Botero et al., 2018). A considerably large body of the literature confirms teachers and learners' overall positive attitude toward the use of mobile devices for language learning (Bradley, Lindström & Hashemi, 2017; Wrigglesworth, 2019). Generally, students successfully engage in mobile tasks and find them useful (Kim et al., 2013; Wrigglesworth, 2019). In particular, mobile devices proved to be useful for accessing information quickly, helping students to easily communicate and collaborate with peers, as well as experiencing multimodal and contextualized learning (Stockwell & Hubbard, 2013; Gikas & Grant, 2013).

The cross-national study conducted by Hsu (2013) with EFL learners, for example, showed how the use of MALL activities in an authentic context allowed students to perform specific tasks in a specific place overcoming time and space issues and increasing students' motivation. On the other hand, Hsu's study (2013) showed evidence of how a different cultural background can influence MALL perceptions and attitudes. For instance, EFL learners from Japan were not particularly impressed by the affordances of mobile devices in the language classroom. Overall, learners defined using mobile devices, particularly smartphones, for language learning as simply being 'more convenient', meaning that it is easier to work with a small easy-to-carry device, instead of using a larger and heavier mobile device such as a tablet, or a portable device such as a laptop (Gikas & Grant, 2013). Control and personalization have also been reported as highly positive factors in using mobile devices for language learning (Edmunds et al., 2012). Moreover, focusing on the role of the teachers and task design, recent studies have shown that when mobile technologies are integrated effectively into the language classroom, they provide a valuable contribution to the learning approaches, also creating a collaborative learning environment (Sabah, 2016). Additionally, a meta-analysis and research synthesis study conducted by Sung, Chang & Liu (2015) also found evidence that learners were more engaged using mobile devices than desktop computers, and that learning was more effective in informal educational settings.

Issues and Concerns on MALL Implementation

Although a large body of research has generally reported positive attitudes, learners and teachers have various concerns toward MALL implementation and practice. Most of the issues seem to be related to the device in itself: the screen or the keyboard were too small, some apps did not work properly (Liu et al., 2015; TELL-OP, 2015), the learning tools were not sufficiently user-friendly to be pedagogically effective (Kim et al., 2013). To support this, a study conducted by Stockwell (2010) with English as foreign language learners in a Japanese university, showed how vocabulary activities performed on mobile devices took longer than those performed on laptop computers, thus confirming that small screen devices often require the learner to scroll up and down the screen to be able to access all the information. Learners also described mobile devices as distracting and showed frustration when working with teachers whom they perceived as not having positive attitudes toward mobile learning (Gikas & Grant, 2013).

In order to empower educators and practitioners with the correct instruments to design and implement formal and informal language learning tasks with the support of mobile devices, it is crucial to gain a thorough understanding of what drives teachers and learners' attitudes and preferences. The present study aims at describing higher education learners and teachers' preferences toward the use of mobile devices for formal and informal language learning in different educational contexts across the world.

METHODOLOGY

Context of the Study

The study was born from the intention of EuroCALL's Mobile-Assisted Language Learning (MALL) Special Interest Group (SIG) board members to share knowledge and expertise with the aim of investigating the field of MALL and giving recommendations for further MALL research. The coordinating team first decided on the survey format of the study and chose to focus on HE teachers and learners' use of MALL. Following a call for contribution distributed at the beginning of 2018 amongst the MALL SIG community of the association, a multinational team of 12 researchers started to collaborate in the research project. The first step consisted in defining the goals of the study and in examining the possible distribution of the survey. The team included researchers from Spain, Belgium, Italy, Sweden and Japan who teach languages to HE students at various levels and who could use their professional networks to distribute the questionnaire to language lecturers in their country. According to the team's various interests in the field of MALL, the survey was designed to explore pedagogically sound practices and provide a better understanding of the current and future role of MALL on language learning, specifically in HE. The questionnaire sought therefore to answer two primary research questions:

RQ1: How do HE learners/teachers use mobile devices for language learning?
RQ2: What are HE learners/teachers' attitudes toward the use of mobile devices for language learning?

Questionnaire Design

The survey includes two parallel questionnaires that were co-created by the team members with the help of online collaborative tools, namely *Google docs* and *Google forms*. Most communication was conducted via e-mails or Skype sessions. The researchers shared their knowledge and expertise at different stages of the questionnaire design. They were first asked to share MALL surveys from previous projects (e.g. TELL-OP, 2015; STRATAPP, 2016) or from the literature (e.g. Wakefield & Whitten, 2006; Cheng, 2015) then to add questions to the questionnaires according to their own research interests (e.g. attitudes, practices, specific tools, target respondents). After the collection of information and the selection of relevant studies to help construct the questionnaire, a subsequent step consisted in structuring

Table 1. Questionnaire design

	Learners' questionnaire	Teachers' questionnaire
Section 1	Educational and language background	Teaching experience and educational background
Section 2	Use of mobile devices	Use of mobile devices
Section 3	Use of mobile devices for educational purposes	Use of mobile devices for educational purposes
Section 4	Attitudes toward the use of mobile devices for language learning	Attitudes toward the use of mobile devices for language learning
Section 5	Background information	Background information

and reviewing the questions with regard to terminology, respondents' cognitive load, relevancy of the items, and adaptation to various contexts. For example, questions related to the language background of the respondents were formulated independently from any specific language proficiency test (e.g. CEFR, TOEIC) since these assessment tools are not of common knowledge among language learners. Finally, the questionnaires were tested by some researchers' colleagues prior to the actual distribution among HE institutions so as to ensure the comprehensibility and the validity of all items. The two questionnaires (see appendices 1 & 2) were created on *Google forms* and were accessible online. The students' survey included 37 items while the teachers' set included 31 items. Both of them would take approximately ten minute to complete and the questions were presented in English.

As can be seen in Table 1, the two questionnaires show a very similar structure which should allow for comparisons. In sections 1 and 5, personal information is gathered to depict the respondents' profile. Sections 2 and 3 include all the items relating to the use of mobile devices, be it for personal or for educational purposes. Several surveys were used to shape this part (e.g. TELL-OP, 2015; Saidouni & Bahloul, 2016*)*. Most items are closed, multiple choice or 5-point Likert-scale questions. This rating scale type of questions is used to measure opinions and attitudes and is used in this section to measure the frequency of use of mobile devices by the respondents ranging from (1) *never* to (5) *very often*. Section 4 addresses respondents' attitudes toward the use of MALL and is primarily based on Venkatesh & Davis' extended Technology Acceptance Model (TAM) (2000). Items regarding perceived usefulness, ease of use and enjoyment have been chosen from various studies (e.g. Cheng, 2015; Al-Emran et al., 2016; Van de Vyver, 2016) to serve our purpose. This section contains 4 open questions and 3 sets of 5-point bipolar Likert-scale questions addressing the respondents' attitudes on MALL, ranging from (1) *totally disagree* to (5) *totally agree,* (3) being a neutral point.

Data Collection and Analysis

Each of the 12 researchers distributed the questionnaire in their country to HE language institutes, language instructors and language students through their institution's Learning Management Systems (LMS), personal networks, professional and personal mailing lists. The respondents could then, on a voluntary basis, answer the 10-minute questionnaire online individually. Japanese researchers and teachers had their students fill in the questionnaire in Japanese in class. The survey took place from the beginning of May to the end of June 2018 with a total of 748 student respondents, 511 of whom were from Japan and the remaining 236 from other countries. As this imbalance in student responses would bias the results, this study only focuses on the latter. Additionally, a total of 127 teacher responses were received.

Right after the collection of the data, a video-conference session was organized with the whole team in order to decide on methods of analysis as well as to form working teams for each of the three data sets. The decision was made to analyse the teachers' and the learners' set separately as a first step. The learners' data set from Japan would be analysed as a third set since the number of respondents was particularly high (n=511) compared to the other learners' set (n=236) but also since the data collection type was different (see previous section).

Data was analysed both quantitatively through descriptive statistics, and qualitatively, based on Glaser & Strauss' Grounded Theory (1967). The majority of the survey items being closed-ended questions, the answers were analysed using descriptive statistics and the teachers' and learners' sets could be compared using the same method. Various Likert items composed the 3 attitudinal constructs from Venkatesh & Davis' extended TAM (2000); Cronbach's alpha values were calculated to measure the internal consistency of these constructs, namely, perceived ease of use, perceived usefulness and perceived enjoyment, before starting the analyses. Table 2 shows that these values are all acceptably high (≥0.7).

Four open-ended questions on practices and attitudes were created to complement the quantitative data collected and required a qualitative analysis. Glaser & Strauss' grounded theory (1967) was used for various reasons. This approach consisted in observing the information provided by the respondents without using any specific criteria from the literature or from any previous experience. This was convenient in this research project for practical reasons since the group did not have to agree on a single framework to follow for analyses. Moreover, having a group of researchers examining the same data could help on making reliable categories. Inter-rater reliability was sought by conducting collaborative discussions among the researchers in order to reach a consensus in terms of data interpretation, which ultimately led to agreed categories. Besides, having patterns emerge from the data was seen to be a pertinent approach for this exploratory study on learners and teachers' perceptions.

Table 2. Students and teachers' attitudes – Cronbach alpha

Construct	Cronbach Alpha STUDENTS	Cronbach Alpha TEACHERS	Likert items
Perceived enjoyment	0.73	0.90	• I find using mobile devices to be enjoyable. • I have fun using mobile devices in my classes. • I like using mobile devices in my classes.
Perceived usefulness	0.82	0.84	• Using mobile devices can help me/my students accomplish language learning tasks more quickly. • Mobile devices allows access to information anywhere anytime. • Using mobile devices can make language learning /teaching more efficient. • Using mobile devices can be useful for language learning/teaching. • Mobile learning allows us to learn wherever we need to. • Mobile learning offers opportunities for communication and teamwork
Perceived ease of use	0.70	0.79	• My interaction with mobile devices is clear and understandable. • I find it easy to get mobile devices to do what I want them to do. • I find it easy to become skillful at using mobile devices.
Intention to use MALL	0.86	0.85	• I would like to learn about how to use mobile devices for language learning/teaching. • I would like to use mobile learning on a regular basis IN the language classroom. • I would like to use mobile learning on a regular basis OUTSIDE the language classroom.

The idea was to "deconstruct the data into manageable chunks in order to facilitate an understanding of the phenomenon in question" (Cohen, Manion & Morrison, 2007, p. 493). The working group in charge of the teachers' set initiated the analyses. The data was first coded manually and patterns that emerged thanks to the coding were identified and categorised into variables. It was decided not to use any software as the researchers did not all have easy access to a text-mining software programme. The second group, in charge of the learners' set, coded the learners' data manually, established categories, then crossed the results with the teachers' set and refined their categorisation so as to account for the two data sets as accurately as possible. Combining a quantitative and a qualitative approach facilitated gaining further understanding of learners and teachers' attitudes toward and uses of mobile devices for language learning in HE.

Participants

The study targeted participants –both HE instructors and HE learners– based in Asia, North America and Europe. The data was collected by means of a large-scale online survey which investigated their practices and attitudes toward the use of mobile devices as a language learning tool in today's educational context. The aim of the study was to gather information from HE learners and teachers worldwide and to get a glimpse of the language learning and teaching communities' practices on mobile-assisted language learning.

Teachers

The results are based on 127 replies from HE teachers, mainly teaching English in Europe (83%). Table 3 shows a substantial participation rate from respondents teaching in Belgium, Spain and Italy, which are in fact some of the researchers' home countries. Teachers in Spain and Belgium mostly originate from these two countries. Teachers in Italy are Italian or English speaking and they all teach English. The respondents teaching in Japan are not Japanese, but come from the UK, Australia or the USA.

77 of the respondents are exclusively English teachers. 26 others teach English and another language (e.g. Dutch, German, Spanish). The 23 other participants teach other languages to English (e.g. Dutch, German, Spanish, French). Table 4 represents the native language of the teachers. When observing the distribution of the participants, it can be seen that most teachers in Belgium are French (27) or Dutch (4) speakers. In Spain, 29 teachers' native language is Spanish. German speakers mostly teach in Spain and in Belgium, while English speakers teach in various countries, namely Japan, Belgium, Israel, Italy, Spain, Ireland or the UK. 20 respondents speak another native language (e.g. Russian, Polish, Ukrainian), the majority of whom teach English (14).

Table 3. Participants per country of teaching and origin

	Belgium	Spain	Italy	Israel	Japan	Ukraine	UK	Germany	Other
Country of teaching	38	35	13	10	6	5	3	1	16
Nationality	33	28	7	8	0	4	13	8	26

Table 4. Teachers' native language

French	Dutch	Spanish	Italian	English	German	Swedish	Other
27	4	32	6	28	9	1	20

Over 70% (N=91) of the teachers have 10 years or more of experience as teachers while 14% (N=18) have between 5 to 10 years of experience. The 18 remaining teachers (14%) have less than 5 years of experience. When asked about the type of education they are teaching in (cf. Table 5), 93% (N=118) state they do face-to-face teaching, among which 64% (N=81) do face-to-face teaching only. 29% (N=37) report teaching in different types of education and a minority teaches in blended learning courses (N=8) or online/distance teaching (N=1) exclusively.

Learners

The questionnaire was completed by 236 HE students from various countries. 511 Japanese HE students also filled in the survey but their responses are not addressed in this chapter. As shown in Table 6, most of our respondents are living in Italy (62%), Spain (17%) and Belgium (13%), which are some of the researchers' home countries. The other participants are living in Ukraine (4%), Greece (1.5%) and other countries (1.5%) such as Germany or the USA. The majority originate from the country where they are living. Still, 7% (N=17) of them come from other countries like Vietnam, China or Poland.

As can be seen in Figure 1, the most common age groups among the respondents are the 18-20 range (N=98, 41.5%), and the 21-23 range (N= 91, 38.5%). They are followed by the 24-26 range (N=25, 10.5%) and finally the respondents aged between 27 and 43 (N=24, 10%). 90% of the learners are, in fact, aged between 18 and 26.

Table 5. Type of education

Face-to-face teaching	81
Face-to-face teaching, Blended learning courses	15
Face-to-face teaching, Blended learning courses, Online/distance teaching	13
Face-to-face teaching, Online/distance teaching	9
Blended learning courses	8
Online/distance teaching	1

Table 6. Learners' country of living and nationality

	Italy	Spain	Belgium	Ukraine	Greece	Other
Country of living	146	40	32	9	4	5
Nationality	143	32	25	9	5	22

Figure 1. Age distribution of the learners

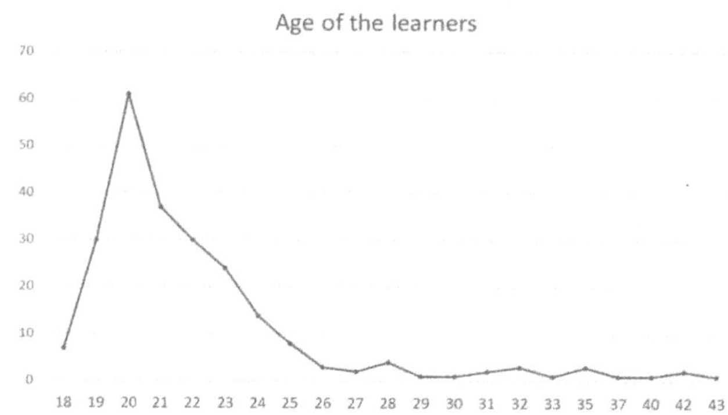

70% of the learners (N=164) are enrolled in a Bachelor's Degree and 29% (N=69) are enrolled in a Master's degree, while the remaining 1% (N=3) are PhD students. The majority of them are following face-to-face courses (N=186, 79%) whereas the others are following blended learning courses (N=35, 15%) or online/distance learning (N=14, 6%). The learners' mother tongue, as can be seen from Table 7, essentially correlates with their nationality (Table 6), at least for Italian, Spanish and French speakers. It can be observed that nearly 10% (N=23) speak other languages as their mother tongue to the languages mentioned in the survey. As for their first foreign language, the results show that the majority of the respondents (84%) have learnt English as a first foreign language whereas some others have first learnt Dutch, French, Italian or other languages like German, Spanish, Arabic or Japanese.

The respondents are currently learning from 1 to 5 languages. Most of them are learning 2 foreign languages (N=137; 59%), 27% of them (N=62) are currently learning one foreign language, 11% (N=25) 3 languages, and the remaining 3% (N=7), 5 languages.

Figure 2 represents the different languages learnt by the students. This figure and further results show that English is the most common foreign language learnt by the respondents either on its own (N=31) or in combination with other languages (N=157) like German, Chinese, Dutch, French and/or Spanish. It is followed by Spanish, for some learnt on its own (N=13), or most often in combination with English (N=45), and by German, learnt most of the time in combination with English and/or Spanish (N=48). French, Chinese, Dutch and Italian follow the trend as they are mostly chosen in combination with English. The additional languages include Japanese, Korean and Swedish.

When asked about the skill(s) they needed to develop at the moment of the survey, 7% of the learners declared having to develop all 4 skills (N=17) whereas 60% (N=144) mentioned one skill only (cf. Figure 3). Among these responses, speaking skills comes first (N=75), followed by writing skills (N=44),

Table 7. Native language of the learners

	Italian	Spanish	French	English	Chinese	German	Dutch	Other
Native language	144	33	24	6	3	2	1	23
First foreign language	3		10	198			15	10

Figure 2. Languages currently learnt by the students

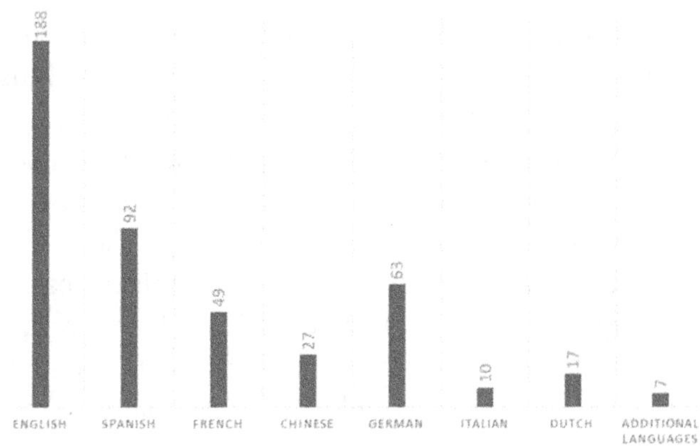

listening comprehension (N=23) and finally, reading comprehension (N=3). 25% choose speaking skills in combination with writing skills (N=33) or listening comprehension (N=30), while the remaining 5% (N=12) make other combinations.

RESULTS AND DISCUSSION

This section seeks firstly to address, through descriptive analyses, the uses of mobile devices reported by the learners and teachers in order to answer the first research question. The second part of this section deals with our second research question and seeks to examine the attitudes of HE learners and teachers toward the use of mobile devices for language learning.

The first part reports on the learners' personal and educational use of mobile devices in terms of frequency and access. The analyses then compare the learners and teachers' perceptions as regards the encouragement toward using mobile devices in the language classroom. The second part categorizes the

Figure 3. Language skill(s) the learners need to develop

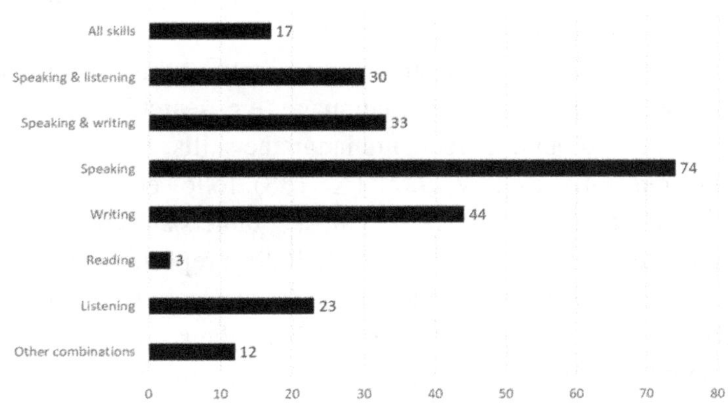

types of apps used that were cited by the learners, and the targeted skills students subsequently reported practising. These results provide an overview of the learners' practices of MALL, as reported by the users themselves, which are later compared with the instructors' responses on related issues. Confronting the two viewpoints leads to interesting findings and a discussion on MALL uses that allow for a number of pedagogical recommendations, listed at the end of the chapter.

This section closes with the results linked to HE learners and teachers' attitudes toward the use of MALL. Open answers from learners and instructors are displayed in categories to observe the rationale behind the choice of using a mobile device for language learning. Answers to Likert questions are then examined to study the perceptions of both learners and teachers in terms of usefulness, ease of use and enjoyment when using mobile devices for language learning. The TAM model (Venkatesh & Davis, 2000) results, confirming an influence of these attitudes on the intention of use of certain technology, leads to the identification of factors of attention for future MALL research and MALL integration in the language classroom.

Comparing Students and Instructors' Use of Mobile Devices

Regarding the frequency of using their smartphone, 90% (N=212) of the students responded doing so, on average, every day, whereas a slightly lower amount of instructors agreed with this option (N=105, 82.6%). A high percentage of the students also estimated using it more than 2 hours on a daily basis (N=190, 80.5%), while this figure decreased to nearly half that amount when teachers were asked (N=54, 42.5%). These first figures already display differences between teachers and learners and demonstrate the potential interest of examining practices and attitudes of the various actors involved in MALL.

Concerning where the learners' most commonly access their mobile devices to complete educational activities like finding information for their courses, studying, working on class assignments (see Figure 4), two thirds of them (N=156, 66.1%) acknowledged that they very often do this at home, whereas just over one third (N=84, 35.6%) report often using them for this purpose at university. If we take into account the number of students who claim using their devices while commuting from one place to another, "very often" (N=22, 9.3%), "often" (N=37, 15.5%) and "sometimes" (N=64, 27.1%) add up to a non-negligent amount of 123 users, that is, 52% –just over half of them. This favourable disposition to make the most of time while on the go, to either learn or revise educational content, should be taken into account by language instructors to design activities and tasks that are suitable to be carried out on small screens with limited access to additional resources to aid in completing them. These findings are aligned with Lai & Zheng's study (2018) on self-directed use of mobile devices beyond the classroom, also for language learning. They can also be linked to the need to implement authentic tasks in authentic contexts as advocated by Hsu (2013).

Figure 5 illustrates how often university students use both mobile devices and desktop or laptop computers as a means toward learning a foreign language (e.g. searching the meaning of words, working on assignments for language courses, practising language skills, etc.). As we can see, computers are still slightly in the lead, frequently used by 79.6% (N=188), followed by smartphones, which are also frequently used by 77% (N=182). However, as we can see, tablets are scarcely reported being used.

However, not even one third of the students (N=73, 30.9%) reported being encouraged by their language teachers to use their mobile devices in the language classroom (Q17), despite the fact that 63.7% (N=81) of the teachers (Q14, cf. Figure 6) reported advising their students to do so as classroom work and the exact same amount as a homework assignment. The discrepancy between the learners and the instructors' responses may reflect a somewhat unclear status of the use of mobile devices in education.

Figure 4. Results for Q24 addressed to students, "When you use mobile devices for educational activities, where do you use them?"

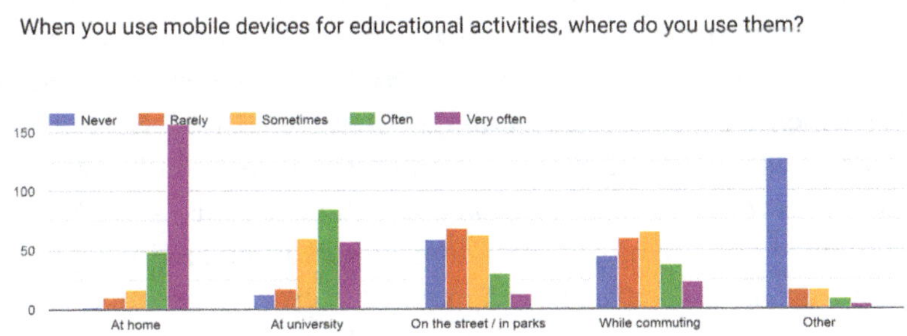

Figure 5. Results for Q14 addressed to students, "How often do you use the following devices for language learning purposes?

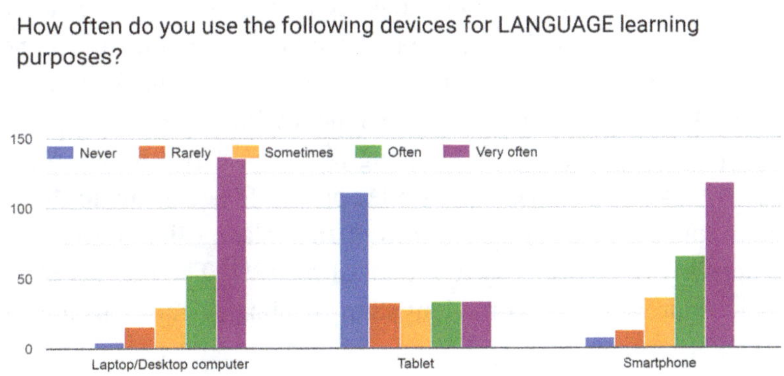

Figure 6. Results for Q14 addressed to teachers, "How often do you recommend your students to use mobile devices for…"

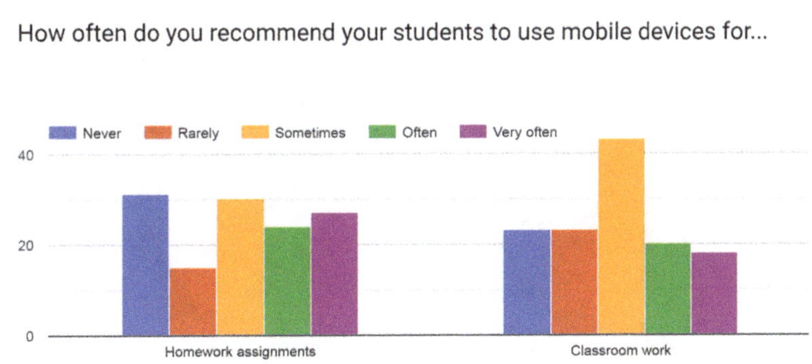

Higher Education Students' Use of Mobile Devices for Language Learning

One of the core questions in the survey queried language learners' practices upon using their mobile device for language learning purposes. As we can see in Figure 7, there was an outstanding majority of respondents who claimed mainly using dictionaries[2] (N=113, 47.88%), followed by gamified tutorial apps (N=48, 20.34%), which are apps such as Duolingo, Memrise, Busuu, Babbel or Lingualeo that offer practice activities –to a greater or lesser extent– in all four language skills (reading, writing, listening and speaking, in addition to vocabulary acquisition and grammar drills) and offer a sequential learning path that goes from less demanding to more demanding, as well as integrating a gamification component. The word "gamification" is understood as the technique of exerting game mechanics into the non-game environment, such as websites and mobile applications. These game elements commonly come in the form of a challenge, which can sometimes translate into increased interaction with the application.

Figure 7 also sheds light as to what students *do not* customarily seek online to improve their language skills, namely, translation tools, which were only mentioned by 12 students (5.08%), dedicated vocabulary practice aids (N=10, 4.24%), pronunciation (N=7, 2.97%) and grammar practice (N=6, 2.54%). This can perhaps be explained because students no longer understand language learning by translating into their mother tongue but, rather, as a communicative process. This is in agreement with the way language learning methods have evolved, where the focus has shifted from accuracy to fluency with its emphasis on meaning, fluency, and real-life communication (Tejada Molina et al., 2005).

This same question (Q23) also provided the researchers with a wealth of information regarding the skills that the learners primarily targeted when using these apps. As shown in Figure 8, HE students report practising all four language skills using mobile apps. Still, according to the categories into which the online tools were organised, vocabulary comes in first place (dictionaries 47.88% and dedicated vocabulary practice apps 4.24%) adding up to a total of 52.12%. The perceived need on behalf of the learners in this study to improve their foreign language vocabulary confirms a general trend of MALL

Figure 7. Results for Q23, what tools (apps or other software) do you use for language learning, what do you use them for? Numbers in brackets represent actual responses, i.e. raw data (N=236).

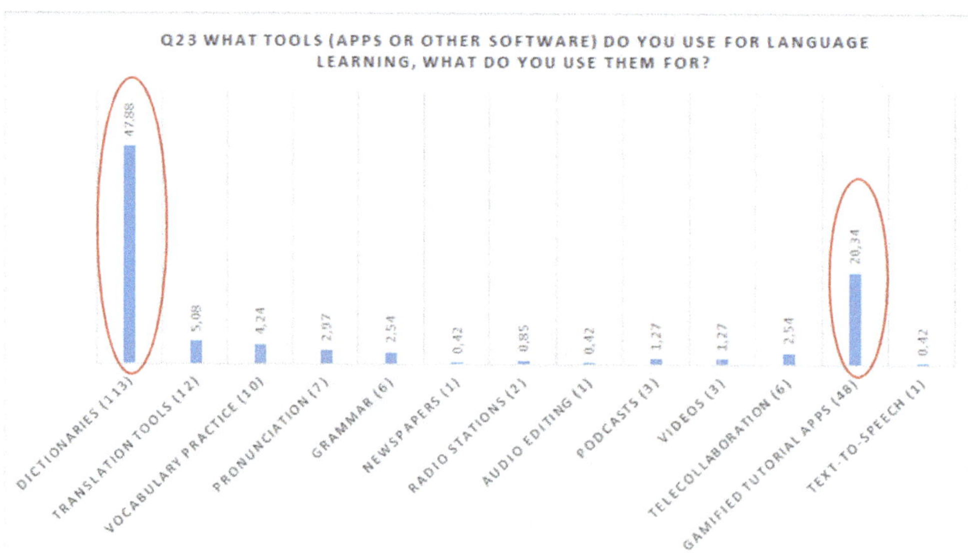

research studies (Li & Cummin, 2019; Stockwell & Liu, 2015). Students are willing to work on vocabulary reinforcement activities using their mobile device. The fact that they are mainly using dictionary apps or self-directed web sources is probably due to the lack of experimental designs and methodologies such as the ones presented in Li & Cummin (2019) or in Hazaea and Alzubi (2016). Far behind in popularity come the other skills, together with grammar practice. In second and third place come productive skills, i.e. Speaking (5.93%) and Writing 5.5%; in fourth place, Listening Comprehension (3.39%), followed by Grammar (2.54%) and Reading Comprehension (0.42%). The combined skills practised in gamified tutorial apps have been excluded from these numbers.

Higher Education Instructors' Use of Mobile Devices for Language Teaching

However, when it comes to comparing student input to what HE language instructors reported, we can observe that there is quite a different viewpoint. As we can see in Figure 9, 70.87% (N=90) of the respondents claimed asking their students to use their mobile devices for vocabulary practice and acquisition, followed by 44.88% (N=57) who reported asking them to conduct activities to improve their listening skills. The use of dictionaries –which as we saw, students reported being their preferred choice of language learning apps– were not contemplated by teachers and came last, with only one person naming this option as a language learning tool.

Like learners, language teachers also reported asking their students to practice all four language skills through mobile devices but, in the latter case, focusing on passive rather than active skills. As displayed in Figure 10, Listening Comprehension was targeted by 44.88% (N=57) of the respondents and Reading Comprehension by 38.58% (N=49), whereas Writing was targeted by 22.83% (N=29), and Speaking practice by 18.90 (N=24).

Over one third of the respondents also claimed recommending the use of grammar (N=45, 35.43%) and pronunciation tools (N=41, 32.28%), contrary to what students claimed to use since they reported scarcely using these tools at all (see Figure 7). Despite the very low numbers, some instructors also mentioned the use of formative assessment tools to conduct tests and quizzes[3] (N=7, 5.51%), as well as requesting students to search for information using their mobile devices (N=5, 3.93%).

Figure 8. Categorised responses and the number of items identified per category. Numbers in brackets represent actual responses (i.e. raw data).

Categorised responses & no. of items identified (213)	% of respondents
Dictionaries (113)	47,88
Translation tools (12)	5,08
Vocabulary practice (10)	4,24
Pronunciation (7)	2,97
Grammar (6)	2,54
Newspapers (1)	0,42
Radio stations (2)	0,85
Audio editing (1)	0,42
Podcasts (3)	1,27
Videos (3)	1,27
Telecollaboration (6)	2,54
Gamified tutorial apps (48)	20,34
Text-to-speech (1)	0,42

Skills
Speaking (5.93%)
Writing (5.5%)
Listening (3.39%)
Reading (0.42%)

Linguistics elements
Vocabulary (47.88%)
Grammar (2.54%)

Skill and linguistic element targeted by the items in each category.

Figure 9. Results arising from Q12 addressed to instructors. If you ask your students to use mobile devices for language learning, which linguistic aspect(s) do you target?

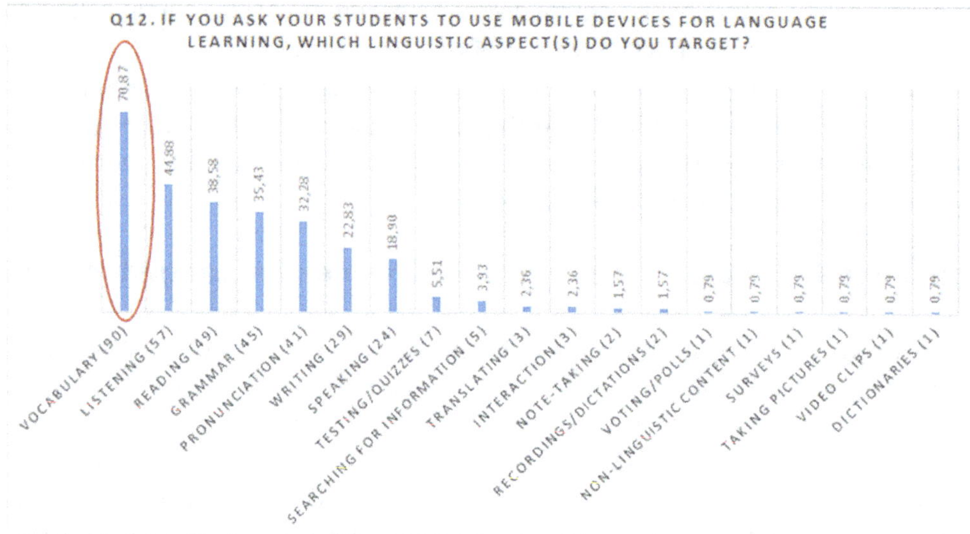

Figure 10. Results of Q12. Categorised responses and the number of items identified per category. Numbers in brackets represent actual responses (N=127), i.e. raw data.

Categorised responses and no. of items identified	(%)
Vocabulary (90)	70,87
Listening (57)	44,88
Reading (49)	38,58
Grammar (45)	35,43
Pronunciation (41)	32,28
Writing (29)	22,83
Speaking (24)	18,90
Testing/quizzes (7)	5,51
Searching for information (5)	3,93
Translating (3)	2,36
Interaction (3)	2,36
Note-taking (2)	1,57
Recordings/dictations (2)	1,57
Voting/polls (1)	0,79
Non-linguistic content (1)	0,79
Surveys (1)	0,79
Taking pictures (1)	0,79
Video clips (1)	0,79
Dictionaries (1)	0,79

Skills
Speaking (55.11)
Listening (45.67%)
Reading (43.3%)
Writing (27.55%)

Linguistic elements
Vocabulary (71.66)
Grammar (35.43%)

When asked to name the apps/resources language instructors used in combination with the linguistic aspect(s) targeted (Q13), nearly half of the respondents said they primarily asked their students to use their mobile devices for formative assessment (N=62, 48.82%), followed by the use of dictionaries and encyclopaedias (N=40, 31.50%). These results represented in Figure 11 correlate with the fact that, like the students themselves, when it comes to practising lexical items, instructors tend to focus mainly on intentional vocabulary development, which can be interpreted as being part of what Schmitt (1997, p. 206) called consolidation strategies[4], i.e. methods that foster the consolidation of a word once it has been encountered. These activities may be understood as reinforcement techniques. In some cases, recommending learners to use online dictionaries or encyclopaedias to look up unknown words and concepts pursues the same outcome as implementing quizzes as a consolidation strategy to aid the acquisition of new lexical items. Moreover, these mobile uses are in line with a socio-constructivist approach that puts the learners at the centre of their learning.

Although considerably behind in popularity, streamed video came in third place (N=16, 12.60%). This result is associated to watching/listening comprehension activities, which correlates with the fact that this skill, according to the respondents, is the second most targeted when it comes to using a mobile device for language learning purposes.

If we have a closer look at the categorised responses and the number of items identified, which amount to a total of 219, there is clear evidence that mobile applications serve a wide spectrum of purposes geared toward improving the students' language learning experience. As can be seen in Figure 12, 29 categories of tools were identified, targeting –as mentioned above– all four language skills, in addition to vocabulary and grammar practice. The 29 categories of tools reported cover a wide range of purposes for using mobile apps, ranging from social media to song lyrics, through collaborative tasks.

Despite the fact that instructors report integrating MALL into their teaching practices, very few provide feedback to their students with a mobile device (See Figure 13). Those that do, however, still use a traditional means such as e-mail. It is a vast minority who responded having provided feedback

Figure 11. Results arising from Q13 addressed to instructors. If you ask your students to use mobile devices for language learning, could you name some apps/resources you use and the linguistic aspect(s) targeted?

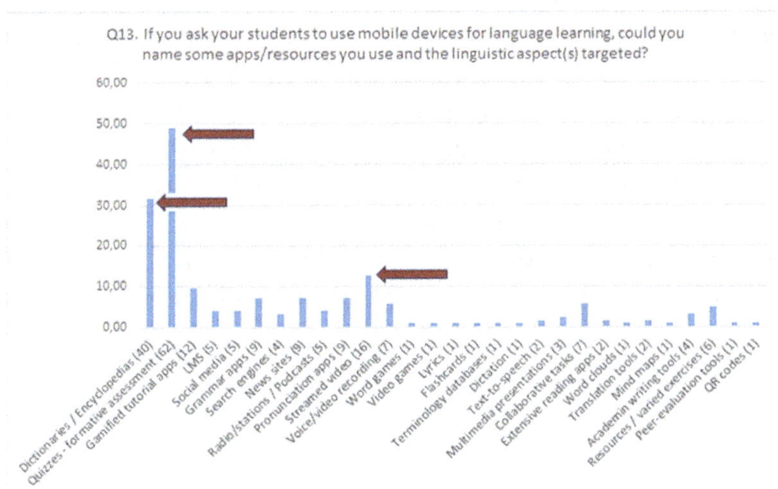

Figure 12. Results of Q13. Categorised responses and the number of items identified per category. Numbers in brackets represent actual responses (N=127), i.e. raw data.

Categorised responses & no. of items identified (219)	% of respondents
Dictionaries / Encyclopaedias (40)	31,50
Quizzes - formative assessment (62)	48,82
Gamified tutorial apps (12)	9,45
LMS (5)	3,94
Social media (5)	3,94
Grammar apps (9)	7,09
Search engines (4)	3,15
News sites (9)	7,09
Radio stations / Podcasts (5)	3,94
Pronunciation apps (9)	7,09
Streamed video (16)	12,60
Voice/video recording (7)	5,51
Word games (1)	0,79
Video games (1)	0,79
Lyrics (1)	0,79
Flashcards (1)	0,79
Terminology databases (1)	0,79
Dictation (1)	0,79
Text-to-speech (2)	1,57
Multimedia presentations (3)	2,36
Collaborative tasks (7)	5,51
Extensive reading apps (2)	1,57
Word clouds (1)	0,79
Translation tools (2)	1,57
Mind maps (1)	0,79
Academic writing tools (4)	3,15
Resources / varied exercises (6)	4,72
Peer-evaluation tools (1)	0,79
QR codes (1)	0,79

Skills
Speaking (14.96%)
Listening (18.11%)
Reading (12.60%)
Writing (12.60%)

Linguistic elements
Vocabulary (35.43%)
Grammar (7.09%)

through a social network, an instant messaging system or an online bulletin board such as *Padlet*. Two respondents reported using *PeerEval* for this purpose, the only currently available peer assessment app that has been created for the sole purpose of providing language learners with an evaluation of their work based on previously designed rubrics.

If we compare these results, however, with those reported by university students in response to Q29, "Do you ever receive feedback from your language teacher with a mobile device? If so, please explain how", there is a clear correlation between the claims, that is, students also acknowledge receiving feedback from their instructor via e-mails on their mobile devices. This is also a clear indication that e-mails are increasingly being accessed (and perhaps replied to) on a mobile device.

As we can see in Figure 14, over two thirds of the students (N=172, 70.49%) either do not respond or answer "no" to this question, which correlates with the information provided by teachers, two thirds of whom (N=77, 66.62%) also deny providing their learners' with feedback through the medium of a mobile device.

Figure 13. Results of Q21 addressed to instructors. Do you ever provide feedback to your students with a mobile device. If so, please explain how.

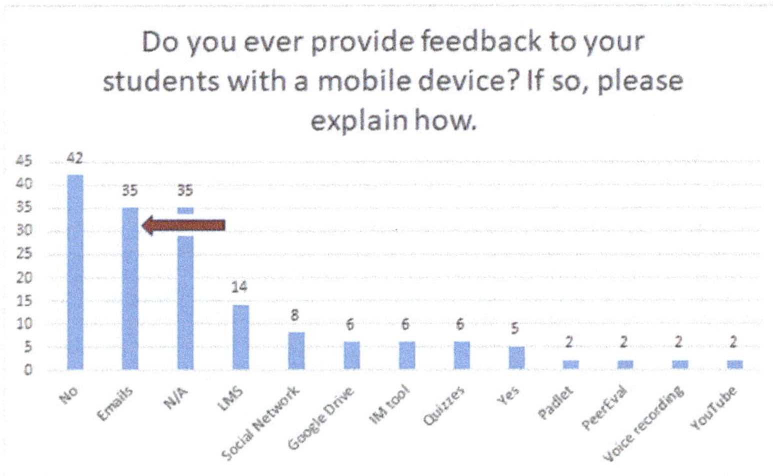

Figure 14. Results of Q29 addressed to university students. Do you ever receive feedback from your language teacher with a mobile device? If so, please explain how.

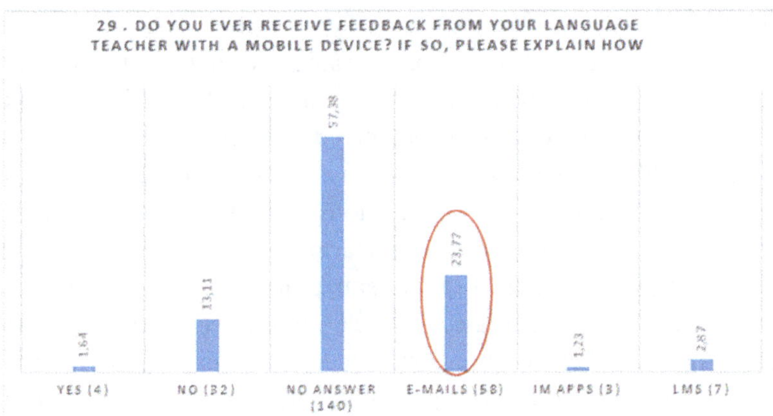

University Students and Instructors' Attitudes toward the Use of Mobile Devices for Language Learning

Many were the attitudes conveyed in the survey toward using mobile devices for language learning on the learners' part. When asked why they sometimes chose to use a mobile device instead of a fixed device for language learning (Q26), the categories that stood out considerably in this open question were those shown in Figure 15. Four main reasons stand out dealing with speed, convenience, ease of use and portability, among other less popular reasons. These are not features that would normally be associated to the fact that we are dealing with language learning, but to the inherent characteristics of the mobile devices themselves.

Figure 15. Results of Q26 addressed to students, "Why do you sometimes choose to use a mobile device instead of a fixed device for language learning?"

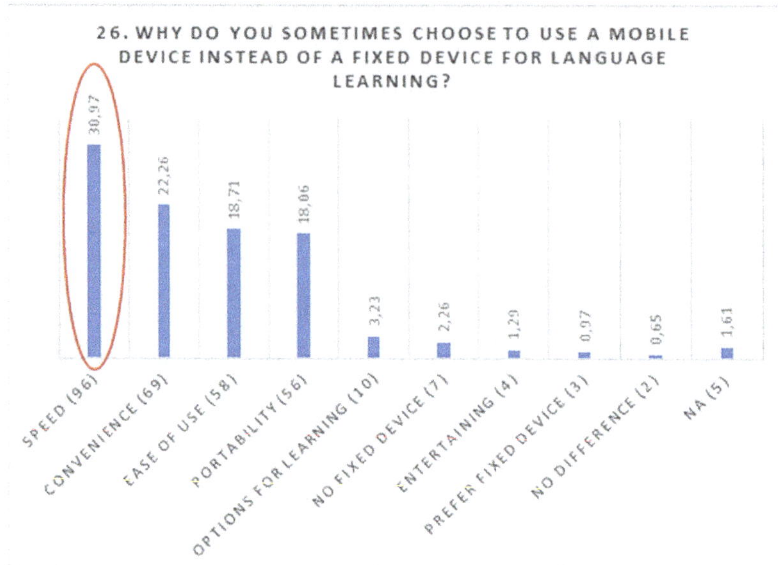

The results are similar when the instructors were asked why they choose a mobile device instead of a fixed device (Q18). As we can see in Figure 16, the categories depicted from the open answers also drew attention to reasons such as convenience and portability, together with motivation, which is a clear indication that using mobile apps and devices is perceived as an incentive for learners.

In terms of likes and dislikes, both students and instructors were asked to give some thought to this, also through open-ended questions. These questions sought to inform their perceptions in terms of the role mobile devices have taken on in language learning. Regarding what they liked about mobile devices for language learning (Q27, Figure 17), students pointed out the many affordances inherent to these devices (39%). Among them, immediate feedback, interactive features, portability, being able to personalise learning, being fun and entertaining, and speed. They also highlighted a number of features that made these devices convenient to use (8%) such as being handy, practical, efficient, helpful, up-to-date and versatile. Other aspects that were considered favourable were the availability of a wealth of apps and resources capable of enriching the learning process (19%) and the variety and usefulness of the content to support learning (11%). These findings reflect those in Stockwell & Hubbard (2013) and Gikas & Grant (2013). However, they are in disagreement with part of the results in Hsu (2013), who argues that for a large group of participants, mainly from Asian countries (e.g. Japan), the affordances of mobile devices were not so relevant for language learning.

When asked what they liked about using mobile devices for language learning (Q19, Figure 18), the instructors who replied (N=108) pointed out reasons such as learner motivation (N=29, 26.8%), ease of access (N=26, 24%) and praised the large variety of apps and activities that are available for this purpose (N=20, 18.5%).

In terms of dislikes (Q28, Figure 19), in addition to 23% who disliked nothing about using their mobile devices for language learning, other students remarked displeasure in technical issues (25%), unreliable or limited content (20%), health and addictions (4%), becoming easily distracted (5%), and also missing face-to-face interaction (5%).

Figure 16. Q18 addressed to teachers, "Why do you sometimes choose to use a mobile device instead of a fixed device for language learning/teaching?"

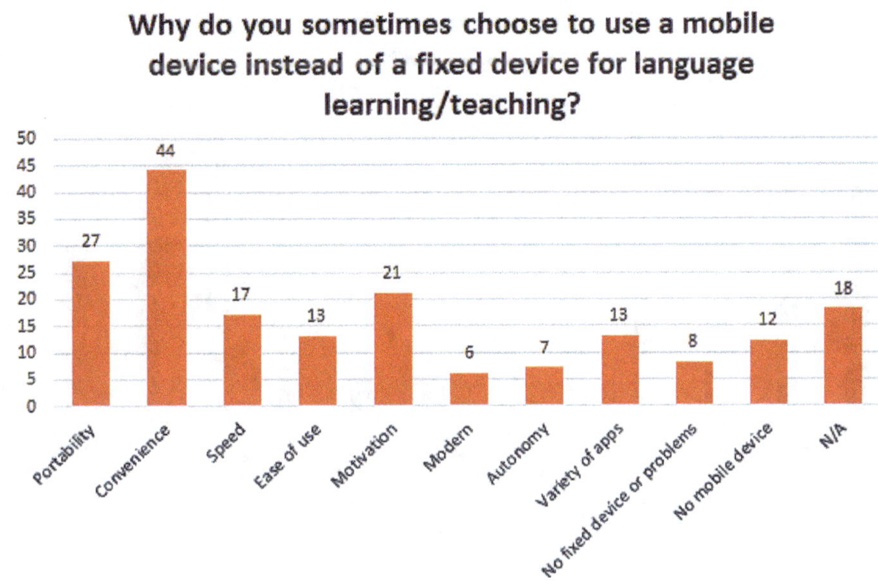

Figure 17. Q27 addressed to students, "What do you like about using mobile devices for language learning?"

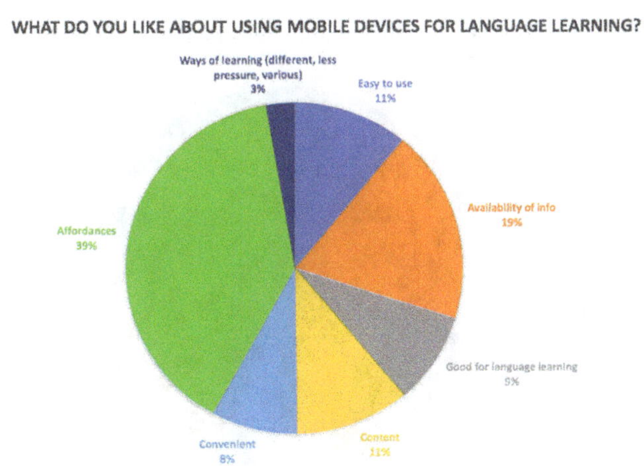

In terms of what teachers indicated disliking (Q20, Figure 20), two clear reasons came to the forefront among those who answered this question (N=103). On the one hand, the many distractions that may arise disturbing the learning process and the focus on the activity or task at hand (42%) and, on the other, technical difficulties (32%). Resembling the students, a certain amount of teachers also said they disliked nothing about them (10%).

In terms of how useful, easy to use and enjoyable students' perceived mobile devices for language learning, six 5-point Likert scale questions were posed (Q30): a) I find using mobile devices for language learning to be enjoyable; b) Using mobile devices can help me accomplish language learning tasks more

Figure 18. Q19 addressed to teachers, "What do you like about using mobile devices for language learning?"

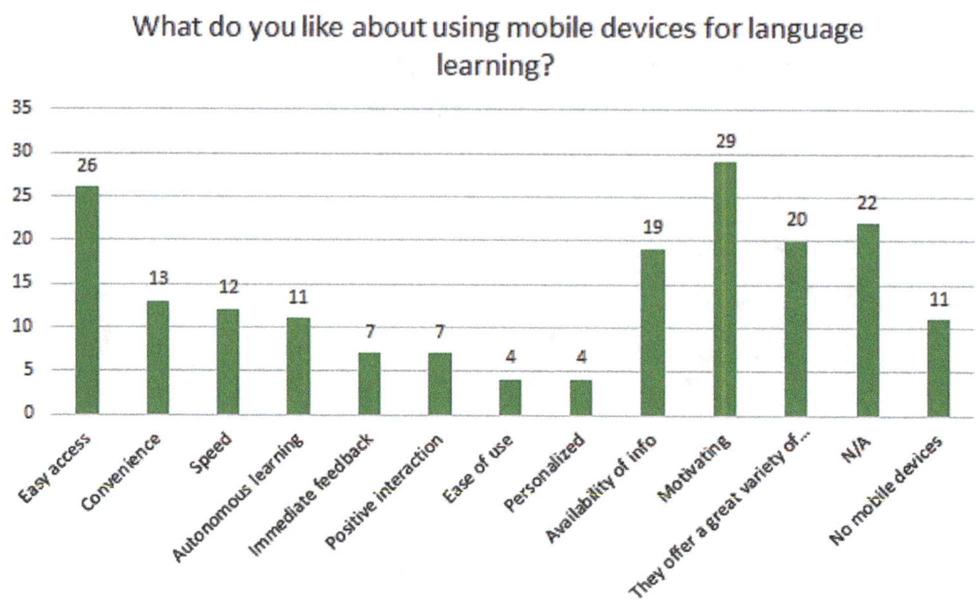

Figure 19. Results of Q28 addressed to students, "What do you dislike about using mobile devices for language learning?"

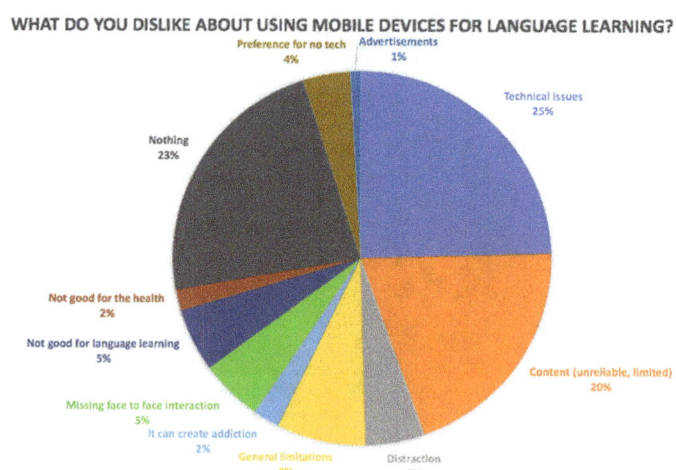

quickly; c) My interaction with mobile devices is clear and understandable; d) Mobile devices allow access to information anywhere anytime; e) I find it easy to get mobile devices to do what I want them to do; and f) Using mobile devices can make language learning more efficient. As shown in Figure 21, overall, students are in favour of using their devices because, to a large extent, if we count the results for "quite agree" and "totally agree", they find them enjoyable (83.4%), useful (83.4%), quick (81.3%), straightforward to handle (76.7%), efficient (66.1%) and ubiquitous (92.3%).

Figure 20. Results of Q20 addressed to teachers, "What do you dislike about using mobile devices for language learning?"

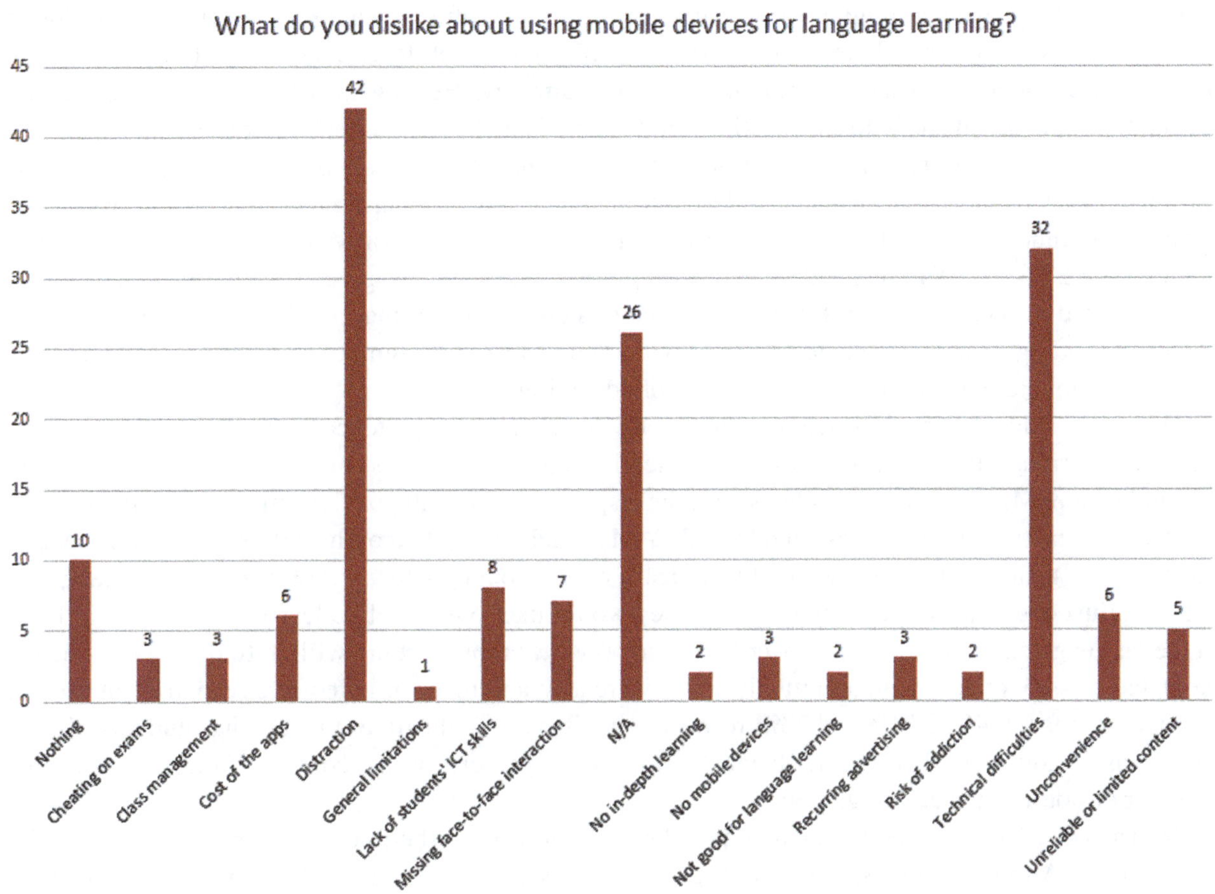

Figure 21. Q30 addressed to students requesting their opinion about mobile devices for language learning

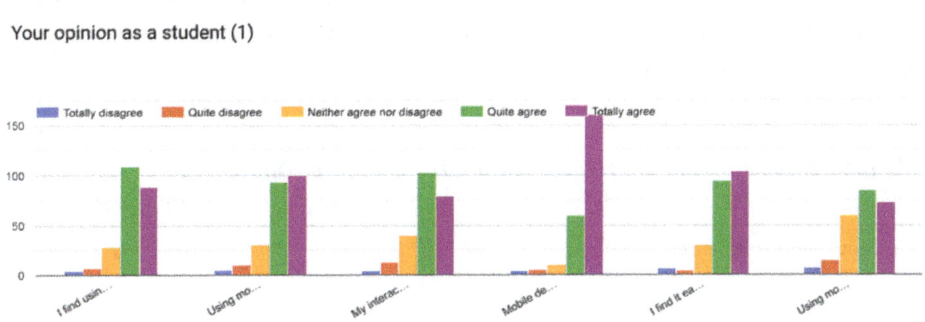

An additional set of six questions were posed to further explore these same parameters (Q31), also on a 5-point Likert scale: a) I have fun using mobile devices in my classes; b) Using mobile devices can be useful for language learning; c) I find it easy to become skillful at using mobile devices; d) Mobile learning allows us to learn wherever we need to; e) I like using mobile devices in my classes; f) Mobile learning offers opportunities for communication and teamwork. Here again, when we take into account the number of respondents who chose either "quite agree" and "totally agree", we observe the learners' favourable opinion toward MALL regarding entertainment (39.4%), usefulness (87.7%), ease of use (70%), ubiquity (83%), enjoyment and collaboration (45%). It is also a fact, however, that a considerable quantity of students do not have a clear opinion about these aspects of MALL. For example, a non-negligible 42.7% (N=101) responded that they "neither agree nor disagree" (cf. Figure 22). This can be interpreted as owing to the fact that, when used in a classroom setting, mobile devices are no longer objects providing pleasure and entertainment but simply another medium of instruction, like a desktop or laptop computer and, as such, at the service of education.

When HE students were queried in connection with their future pretences to use mobile devices for language learning purposes (Q32), as we can see in Figure 23, the vast majority say that they would willingly adopt MALL, which correlates with the respondents' overall positive attitude toward the use of MALL, as confirmed by Venkatesh and Davis' TAM model (2000). Out of the 236 respondents, over two thirds (72.4%) claimed that they would like to learn more about how to use mobile devices for language learning, just over half of them (54.6%) said they would like to use mobile learning on a regular basis in the language classroom, and just over two thirds (68.6%) replied being willing to do this outside the language classroom although, admittedly, there were also a number of indecisive responses in each of these questions (19.4%, 27.1% and 22%, respectively). This data illustrates the fact that students still see MALL as an add-on to their regular language tuition and perceive it in a context of informal learning, whether or not it is a class assignment.

Instructors were also presented with equivalent questions from their point of view, also on a 5-point Likert scale. Although the responses were generally less enthusiastic than those of students, there is nevertheless also an evident positive attitude toward MALL when measured according to the following parameters: a) I find using mobile devices for language teaching enjoyable; b) Using mobile devices can help my students accomplish language learning tasks more quickly; c) My interaction with mobile devices is clear and understandable; d) Mobile devices allow access to information anywhere anytime; e) I find it easy to get mobile devices to do what I want them to do, and f) Using mobile devices can make language teaching more efficient.

Figure 22. Q31 addressed to students requesting their opinion about MALL

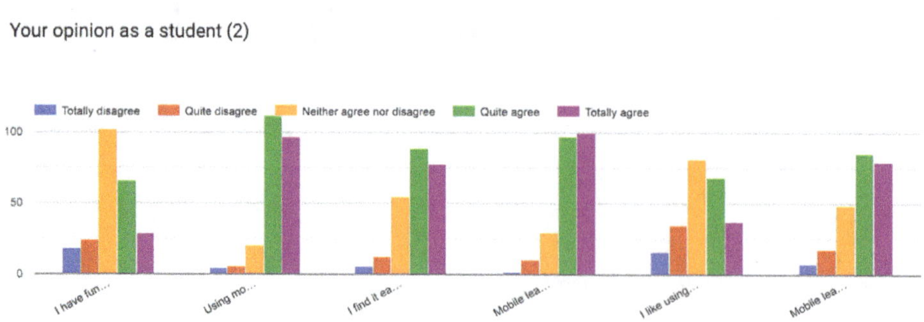

Figure 23. Q32 addressed to students requesting their opinion about their future use of MALL

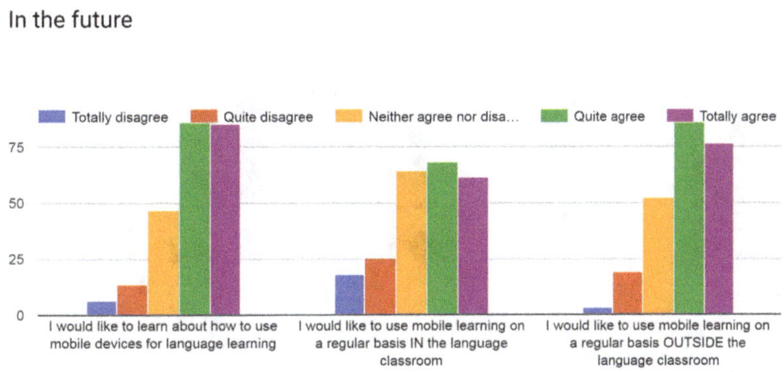

The one response that stands out is the ubiquitous nature of MALL, i.e. its usefulness to ensure access to information anywhere, anytime. As we can see in Figure 24, if we take into account both favourable categories, i.e. "quite agree" and "totally agree", this adds up to 95% of the respondents (N=121 out of 127). In terms of enjoyment, two thirds (67%) agreed with statement a) above. Regarding the speed with which students could accomplish tasks, two thirds of the teachers also had a favourable opinion, i.e. 66% in all. As regards smooth interaction with mobile devices, nearly two thirds (65%) agreed positively with this and two thirds (66%) also thought that they take control over their mobile devices (66%). A slightly higher percentage (70%) agreed that MALL could make teaching a foreign language more efficient. As with the students, teachers also believed that using mobile devices could be useful for language teaching (87.4%) –cf. Figure 25; over two thirds thought that it is easy to become skillful using them (71.6%). A further 85% praised their ubiquitous nature and just under two thirds (63%) stated liking applying MALL in their classes; 71.6% of the respondents also agreed that mobile learning offers opportunities for learners to intercommunicate and collaborate. The lowest rate was given to perceiving mobile devices as being fun within a classroom context. Only just over half of the respondents (54.3%) agreed to this.

Figure 24. Q22 addressed to teachers requesting their opinion about mobile devices for language teaching

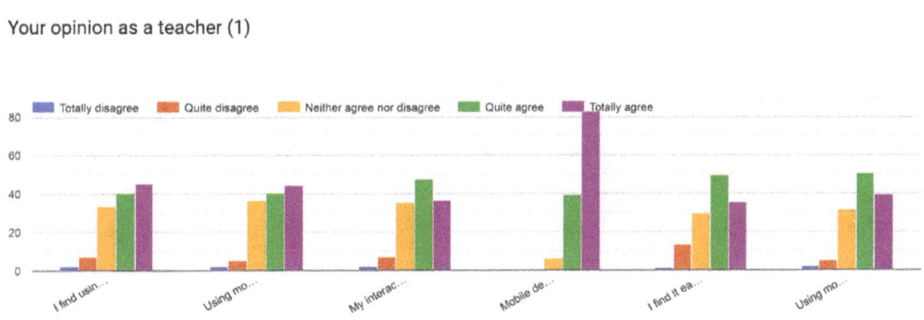

Figure 25. Q23 addressed to teachers requesting their opinion about MALL

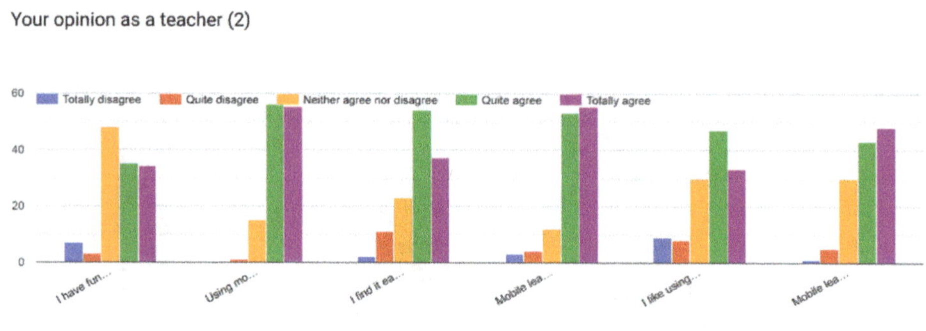

Regarding how language instructors foresee applying mobile devices into language teaching, a majority (80%) stated that they would like to learn more about how to integrate MALL into their teaching practice although they would like to integrate MALL into their classes to a lesser extent (67.7%). It was 76.3% of the instructors who claimed they would advise their students to use their mobile devices for language learning outside the class (see Figure 26).

To summarise students' attitudes, descriptive statistics enabled us to determine the extent to which the four elements that comprise the Technology Acceptance Model (Venkatesh & Davis, 2000) were perceived by the respondents. As we can see in Table 8, the learners' overall perceptions toward MALL are positive. Perceived Ease of Use and Perceived Usefulness are the highest scores, followed by the Intention to use MALL. This suggests that learners find mobile devices useful and easy to use, and that they consider them something valuable for language learning. The score for Perceived Enjoyment is also above the mid-point of the scale, but slightly lower than the rest.

To summarise teachers' attitude, overall, the teachers who responded to the survey have a positive attitude toward MALL. In a 5-point Likert scale, where the mid-point would be 3, the mean responses for the 4 constructs measured are all above 3.7. The highest score was obtained for Perceived Usefulness, followed by the Intention to use MALL. The score for Perceived Ease of Use and Perceived Enjoyment were also positive, although not as high as that of the perceived usefulness or the intention of use.

Figure 26. Q24 addressed to teachers requesting their opinion about their future use of MALL

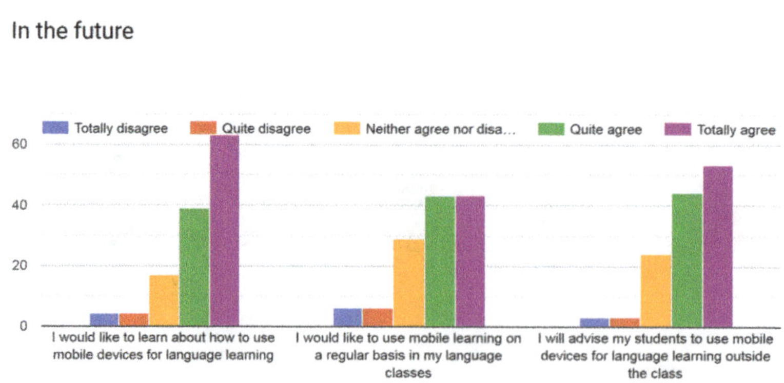

Table 8. Students and teachers' attitudes according to the TAM

	Number of Likert items	Students' set Mean score	Students' set Standard Deviation	Teachers' set Mean score	Teachers' set Standard Deviation
Perceived enjoyment	3	3.57	1.00	3.77	0.97
Perceived usefulness	6	4.14	0.91	4.17	0.64
Perceived ease of use	3	4.05	0.93	3.85	0.82
Intention to use MALL	3	3.81	1.06	4.06	0.89

As we can see, the students' mean scores are generally slightly lower than those obtained for teachers are. The mean score for the learners' Perceived Ease of Use is slightly higher than that for teachers, which may be due to a possible higher familiarity with mobile devices.

SOLUTIONS AND RECOMMENDATIONS

For Teachers

The use of mobile devices inside and outside standard learning environments is expanding and evolving rapidly. Teachers, second language acquisition researchers and educators cannot ignore these changes, as current and future HE learners are expecting them to deliver new ways of learning to improve motivation and to provide innovative learning tools. This study aimed at providing possible directions for teachers. More specifically, it identified a series of suggestions that could be explored in the context of this study (HE in Europe), but also in other language learning contexts.

It is crucial to raise awareness on existing practices, which, as we have seen, are quite varied, by developing action research projects and sharing best practices. Technology-mediated tasks have a strong potential in formal and informal language learning. Learners expect to work on tasks that have been designed to develop specific skills with the support of a specific and tested technology/app. Moreover, it is important to use mobile devices for developing all language skills and linguistic elements.

Due to the disparate use that students claim in terms of MALL in informal contexts and instructors' beliefs regarding how to implement the use of mobile devices in the language classroom, it stands out that there should be more communication and exchanges of information between teachers and learners regarding their mobile practices and the underlying rationale. Likewise, in order to improve successful MALL practices teachers need to be trained on technology-mediated task-based language teaching and learning. Raising their awareness on the multiple MALL affordances could make them make relevant choices in terms of tools and activities and increase the usefulness and ease of use of mobile devices for language learning and teaching.

As just over half of the students in the survey reported using their smartphones for language learning while commuting to and from university, teachers should take advantage of this and design activities and tasks that are suitable to conduct while on the go on a small screen, bearing in mind that the number of additional resources at hand to aid in completing them may be limited.

For Learners

Although limited to the European HE context, this study offers a general overview of the attitudes of university students toward MALL and, as such, has important implications. Based on the reported use of smartphones and other mobile devices, it has become fundamental for language specialists to curate apps and online resources for language learning so that learners can make informed choices and access high quality second language multimedia content to suit their learning needs. Institutions, policy makers and curriculum designers can take advantage of learners' favourable attitude toward MALL and increase the number of apps and resources integrated into the curriculum and the second language syllabus. The fact that often learners did not answer questions on specific uses of mobile devices for language learning probably shows some uncertainty about how to use these to develop specific abilities and skills. Specific training and support from the institutions would certainly help learners overcome these issues.

Lastly, this study provides evidence that university teachers and learners are willing to implement MALL practices when supported by the use of tools and apps specifically designed for language learning and guided by SLA methodologies.

CONCLUSION

The results of the large-scale online survey revealed general positive attitudes toward the use of mobile devices for language learning purposes. In the context of Europe, this study showed how HE teachers and learners are willing to integrate MALL tools and methods in standard educational settings and even more in informal environments, confirming the trends in Sung et al. (2015).

This study enabled a useful comparison between teachers and learners' uses and attitudes toward MALL. Overall, learners reported a more frequent use of these devices than teachers did, and this probably explains the learners' positive stance toward the use of mobile devices for learning. Evidence also shows that most learners use mobile devices to practise and develop a foreign language outside the educational context revealing a growing trend toward a self-directed use of MALL that can inevitably influence educational practices (see also Lai et al., 2018). In particular, students make use of gamified tutorial apps such as Duolingo, etc. as a means of informal learning to supplement their regular language tuition. Additionally, the fact that 188 learners state that they are learning English as a foreign language (EFL) is an indication that MALL in the European context is dominated by learners of English, probably because there is more demand to learn English due to its *lingua franca* nature, which in turn pushes app providers to offer EFL more than any other language (see Figure 2).

According to learners' responses, speaking is the foreign language skill they need to develop the most. However, they report mainly using vocabulary practice or dictionary apps and gamified tutorial apps. This is probably due to the fact that students are not aware of how to use mobile devices to develop speaking skills to their full potential. In educational contexts students seem to perceive their smartphone more as a motivational tool than as an educational tool, therefore it is crucial to provide them with guidance as to how to use it for language learning purposes.

Teachers encourage the use of mobile devices mainly for vocabulary practice and acquisition, recommending apps primarily for reinforcement activities. Both teachers and students see the potential of using mobile devices/apps for vocabulary development and consolidation. Language instructors also make use of streamed video, mainly to develop listening skills and encourage student engagement. Motivation is, in fact, one of the key drivers for teachers, while MALL for learners is mainly driven by perceived usefulness and functionality, as already reported in Edmunds et al. (2012).

Findings also show some key issues related to MALL implementation. Learners focus on the importance of device and network functionality, reporting issues such as small screen and keyboard, slow Wi-Fi connection, confirming the results in Liu et al. (2015). However, the study also shows issues relating to task management and the role of the teachers. The discrepancy between learners claiming not being encouraged by the teacher to use a mobile device for language learning and the opposite claim put forth by teachers confirms a certain frustration on behalf of the learners when dealing with unmotivated teachers (Gikas & Grant, 2013). This also confirms the idea that it is not the technology in itself that will change learning, but how it is methodologically integrated into the language curriculum. Evidence also shows that providing immediate and personalized feedback is an issue for both teachers and learners. Although university language instructors are slightly more skeptical about MALL than university students, evidence suggests that both see important implications for language teaching and learning.

ACKNOWLEDGMENT

This research was supported by EuroCALL, the European Association for Computer-Assisted Language Learning and conducted by members of its MALL Special Interest Group.

REFERENCES

Al-Emran, M., Elsherif, H. M., & Shaalan, K. (2016). Investigating attitudes towards the use of mobile learning in higher education. *Computers in Human Behavior*, *56*, 93–102. doi:10.1016/j.chb.2015.11.033

Andujar, A. (2016). Benefits of mobile instant messaging to develop ESL writing. *System*, *62*, 63–76. doi:10.1016/j.system.2016.07.004

Baleghizadeh, S., & Oladrostam, E. (2010). The effect of mobile assisted language learning (MALL) on grammatical accuracy of EFL students. *Mextesol Journal*, *34*(2), 1–10.

Bebell, D., & O'Dwyer, L. M. (2010). Educational outcomes and research from 1:1 computing settings. *The Journal of Technology, Learning, and Assessment*, *9*(1), 5–15.

Botero, G. G., Questier, F., Cincinnato, S., He, T., & Zhu, C. (2018). Acceptance and usage of mobile assisted language learning by higher education students. *Journal of Computing in Higher Education*, *30*(3), 426–451. doi:10.100712528-018-9177-1

Bradley, L., Lindström, N. B., & Hashemi, S. S. (2017). Integration and language learning of newly arrived migrants using mobile technology. *Journal of Interactive Media in Education*, *2017*(1), 3. doi:10.5334/jime.434

Burston, J. (2015). Twenty years of MALL project implementation: A meta-analysis of learning outcomes. *ReCALL*, *27*(01), 4–20. doi:10.1017/S0958344014000159

Byrne, J., & Diem, R. (2014). Profiling mobile English language learners. *JALT CALL Journal, 10*(1), 3–19.

Castrillo, M. C., Martín-Monje, E., & Bárcena, E. (2014). Mobile-based chatting for meaning negotiation in foreign language learning. In *Proceedings of the 10th International Conference on Mobile Learning*. (pp. 49-59). International Association for Development of the Information Society.

Ch'ng, L. K., & Samsudin, Z. (2013). Integration of mobile devices into ubiquitous learning by the 21st century teenagers. *Education, 3*(6), 362–374. doi:10.5923/j.edu.20130306.12

Chen, I. J., & Chang, C. C. (2011). Content presentation modes in mobile language listening tasks: English proficiency as a moderator. *Computer Assisted Language Learning, 24*(5), 451–470. doi:10.1080/09588221.2011.577749

Cheng, Y. (2015). Towards an understanding of the factors affecting m-learning acceptance: Roles of technological characteristics and compatibility. *Asia Pacific Management Review, 20*(3), 109–119. doi:10.1016/j.apmrv.2014.12.011

Cohen, L., Manion, L., & Morrison, K. (2013). *Research Methods in Education*. London, UK: Routledge. doi:10.4324/9780203720967

Edmunds, R., Thorpe, M., & Conole, G. (2012). Student attitudes towards and use of ICT in course study, work and social activity: A technology acceptance model approach. *British Journal of Educational Technology, 43*(1), 71–84. doi:10.1111/j.1467-8535.2010.01142.x

Fleischer, H. (2012). What is our current understanding of one-to-one computer projects: A systematic narrative research review. *Educational Research Review, 7*(2), 107–122. doi:10.1016/j.edurev.2011.11.004

Gikas, J., & Grant, M. M. (2013). Mobile computing devices in higher education: Student perspectives on learning with cellphones, smartphones & social media. *The Internet and Higher Education, 19*, 18–26. doi:10.1016/j.iheduc.2013.06.002

Glaser, B. G., & Strauss, A. L. (1967). *The discovery of grounded theory: Strategies for qualitative research*. New Brunswick, NJ: Transaction Publishers. [2006 reprint]

Gutiérrez-Colon Plana, M., Gimeno-Sanz, A., Appel, C., & Hopkins, J. (2015). Improving learners' reading skills through instant short messages: A sample study using WhatsApp. In A. Gimeno-Sanz, M. Levy, D. Barr, & F. Blin (Eds.), *WorldCALL: Sustainability and computer-assisted language learning* (pp. 266–281). London, UK: Bloomsbury.

Hazaea, A. N., & Alzubi, A. A. (2016). The effectiveness of using mobile on EFL learners' reading practices in Najran University. *English Language Teaching, 9*(5), 8–21. doi:10.5539/elt.v9n5p8

Hoven, D., & Palalas, A. (2013). The design of effective mobile-enabled tasks for ESP students: A longitudinal study. *CALICO Journal, 30*, 137–165.

Hsu, C. K., Hwang, G. J., & Chang, C. K. (2013). A personalized recommendation-based mobile learning approach to improving the reading performance of EFL students. *Computers & Education, 63*, 327–336. doi:10.1016/j.compedu.2012.12.004

Hsu, L. (2013). English as a foreign language learners' perception of mobile assisted language learning: A cross-national study. *Computer Assisted Language Learning*, *26*(3), 197–213. doi:10.1080/09588221.2011.649485

Kim, D., Rueckert, D., Kim, D.-J., & Seo, D. (2013). Students' perceptions and experiences of mobile learning. *Language Learning & Technology*, *17*(3), 52–73.

Kukulska-Hulme, A. (2012). Chapter One: Language learning defined by time and place: A framework for next generation designs. In J. Díaz-Vera (Ed.), *Left to my own devices: Learner autonomy and mobile-assisted language learning* (pp. 1–20). BRILL.

Kukulska-Hulme, A., & Viberg, O. (2018). Mobile collaborative language learning: State of the art. *British Journal of Educational Technology*, *49*(2), 207–218. doi:10.1111/bjet.12580

Lai, C., & Zheng, D. (2018). Self-directed use of mobile devices for language learning beyond the classroom. *ReCALL*, *30*(3), 299–318. doi:10.1017/S0958344017000258

Li, J., & Cummins, J. (2019). Effect of using texting on vocabulary instruction for English learners. *Language Learning & Technology*, *23*(2), 43–64.

Liu, G. Z., Kuo, F. R., Shi, Y. R., & Chen, Y. W. (2015). Dedicated design and usability of a context-aware ubiquitous learning environment for developing receptive language skills: A case study. *International Journal of Mobile Learning and Organisation*, *9*(1), 49–65. doi:10.1504/IJMLO.2015.069717

Lys, F. (2013). The development of advanced learner oral proficiency using iPads. *Language Learning & Technology*, *17*(3), 94–116.

Pegrum, M. (2014). *Mobile learning: Languages, literacies and cultures*. Basingstoke, UK: Palgrave Macmillan. doi:10.1057/9781137309815

Pegrum, M. (2016). Three agendas for MALL. In A. Palalas & M. Ally (Eds.), The international handbook of mobile-assisted language learning (pp. 86-112). China Central Radio & TV University Press Co.

Read, T., Bárcena, E., & Kukulska-Hulme, A. (2016). Exploring the application of a conceptual framework in a social MALL app. In A. Pareja-Lora, C. Calle-Martínez, & P. Rodríguez-Arancón (Eds.), New perspectives on teaching and working with languages in the digital era (pp. 223-232). Research-Publishing.net. doi:10.14705/rpnet.2016.tislid2014.436

Roschelle, J., Rafanan, K., Estrella, G., Nussbaum, M., & Claro, S. (2010). From handheld collaborative tool to effective classroom module: Embedding CSCL in a broader design framework. *Computers & Education*, *55*(3), 1018–1026. doi:10.1016/j.compedu.2010.04.012

Sabah, N. M. (2016). Exploring students' awareness and perceptions: Influencing factors and individual differences driving m-learning adoption. *Computers in Human Behavior*, *65*, 522–533. doi:10.1016/j.chb.2016.09.009

Saidouni, K. & Bahloul, A. (2016) Teachers and students' attitudes towards using mobile-assisted language learning in higher education, Arab World English Journal Special Issue on CALL No. 3 July 2016 (pp.123-140).

Schmitt, N., & McCarthy, M. (Eds.). (1997). *Vocabulary: Description, acquisition and pedagogy*. Cambridge, UK: Cambridge University Press.

Steel, C. (2015). Students' perspectives on the affordances and constraints of using mobile devices and applications for learning languages. In A. Gimeno-Sanz, M. Levy, D. Barr, & F. Blin (Eds.), *WorldCALL: Sustainability and computer-assisted language learning* (pp. 230–243). London, UK: Bloomsbury.

Stockwell, G. (2010). Using mobile phones for vocabulary activities: Examining the effect of platform. *Language Learning & Technology*, *14*(2), 95–110.

Stockwell, G. & Hubbard, P. (2013). Some emerging principles for mobile-assisted language learning. *The International Research Foundation for English Language Education,* 1-15.

Stockwell, G., & Liu, Y. C. (2015). Engaging in mobile phone-based activities for learning vocabulary: An investigation in Japan and Taiwan. *CALICO Journal*, *32*(2), 299–322. doi:10.1558/cj.v32i2.25000

STRATAPP. Gamifying academic English skills in higher education: Reading academic English app (2016-2019). Erasmus + Project. Available at http://stratapp.eu

Sung, Y. T., Chang, K. E., & Liu, T. C. (2016). The effects of integrating mobile devices with teaching and learning on students' learning performance: A meta-analysis and research synthesis. *Computers & Education*, *94*, 252–275. doi:10.1016/j.compedu.2015.11.008

Tejada Molina, G., Pérez Cañado, M. L., & Luque Agulló, G. (2005). Current approaches and teaching methods. *Bilingual programmes. TEFL in Secondary Education.* Granada, Spain: Editorial Universidad de Granada. Available at http://www4.ujaen.es/~gluque/Chapter4HANDBOOKDEFINITIVO.pdf

TELL-OP. Transforming European learner language into learning opportunities. (2015-2017). OER survey: Language processing technologies. Erasmus + Project. Available at http://www.tellop.eu/wp-content/uploads/2015/09/TELL-OP-Survey-ENG.pdf

Tuttle, H. G. (2013). Improving students' modern language speaking skills through mobile learning. In Z. L. Berge & L. Y. Muilenburg (Eds.), *Handbook of mobile learning*. New York, NY: Routledge.

United Nations Educational, Scientific and Cultural Organization. (2012). Mobile learning and policies. Key issues to consider. Paris, France: UNESCO. Available at https://unesdoc.unesco.org/ark:/48223/pf0000217638

United Nations Educational, Scientific and Cultural Organization. (2013). The future of mobile learning. Implications for policy makers and planners. Paris, France: UNESCO. Available at https://unesdoc.unesco.org/ark:/48223/pf0000219637

Van de Vyver, J. (2016). Mobile-assisted language learning community and culture in French-speaking Belgium: the teachers' perspective. In S. Papadima-Sophocleous, L. Bradley, & S. Thouësny (Eds.), CALL communities and culture – short papers from EUROCALL 2016, (pp. 450-455). Research-Publishing.net.

Venkatesh, V., & Davis, F. D. (2000). A theoretical extension of the technology acceptance model: Four longitudinal field studies. *Management Science*, *46*(2), 186–204. doi:10.1287/mnsc.46.2.186.11926

Wakefield, R., & Whitten, D. (2006). Mobile computing: A user study on hedonic/utilitarian mobile device usage. *European Journal of Information Systems*, *15*(3), 292–300. doi:10.1057/palgrave.ejis.3000619

Warschauer, M. (2007). A teacher's place in the digital divide. *Yearbook of the National Society for the Study of Education*, *106*(2), 147–166. doi:10.1111/j.1744-7984.2007.00118.x

Wrigglesworth, J. (2019). Using smartphones to extend interaction beyond the EFL classroom. *Computer Assisted Language Learning*, 1–22. doi:10.1080/09588221.2019.1569067

Zucker, A. A., & Light, D. (2009). Laptop programs for students. *Science*, *323*(5910), 82–85. doi:10.1126cience.1167705 PMID:19119224

ENDNOTES

1 European Association for Computer-Assisted Language Learning. For further information, refer to http://www.eurocall-languages.org/.

2 The dictionaries that were mentioned were Reverso, Oxford-Hachette, Wordreference, Merriam Webster, Abbyy Lingvo, Linguee, Collins Dictionary, Van Dale (for Dutch language), PONS, Cambridge Dictionary, Dictionary.com, Thesaurus.com, Multitran, Real Academia Española de la Lengua, and Bab.la.

3 The tools mentioned in this category included: Quizlet, Kahoot!, Socrative, Wooclap, Plickers, Quizizz, and Mentimeter.

4 According to Schmitt's (1997) taxonomy, there are two main groups of vocabulary learning strategies: discovery strategies and consolidation strategies. Discovery strategies are those strategies that are used in discovering the meaning of a new word whereas consolidating strategies deal with the consolidation of a word once it has been encountered. The former consists of determination strategies and social strategies, whereas the latter includes social strategies, memory strategies, cognitive strategies and metacognitive strategies.

APPENDIX

Babbel: https://www.babbel.com
Busuu: https://www.busuu.com/en
Duolingo: https://en.duolingo.com
Lingualeo: https://lingualeo.com
Memrise: https://www.memrise.com
Padlet: https://padlet.com
PeerEval: https://peereval.mobi
MALL student questionnaire: https://tinyurl.com/y2wzctbt
MALL teacher questionnaire: https://tinyurl.com/yy7ywl7w

Chapter 2
Twenty-Five Years of MALL Experimental Implementation Studies:
What Do We Really Know About It?

Jack Burston
https://orcid.org/0000-0003-2905-5585
Cyprus University of Technology, Cyprus

Androulla Athanasiou
https://orcid.org/0000-0002-0125-8033
Cyprus University of Technology, Cyprus

ABSTRACT

This chapter offers a comprehensive, up-to-date overview of Mobile-Assisted Language Learning (MALL) studies, from the first in 1994 through 2018, relating to the experimental implementation of mobile-based affordances and applications in the teaching and learning of languages, L1 as well as L2. It derives from the analysis of over 2000 MALL studies that have appeared in almost 500 distinct journals and presentations in conferences given by nearly 300 different professional organizations, the great majority of both sources being unrelated to language learning and teaching. It provides background information relating to these studies, documents their profoundly atheoretical nature, details their pedagogical features, and assesses their learning outcome effectiveness.

MALL STUDY SOURCES

MALL studies have been the focus of many reviews. In fact, thirty-two have been published, ten in the past two years alone. The first MALL review was that of Chinnery (2006), which included 22 studies published between 2001-2005. Unlike the latter, which treated the field broadly, most of the remaining 31 overviews have had a restricted focus, with the greatest attention being paid to vocabulary acquisi-

DOI: 10.4018/978-1-7998-1097-1.ch002

tion (Derakhshan & Khodabakhshzadeh, 2011; Afzali, et al., 2017; Mahdi, 2017; Çelik & Yavuz, 2018; Elgort, 2018; Lin & Lin, 2019). Other areas of specific interest were English as the target language (AbuSa'aleek, 2014; Hassan Taj, et al., 2016; Ok & Ratliffe, 2017), authentic environments (Shadiev, Hwang & Huang, 2017; Liu, Shadiev & Hwang, 2017; Shadiev, Liu & Hwang, 2019), collaborative language learning (Cong-Lem, 2018; Kukulska-Hulme & Viberg, 2018), location-based applications (Wang, Liu, & Hwang, 2017), speaking and listening (Kukulska-Hulme & Shield 2007), qualitative research (Yükselir, 2017), Activity theory (Lin *et al.,* 2019) and mobile phone usage (Darmi, & Albion, 2014). Another twelve meta-analyses have comprehensively covered the field of experimental implementations (Kukulska-Hulme & Shield, 2008; Viberg & Grönlund, 2012; Yang, 2013; Lee, et al., 2014; Bozdoğan, 2015; Burston, 2015; Duman, Orhon & Gedik, 2015; Sung, Chang, & Yang, 2015; Al-Qudaimi, 2013; Persson & Nouri, 2018; Cho, et al., 2018; Metruk, 2019).

With so many, and so recent, published MALL overviews, it is legitimate to ask what this chapter has to offer that previous reviews have not. The simple answer to that question is that all previous overviews have been based on very incomplete data. The number of experimental implementations analyzed in existing overviews is remarkably small. Half of the above analyses consisted of less than 25 studies, and another twelve between 25 and 45. Only four reviews involved more than fifty studies, the two most extensive 69 and 82.

The starting point of the present MALL overview was a general 1994-2012 MALL bibliography of some 575 studies underlying Burston 2013, which contained nearly 350 references to experimental implementations. This in itself was considerably more extensive than anything published before then. This general MALL bibliography derived primarily from the most prominent CALL and educational technology journals (*CALICO, CALL, Language Learning & Technology, ReCALL, British Journal of Educational Technology*) and associated conference proceedings. For reasons explained in Burston 2020 (forthcoming), however, reliance on such sources left many MALL studies unreported. Based on a process of systematic bibliographic mining, it turns out that some 876 MALL studies actually appeared during the 1994-2012 period, which is to say nearly a third were missing from the original underlying bibliography.

Rather than relying upon keyword searches of journal and conference publications or academic databases, systematic bibliographic mining started with publications listed in existing MALL overviews. These were then all directly consulted in order to extract from their references any potentially relevant publications not included in the overview bibliographies. In turn, the references in the new finds were consulted and the process systematically continued until no further new references were discovered in any further searches. A second important source for bibliographic mining was *ResearchGate*, which automatically notifies authors when their publications are cited in other papers. Given the frequent citation of the annotated MALL bibliography in Burston 2013, *ResearchGate* notifications provided an ongoing and up-to-date source of new MALL publications since 2012. These in turn were systematically mined for bibliographical references.

On the basis of the journals and conferences cited in the mined bibliographies, a manual search was then undertaken of the article titles and abstracts in these journals and conference proceedings. Where relevant, new papers were consulted and their references mined.

As a result of this systematic bibliographic mining, a total of 2072 MALL studies were identified between 1994 and 2018 (Table 1).

A little more than half (53%) of all MALL studies appeared as journal articles. Amazingly, these were distributed over 485 distinct journals. Equally surprising, two-thirds of these journals had no focus on language teaching, language learning or applied linguistics in any form. The second greatest source of

Table 1. MALL study sources

Sources	Occurrence	
Journals	1108	53.47%
Conference presentations	593	28.62%
Book chapters	167	8.06%
Masters theses	43	2.08%
Doctoral dissertations	34	1.64%
Websites	34	1.64%
Books	11	0.53%
Newspaper articles	10	0.48%
Other	72	3.47%

MALL studies was conference presentations, which accounted for 29% of the total. These appeared in conferences organized by 298 different professional groups, nearly three-quarters (74%) of which were devoted to non-language related topics. In distant third place, book chapters represented 8% of the total. The remaining 10% of total MALL publications included a variety of sources: masters/doctoral dissertations, websites, books, newspaper articles and other miscellaneous publications (e.g., project reports, in-house newsletters, undergraduate theses, papers of undisclosed publication status, etc.).

Of the total 2072 MALL studies, about 60% concerned research topics having nothing to do with experimental implementations: mobile device ownership, software reviews, technical design, pedagogical framework, learning theory, institutional infrastructure, teacher training, etc. Only 851 studies, contained in 789 articles, involved experimental implementations. On closer inspection, however, it turned out that 64 of the studies were in fact repetitions, either completely or substantially the same. For the most part, this was the result of conference presentations subsequently being published as journal articles. Removing these redundancies resulted in the database upon which this study is based containing 787 distinct MALL experimental implementation studies. Of these, full information is known from 699 available studies (i.e., 89%). For the remaining 88, although the actual texts could not be consulted, it was nonetheless possible to glean essential information (e.g., target language, linguistic focus, country of study, learning outcomes, etc.) from titles, abstracts and descriptions in literature reviews. To maximize comprehensiveness, this information has been included in the following analyses, though this results in the numerical basis of each analysis varying slightly from topic to topic.

GENERAL BACKGROUND

Year of Publication

As can be seen in Table 2, very few MALL implementation studies appeared prior to 2003, and even then the annual output did not surpass 20 before 2007. The low number of MALL studies in the early years was a direct consequence of the lack of mobile-device accessibility and the need to provide participants with them. As the personal ownership of mobile phones has increased, especially among

Table 2. MALL study publication years

Year	MALL Implementation Studies	Year	MALL Implementation Studies
1994	1	2007	26
1995	1	2008	33
1996	0	2009	40
1997	2	2010	55
1998	0	2011	99
1999	3	2012	68
2000	1	2013	69
2001	2	2014	63
2002	1	2015	80
2003	13	2016	78
2004	14	2017	70
2005	19	2018	44
2006	5		

university students, so, too, has the annual number of MALL implementation studies. The last mention of loaning phones to participants (Wong *et al.*, 2010) involved the provision of smartphones to primary school children. Some recent studies (Rachels & Rockinson-Szapkiw, 2018), however, still report loaning tablet computers to students, but again to primary school participants. Lastly, it is to be noted that the decrease in MALL studies evidenced in 2018 is almost certainly a result of the time lag involved in collecting bibliographical references and the total number can be expected to increase at least to the level of 2017. In any event, over 50 MALL implementation studies have already been published during the first half of 2019.

Country of MALL Study

As Table 3 attests, experimental MALL studies are indeed a worldwide phenomenon. In fact, MALL implementation studies have been undertaken in 61 countries around the world. That being said, however, it is equally apparent that the distribution of MALL studies around the world is very uneven. Two countries, Taiwan and Japan, account for the greatest number of studies, over 30% between them, more than the next four countries (USA, Iran, Korea, China) combined. Together, the top 14 countries account for over 82% of all experimental MALL studies. The remaining 47 countries each represent less than 1% of all MALL implementations. It is to be noted that, contrary to what might be assumed, this disparity in the distribution of experimental MALL studies is not simply a matter of the level of economic development in a country. Turkey and Malaysia, for example, are on a par with Singapore, Spain and Australia in terms of MALL implementations. Likewise, a number of very developed countries (e.g., Germany, Ireland, Sweden, Belgium) are at the bottom of the list of MALL studies.

Table 3. Country of MALL study

Country	MALL Studies	
Taiwan	111	15.33%
Japan	108	14.92%
USA	74	10.22%
Iran	59	8.15%
Korea	40	5.52%
China	30	4.14%
Canada	26	3.59%
UK	26	3.59%
Saudi Arabia	25	3.45%
Turkey	22	3.04%
Malaysia	21	2.90%
Singapore	21	2.90%
Spain	19	2.62%
Australia	15	2.07%
47 Other countries	1-7	<1% each

Institutional Level of MALL Studies

The great majority, nearly three-quarters (72%), of MALL studies has been conducted with students at post-secondary level, i.e. colleges and universities (Table 4). Primary school represents about 11% and middle/high school combined about the same. The remainder of MALL studies, about 7%, have been undertaken in pre-schools, adult education centers and in the workplace. The high concentration of MALL implementation studies at tertiary level is attributable to two factors. Firstly, universities are the natural domain for research to be conducted in any field and thus investigators are inclined to work with their own students at this level. The second factor pertains to mobile device access and, since the earliest experimental MALL studies, university students have led the way in the ownership of mobile phones.

Table 4. Institutional level of MALL studies

Institution	Number	
University/College	468	71.67%
Primary school	71	10.87%
Middle school	37	5.67%
High school	29	4.44%
Adult Ed	27	4.13%
Pre-school	18	2.76%
Workplace	3	0.46%

Mobile Devices

The mobile devices used in MALL studies very much follow technological developments (Burston, 2015). Palmtop computers, otherwise known as Personal Digital Assistants (PDAs), were the first devices used in MALL studies (Callan, 1994). Owing to their expense, PDAs always had to be loaned to MALL study participants. Although the mainstay PALM computer went out of production in 2009, references to the use of PDAs in MALL studies persist as late as 2016 (Hsu, 2016).

Widespread use of handheld e-dictionaries among Japanese university students facilitated MALL studies based on these devices from 1995 (Sharpe, 1995). The relative infrequency of PC ownership among Japanese university students, compared to their ubiquitous ownership of basic mobile phones, prompted the use of the latter in MALL studies from 2001 (Houser, Thornton, Yokoi & Yasuda, 2001). The advent of the iPod in 2005 introduced the use of MP3 players into MALL studies (Belanger, 2005), though as with PDAs earlier, these initially had to be provided to students. Shortly after, more sophisticated feature phones, which incorporated color screens, cameras, mp3 players and allowed Internet access, began to appear in MALL studies (Alabbadi, 2007).

Following the release of the iPhone (2007), smartphones started making their appearance in MALL studies in 2010 (de Jong, Specht & Koper, 2010; Summet, 2010). Studies based on the iPad followed suit the next year (Sheppard, 2011). Again, initially participants had to be provided with these devices. Although the worldwide sale of smartphones surpassed that of basic/feature mobile phones in 2013 (Bilton, 2014), MALL studies continue to refer to the use of "mobile phones" through 2018. This, however, is almost certainly the result of equating all mobile phones with smartphones since in the countries where MALL studies are the most frequent, this has been the case since at least 2016 when globally feature phones accounted for only 21% of mobile phone sales (Nekkanti, 2018). In any event, where they are identified, nearly two-thirds of all mobile devices used in experimental MALL studies since 2010 have been some type of mobile phone. In comparison, tablet computers account for only 16% of all identified mobile devices during this period. The actual percentage of mobile phone, especially smartphone, usage could in fact be in excess of 75% since some 15% of MALL studies since 2010 do not identify the mobile device used and mobile phones would doubtless represent the majority of unreported devices.

The fact that nearly 100 MALL studies between 2010 and 2018 do not identify the devices used is indicative of the extent to which mobile device accessibility has become commonplace in the past decade. If a MALL study is based on mobile phone usage, it can easily be assumed that participants all have their own smartphones, thus there is no need to state the obvious. So, too, many MALL implementations during this period involve the use of affordances (e.g., audio/video recording and playback) common to a range of mobile devices, so the device used does not attract particular attention. Likewise, where multi-platform MALL apps are involved, which for some time now has been the case, the device used to access them is not deemed worthy of mention.

Mobile Apps and Affordances

Mobile devices are used in two main ways in experimental MALL implementations, to run apps (e-dictionaries, tutorials, games, messaging, social networking) and to exploit their affordances (audio/video playback & recording, Internet access, GPS, RFID tag readers, e-text readers). Unlike mobile device usage, where mobile phones clearly dominate, no app or affordance represents the lion's share

Table 5. Most frequent mobile apps and affordances

Apps	Occurrence		
SMS	82	11.97%	
MMS	19	2.77%	14.74%
Audio/Video Broadcasts	60	8.76%	
Audio/Video Recording	18	2.63%	
Pictures	14	2.04%	13.43%
Internet	66	9.64%	**37.81%**
Other	426	**62.19%**	
Total	685		

in MALL studies. Together, the six most frequent types account for less than 38% of the 685 apps/affordance usage that are encountered (Table 5).

The most frequently occurring app, SMS, represents only about 12%, with MMS bringing the total to 15%. SMS/MMS usage increased substantially between 2008 and 2013, then rapidly declined back to early levels. This decline coincides with the increasingly ubiquitous ownership of smartphones among university students, the largest group of MALL study participants, during the latter period. Notwithstanding, it is interesting to note that SMS/MMS-based MALL studies continue to be undertaken in countries like Canada, Japan, Taiwan and especially Iran, where student-owned smartphones are certainly not lacking.

The advent of audio/video recording and playback facilities, beginning with iPods and extending through incorporation of these affordances into feature phones and smartphones, gave rise to MALL studies based on audio/video broadcasts, most notably pod/vodcasts. In total, however, these account for less than 9% of all MALL implementations. Even adding implementations that required participants to record their own audio, video or still pictures only brings this up to 13%.

Internet-based social networking messaging represents the third largest area of MALL implementations, albeit with less than 10% of the total. Mobile access to these apps is only possible with smartphones. This field is dominated by four apps: *WhatsApp*, *LINE APP*, *WeChat* and *Kakao Talk*. Whereas *WhatsApp* is used quite widely, the use of *LINE APP* occurs primarily in Taiwan, *WeChat* in China and *Kakao Talk* in South Korea.

The remaining 62% of MALL studies employ a wide range of mobile-accessible apps including e-books, e-mail, Facebook, Skype, discussion forums, Google Docs, Automatic Speech Recognition, commercial apps, custom-designed apps, subtitling apps, translation apps, etc., etc.

THEORETICAL FOUNDATIONS OF MALL

There are two distinct types of MALL implementation studies, those that target language learning outcomes and those that do not. Those that do not are concerned rather with student perceptions about mobile device ownership and usage and attitudes regarding MALL pedagogic applications, their perceived convenience, accessibility, effectiveness and factors such as motivation and stress, etc. Whether or not they focus on language learning outcomes, MALL implementations are further distinguished according to whether they involve an L1 or an L2.

In analyzing the theoretical foundations of MALL implementations, whatever their main categorization or L1/L2 focus, the most striking characteristic that emerges is their profoundly atheoretical nature. Nearly two-thirds (65%) of all MALL experimental studies make no reference at all to any guiding theory. While many MALL studies include a literature review that discusses theoretical considerations, quite a number of these do not actually relate these theories to their own MALL implementations. In all, only 29% of MALL implementations are explicitly informed by underlying theories of any sort. However, in a number of cases, participant activities are described in a way that implies the guiding influence of an underlying theory. References to repetition and reinforcement, for example, point to a behavioristic learning framework. Similarly, a focus on the collaborative completion of tasks can be taken to imply a social-constructivist approach. So, too, particular importance accorded to operating in a real-life environment is consistent with a theory of situated learning. Notwithstanding, all together, such implicit theoretical underpinnings add only another 6% to the tally, i.e., even the most generous interpretation of what constitutes a theoretical foundation does not surpass 35% of all MALL implementation studies.

It is to be noted that within MALL experimental implementations not all referenced theories relate to learning in general or language acquisition in particular. In reality, cited theories cover a very broad range including theory construction (Grounded Theory), teaching competence (TPCK), testing (Item Response Theory), motivation (Self-determination Theory), technology acceptance (TAM), to mention only a few.

Within L1-based MALL implementations, social constructivism is by far the most frequently referenced theory (Table 6). Aside from context-based learning, all the other theoretical references in L1 MALL studies occur only once or twice.

Table 6. Theories in L1 MALL studies

Theory	Studies
Social Constructivism	11
Context-based Learning	5
Processing theory	2
Reader Response theory	2
Motivation theory	2
Engagement theory	2
Negotiated meaning	1
Phonological coding	1
Processing theory	1
Experiential Learning	1
Flow theory	1
Grounded theory	1
Self-Determination theory	1
Self-directed Learning	1
Situated Learning	1
TPCK	1

As with L1-based MALL implementations, social constructivism is far and away the most frequently cited theory when working within an L2 context (Table 7). In fact, it is referenced more than three times more frequently than the second place Situated Learning. Four other theories (Dual Coding, Cognitive Load, Behaviorism, Task-based Learning) are closely grouped in what could be considered third place. Beyond this, what is most notable about the theoretical basis of L2 MALL studies, where this is attested, is its enormous variety. A total of 79 other theories are invoked, of which 60 only appear once or twice.

In considering the impact of theory upon MALL experimental implementations, it is instructive to observe how the proportion of (explicit or implicit) theory-based studies has varied over the years (Table 8). The first MALL implementation with a theoretical basis appeared in 2003 (Kim, 2003). Since eight MALL implementations were reported that year, this represents 12.5% of the studies. The relevance of the proportion of theoretically informed MALL implementations very much depends on the number of studies published in a given year. Thus, with only 13 MALL studies published in 2004, the enormous jump of theory-based MALL implementations to nearly 62% (8/13) is really not all that significant.

The proportions of theory-based MALL implementations that are the most significant occur from 2011 onwards when the annual number of published MALL studies remains in excess of fifty. What is most obvious about the proportion of theory-informed MALL studies is that there is no trend for it to increase with time. Except for two peaks around 50% in 2011 and 2012, since 2010 it has consistently ranged about 25%-35%.

Table 7. Theories in L2 MALL studies

Theory	Studies	Theory	Studies
Social Constructivism	73	Input Hypothesis	4
Situated Learning	23	Cognitive theory	3
Dual Coding theory	16	ARCS	3
Cognitive Load	14	Collaborative Learning	3
Behaviorism	13	Comprehensible Input	3
Task-based Learning	13	Connectionism	3
Spaced repetition	10	Grounded theory	3
Authentic Learning	9	Ecological Social Constructivism	3
Activity theory	7	Interactionist theory	3
Context-based Learning	7	IRT	3
Output Hypothesis	7	Lexical Quality hypothesis	3
Self-Determination theory	5	Inquiry-based Learning	3
TAM	5	+60	1-2

Table 8. Theoretical basis of mall studies over time

Year	Theory-based MALL	Total MALL studies	Proportion
2003	1	8	12.50%
2004	8	13	61.54%
2005	2	18	11.11%
2006	1	4	25.00%
2007	5	24	20.83%
2008	13	32	40.63%
2009	16	32	50.00%
2010	13	48	27.08%
2011	41	87	47.13%
2012	30	64	46.88%
2013	19	64	29.69%
2014	18	52	34.62%
2015	18	74	24.32%
2016	26	76	34.21%
2017	21	66	31.82%
2018	15	41	36.59%

PEDAGOGICAL FEATURES OF MALL STUDIES

Targeted Languages

Although totally ignored in previous MALL overviews, experimental MALL studies have in fact been reported for eleven L1, involving a total of 53 implementations (Table 9).

Admittedly, this is very small compared to the 39 L2 languages that have been the focus of 715 MALL studies (Table 10).

Table 9. L1 MALL implementation languages

L1 Language	Studies	
English	31	58.49%
Chinese	9	16.98%
Spanish	4	7.55%
Urdu	2	3.77%
Farsi	1	1.89%
Finnish	1	1.89%
German	1	1.89%
Hausa	1	1.89%
Macedonian	1	1.89%
Malay	1	1.89%
Zarma	1	1.89%

Table 10. L2 MALL implementation languages

L2 Language	Occurrence		L2 Language	Occurrence		L2 Language	Occurrence	
English	571	75.53%	Norwegian	4	0.53%	Albanian	1	0.13%
French	36	4.76%	Irish	4	0.53%	American Sign Language	1	0.13%
Chinese	32	4.23%	Korean	4	0.53%	Arabic	1	0.13%
German	21	2.78%	Swedish	3	0.40%	Basque	1	0.13%
Japanese	21	2.78%	Afrikaans	2	0.26%	ciNyanja	1	0.13%
Spanish	16	2.12%	Turkish	2	0.26%	Croation	1	0.13%
Italian	13	1.72%				Danish	1	0.13%
						Dutch	1	0.13%
						Finnish	1	0.13%
						Greek	1	0.13%
						Greenlandic	1	0.13%
						Hindi	1	0.13%
						Hungarian	1	0.13%
						Icelandic	1	0.13%
						Indonesian	1	0.13%
						Kiswahhil	1	0.13%
						Latin	1	0.13%
						Lithuanian	1	0.13%
						Luxembourgish	1	0.13%
						Māori	1	0.13%
						Ojibwe	1	0.13%
						Portuguese	1	0.13%
						Sanscrit	1	0.13%
						Sepedi	1	0.13%
						Serbian	1	0.13%
						Tagalog	1	0.13%

What is most striking in both cases is the preponderance of English as the target language, nearly 60% as an L1 and over 75% as an L2. Noticeable also is the very quick rate at which the proportion of other languages falls off. Surprisingly, commonly taught L2 languages like German and Spanish account for less than 5% between them. Seven of the L1 and twenty-six of the L2 are represented by a single language. That being said, less commonly taught languages are well represented both as an L1 and L2.

Targeted Proficiency Levels

With the exception of one study (Higgins & Raskind, 2004) that targeted participants with cognitive impairments, all L1 MALL implementations involved native speakers with normal (or at most, delayed) language development. As a consequence, with children and adolescents, language competence level has generally just been equated with the age of the participants. In only one instance (Fotouhi-Ghazvini, Earnshaw & Haji-Esmaeili, 2009) was the ability level of the participants explicitly identified, and this only in general terms, i.e. low. In the case of adults, L1 MALL implementations presuppose native speaker oral competence though, with the exception of two studies that targeted university students (Schugar, Schugar & Penny 2011; Acheson, Cason Barratt & Balthazor, 2013), only basic level literacy.

As regards the linguistic competence level of participants, what is most notable about L2 MALL implementation studies is the lack of information provided about this. Of the 715 studies that were available for analysis, only 353 gave any indication at all about language level. Among these, many times the information provided was too vague to allow the determination of competence level, e.g., four years of previous study, first-year university students, graduate students, etc.). In all, only 332 studies (46%) gave enough information to be able to classify students by competency level. Of these, only 25 (7%) were substantiated by reference to any objective standardized proficiency assessment, and these only for English (e.g., TOEFL, TOEIC, IELTS, PET, etc.). Moreover, there is no consistency in the terminology used to describe competency levels. Sometimes language learner proficiency is indicated relative to the CEFR (e.g., A2, B1, etc.) or ACTFL (e.g., Beginner-mid, Intermediate-low, etc.); but usually only very general terms are used (Beginner, Intermediate, Advanced).

Within these constraints, it is only possible to classify learners into four broad levels: Pre-literate, Beginner, Intermediate, Advanced. The Pre-literate level applied to participants just learning to recognize letters and decipher simple words.

As can be seen in Table 11, although the competence levels reported cover the whole range of proficiency, the great majority (84%) of L2 MALL implementation studies targets learners below the advanced level.

MALL Activity Types

For both L1 and L2 MALL studies, the most striking characteristic of activity types is their very uneven distribution. Among the 52 L1 MALL studies, incorporating a total of 62 activities, the two most frequent (tutorials and e-Reading) together account for more than the next four combined (Table 12). So, too, the top ten account for over 87% of all activity types.

The distribution of activity types within L2 MALL implementations is even more skewed in the direction of tutorial exercises. Of the 809 activities within 665 studies, tutorials on their own account for 27% (Table 13) of the total. This is as much as the six next most frequent combined. Aside from their uneven distribution, L2 MALL activity types are noteworthy for their enormous diversity. In all, 84 different types are represented, with 63 occurring less than one percent of the time.

Table 11. L2 linguistic competence level

Level	Occurrence	
Pre-literate	7	2.11%
Beginner	134	40.36%
Intermediate	139	41.87%
Advanced	52	15.66%

Table 12. L1 MALL study activity types

Activity Type	Occurrence			Activity Type	Occurrence	
Tutorial	11	17.74%		Discussion	1	1.61%
e-Reading	11	17.74%		Social networking	1	1.61%
Game	6	9.68%		MMS	1	1.61%
SMS	6	9.68%		Pre-Reading	1	1.61%
Word processing	5	8.06%		Text deciphering	1	1.61%
Photo taking	4	6.45%		Story telling	1	1.61%
Wiki	4	6.45%		Picture viewing	1	1.61%
E-dictionary	3	4.84%		Identify morphological categories	1	1.61%
Quizzes	2	3.23%				
Word construction	2	3.23%				
		87.10%				**12.90%**

Table 13. L2 MALL study activity types

L2 Activity Type	Occurrence			Occurrence	
Tutorial	218	26.95%	Video creation	19	2.35%
Game	49	6.06%	Communication	18	2.22%
SMS	45	5.56%	Discussion	18	2.22%
e-Reading	43	5.32%	Video viewing	17	2.10%
Audio listening	39	4.82%	MMS	16	1.98%
Audio recording	24	2.97%	Written composition	15	1.85%
Quiz	21	2.60%	Podcast listening	14	1.73%
Word look-up	21	2.60%	Flashcards	12	1.48%
Blog	20	2.47%	Chat	9	1.11%
Information retrieval	20	2.47%	QR scans	9	1.11%
Photo taking	20	2.47%	**+ 63 Other Types**		**< 1% each**
		82.45%			**17.55%**

Over time, the combined L1/L2 frequency of occurrence of the four most favored activity types (tutorials, games, e-reading, SMS) evidences considerable variation. As can be seen in Figure 1, which tracks year by year on the horizontal axis the total occurrences of each activity type, the distribution of these activities is for the most part clustered in the last ten years. Before then, due to technological constraints, SMS-based activities were the dominant paradigm. Usage of SMS more than doubled between 2005 and 2012, but has since essentially disappeared from the scene. Tutorials only begin to appear in any number from 2008. Their occurrence then fairly regularly increases until 2017, after which they fall back to about their 2012 level. However, as indicated previously with regard to yearly publication statistics, this decline may only be a consequence of the time lag involved in collecting data about the most recent MALL studies. Game-based apps occur as early as 2005 (Yang, Lai & Chu, 2005) and climb

Figure 1. Activity types distribution of occurrence

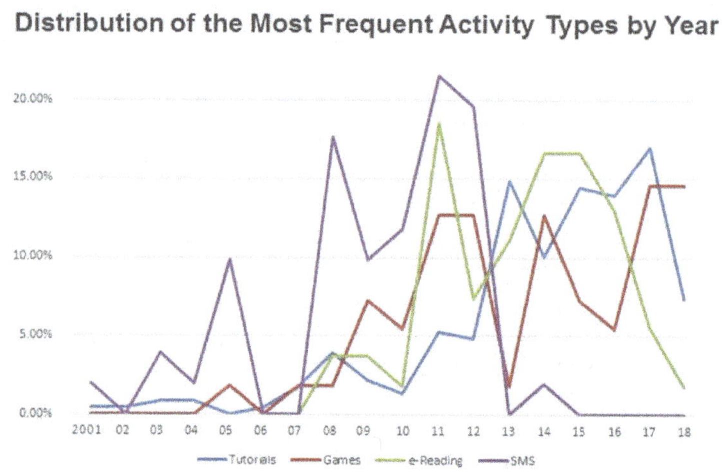

regularly in number until 2012. More recent years are characterized by variations in their occurrence, though their use appears to be stabilizing since 2017. The occurrence of e-reading activities is only attested since 2008 (Chen & Hsu, 2008). Its use since then has evidenced considerable ups and downs, but has been in free-fall since 2015.

Individual vs Collaborative Studies

In line with the preponderance of tutorial exercises, not surprisingly MALL implementations have been overwhelmingly oriented towards individual activities. These account for about 70% of all MALL studies with a language learning focus. As can be seen in Figure 2, which tracks on the vertical axis the yearly proportion of orientation types, prior to 2003 all experimental MALL implementations involved individual activities. However, up until 2006 the number of annual experimental MALL studies was quite small, so the relative percentages of orientation types in the early years need to be considered accordingly. Since 2007, the annual number of published MALL implementation studies has been high enough (31-87) to regard the relative proportion of orientation types with confidence. Accordingly, a very clear decrease in the proportion of individually-oriented MALL implementations is observable, falling from 88% in 2007 to 67% in 2018. During the same period, collaborative-based MALL implementations (with or without an individual component) correspondingly increased from 15% to over 41% in 2016. They then dropped back to pre-2016 levels with individually-oriented MALL implementations picking up the difference.

Linguistic Focus

The linguistic focus of L1 MALL studies is very much on literacy skills, with 69% of all implementations involving reading in one form or another (Table 14). The rest deal with vocabulary, grammar and writing. Studies explicitly targeting literacy are of four types. Pre- literacy targets the sound/symbol correspondence associated with alphabet learning and primarily involves pre-school children. Early literacy involves learners of all ages and focuses on the reading of words and simple sentences. Basic

Figure 2. Individual vs collaborative studies by year

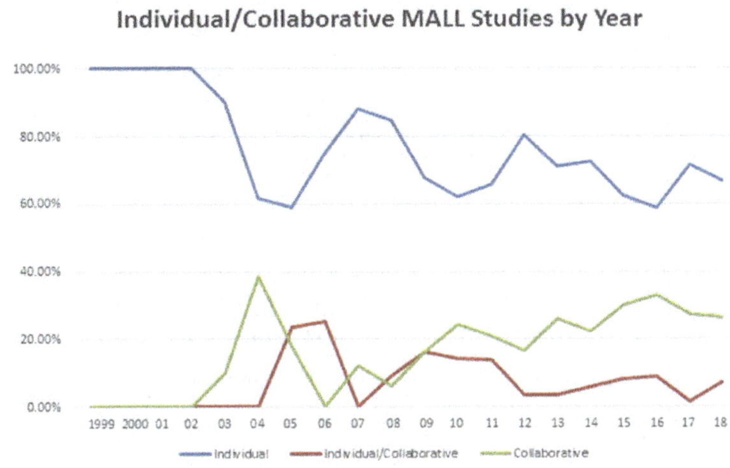

literacy involves the reading of short texts as undertaken by school children and adult education learners. Advanced literacy targets higher levels of literacy, typically in high school environments. Nearly half of the language focus categories (47%) fall outside the domain of literacy; but each has a very low rate of occurrence (i.e., 1.28%). Not surprisingly, with such emphasis being put on reading, interpersonal communication is notable by its relegation to this marginal linguistic focus.

Table 14. MALL studies L1 language focus

Language Focus	Studies	
Reading	22	28.21%
Vocabulary	16	20.51%
Pre-literacy	8	10.26%
Writing	8	10.26%
Advanced Literacy	5	6.41%
Early literacy	4	5.13%
Spelling	3	3.85%
Grammar	2	2.56%
Basic literacy	2	2.56%
Listening	1	1.28%
Note taking	1	1.28%
Speaking	1	1.28%
Test preparation	1	1.28%
Rhetoric	1	1.28%
Communication	1	1.28%
Handwriting	1	1.28%
Sentence building	1	1.28%

Table 15. MALL studies L2 language focus

Language Focus	Studies		Language Focus	Studies	
Vocabulary	294	33.07%	Community	7	0.79%
Listening	125	14.06%	Translation	4	0.45%
Speaking	115	12.94%	Verb conjugation	3	0.34%
Reading	111	12.49%	Oral reading	3	0.34%
Writing	98	11.02%	Spelling	2	0.22%
Grammar	40	4.50%	Signing	1	0.11%
Pronunciation	29	3.26%	Business presentations	1	0.11%
Communication	14	1.57%	Task-based language usage	1	0.11%
Test preparation	13	1.46%	Oral sentence construction	1	0.11%
Culture	8	0.90%	Phonetics	1	0.11%
Character R&W	8	0.90%	Chanting	1	0.11%
Literacy	8	0.90%	Sentence building	1	0.11%

Even more than with L1 MALL studies, the linguistic focus of L2 implementations covers a very broad range, twenty-four categories in all, and an even greater proportion of these (63%) is relegated to marginal status, i.e., each below 1% of all occurrences (Table 15). By far, the main linguistic focus of L2 MALL studies is on vocabulary acquisition, which by itself accounts for a third of all studies. Combined, a little over half of the MALL implementations target listening, speaking, reading and writing in fairly equal measure. Interest in other language areas drops markedly after this, with grammar and pronunciation together accounting for less than 8%. Again, at less than 2%, it is to be noted that interpersonal communication fares little better in L2 than in L1 studies.

As can be seen in Figure 3, which tracks on the vertical axis the relative yearly proportion of the five most frequent language focuses, prior to 2006 vocabulary totally dominated L2 MALL studies. Though displaced from top position by listening in 2006, vocabulary has since then consistently accounted for over a third of all L2 MALL implementations. The remaining four language focuses (reading, listening, speaking and writing) have on average fluctuated within a fairly narrow range of about 14%-18% of the total per year.

MALL Curriculum Integration

The degree of curricular integration of MALL can be classified into five categories: Pilot testing, In-class trialing, Experimental supplements, Experimental complements, Fully integrated. Overall, some 28% of all MALL implementations have been of very short duration (pilot tests/class trials), with more than half of these taking place over 1-2 sessions lasting at most a total of three hours. Some were as short as five minutes. Pilot tests were intended to get feedback on MALL apps as part of the development process. In-class trials took place over 1-3 weeks and were similarly primarily concerned with operational feedback and student perceptions, though some learning outcomes are reported.

Figure 3. MALL studies L2 language focus by year

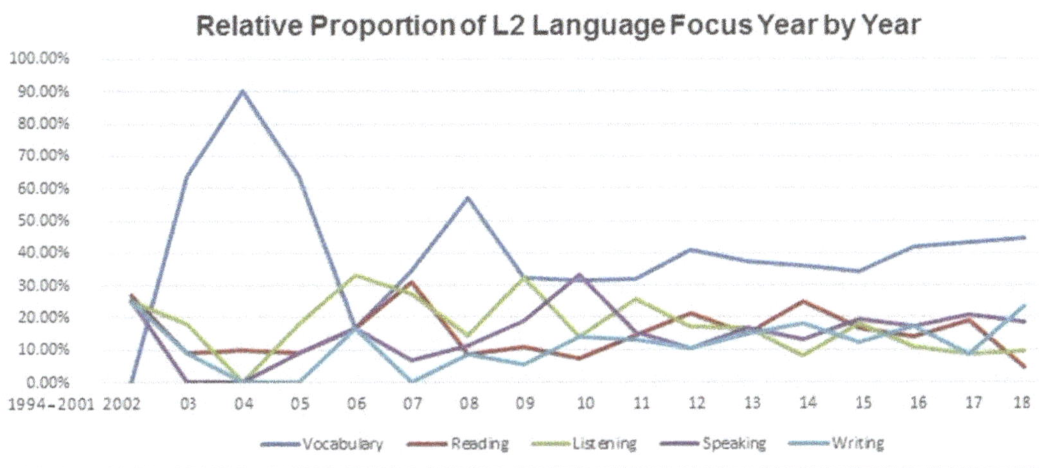

Both pilot testing and in-class trials were more prevalent in the early years 1999-2006, at times reaching nearly 40% for pilot testing and 50% for in-class trialing. However, as indicated previously, the number of MALL implementations during these years was quite low, so these percentages are not as relevant as they might seem. As can be seen in Figure 4, which tracks on the vertical axis the relative yearly proportion of the integration levels, short-term studies have progressively decreased since 2007 and in the last two years have accounted for only about 10%-15% of all MALL implementations. Experimental class supplements, which were voluntary in nature and did not enter into course grade calculations, typically lasted only a few weeks. Like pilot testing and in-class trialing, these occurred more frequently during the early MALL years and their relative proportion has gradually decreased since 2007, though they continue to represent about 20% of all MALL implementations. The proportion of experimental complements, involving compulsory (and usually assessed) course components and mostly lasting an entire term/semester (i.e., 10 weeks +), has steadily increased since 2007 and in the past two years has represented over 50% of all MALL studies. An indication of the increasing maturity of MALL implementations can be seen in the number of studies in which MALL is regarded as a fully integrated course component as opposed to an experimental constituent. Since 2011, these studies, which last for one year or more, have gradually increased to the point where they now regularly represent 15%-20% of all MALL studies. The proportion of these dropped considerably in 2018, but again this result is very likely a consequence of incomplete data for this year.

Learning Outcomes

Language learning outcome studies fall into two distinct groups, those that substantiate results with objective evidence (i.e., formal targeted tests) and those that rely only on subjective evidence (e.g., student self-reports, instructor impressions) or general results such as mid-term, final or overall course grades. Retaining only MALL implementations with objective learning outcome results, there are 414 studies that can be considered when evaluating MALL language learning effectiveness. Three types of consistent results are reported:

Figure 4. Level of MALL curricular integration

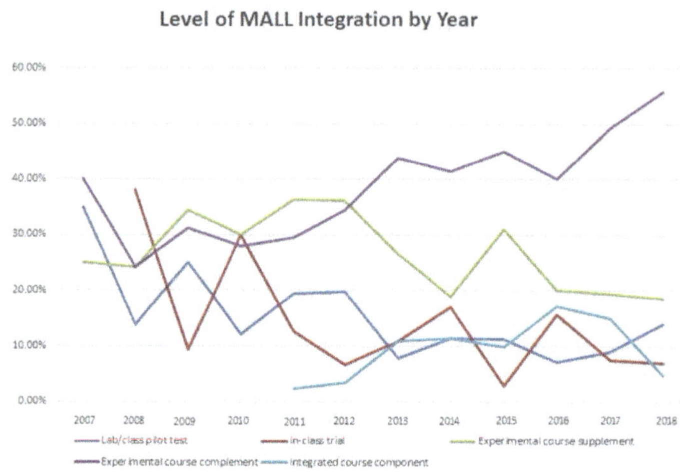

1. those which are entirely positive for the experimental MALL group (+EXP),
2. those with no significant difference between the experimental MALL group and the control (~EXP)
3. those where the results of the experimental MALL group were negative compared to the control (–EXP).

Mixed results are also indicated (EXP), e.g., positive for vocabulary and grammar, but no difference for writing.

As can be seen in Table 16, nearly two-thirds (64%) of the substantiated learning outcome studies favored the MALL condition. Studies that showed no difference or had mixed results were about the same at 16%-17% of the MALL studies. Negative MALL-based learning outcomes are reported in less than 3% of the studies.

In considering the high level of reported MALL successes, it needs to be kept in mind that some studies indicate the number of participants involved (N) but not the duration of the treatment (T) and vice versa. Other studies specify neither the N nor the T. Moreover, although the N may be specified, in about ten percent of the cases it is no more than nine and can even be as low as one participant. Likewise, even when T is specified, it is not always clearly indicated to what this corresponds, i.e., authors

Table 16. Learning outcomes in MALL studies

Language learning outcomes	Results	Non-objective evidence		Objective evidence	
500		**86**	17.20%	**414**	82.80%
	+EXP	48	55.81%	264	63.77%
	~EXP	5	5.81%	71	17.15%
	-EXP	7	8.14%	12	2.90%
	EXP	26	30.23%	67	16.18%

may indicate that their study lasted 4 weeks, but fail to specify the number of contact sessions and their duration. So, too, in nearly 15% of MALL implementations T represents a single session, sometimes as short as 5 minutes.

In determining the validity of research results, minimal conditions must be met regarding the number of treatment participants (N) and the duration of the treatment (T). Although there are no hard rules concerning a minimal N and T, it is reasonable to assume a minimal N of 10 and T of one month (i.e., four weeks x 2 hours per week). On this basis, 257 of the original 414 studies (63%) were retained as the basis of analysis.

As can be seen in Table 17, though much fewer in number, the proportion of favorable MALL outcomes is slightly higher when minimal N/T conditions are applied to the analysis. Positive MALL outcomes rise to over 65% accompanied by a decrease in the incidence of implementations showing no difference and a decline of negative results below 2%.

Finally, in examining the percentage of valid N-T objective MALL language outcome studies on a yearly basis (Figure 5), it can be seen that, with a couple of dips, it has climbed steadily since 2006 to over 70%. Once again, it needs to be kept in mind that percentages before 2007 are based on a relatively low number of studies and the apparent decrease in 2018 must be considered in light of the incomplete data for this year.

Table 17. N-T valid objective mall language learning outcomes

MALL results	Studies	
+EXP	168	65.37%
~EXP	41	15.95%
-EXP	5	1.95%
EXP	43	16.73%

Figure 5. Valid N-T objective MALL language outcome studies by year

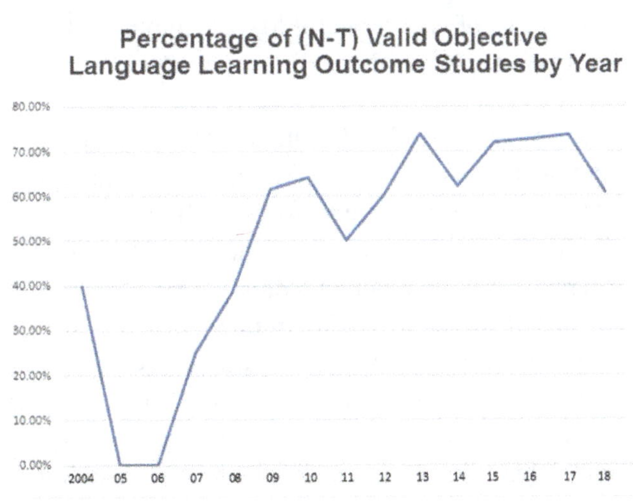

CONCLUSION

As may be judged by the preceding discussion, this overview of experimental MALL implementations is noteworthy in two fundamental respects. Firstly, based on a maximally comprehensive review of the literature, it confirms a number of observations in previous, much more limited, meta-analyses. Secondly, and more importantly, it brings to light new facts about MALL implementations which have hitherto not been reported.

In conformity with previous MALL overviews, this study has shown Taiwan and Japan to be the countries in which the greatest number of experimental MALL projects have been implemented. Likewise, in whatever country, it is at university level where MALL studies have been the most heavily concentrated. So, too, student-owned mobile phones are by far the device of choice for MALL implementations. As many other reviews have shown, MALL studies are dominated by a focus on L2 English below the advanced level. Whatever the targeted L2, tutorial applications with an emphasis on vocabulary acquisition have been the mainstay of experimental MALL studies. Moreover, the use of MALL to support interpersonal communication has been virtually non-existent. Lastly, this analysis has confirmed the longstanding and overwhelming claims about successful MALL learning outcomes. That being said, however, it has also once again demonstrated serious design flaws in the determination of such results, most notably the absence of objective evidence, small participant numbers, short treatment times and/or failure to even specify one or the other.

The new findings emerging from this review are numerous and significant. Firstly, it has demonstrated the effectiveness of bibliographical mining in identifying MALL studies in general and experimental implementations in particular, over 2000 for the former and nearly 800 for the latter. In the process, it has shown that MALL studies have appeared in nearly 500 different journals, the great majority outside the domain of those focused on language learning and teaching. Likewise, with regard to conference presentations, it has documented that out of nearly 300, three-quarters were given at non-language related symposiums. So, too, it has shown the truly worldwide spread of MALL studies and the impressive number of languages to which it has been applied, L1 as well as L2. Although mentioned in some previous reviews, the real extent of the profoundly atheortetical nature of experimental MALL studies has been documented and tracked annually for the first time, the results of which demonstrate no upward trend. Regarding the theoretical foundations of experimental MALL implementations, a striking contrast has also been shown between the small number of theory-informed studies and the great variety of theories that are invoked.

With respect to the linguistic parameters of MALL implementations, this study is the first to document the existence of L1 studies and their focus on literacy. While it has been reported for some time that the great majority of experimental L2 MALL implementations is directed towards beginning and intermediate level learners, this study has also demonstrated for the first time the very limited extent to which the proficiency level of learners is actually documented, not even half the time. Although tutorials are well known to be the activity type of choice in MALL implementations, only in this study has the great variety of activity types within MALL been shown. This study is also unique in tracking the distribution of activity types and language focus over time. Though, overall, among the most frequently occurring activity types, SMS and e-reading have all but disappeared in recent years. Likewise, while vocabulary acquisition still holds sway in L2-based studies, reading, listening, speaking and writing taken together have over the last decade carved out about a 60% representation. In addition, while representing only a small fraction of the total, a considerable variety of language focuses is attested in MALL implementations. In line with the demonstrated dominance of tutorial exercises with their focus on reading, listening, and writing, individually-oriented activities are shown to be the norm, though decreasingly so over the past ten years. Finally, a diachronic analysis during this period has also demonstrated the increasing level of integration of MALL implementations into the curriculum and a very substantial improvement in the proportion of study outcome results based on objectively supported and valid N/T data.

REFERENCES

AbuSa'aleek, A. (2014). A review of emerging technologies: Mobile Assisted Language Learning (MALL). Asian Journal of Education and e-Learning, 2(6), 469-475. Retrieved from https://www.academia.edu/9844553/A_Review_of_Emerging_Technologies_Mobile_Assisted_Language_Learning_MALL_

Afzali, P., Shabani, S., Basir, Z., & Ramazani, M. (2017). Mobile-assisted vocabulary learning: A review study. *Advances in Language and Literary Studies, 8*(2), 190-195. doi:10.7575/aiac.alls.v.8n.2p.190

Al-Qudaimi, K. (2013). Mobile-Assisted Language Learning: A literature review. 1-15. Unpublished. Retrieved from https://www.academia.edu/6369026/Mobile-Assisted_Language_Learning_A_Literature_Review

Bozdoğan, D. (2015). MALL revisited: Current trends and pedagogical implications. *Procedia: Social and Behavioral Sciences, 195*, 932-939. doi:10.1016/j.sbspro.2015.06.373

Burston, J. (2015). 20 years of MALL project implementation: A meta-analysis of learning outcomes. *ReCALL, 27*(1), 4-20. doi:10.1017/S0958344014000159

Çelik, Ö., & Yavuz, F. (2018). An extensive review of literature on teaching vocabulary through mobile applications. *Bilecik Şeyh Edebali University Journal of Social Sciences Institute*, 3(1), 56-91. Retrieved from https://www.researchgate.net/publication/325540454_An_Extensive_Review_Of_Literature_On_Teaching_Vocabulary_Through_Mobile_Applications

Chinnery, G. (2006). Going to the MALL: Mobile Assisted Language Learning. *Language Learning & Technology, 10*(1), 9-16. Retrieved from http://llt.msu.edu/vol10num1/emerging/default.html

Cho, K., Lee, S., Joo, M.-H., & Becker, B. (2018). The effects of using mobile devices on student achievement in language learning: A meta-analysis. *Education in Science, 8*(3), 1-16. doi:10.3390/educsci8030105

Cong-Lem, N. (2018). A review of research in mobile-assisted collaborative language learning. *VNU Journal of Foreign Studies*, 34(4), 71–82. Retrieved from https://www.researchgate.net/publication/327315618_A_REVIEW_OF_RESEARCH_IN_MOBILE_ASSISTED_COLLABORATIVE_LANGUAGE_LEARNING

Darmi, R., & Albion, P. (2014). A review of integrating mobile phones for language learning. In *Proceedings of 10th International Conference Mobile Learning* (pp. 93–100). International Association for Development of the Information Society; Retrieved from http://files.eric.ed.gov/fulltext/ED557201.pdf

Derakhshan, A., & Khodabakhshzadeh, H. (2011). Why CALL not MALL: An in-depth review of text-message vocabulary learning. *Theory and Practice in Language Studies, 1*(9), 1150-1159. doi:10.4304/tpls.1.9.1150-1159

Duman, D., Orhon, G., & Gedik, N. (2015). Research trends in mobile assisted language learning from 2000 to 2012. *ReCALL, 27*(2), 197-216. doi:10.1017/S0958344014000287

Elgort, I. (2018). Technology-mediated second language vocabulary development: A review of trends in research methodology. *CALICO Journal, 35*(1), 1-29. Retrieved from https://journals.equinoxpub.com/index.php/CALICO/article/view/34554/pdf

Hassan Taj, I., Sulan, N., Sipra, M., & Ahmad, W. (2016). Impact of Mobile-Assisted Language Learning (MALL) on EFL: A meta-analysis. *Advances in Language and Literary Studies, 7*, 2. Retrieved from http://www.journals.aiac.org.au/index.php/alls/article/view/2164/1914

Kukulska-Hulme, A., & Shield, L. (2007). An overview of Mobile Assisted Language Learning: Can mobile devices support collaborative practice in speaking and listening? EuroCALL Virtual Strand. Retrieved from http://vsportal2007.googlepages.com/collaborativepractice

Kukulska-Hulme, A., & Shield, L. (2008). An overview of Mobile Assisted Language Learning: From content delivery to supported collaboration and interaction. *ReCALL, 20*(3), 271-289. doi:10.1017/S0958344008000335

Kukulska-Hulme, A., & Viberg, O. (2018). Mobile collaborative language learning: State of the art. *British Journal of Educational Technology, 49*(2), 207-218. doi:10.1111/bjet.12580

Lee, Y.-S., Sung, Y.-T., Chang, K.-E., Liu, T.-C., & Chen, W.-C. (2014). A meta-analysis of the effects of learning languages with mobile devices. Lecture Notes in Computer Science, 8699, 106-113. Retrieved from https://link.springer.com/chapter/10.1007%2F978-3-319-13296-9_12

Lin, C.-C., Lin, V., Liu, G.-Z., Kou, X., Kulikova, A., & Lin, W. (2019). Mobile-assisted reading development: A review from the Activity Theory perspective. Computer Assisted Language Learning. Retrieved from https://www.tandfonline.com/doi/pdf/10.1080/09588221.2019.1594919?needAccess=true

Lin, J.-J., & Lin, H. (2019). Mobile-assisted ESL/EFL vocabulary learning: A systematic review and meta-analysis. *Computer Assisted Language Learning*, 1-42. Retrieved from https://www.tandfonline.com/doi/full/10.1080/09588221.2018.1541359?scroll=top&needAccess=true

Liu, T., Shadiev, R., & Hwang, W.-Y. (2017). Review of research on MALL in familiar authentic environments. *Computer Assisted Language Learning, 30*(3-4), 284-303. doi:10.1080/09588221.2017.1308383

Mahdi, H. (2017). Effectiveness of mobile devices on vocabulary learning: A meta-analysis. *Journal of Educational Computing Research, 55*(3), 1-21. Retrieved from https://kku-sa.academia.edu/HassanMahdi

Metruk, R. (2019). The call of the MALL: The use of smartphones in higher education. A literature review. Revista Dilemas Contemporáneos: Educación, Política y Valores, 6(3), np. Retrieved from https://www.academia.edu/38990631/The_Call_of_the_MALL_the_Use_of_Smartphones_in_Higher_Education._A_Literature_Review?email_work_card=view-paper

Ok, M., & Ratliffe, K. (2017). Use of mobile devices for English language learner students in the United States: A research synthesis. *Journal of Educational Computing Research, 56*(4), 538-562. doi:10.1177/0735633117715748

Persson, V., & Nouri, J. (2018). A systematic review of second language learning with mobile technologies. *International Journal of Emerging Technologies in Learning*, *13*(2), 188-210. doi:10.3991/ijet.v13i02.8094

Shadiev, R., Hwang, W.-Y., & Huang, Y.-M. (2017). Review of research on mobile language learning in authentic environments. *Computer Assisted Language Learning*, *30*(3-4), 284-303. doi:10.1080/09588221.2017.1308383

Shadiev, R., Liu, T., & Hwang, W.-Y. (2019). Review of research on mobile-assisted language learning in familiar, authentic environments: Review of research on MALL in familiar contexts. *British Journal of Educational Technology*, 1-12. Retrieved from https://onlinelibrary.wiley.com/doi/pdf/10.1111/bjet.12839

Sung, Y.-T., Chang, K.-E., & Yang, J.-M. (2015). How effective are mobile devices for language learning? A meta-analysis. *Educational Research Review*, *16*, 68-84. doi:10.1016/j.edurev.2015.09.001

Viberg, O. & Grönlund, Å. (2012). Mobile Assisted Language Learning: A literature review. *Proceedings of 11th World Conference on Mobile and Contextual Learning, mLearn 2012* (pp. 1-8). Retrieved from http://du.diva-portal.org/smash/get/diva2:549644/REFERENCES01

Wang, H.-Y., Liu, G.-Z., & Hwang, G.-J. (2017). Integrating socio-cultural contexts and location-based systems for ubiquitous language learning in museums: A state of the art review of 2009-2014. *British Journal of Educational Technology*, *48*(2), 653-671. doi:10.1111/bjet.12424

Yang, J. (2013). Mobile-Assisted Language Learning: Review of the recent applications of emerging mobile technologies. *English Language Teaching*, *6*(7), 19-25. doi:10.5539/elt.v6n7p19

Yükselir, C. (2017). A Meta-synthesis of qualitative research about Mobile-Assisted Language Learning (MALL) in foreign language teaching. *Arab World English Journal*, *8*(3), 302-318. doi:10.24093/awej/vol8no3.20

ADDITIONAL READING

Acheson, P., Barratt, C., & Balthazor, R. (2013, December). Kindle in the Writing Classroom. *Computers and Composition*, *30*(4), 283-296. doi:10.1016/j.compcom.2013.10.005

Alabbadi, M. (2007). Learner's acceptance based on Shackell's usability model for supplementary mobile learning of an English course. Science and Technology, 121-128. Retrieved from http://www.scitepress.org/DigitalLibrary/Link.aspx?doi=10.5220/0002810101210128

Belanger, Y. (2005). Duke University iPod First Year Experience Final Evaluation Report. Center for Instructional Technology. Duke University. Retrieved from http://cit.duke.edu/pdf/reports/ipod_initiative_04_05.pdf

Bilton, N. (2014). Smartphone sales beat feature phones in 2013. BITS Business, Innovation, Technology, Society. Retrieved from https://bits.blogs.nytimes.com/2014/02/13/smartphone-sales-beat-feature-phones-in-2013/

Burston, J. (2013). Mobile-Assisted Language Learning: A selected annotated bibliography of implementation studies 1994-2012. Language Learning & Technology, Special volume on Mobile-Assisted. *Language Learning*, *17*(3), 157–224. Retrieved from http://llt.msu.edu/issues/october2013/burston.pdf

Burston, J. (2020). MALL: What do we really know about it? In A. Kukulska-Hulme & V. Morgana (Eds.), *Selected papers from the 2018 CALLmi Conference*, May 23-24, 2019, Milan, Italy. The Routledge Focus Series.

Callan, S. (1994). Can the use of hand-held personal computers assist transition students to produce written work of excellent quality? Ontario, Canada: Wentworth County Board of Education.

Chen, C.-M., & Hsu, S.-H. (2008). Personalized intelligent mobile learning system for supporting effective English learning. *Journal of Educational Technology & Society*, *11*(3), 153-180. Retrieved from http://www.ifets.info/journals/11_3/12.pdf

de Jong, T., Specht, M., & Koper, R. (2010). A study of contextualised mobile information delivery for language learning. *Journal of Educational Technology & Society*, *13*(3), 110-125. Retrieved from http://www.ifets.info/journals/13_3/11.pdf

eMarketer (2015). Mobile phone users in Japan upgrade to smartphones, finally. Retrieved from https://www.emarketer.com/Article/Mobile-Phone-Users-Japan-Upgrade-Smartphones-Finally/1012880

Fotouhi-Ghazvini, F., Earnshaw, R., & Haji-Esmaeili, L. (2009). Mobile-Assisted Language Learning in a developing country context. *International Conference on CyberWorlds*, 391-397. Retrieved from http://www.computer.org/portal/web/csdl/doi/10.1109/CW.2009.28

Higgins, E., & Raskind, M. (2004). The compensatory effectiveness of the *Quicktionary Reading Pen II* on the reading comprehension of students with learning disabilities. *Journal of Special Education Technology*, *20*(1), 31-40. doi:10.1177/016264340502000103

Houser, C., Thornton, P., Yokoi, S., & Yasuda, T. (2001). Learning on the move: Vocabulary study via mobile phone email. ICCE 2001 Proceedings, 1560-1565. Retrieved from https://www.learntechlib.org/noaccess/9065/

Hsu, T.-C. (2016). Enhancing listening comprehension and vocabulary recall in mobile-assisted language learning. In A. Palalas & M. Ally (Eds.), *International Handbook of Mobile-Assisted Language Learning* (pp. 608-632). China Central Radio & TV University Press, Co., Ltd.

Kim, I.-S. (2003). The development of the PDA-based multimedia contents for teaching an English syntax course. Multimedia-Assisted Language Learning, 6(2), 9-34. Retrieved from http://kmjournal.bada.cc/wp-content/uploads/2013/05/6-2-1Kim.pdf

Nekkanti, N. (2018). Samsung loses top spot in feature phones to Reliance Jio in India. SAMMobile. Retrieved from https://www.sammobile.com/news/samsung-loses-top-spot-in-feature-phones-to-reliance-jio-in-india/

Rachels, J., & Rockinson-Szapkiw, A. (2018). The effects of a mobile gamification app on elementary students' Spanish achievement and self-efficacy. *Computer Assisted Language Learning*, *31*(1-2), 72-89. Retrieved from http://www.tandfonline.com/doi/pdf/10.1080/09588221.2017.1382536. doi:10.1080/09588221.2017.1382536

Schugar, J., Schugar, H., & Penny, C. (2011). A Nook or a book: Comparing college students' reading comprehension level, critical reading, and study skills. *International Journal of Technology in Teaching and Learning*, *7*(2), 174-192. Retrieved from http://www.sicet.org/journals/ijttl/issue1102/6_Schugar.pdf

Sharpe, P. (1995). Electronic dictionaries with particular reference to the design of an electronic bilingual dictionary for English-speaking learners of Japanese. *International Journal of Lexicography*, *8*(1), 39-54. Retrieved from http://ijl.oxfordjournals.org/content/8/1/39.full.pdf+html. doi:10.1093/ijl/8.1.39

Sheppard, D. (2011). Reading with iPads - The difference the difference makes. *Education Today*, 3, 12-17. Retrieved from http://www.minnisjournals.com.au/articles/ipads%20et%20t3%2011.pdf

Summet, V. (2010). Facilitating Communication for Deaf Individuals with Mobile Technologies. PhD dissertation, Georgia Institute of Technology. Retrieved from https://smartech.gatech.edu/bitstream/handle/1853/33878/summet_valerie_h_201005_phd.pdf

Wong, L.-H., Chin, C.-K., Tan, C.-L., Liu, M., & Gong, C. (2010). Students' meaning making in a mobile assisted Chinese idiom learning environment. Learning in the Disciplines: ICLS 2010 Conference Proceedings - 9th International Conference of the Learning Sciences 1 (pp. 349-356). Retrieved from https://www.researchgate.net/publication/220934587_Students%27_meaning_making_in_a_mobile_assisted_Chinese_idiom_learning_environment

Yang, J.-C., Lai, C.-H., & Chu, Y.-M. (2005). Integrating speech technologies into a one-on-one digital English classroom (pp. 159-163). *Wireless and Mobile Technologies in Education*; Retrieved from http://ieeexplore.ieee.org/xpl/articleDetails.jsp?arnumber=1579256

Chapter 3
WhatsApp and Jitsi to Foster Student Engagement in an American-Spanish Telecollaboration Exchange

Alberto Andujar
https://orcid.org/0000-0002-8865-9509
University of Almeria, Spain

Jose M. Franco Rodriguez
Fayetteville State University, USA

ABSTRACT

This chapter explores students' engagement in a telecollaboration project between a Spanish and an American university. Students' cognitive, emotional, and behavioral engagement were evaluated throughout the project. A total of 53 students participated in an online exchange during one and a half months through two different applications, WhatsApp representing the text-based environment and Jitsi representing the synchronous videoconferencing platform. The engagement construct was explored using pre and post measures as well as tracking students' conversation in the platforms. Results yielded high levels of cognitive engagement as a result of the interaction. Values for emotional engagement were found to be higher in the instant messaging platform and behavioral engagement did not present significant values. Implications and recommendations for future research were drawn.

INTRODUCTION

Telecollaboration processes, which involve the interaction of students that are geographically distant from each other (Dooly, 2008), have normally made use of the available technological resources to develop virtual exchanges between classes with similar aims. From an early use of email exchanges (O'Dowd, 2005) to the development of gamified environments and videoconferencing tools to simulate real-life

DOI: 10.4018/978-1-7998-1097-1.ch003

interaction (Andujar, 2019; Melchor-Couto, 2019), telecollaboration processes have experienced a gradual change. This evolution of the technological choice in telecollaboration environments has contributed to enhancing the possibilities of foreign language interaction between peers which can access these virtual environments anytime and anywhere. Likewise, the possibilities of developing students' linguistic competence have been also enhanced due to the different channels of communication that can be implemented in telecollaboration projects. However, this technological evolution has also entailed teachers and students' adaptation to the new virtual environments in which the projects take place. So as to guarantee an adequate adoption to the new tools in use, researchers as well as language practitioners need to further understand students' behaviour in technology-mediated learning and how the use of certain technological tools instead of others can modify the quality and perceptions about a telecollaboration project. Notwithstanding that factors such as the task design, the type of interaction and the aims of the virtual exchange play a fundamental role in the overall perception of the telecollaboration process, students' use of technology as well as their degree of engagement with the virtual environment may also be of great interest for the success of this type of projects.

Students' engagement has been scarcely investigated in the telecollaboration literature which have normally focused on factors such as developing language skills and intercultural competence (e.g. Helm & Guth, 2016; Gimeno-Sanz, 2018; Yang, 2018), or tutoring language learners (e.g. Angelova & Zhao, 2016) among others. Likewise, telecollaboration projects have frequently made use of a single virtual platform for the development of the interaction, in some cases, limiting students' possibilities of using oral and written modes of communication. Nevertheless, advancements in technology allow students to access a videoconference through the browser or keep a text-based conversation through a ubiquitous device. These new possibilities offer a fertile ground for language teachers willing to connect classes and implement telecollaboration exchanges through the Internet. Mobile instant messaging services such as WhatsApp, Wechat, Line, or Facebook messenger allow participants to interact through text messages, voice recording and video calls allowing participants to interact ubiquitously through different channels. Similarly, applications such as Jitsi provide a private videoconference room in the browser through the Web Real-time communication protocol (WebRTC). Students can access the room and start the interaction through a link without the need to register or download any specific software, avoiding difficulties of previous videoconferencing platforms. In this context, understanding students' patterns when using the aforementioned technological tools and measuring the degree of engagement generated in students seems fundamental to develop successful telecollaboration processes. This greater understanding may contribute to the subsequent integration of telecollaboration projects into the language curriculum, which may benefit language learners at all levels.

LITERATURE REVIEW

Virtual Exchange

Telecollaboration between learners has taken place since the late 1800s with occasional developing spurts, but it was the advent of personal computers and the Internet in the 1990s that led to the current surge in this type of instructional practices, also known as virtual exchange (VE) (Dooly and O'Dowd, 2018, pp. 11-15).

Intercultural exchange and language learning have been the driving force of the plethora of studies on telecollaboration produced in the last two decades. From a top-down perspective, governments have seized upon the momentum to develop VE in education as an agent of social change internationally. The interweaving of VE with soft power has triggered the interest of policy makers since 1988, when the New York State-Moscow Schools Telecommunications Project started, and continues today with initiatives such as the current Erasmus+ Virtual Exchange pilot project implemented by the European Commission (Helm, 2018; O'Dowd, 2018) and the launch of the *Journal of Virtual Exchange* in 2018.

Lewis and O'Dowd (2016) examined studies on telecollaboration from 1996 to 2015 and Çiftçi and Savaş (2018 [2017]) surveyed the period between 2010 and 2015. Their findings show that telecollaboration projects contribute to the development of second language acquisition/learning, intercultural competence, and learner autonomy at the college level, save those that involve only tandem language exchanges. Participants largely reported positive experiences with both their partners and the technology. There were recognizable language learning gains, greater opportunities for authentic interaction, and meaningful feedback. Intercultural competence was enhanced in terms of knowledge, interest, and awareness of both participants' own cultural background and their peers' perspectives through exchanges that were mainly factual, rather than critical, but became more elaborate over time. Recent studies confirm most of these advantages (Gimeno-Sanz, 2018; Andújar and Medina-López, 2019) and emphasize the potential of addressing diversity in the classroom, improving communicative skills, and providing some content knowledge (Pennock-Speck and Clavel-Arroitia, 2018).

As VEs escalate not only in number but also in complexity, with new approaches and applications, scholars face the need to address the intricacies and challenges of this educational practice. O'Dowd (2018) and Dooly and O'Dowd (2018) bring some order by providing an overview of terminology, models, current trends, developments, and open pedagogical issues. Lewis and O'Dowd (2016) and Çiftçi and Savaş (2018 [2017]) observe a prevalence of asynchronous projects, at the expense of underrepresented communication over Internet protocol telephony and videoconferencing, and a decline in the use of virtual worlds despite the potential of immersive environments to be an optimal, neutral space to minimize intercultural challenges and develop learner autonomy. Prevalence occurs also in VE projects at the college and elementary school levels in comparison to secondary education (Jauregi and Melchor-Couto, 2017). The type of partnership may require additional consideration. Pairing groups from different disciplines with distinct goals and objectives, dissimilar academic expectations, and an uneven global status of each cultural community may impact the interaction among peers and highly influence learners' perceptions of the VE (Gimeno-Sanz, 2018, pp. 11-12; Çiftçi and Savaş, 2018 [2017]).

A recurrent claim is the insufficient training and coaching on VE received by both instructors and learners in their respective roles (Pennock-Speck and Clavel-Arroitia, 2018; Çiftçi and Savaş, 2018 [2017]; Jauregi and Melchor-Couto, 2017). Lack of familiarity with the technology may lead to participants either abandoning a tool that they had difficulty managing or replacing it with an alternative one to maintain the communication (Antoniadou, 2011). Students need guidance not only on task completion and procedures, but also on their participation as "experts" and tutors, since peer corrective feedback seems to be more productive that teacher feedback in the traditional classroom setting (Lewis and O'Dowd, 2016).

One of the most critical issues is task design. Timing, scheduling, and consistent participation in peer interaction are often problematic for participants (Sevilla-Pavón & Haba-Osca, 2016; Gimeno-Sanz 2018). Some authors suggest monitoring the exchanges closely and stimulating motivation with the inclusion of participation as part of assessment (Çiftçi and Savaş, 2018 [2017]) and the implementation of collaborative project-based learning so that the target language becomes the means to the end (Dooly and O'Dowd, 2018). Controversial topics are preferable because they can help develop critical thinking and thoughtful dialogue (Gimeno-Sanz, 2018, pp. 11-12). Likewise, contextualization of the task is key to successful intercultural communication (Çiftçi and Savaş, 2018 [2017]).

The usual short duration of the experience and limited ethnographic training curtail the learners' development of cultural awareness and hinder their suspension of judgement. The result is widespread cases of cultural misunderstanding and miscommunication. Student engagement is essential in this case since intercultural communicative competence development is determined by prior experience and degree of participation (Lewis and O'Dowd, 2016). O'Dowd (2019, p. 11) notes that "the engagement between learners in e-tandem and telecollaborative models is often superficial and involves tasks which rarely lead to genuine negotiation of cultural meaning." Thus, the author articulates a ground-breaking approach to VEs that expands current models with the principles of global citizenship instruction and sets aside the customary focus on second language learning.

Foreign/Second Language Education and Mobile Tools

Literature reviews on collaboration in Mobile-Assisted Language Learning in the period 1993-2012 reveal a greater focus on individual learning rather than collaborative learning and show equivocal results regarding these devices' effectiveness in enhancing students' learning outcomes. Research on mobile technologies applied to collaborative language learning in the period 2012-2016 attests to the multiple advantages of this type of learning technology, including higher levels of engagement (Kukulska-Hulme and Viberg, 2018).

Shadiev, Hwang, and Huang (2017) elaborated a review study of authentic mobile language learning from 2007 to 2016. They report students' positive perceptions of the learning tools, gains in linguistic proficiency, learning behavior, and learning differences. Smartphones superseded all other technology and their use was associated with higher motivation and better performance. Only one fourth of the learning activities were designed based on a student-centered approach and most of them had a short duration and a small sample size.

Akiyama and Cunningham (2018) also reviewed research on virtual second language learning at the college level between 1996 and 2016 with a specific focus on synchronous telecollaboration. They categorized all practices into six different models and explained their evolution as an expected result of technology developments. The authors highlight the need for longitudinal studies, a wider range of proficiency levels, and comprehensive reports that include type of tasks, proficiency level, and use of synchronous and asynchronous tools.

Whatsapp and Jitsi as Teaching-Learning Tools

This literature review does not aim to be comprehensive, but rather illustrative of the rationale that supports why these communication technologies were selected for this study.

Whatsapp has become a desired didactic tool in disparate disciplines worldwide. In teacher education, it has been used for student-teacher and student-student interaction based on its functionality and accessibility. Its integration enhances academic performance and generates positive attitudes (So, 2016). In medicine, it has been effective in improving content knowledge, delivering content, fostering academic discussion, and increasing student participation, and interaction among students. As an expected drawback, instructors had to cope with off-topic, unneeded and uncouth posts (Sopanrao et al., 2018).

Timing was not an issue in either case because learners enjoyed being able to access content anytime, anywhere. *Whatsapp* has proved to promote effective collaborative student engagement and uninhibited interaction among all participants (Rambe and Chipunza, 2013). Like other Internet-based technologies, its successful application is contingent upon instructor and student training (Oriji and Fanny, 2019).

The use of *Whatsapp* has been especially prolific in English teaching and studies have shown that most participants report positive attitudes towards this learning tool and the course. Samaie, Nejad Mansouri, and Qaracholloo (2018) used it for self- and peer-assessments of EFL learners' oral language proficiency, albeit participants found the assessment component to be cumbersome and an inhibitor of spontaneity and natural talk. In conjunction with other Internet-based tools, *Whatsapp* produced higher learning outcomes than *Facebook*, but received some negative comments regarding timing and off-topic or needless messages (Çetinkaya, Levent and Sütçü, 2018). Written and oral discussions scheduled out of school hours via Whatsapp improved learners' self-efficacy beliefs (Can, Yafes & Çam Aktaş, 2019). In some cases, this application has been integrated merely as a teacher-student communication tool (Gutiérrez-Colón et al. 2013).

Unlike *Whatsapp*, *Jitsi* is not as widespread, but it is considered a technically promising educational tool (Petrangeli, 2019) that delivers high quality video with low resources. It is simple to use and secure and incorporates joint document editing and screen and document sharing. *Jitsi* is a free, open-source application based on Web real-time communication (WebRTC) technology. It provides voice, video-conferencing, and instant messaging. An account of the very few online learning tools that integrate WebRTC protocol is provided by Andújar and Medina-López (2019), the only scholars who have used *Jitsi* for this purpose. Their preliminary findings indicate a positive impact on student engagement and manifest growth in intercultural and linguistic competencies.

Student Engagement

Most research on telecollaboration projects label these practices as positive, engaging experiences based on participants' reports regarding two main aspects: their interaction with distant peers and the use of network-based technologies (Çiftçi and Savaş, 2018 [2017], p. 285). This information is mainly collected through surveys on their perceptions and attitudes about the project as a whole or selected components of the project. However, the quality or level of engagement with the digital tool is hardly addressed or not addressed at all.

Nonetheless, the effects of mobile technology on student engagement have been described by researchers using qualitative and quantitative data. Some studies base their conclusions on learners' responses to open questions or correlations between participation and performance. Thus, Ogunduyile (2013, p. 1152) tags SMS text messaging and 2go voice chat as "very engaging," Hwang et al. (2014) assert that the degree of engagement in the written task determines the effective learning results, and Ibáñez Moreno and Vermeulen (2015, 1355) affirm that the VISP app is "a motivating and engaging learning activity." Other studies measure engagement by its duration. Chen (2013, p. 26) notes that "the proportion of time spent learning English [on the tablet] increased by 11.4%." Kohn & Hoffstaedter (2017) used the BigBlueButton videoconference platform and Moodle written chat for intercultural communication and measured student engagement based on the length of interactions and language production. The novel component of their engagement analysis is the differentiation between greetings/procedural issues, assigned topics, and self-initiated topics. Some researchers use discourse analysis to determine the quantity and quality of the interaction. Oskoz, Gimeno-Sanz, and Sevilla-Pavón (2018) gauged engagement through students' utterance analysis based on written discussions on the online forum in Google+. Chun (2011) used discourse analysis to assert that students produce more elaborated language in asynchronous forum postings than in synchronous chats, but they are much more engaged in the latter. Finally, some scholars follow one or more existing models of analysis designed for this purpose. Çam Aktaş and Can (2019, p. 250) describe the interaction in *Whatsapp* as "moderate" according to a rubric for quality of the online learning environment. Pellas (2013) applied a research model that includes length of interaction and survey questions that assess behavioral, emotional and cognitive engagement to examine the impact of the implementation of virtual worlds. He concluded that there is a noticeable engagement increase when the virtual world involves meaningful interaction between peers. Andújar and Molina-López (2019, 211-212) also measured those same three types of engagement and reported a "high degree of engagement" in the application.

Case Study

The present case study involved 53 students from two universities, Fayetteville State University (U.S.) and the University of Almería (Spain). Thirty students from the American university and twenty-three students from the Spanish university participated in a telecollaboration project that lasted for a month and a half, from mid-February to the end of April, 2019. Students at Fayetteville State University were English native speakers and were enrolled in a course on Spanish culture and civilization. On the other hand, participants at the University of Almeria were Spanish speakers who were taking an English phonetics course. In order to develop the telecollaboration project between the two universities, both teachers required institutional permission as well as students' permission to carry out the project. Two different types of telecollaboration platforms were proposed by the teachers to carry out the virtual exchange. Firstly, the mobile instant messaging service *Whatsapp* was used to develop the text-based interaction between students in both classes. This tool allowed students to interact with their peers ubiquitously and carry out the task proposed by the teachers. Likewise, the tool was also used to schedule the subsequent videoconferences with their peers. These online videoconferences were carried out through the application *Jitsi* that allows P2P videoconferencing through the browser and does not require any software or plug-in installation. Students only had to share a link with their peers to start the videoconference. Participants had a minimum requirement of 4 interactions per week, two on the text messaging service and two on the videoconferencing platform. Each participant had to speak with two of their peers using their mother tongue in two interactions and the target language in the other two. This type of interaction has normally been referred to as "teletandem" in the literature (Telles, 2015) where students exchange their role of native speaker to benefit from each other's mother tongue. Due to the different aims of the two courses, the language used in the videoconferencing tool was English whereas Spanish was used in the text-based application.

Regarding task design, social constructivism was taken into consideration when developing the different tasks for the project. In this sense, the possibility of students co-constructing knowledge was considered from the beginning of the project. A series of topics related to the contents explained in each of the modules were proposed by participants, following the guidance of the two teachers responsible for the groups. This guidance considered previous literature in the telecollaboration field such as language learning potential, practicality, authenticity or meaning-focus (Chapelle, 2003; Jauregi, De Graaff & van den Bergh, 2012). In addition, the interaction also included different types of tasks such as information gap activities, pre-readings, debates or video-sharing among others. Thus, the topics and the type of tasks varied throughout the telecollaboration project. Among the topics used for the interaction, cultural aspects of each country such as education, habits, and traditions were included in the tasks. Also other current issues such as feminism or anorexia were used in the interaction.

For each of the tasks, students had to interact a minimum time of 10 minutes on each of the platforms. However, no time limitation was set by the teacher for each of the topics. To track the interactions on the videoconference platform, a virtual log was provided to participants in the exchange in order to register the time spent on the assignment and the date of every interaction. The conversation on the videoconference platform was not recorded in order to maintain students' privacy. Nevertheless, students were asked to write a brief report after each conversation. The data obtained from the text-based interaction was collected using the web version of the instant messaging service at the end of the project.

Practical considerations included explaining the project to participants in a 45-minute lesson, guaranteeing that all the participants could have access to a smartphone or a laptop with a built-in camera and microphone, exchanging students' contact details, controlling the technical problems that may occur during the process, and dealing with potential issues resulting from the interaction.

METHODOLOGY

As the main aim of this case study was to measure students' degree of engagement with the telecollaboration project as well as the tools used for the development of the interaction, quantitative methods together with the tracking of the conversation in each application were considered. In order to explore students' engagement, indications by Henrie, Halverson and Graham (2015), experts in measuring students' engagement in technology-mediated learning, were followed. Three different data collection instruments were used to assess the engagement construct. Firstly, the Technology Category Adopter Index (Dugas, 2005) was modified incorporating aspects related to telecollaboration environments. This scale attempted to explore students' use of technology and previous experiences in telecollaboration environments. The aim of these data collection instruments was to further understand students' use of technology and its possible relation with the time spent in the telecollaboration platforms used in the experiment. Secondly, a log was used during the project to collect the information regarding the time spent in each interaction as well as the date and the participants involved. This log was used to track students' engagement throughout the project, as previous research in the field (Fredricks, Blumenfeld & Paris, 2004; Henrie, Bodily, Manwaring, & Graham, 2015) highlighted the need to measure variations in the engagement construct. Lastly, Henrie, Larsen, Manwaring, Halverson & Graham's (2016) scale to measure students' engagement was given to participants at the end of the project, as the aim was to measure students' engagement at activity level. This longitudinal scale measured the engagement construct taking into consideration Fredricks et al.'s, (2004) distinction about the different types of engagement: behavioral, emotional and cognitive engagement. The first refers to those behaviors which are normally related with academic success, active participation, high attendance or homework completion. Emotional engagement refers to students' feelings about the learning experience. Particularly, factors such as interest, boredom, apathy or frustration. With regard to cognitive engagement, it deals with students' efforts to understand teachers' explanations as well as self-regulation processes and metacognitive behaviors. Thus, this data collection instrument addressed the three types of engagement described by Fredricks et al., (2004).

Internal threats to validity were minimized as the instruments had been used in previous research and the methodology followed indications of experts in the field. However, limitations with regard to the number of participants (n = 53) may affect the generalizability of the results.

RESULTS

The results presented in this case study were analysed quantitatively through the use of the statistical software SPSS in order to explore significant differences in the data collected. On the other hand, open-ended items in the data collection instruments given to participants were analysed qualitatively.

Firstly, the TACI index (Dugas, 2005) explore students' use of technology and experience in telecollaboration environments. Questions 1-7 consisted of a 5-point likert scale going from "A little" to "A lot". Yes/no questions were used in items 8-12 in which "Yes" was coded as 1 and "No" was coded as 2. Table 1 presents the descriptive measures obtained in this scale at the beginning of the course.

As observed in Table 1, the mean score for items 3 and 5 indicated that participants were heavy internet users and they made use of it for studying as well as in their private lives. Similarly, the mean scores obtained in items 4 and 7 highlighted a high use of chat-based communication tools in their private lives as well as at the university. This factor became a predictor of students' engagement when using chat-based communication tools for learning. Participants were already familiarized and, to a certain extent, engaged with this type of instant messaging services which may have an impact on students' level of engagement. Virtual learning environments were also used by a majority of the students in the telecollaboration exchange to study the different modules at their respective universities. These findings support the idea of students as "digital natives" (Prensky, 2001) as they have already mastered the use of the Internet, communication tools, and virtual environments. With regard to their previous experience in telecollaboration projects as well as their participation in eTandem exchanges, 62% of the participants,

Table 1. Descriptive measures of the Technology Category Adopter Index (TACI)

Items	N	Mean	SD
1.Are you familiar with ICT tools?	53	2.96	1.54
2.Do you use virtual environments for your studies?	53	3.64	1.42
3.Do you use the Internet for your study?	53	4.72	.69
4.Do you use chat communication tools at your university?	53	3.58	1,29
5.Are ICT tools used for the study at your university?	53	3.11	1.45
6.Do you use Internet in your private life?	53	4.85	.63
7.Do you use video chats to communicate in your private life?	53	3.62	1.40
8.Have you ever used Jitsi or any videoconferencing tool before?	53	1.43	.50
9.Have you ever used WhatsApp or any mobile instant messaging tool before?	53	1.38	.48
10.Have you ever participated in a telecollaboration project?	53	1.38	.48
11.Have you ever participated in an online language exchange?	53	1.51	.50
12.Have you ever practiced etandem before?	53	1.28	.45

most of them American, responded positively. However, as for Spanish students, it was the first time they were involved in this virtual exchange. This may have an impact on the engagement results at the end of the telecollaboration project as some students were not familiar with this type of interaction.

At the end of the course, Henrie et al.'s (2016) longitudinal scale of engagement was distributed to participants. This scale designed to measure students' emotional, cognitive and behavioral engagement was slightly modified in Items 1 and 2 to address specific questions of this investigation. Table 2 presents descriptive measures of the longitudinal scale of engagement.

Table 2. Descriptive measures of the longitudinal engagement scale

Item	N	Mean	SD
1. Did you enjoy using jitsi?	53	1.91	1.13
2. Did you enjoy using WhatsApp?	53	4.11	.82
3. Was this virtual exchange interesting?	53	4.00	.98
4. Did you wish you had been doing something else?	53	3.21	1.23
6. How well were you concentrating?	53	3.74	.90
9. I believe I will receive an excellent grade in this class	53	4.72	1.45
10. I'm confident I can understand the most complex material in this course.	53	4.87	1.31
11. I'm confident I can do an excellent job on the assignments and tests in this course.	53	5.04	1.25
12. Considering the difficulty of this course, the teacher, and my skills, I think I will do well in this class.	53	5.13	1.25
13. I often feel so lazy or bored when I study for my classes that I quit before I finish what I planned to do.	53	3.21	1.35
14. I work hard to do well in my classes even if I don't like what we are doing	53	5.45	1.57
15. When coursework is difficult, I give up or only study the easy parts	53	2.38	1.28
16. I like the subject matter of this course.	53	5.09	1.45
17. I am very interested in the content area of this course.	53	5.23	1.25
18. Even when course materials are dull and uninteresting, I manage to keep working until I finish.	53	5.40	1.33
19. Understanding the subject matter of this course is very important to me.	53	6.02	1.10
20. I am capable of solving, or getting help to solve, my computer-related problems.	53	5.79	1.30
21. I am very comfortable doing class work that is online	53	5.60	1.57
22. I am capable of using the internet to find the information I need.	53	6.28	1.27
23. My aim is to pass the course while doing as little work as possible.	53	2.43	1.42
24. I only study seriously what's given out in class or in the course outlines	53	3.08	1.12
25. I find I can get by in most tests by memorizing key sections rather than trying to understand them	53	2.49	1.12
26. I see no point in learning material which is not likely to be in the examination.	53	2.19	1.00
27. I often spend extra time trying to obtain more information about class topics	53	2.83	.99
28. I test myself on important topics until I understand them completely	53	3.30	1.11
29. I come to most classes with questions in mind that I want answering	53	2.51	1.08
30. I make a point of looking at most of the suggested readings that go with lectures.	53	3.15	1.19
31. Before this semester, had you ever taken an online course?	53	1.64	.48
32. Before this semester, had you ever taken a hybrid/blended course?	53	1.74	.44

In order to analyse the results of the engagement scale, the one-sample Wilcoxon ranked test was used to explore significant differences in the data collected. Items 1-4 explored students' emotional engagement through a 5-point likert scale going from "Not at all" to "Very much". Likewise, emotional engagement was also observed in Item 5 in which students responses went from "Excited" to "Bored". The results regarding the degree of enjoyment varied from one tool to the other, being *WhatsApp* the application selected by participants as the most enjoyable, where significant differences ($Z= 6.32$, $p < 0.01$, $r = .74$) were found. On the contrary, the enjoyment in *Jitsi* seemed to be very low due to the technical difficulties such as delays in the conversation and internet problems students encountered during the interaction ($Z= 3.6$, $p < 0.01$, $r = .49$). Responses to Item 4 and 5 indicated that participants found the virtual exchange to be significantly interesting ($Z= 6.06$, $p < 0.01$, $r = .83$). However, their degree of excitement did not differ significantly from the hypothesized mean ($Z= .30$, $p > 0.01$). These findings indicated that although students' level of emotional engagement was high, results did not yield significant differences in all the items addressing this type of engagement.

The second part of the engagement scale explored students' cognitive engagement through a seven-point semantic differential response scale. The positive end in this scale indicated an aspect of engagement whereas the negative end indicated disengagement. Table 3 presents students' responses to the cognitive engagement items in the scale.

As shown in Table 3, students' level of cognitive engagement varied from Item 7 to Item 8. Most of the students described their mood during the exchange as "Neutral". Others indicated feeling to a certain extent "Active" and others to a certain extent "Passive". This may be due to the fact that the interaction took place not only through videoconference but through text-based conversation, which may have affected participants' mood. Thus, there were no clear distinctions with regard to Item number 7. However, a majority of participants highlighted feeling "Focused" during the virtual interaction ($Z= 6.18$, $p < 0.01$, $r = .84$) in line with previous studies exploring cognitive engagement in telecollaboration projects (Andujar & Medina-López, 2019). The need to understand the language of their peers and keep a constant thread of conversation during the interaction may be possible explanations of this high level of concentration.

Taking into consideration the transactional model of students' engagement (Eccles & Wang, 2012; Henrie et al., 2016; Lawson & Lawson, 2013), which includes the learning environment as well as learners' characteristics as indispensable elements of students' engagement, the rest of the elements in the scale addressed these characteristics. Due to the amount of items in the scale, the results were grouped in order to provide a visual representation of students' characteristics. Items 9 to 12 and Item 14 explore students' positive attitude towards the course and the module as opposed to Items 13 and 15 which explored participants' negative attitude towards the aforementioned aspects. Figure 1 shows a visual representation of students' characteristics with regard to these two factors.

Table 3. Students' cognitive engagement throughout the exchange (%)

	Max	Medium	Min	Neutral	Min	Medium	Max	
Active	0	0	28.3	34	28.3	7.5	1.9	**Passive**
Focused	30.2	39.6	22.6	7.5	0	0	0	**Distracted**

Figure 1. Students' positive and negative attitude towards the course and the module

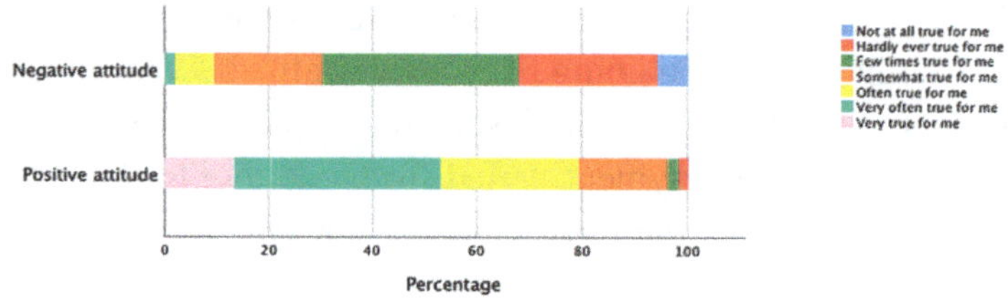

As displayed in Figure 1, about 55% of participants emphasized having a positive attitude towards the course and the module and 25% of the students indicated that having a positive attitude was "Often true" for them. On the contrary, responses with regard to students' negative attitude highlighted that it was not very frequent for students to experience these feelings and most of the participants indicated "Few times true for them" (38%) or "Hardly ever true for them" (27%). Thus, the overall results of the two abovementioned factors indicated a tendency towards a positive attitude of students in the course and the module.

Another aspect explored in the scale was to what extent students gave importance to the subject matter of the module. In this sense, items 16 to 19 dealt with this parameter and were grouped for analysis purposes. Figure 2 presents the results obtained from the data collected in the aforementioned items.

As observed in Figure 2, students gave a high degree of importance to the subject matter with 34% of participants indicating "Important" and 32% indicating "Very important". These results could have contributed to students' degree of engagement in the telecollaboration project and highlighted that students gave careful attention to the contents of the course. These contents were selected by the two teachers in the virtual exchange, addressing concerns of researchers in the telecollaboration field (Jauregi, De Graaff, van den Bergh, & Kriz, 2012) in order to improve students' engagement in the project.

Lastly, students' apathy and idleness as well as their eagerness to learn were investigated. Participants' mood towards learning and passing each of the modules at their respective universities was observed in Items 22 to 29. In order to observe students' responses for the items addressing the aforementioned factors, results were grouped to provide a visual representation of the data.

As shown in Figure 3, a high percentage of participants (53%) indicated that feeling lazy to learn new contents and subjects was "Sometimes true for them". Slightly lower percentages were found with regard to students' responses to their eagerness to learn. Thus, 41% indicated "Sometimes true for me"

Figure 2. Students' perceptions of the importance of the subject matter

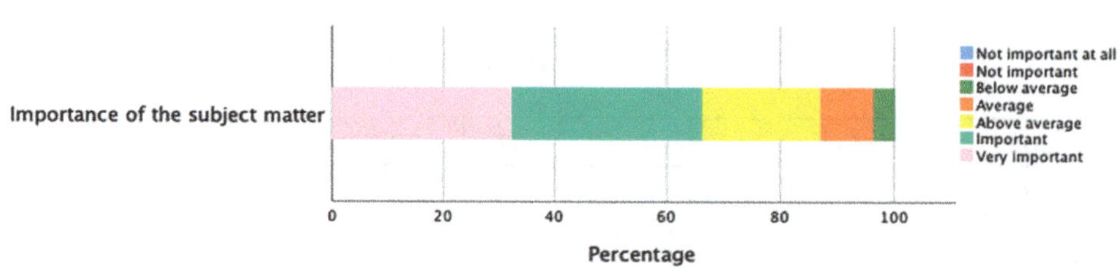

Figure 3. Students' apathy and eagerness to learn

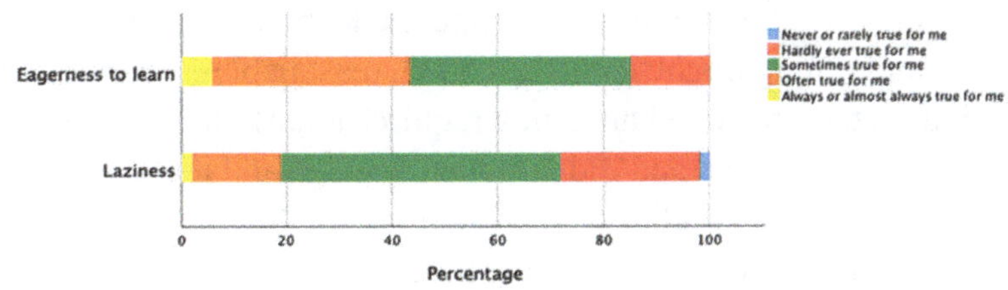

and 38% marked "Often true for me" when referring to their eagerness to learn. The results of these two factors emphasized that students did not present a very positive attitude towards learning but neither a negative one. Thus, students' engagement could not be directly related to their eagerness to learn as well as their level of disengagement could not be directly explained because of their idleness.

Overall, the results presented in this section arose significant results particularly in terms of emotional and cognitive engagement. As for emotional engagement, the telecollaboration project as well the chat-based application used for its development seemed to generate high levels of engagement. Nevertheless, the videoconferencing platform used for this project did not generate the same level of engagement due to the technical problems and conversation delays experienced by participants. It is also worth mentioning students' level of cognitive engagement that was found to be very high and required participants to be focused throughout the exchange. This may be explained because of the efforts made by participants to understand their peers as well as the topics proposed by the teachers. With regard to behavioral engagement no significant differences were found in the information collected and thus, students' engagement in the telecollaboration project could not be directly related to their individual behavior, although clearly contributed to students' engagement in the project.

DISCUSSION

This study adopts the conceptual framework of student engagement developed by Fredricks, Blumenfeld, & Paris (2004) in an attempt to explain the effectiveness of *Jitsi* and *Whatsapp* as learning tools in a telecollaboration project. These scholars understand engagement as a threefold notion that comprises behavior, emotion, and cognition and is linked to positive academic outcomes. This personal involvement or commitment springs not only from intrinsic motivation, but also from external factors. Thus, they found that "engagement is higher in classrooms with supportive teachers and peers, challenging and authentic tasks, opportunities for choice, and sufficient structure" (Blumenfeld, & Paris, 2004, p. 87). Since virtual exchanges can foster and help generate that type of classroom environment, this investigation examines the level of emotional, cognitive, and behavioral engagement in order to gauge the effectiveness of the online learning tools used in our telecollaboration project.

Jitsi and *Whatsapp* are mobile and desktop applications that comprise most highly desired features for optimal telecollaboration projects (Abernathy Marsh-Soloway & Michel Delers, 2019), notably *Jitsi*'s ability to safeguard privacy. Students' responses to the questionnaire on the use of technology and experience with virtual environments show their high competence and readiness to work with both platforms.

The results validate findings from previous studies indicating that this type of learning tools enhances student engagement (Khlaisang & Mingsiritha, 2016, p. 111). However, the data on emotional engagement reveals a strong preference for *Whatsapp*, despite the technological and interpersonal benefits of *Jitsi*. Technical glitches and timing might seem the initial, reasonable explanation for this disparity, but we need to bear in mind that, on the one hand, *Jitsi* requires a fairly low usage of processing resources and, on the other hand, scheduling can and did affect both video and text chat. Another disengagement factor may be the participants' low familiarity with *Jitsi*, given that introducing a new virtual platform can negatively affect students' satisfaction (Litzler, Huguet-Jérez, & Bakieva, 2018) and performance (Batardière & Helm, 2016, pp. 315-316), but students informed that they were similarly unacquainted with *Whatsapp*. More compellingly perhaps, some research indicates that, when other options are available, videoconferencing may become the least chosen means of communication (Jilg & Southgate, 2017), be dropped altogether, or substituted with one the learners know well (Antoniadou, 2011, p. 15).

As for the cognitive aspect, participants' responses point to a high level of engagement: most students stated that they were focused during the activity. A conducive factor attributed to this success was appropriate task design. In addition to an arranged week-long period to familiarize students with the virtual environments, task procedures, and expectations, both instructors agreed on a set of conversation topics. The high interest in course content (Item 17) suggests that the selected topics addressed learners' demand for discussions on issues relevant to their lives (El-Hariri, 2016). This relevance was likely heightened by the partnership between students of similar age and reciprocity in the use of languages (Akiyama, 2015). The effective interaction time recorded in the logs, in the form of continuous communication, may be associated with high cognitive engagement. Moreover, this interaction seems to have been triggered by extrinsic and intrinsic motivation, which positively affect academic motivation (Tasgin & Cosku, 2018) and, consequently, involvement in the activity. On the one hand, task design favored cognitive engagement with a 10-minute minimum interaction time requirement. Intellectual attention to the task was paramount to maintain a congruent, meaningful conversation. On the other hand, there is evidence of students' personal interest in the task as demonstrated by exchanges longer than the required time, genuine negotiations of meaning, and actively monitoring comprehension content. Since this last point is based on anecdotal, unscientific evidence, further data analysis of students' interactions would be very valuable to better assess the level of cognitive engagement.

The analysis of behavioral engagement shows positive attitudes towards the course and the telecollaboration module as well as high importance of the subject matter. These favorable results do not seem to be directly associated with the reported moderate eagerness to learn, which is also neutralized by similar results on apathy/idleness. Effective course organization and task design as extrinsic motivations and students' intellectual preference as intrinsic motivation seem to explain these results. Although individual behavioral engagement may not be unequivocally linked to overall engagement in the telecollaboration project, a look into the intertwined linkage between all three types of engagement may explain the reported passiveness and platform preference. Low emotional engagement has been stated to have a negative impact on behavioral intentions and vice versa (Al-Adwan, Al-Madadha, & Zvirzdinaite, 2018). Thus, poor self-management of learning in terms of idleness could have been an added reason for participants' dislike of *Jitsi* and consequently disregard it. Conversely, unfavorable perceptions of the learning tool (i.e. self-efficacy, ease of use, and usefulness) may have impacted participants' attitudes and their behavioral intentions (Park, Nam, & Cha, 2012) and hence also yielded relatively low cognitive engagement in terms of passiveness.

As a closing note, it is worthwhile to revisit the issue of task design since it seems to be an underlying aspect with the potential to affect all types of engagement. This project might have been enhanced with more cognitively engaging tasks, such as those that require participants' active roles to solve problems or create joint final products (Khlaisang & Mingsiritha, 2016). Likewise, the acceptance of and involvement in a new platform like *Jitsi* might have been increased with further planning, preparation, and monitoring. In particular, further instructor guidance would have been beneficial, paradoxically, to increase student independence and self-efficiency in the use of the virtual learning tools (Jilg & Southgate, 2017; Litzler, Huguet-Jérez, & Bakieva, 2018).

CONCLUSION

Overall, the findings of this investigation shed light on the impact of telecollaboration processes in the engagement construct. Notwithstanding that specific types of engagement such as students' level of cognitive engagement are found to be very present during online virtual exchanges, it becomes fundamental to put into practice telecollaboration tools that can also help foster students' emotional engagement during the learning process. The understanding of how these tools affect learners' attitude and behavior during the virtual language exchange seems of great interest to appropriately develop telecollaboration environments where students can maximize the opportunities for language learning and practice. Furthermore, the possibilities that virtual platforms offer for tracking variations in the engagement construct need to be considered by language teachers and practitioners who are willing to assess the outcomes of telecollaboration projects. In the same vein, the assessment of the interaction between participants in these projects seems fundamental to explore the degree of engagement achieved by the different tasks carried out in the telecollaboration environment.

This type of telecollaboration projects do not constitute a great increase of the teacher's workload as the interaction takes place between peers, and once the process is fully organized, there is no need for the teacher to control or monitor in detail the language exchange. Simply by tracking the interaction in the telecollaboration platform and solving the possible issues between participants, the language exchange takes place without further requirements. This ease of use may lead to a possible integration of telecollaboration projects into the language curriculums, which may maximize students' opportunities for receiving input in the target language and thus accelerate the language learning process of the participants.

Further research is sought to delve into students' engagement, not only from a technological point of view, but also from an interactionist perspective in which conversation analysis may yield relevant data with regard to students' engagement in this type of processes. Similarly, further cross-cultural collaboration between institutions may help develop the implementation of telecollaboration projects that adequately meet the needs of second language learners and teachers. The new possibilities of online communication between peers become a fertile ground for language interaction, intercultural competence development and social gathering through the Internet. These factors may need to be taken into consideration in second language environments, as technology is fast-evolving and the arrival of 5G connectivity may help overcome many of the challenges found in this and previous telecollaboration studies (Ware & Kessler, 2016; Andujar, 2019), facilitating and fostering a better implementation of these projects. Different technical difficulties such as conversation delays and problems associated with the speed of the Internet will gradually disappear and the constraints of traditional classroom environments such as limited in-class time or attending class at a particular time and place may be overcome in the short-term. Thus, it seems fundamental to further understand the elements at play during the development of telecollaboration projects that could contribute to students' engagement, from the task design to the tools implemented, in order to fully exploit their potential in the second language learning field.

REFERENCES

Abernathy Marsh-Soloway, M., & Michel Delers, O. (2019). New directions in eTandem: An expanded vision of capabilities and practices. *Revista do GEL, 15*(3), 176–193. doi:10.21165/gel.v15i3.2395

Akiyama, Y. (2015). Task-based investigation of learner perceptions: Affordances of video-based eTandem learning. In E. Dixon & M. Thomas (Eds.), *Researching language learner interaction online: From social media to MOOCs, CALICO Monograph Series* (Vol. 13, pp. 149–170). San Marcos, TX: CALICO.

Akiyama, Y., & Cunningham, J. (2018). Synthesizing the practice of SCMC-based telecollaboration: A scoping review. *CALICO Journal, 35*(1), 49–76. doi:10.1558/cj.33156

Al-Adwan, A. S., Al-Madadha, A., & Zvirzdinaite, Z. (2018). Modeling students' readiness to adopt mobile learning in higher education: An empirical study. *International Review of Research in Open and Distributed Learning, 19*(1). doi:10.19173/irrodl.v19i1.3256

Andujar, A. (2019). Shaping the Future of Telecollaboration: Web RTC. In Educational Technology and Resources for Synchronous Learning in Higher Education (pp. 151-172). Hershey, PA: IGI Global. doi: 10.4018/978-1-5225-7567-2.ch008

Andujar, A., & Medina-López, C. (2019). Exploring new ways of eTandem and telecollaboration through the WebRTC protocol: Students' engagement and perceptions. *International Journal of Emerging Technologies in Learning, 14*(5), 200–217. doi:10.3991/ijet.v14i05.9612

Angelova, M., & Zhao, Y. (2016). Using an online collaborative project between American and Chinese students to develop ESL teaching skills, cross-cultural awareness and language skills. *Computer Assisted Language Learning, 29*(1), 167–185. doi:10.1080/09588221.2014.907320

Antoniadou, V. (2011). Using activity theory to understand the contradictions in an online transatlantic collaboration between student-teachers of English as a foreign language. *ReCALL, 23*(3), 233–251. doi:10.1017/S0958344011000164

Batardière, M.-T. & Helm, F. (2016). Fostering students' engagement with topical issues through different modes of online exchange. In S. Jager, M. Kurek, & B. O'Rourke (Eds.), *New directions in telecollaborative research and practice: Selected papers from the second conference on telecollaboration in higher education*, 313-319. Research-publishing.net. doi:10.14705/rpnet.2016.telecollab2016.52

Çam Aktaş, B., & Can, Y. (2019). The effect of Whatsapp usage on the attitudes of students toward English self-efficacy and English courses in foreign language Education outside the school. *International Electronic Journal of Elementary Education, 11*(3), 247–256. doi: 10.26822/iejee.2019349249

Çetinkaya, L., & Sütçü, S. S. (2018). The effects of Facebook and WhatsApp on success in English vocabulary instruction. *Journal of Computer Assisted Learning, 34*(5), 504–514. doi:10.1111/jcal.12255

Chapelle, C. A. (2003). *English language learning and technology.* Amsterdam, The Netherlands: John Benjamins. doi:10.1075/lllt.7

Chen, X.-B. (2013). Tablets for informal language learning: Student usage and attitudes. *Language Learning & Technology, 17*(1), 20–36.

Chun, D. M. (2011). Developing intercultural communicative competence through online exchanges. *CALICO Journal, 28*(2), 392–419. doi:10.11139/cj.28.2.392-419

Çiftçi, E. Y., & Savaş, P. (2018). [2017]). The role of telecollaboration in language and intercultural learning: A synthesis of studies published between 2010 and 2015. *ReCALL, 30*(3), 278–298. doi:10.1017/S0958344017000313

Dooly, M. (Ed.). (2008). *Telecollaborative language learning: A guidebook to moderating intercultural collaboration online.* Peter Lang.

Dooly, M. A., & O'Dowd, R. (2018). Telecollaboration in the foreign language classroom: A review of its origins and its application to language teaching Practices. In M. A. Dooly & R. O'Dowd (Eds.), *In this together: Teachers' experiences with transnational, telecollaborative language learning projects* (pp. 11–34). New York, NY: Peter Lang. doi:10.3726/b14311

Dugas, C. A. (2005). *Adopter characteristics and teaching styles of faculty adopters and nonadopters of a course management system.* (Unpublished doctoral dissertation), Indiana State University, Terre Haute, IN.

Eccles, J., & Wang, M. T. (2012). Part I commentary: So what is student engagement anyway? In *Handbook of research on student engagement* (pp. 133–145). Boston, MA: Springer; doi:10.1007/978-1-4614-2018-7_6

El-Hariri, Y. (2016). Learner perspectives on task design for oral–visual eTandem language learning. *Innovation in Language Learning and Teaching, 10*(1), 49–72. doi:10.1080/17501229.2016.1138578

Maske, S. S., Kamble, P. H., Kataria, S. K., Raichandani, L., & Dhankar, R. (2018). Feasibility, effectiveness, and students' attitude toward using WhatsApp in histology teaching and learning. *Journal of Education and Health Promotion, 7,* 158. doi: 10.4103/jehp.jehp_30_18 PMID:30693295

Fredricks, J. A., Blumenfeld, P. C., & Paris, A. H. (2004). School engagement: Potential of the concept, state of the evidence. *Review of Educational Research, 74*(1), 59–109. doi:10.3102/00346543074001059

Gimeno-Sanz, A. (2018). Learner expectations and satisfaction in a US-Spain intercultural telecollaboration project. *Bellaterra Journal of Teaching & Learning Language & Literature, 11*(3), 5–38. doi:10.5565/rev/jtl3.776

Gutiérrez-Colón Plana, M., Gibert-Escofet, M. I., Triana Figueras, I., Gimeno, A., Appel, C., & Hopkins, J. (2013). *Improving learners' reading skills through instant short messages: A sample study using WhatsApp. Global perspectives on Computer-Assisted Language Learning* (pp. 80–84). Glasgow, UK: WorldCALL.

Helm, F. (2018). The long and winding road.... In C. Fuchs, C. Kennedy, Satar, & H. Müge (Eds.), Journal of Virtual Exchange, 1, 41-63. Research-publishing.net. doi: 10.14705/rpnet.2018.jve.3

Helm, F. & Guth, S. (2016). Telecollaboration and language learning. The Routledge Handbook of Language Learning and Technology, 241-254.

Henrie, C. R., Bodily, R., Manwaring, K. C., & Graham, C. R. (2015). Exploring intensive longitudinal measures of student engagement in blended learning. *The International Review of Research in Open and Distributed Learning*, *16*(3). doi:10.19173/irrodl.v16i3.2015

Henrie, C. R., Larsen, R., Manwaring, K., Halverson, L. R., & Graham, C. R. (2016). Validation of a longitudinal activity-level measure of student engagement. *Contemporary Educational Psychology.*

Hwang, W.-Y., Chen, H. S. L., Shadiev, R., Huang, R. Y.-M., & Chen, C.-Y. (2014). Improving English as a foreign language writing in elementary schools using mobile devices in familiar situational contexts. *Computer Assisted Language Learning*, *27*(5), 359–378. doi:10.1080/09588221.2012.733711

Ibáñez Moreno, A., & Vermeulen, A. (2015). Profiling a MALL app for English oral practice: A case study. *International Journal of Computers and Technology*, *21*(10), 1339–1136. doi: 10.3217/jucs-021-10-1339

Jauregi, K., De Graaff, R., van den Bergh, H., & Kriz, M. (2012). Native/non-native speaker interactions through video-web communication: A clue for enhancing motivation? *Computer Assisted Language Learning*, *25*(1), 1–19. doi:10.1080/09588221.2011.582587

Jauregi, K., & Melchor-Couto, S. (2017). The role of coaching in teacher competence development for telecollaboration. *Alsic*, *20*(2). doi: 10.4000/alsic.3218

Jilg, T., & Southgate, M. (2017). Students helping students: A learning partnership initiative for distance language learners. *Language Learning Journal*, *45*(2), 245–262. doi:10.1080/09571736.2013.877067

Khlaisang, J., & Mingsiritha, K. (2016). Engaging virtual learning environment system to enhance communication and collaboration skills among ASEAN higher education learners. *International Journal of Emerging Technologies in Learning*, *11*(4), 103–113. doi:10.3991/ijet.v11i04.5503

Kohn, K., & Hoffstaedter, P. (2017). Learner agency and non-native speaker identity in pedagogical lingua franca conversations: Insights from intercultural telecollaboration in foreign language education. *Computer Assisted Language Learning*, *30*(5), 351–367. doi:10.1080/09588221.2017.1304966

Kukulska-Hulme, A., & Viberg, O. (2018). Mobile collaborative language learning: State of the art. *British Journal of Educational Technology*, *49*(2), 207–218. doi:10.1111/bjet.12580

Lawson, M. A., & Lawson, H. A. (2013). New conceptual frameworks for student engagement research, policy, and practice. *Review of Educational Research*, *83*(3), 432–479. doi:10.3102/0034654313480891

Lee, L., & Markey, A. (2014). A study of learners' perceptions of online intercultural exchange through Web 2.0 technologies. *ReCALL*, *26*(3), 281–297. doi:10.1017/S0958344014000111

Lewis, T., & O'Dowd, R. (2016). Online intercultural exchange and foreign language learning: A systematic review. In R. O'Dowd & T. Lewis (Eds.), *Online intercultural exchange: Policy, pedagogy, and practice* (pp. 21–66). New York, NY: Routledge.

Litzler, M. F., Huguet-Jérez, M., & Bakieva, M. (2018). Prior experience and student satisfaction with e-tandem language learning of Spanish and English. *International Journal of Interactive Mobile Technologies*, *12*(4), 4–20. doi:10.3991/ijim.v12i4.9196

Melchor-Couto, S. (2019). Virtual worlds and language learning. *Journal of Gaming & Virtual Worlds*, *11*(1), 29–43. doi:10.1386/jgvw.11.1.29_1

O'Dowd, R. (2005). Negotiating sociocultural and institutional contexts: The case of Spanish–American telecollaboration. *Language and Intercultural Communication*, *5*(1), 40–56. doi:10.1080/14708470508668882

O'Dowd, R. (2018). From telecollaboration to virtual exchange: State-of-the-art and the role of UNICollaboration in moving forward. *Journal of Virtual Exchange*, *1*, 1–23. doi:10.14705/rpnet.2018.jve.1

O'Dowd, R. (2019). A transnational model of virtual exchange for global citizenship education. *Language Teaching*, 1–14. doi:10.1017/S0261444819000077

O'Rourke, B. & Stickler, U. (2017). Synchronous communication technologies for language learning: Promise and challenges in research and pedagogy. *Language Learning in Higher Education, 7*(1), 1-20. http://dx.doi.org.uncfsu.idm.oclc.org/10.1515/cercles-2017-0009

Ogunduyile, A. O. (2013). Towards the integration of mobile phones in the teaching of English language in secondary schools in Akure, Nigeria. *Theory and Practice in Language Studies*, *3*, 1149–1153. doi: 10.4304/tpls.3.7.1149-1153

Oriji, A., & Fanny, A. (2019). Social media in teaching-learning process: Investigation of the seu of Whatsapp in teaching and learning in University of Port Harcourt. *European Scientific Journal*, *15*(4), 15–39. doi: 10.19044/esj.2019.v15n4p15

Oskoz, A., Gimeno-Sanz, A., & Sevilla-Pavón, A. (2018). Examining L2 learners' use of engagement strategies in telecollaborative written interactions. In B. Mousten, S. Vandepitte, E. Arnó, & B. Maylath (Eds.), *Multilingual writing and pedagogical cooperation in virtual learning environments* (pp. 200–220). Hershey, PA: IGI-Global; doi:10.4018/978-1-5225-4154-7.ch008

Park, S. Y., Nam, M., & Cha, S. (2012). University students' behavioral intention to use mobile learning: Evaluating the technology acceptance model. *British Journal of Educational Technology*, *43*(4), 592–605. doi:10.1111/j.1467-8535.2011.01229.x

Pellas, N., & Kazanidis, I. (2013). On the value of Second Life for students' engagement in blended and online courses: A comparative study from the Higher Education in Greece. *Education and Information Technologies*, *20*(3), 445–466. doi:10.100710639-013-9294-4

Pennock-Speck, B. & Clavel-Arroitia, B. (2018). Teachers' perspectives on telecollaboration in secondary school foreign language education. *LEA - Lingue e letterature d'Oriente e d'Occidente, 7*, 483-509. doi: 10.13128/LEA-1824-484x-24405

Petrangeli, S., Pauwels, D., van der Hooft, J., Ziak, M., Slowack, J., Wauters, T., & De Turck, F. (2019). A scalable WebRTC-based framework for remote video collaboration applications. *Multimedia Tools and Applications*, *78*(6), 7419–7452. doi:10.100711042-018-6460-0

Prensky, M. (2001). Digital natives, digital immigrants part 1. *On the Horizon*, *9*(5), 1–6. doi:10.1108/10748120110424816

Rambe, P., & Chipunza, C. (2013). Using mobile devices to leverage student access to collaboratively-generated resources: A case of WhatsApp instant messaging at a South African University. *British Journal of Educational Technology*, *44*(4), 544–561. doi:10.1111/bjet.12057

Samaie, M., Nejad Mansouri, A., & Qaracholloo, M. (2018). An inquiry into the efficiency of WhatsApp for self- and peer-assessments of oral language proficiency. *British Journal of Educational Technology*, *49*(1), 111–112. doi:10.1111/bjet.12519

Sevilla-Pavón, A., & Haba Osca, J. (2016). Te das cuenta de que el mundo puede ser tan distinto y similar al mismo tiempo": Telecolaboración y desarrollo de la competencia intercultural en la educación superior. *Didáctica. Lengua y Literatura*, *28*, 263–284. doi: 10.5209/DIDA.54088

Shadiev, R., Wu-Yuin, H., & Yueh-Min, H. (2017). Review of research on mobile language learning in authentic environments. *Computer Assisted Language Learning*, *30*(3-4), 284–303. doi:10.1080/09588221.2017.1308383

So, S. (2016). Mobile instant messaging support for teaching and learning in higher education. *The Internet and Higher Education*, *31*, 32–42. doi:10.1016/j.iheduc.2016.06.001

Tasgin, A., & Cosku, G. (2018). The relationship between academic motivations and university students' attitudes towards learning. *International Journal of Instruction*, *11*(4), 935–950. doi:10.12973/iji.2018.11459a

Telles, J. A. (2015). Teletandem and performativity. *Revista Brasileira de Lingüística Aplicada*, *15*(1), 1–30. doi:10.1590/1984-639820155536

KEY TERMS AND DEFINITIONS

CALL: Field of study which explores the use and application of technological tools and automatic processes to develop language learning.

eLearning: Learning process that usually takes places through the internet or through the use of electronic devices.

Engagement: Refers to students' attitudes, emotions or behaviors while they are involved in certain task.

eTandem: Virtual interaction in which participants swap the role of the native speaker so they can benefit from each other's mother tongue.

Telecollaboration: Virtual exchange between classes which are geographically distant from each other.

Videconferencing: Process that uses a virtual platform to connect two devices which are geographically distant from each other using the Internet.

Virtual Exchange: Online process that connects students from different classes to foster the development of language skills.

Chapter 4
Desirability of Multimedia Hyperlinks in Fiction to Foster Pupil Reading Motivation and Immersion:
Reading Tools for MALL

Claudio Vanhees
University of Antwerp, Belgium

Mathea Simons
University of Antwerp, Belgium

Vanessa Joosen
University of Antwerp, Belgium

ABSTRACT

Reading from digital screens has become increasingly common practice in educational and recreational reading. The response to this digital shift has been twofold. Some suspect it will harm children's ability to perform deep reading; others highlight its potential to support reading among different groups. Digital reading tools, such as fiction with multimedia hyperlinks, could engage particularly reluctant readers or children from low-literate families. This chapter presents the results of an experimental, mixed-method study that identifies hyperlink type and frequency desirability in literary texts. A comparative analysis of respondent perspectives revealed that teachers mark on average more explanatory and enriching hyperlinks than pupils. Pupil and teacher hyperlink type desirability are significantly influenced by respectively literary genre and reading motivation, and importance of pupil reading motivation and media use. Pupil and teacher explanatory hyperlink frequency are significantly influenced by respectively literary genre, and importance of pupil reading motivation.

DOI: 10.4018/978-1-7998-1097-1.ch004

INTRODUCTION

As society moves from a print-based to a digital culture, reading inevitably evolves too (Wolf & Barzillai, 2009). In addition to reading on paper, reading on digital screens such as computers, e-readers, tablets and smartphones has become increasingly common in both educational and recreational reading (Hyman, Moser, & Segala, 2014; Barzillai & Thomson, 2018). The response to this shift has been two-fold. On the one hand, it has caused concern about the potential harm to pupils' ability to perform deep reading (Carr, 2011; Baron, 2015; Wolf, 2016). On the other hand, this digital evolution has equally elicited optimism thanks to the potential of technology to support reading among different groups, such as reluctant readers or children from low-literate families (Maynard, 2010; Miranda et al., 2011; Picton, 2014), especially within the current trend towards more autonomous learning (Warschauer & Liaw, 2011). When it comes to literary reading, hypermedia fiction allows learners to read digital texts enriched with multimedia hyperlinks that lead them to supportive or comprehensive information. This chapter reports on an experimental study that was carried out to assess pupils' and teachers' perspectives on the desirability of such multimedia hyperlinks in literary texts in terms of type and frequency.

BACKGROUND

There is a growing body of literature that recognises the importance of both print and digital environments for children's reading development. Barzillai et al. (2018) propose a model of reading displaying the mutual influence of the characteristics of the reader, the text, and the reading goals/tasks (Figure 1).

The model shows that **reading goals/tasks** in digital environments greatly influence children's reading behaviour. For instance, skimming short, online texts invokes shallower reading behaviour than reading a captivating story posted by a loved one (Fitzsimmons, 2016). The way in which schools introduce digital texts, design assignments, and model digital reading behaviour can also influence children's reading goals (Mifsud & Petrová, 2018). However, Barzillai et al. (2018) underline that there is still a gap between children's reading in school, where they mostly read linear texts on paper, and outside school, where they read more typically digital non-linear text formats, often on digital reading devices.

As far as **text properties and content** are concerned, comprehension is influenced by the digital text format which includes text window sizes on different reading devices, text that flows to fit the parameters of devices, and flipping pages versus scrolling. The mere structure of the text and the presence or absence of hyperlinks also play a role. Literary digital texts can for instance be structured as non-linear hypertexts in which readers navigate from the original text to other documents. Traditional hyperfiction is the term used for non-sequential digital novels, solely composed of written words (Landow, 1997) where

Figure 1. A triangle model of research on the development of digital reading

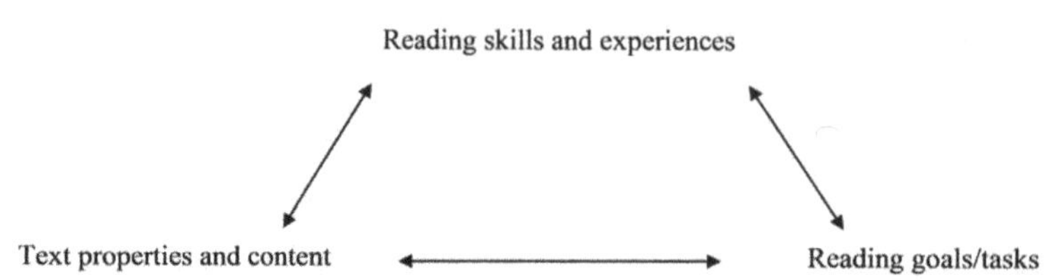

different degrees of reader participation are possible (Gutiérrez, 2014): from constructive hyperfiction (where the reader almost completely constructs the story and partly becomes its author), to exploratory hyperfiction (where the reader can follow different pathways previously created by the author). However, when hyperfiction contains hyperlinks to multimedia, such as sound and dynamic visualisations that are not found in traditional print media, it becomes hypermedia fiction. In exploratory hypermedia fiction the reader moves through the story and can consult several types of supporting or enriching media materials along the way by clicking or tapping on hyperlinks (Authors, 2014). Engaging with more than one text or source in one reading session requires advanced digital reading competencies that must be taught and addressed in schools (Salmerón et al., 2018). Such competencies include focused searching and navigating in hypertext and multimedia sources, integrating multiple sources of information, and critically evaluating the quality of claims made (Afflerbach & Cho, 2010). Multiple cognitive skills (i.e. working memory, attention, executive function and language skills) play a role in these competencies. For instance, greater visuospatial working memory capacity is related to more efficient navigation in hypermedia environments (Salmerón et al., 2018).

With regard to **reader skills and experiences,** Barzillai et al. (2018) highlight that besides the cognitive skills mentioned above - which influence the ease and efficiency of reading - emotions also play an important role in reading development in digital contexts. When it comes to how emotions may impact the digital reading experience, Kaakinen et al. (2018) list the readers' attitudes and prior experience with digital reading environments, and their motivational orientation as possible factors. Readers' attitudes are also influenced by age and gender differences: secondary school children prefer digital texts (McGeown et al., 2016), and boys are more likely to prefer e-readers than girls, although avid readers still prefer printed books (Tveist & Mangen, 2014). Furthermore, readers can be intrinsically or extrinsically motivated (Sweet, Guthrie, & Ng, 1998): the former deliberately seek opportunities to engage in reading, whereas the latter also require external factors for engagement. Their motivational orientation may once again influence whether readers prefer to read on paper or digitally (Larson, 2015). As far as emotional experiences during literary reading are concerned, Miall and Kuiken (2002) distinguish four levels: evaluative, aesthetic, narrative and self-modifying feelings. Evaluative feelings are induced by the interpreted text and the reading activity itself: pleasure, enjoyment and satisfaction of reading. Aesthetic feelings are triggered by formal aspects of text, and serve to capture and keep the readers' attention. Narrative feelings derive from comprehending the plot or identifying with the characters. Self-modifying feelings, however, can only be experienced when readers of literary texts become so absorbed that they feel *transported* to the fictional world, and become *immersed* in the story.

Gerrig (1993) first introduced the concept of *transportation*, roughly defined as *immersion* or *engagement* in (fictional or nonfictional) narratives (see also Busselle & Bilandzic, 2008; Green & Brock, 2000; Nell, 1988), which implies the reader becomes partially and temporarily isolated from the physical and temporal correlates of the real world (Mangen & Kuiken, 2014). Immersion may vary as a function of modality: digital texts can be presented on computer screens, tablets, e-readers, smartphones, or even via virtual reality goggles (Kaakinen et al., 2018), and these technologies may change the perceptual and haptic experience of reading, which again may influence immersion (Mangen, 2008). Consequently, certain hypertext environments that upset the text's structure, such as constructive hyperfiction, may not support immersion (Miall & Dobson, 2001; Miall, 2004) or lead to immersion based on different mechanisms than that in printed text (Douglas and Hargadon, 2001). When hyperfiction includes sound or dynamic visual images, and becomes hypermedia fiction, it is potentially more immersive than traditional print (Schlochtermeier et al., 2015).

So far, research has mostly been confined to cognitive aspects of digital reading and reading comprehension rather than to the emotional and motivational aspects typically related to digital literary reading. Immersion in digital literary texts has remained largely unexplored. Only Mangen and Kuiken (2014) have conducted empirical research on immersion in literary texts on paper and on tablets. However, the texts on paper in their study were limited to five pages and presented as handouts rather than novels, whereas the digital texts did not contain any hyperlinks. It is thus interesting to further investigate immersion in literary texts on paper and on digital devices through full length novels in different genres, but also to examine the role multimedia hyperlinks could play in the digital reading process of literary texts. The latter have so far only been included in research on expository (web)texts, often with a focus on reading comprehension, but could also play a significant role in the emotional aspects of digital literary reading in general, and in story world immersion in particular.

AIM AND RESEARCH QUESTIONS

In order to shed light on the relationship between multimedia hyperlinks and immersion in fiction, this empirical study investigates how pupils and teachers perceive the application of such links in literary texts, based on the following research question:

RQ0: Which hyperlinks do adolescent readers and their language teachers find desirable in literary texts?

Hyperlink type and frequency were also addressed in two subsequent RQs:

RQ1: Which hyperlink *types* do adolescent readers and their language teachers find desirable in literary texts?
RQ2: At what *frequency* are the hyperlinks adolescent readers and their teachers find desirable marked in literary texts?

The meta-reflective study described in this chapter is the first of two studies in a larger transdisciplinary research project on digital reading. Two subsequent empirical studies are carried out on 12- to 13-year-old pupils and their language teachers to evaluate the potential of multimedia hyperlinks for Mobile-Assisted Language Learning (MALL) and its applications within the classroom environment. This research project aims to offer some insight into whether and how multimedia hyperlinks affect the reading experience of fiction and how they can potentially be applied in literary texts to foster immersion.

RESEARCH DESIGN

Method

In this mixed-methods study data were collected for two respondent groups, i.e. pupils and teachers, with three different research instruments: annotated novels, surveys, and qualitative data derived from focus groups. This section first discusses the participants, then the research instruments, and finally the general procedure for data collection and analysis.

Participants

The initial goal of the study was to include 30 pupils (6 per novel) and 30 teachers (6 per novel). To recruit the participants a study proposal was first submitted to the principals of 100 schools in the Flemish region (Belgium). A convenience sample was compiled, which was subsequently expanded through snowballing. The interested principals supported the research and informed their teachers, who were then given the choice to participate with or without their pupils during a meeting with the researchers. Parental permissions for their pupils' participation were also obtained before data collection in the classroom. All respondents gave informed consent and were anonymised with code names to maintain confidentiality (e.g., P001 for pupil 1, T002 for teacher 2, etc.). Thanks to the high response rate, the sample was expanded to 102 pupils and 44 teachers of 42 Flemish secondary schools, as described below.

Participants in the study were **102** seventh- and eighth-grade **pupils** (i.e. 12-14 years old) attending general secondary education (A-stream) in seven schools in Flanders. Three of the schools were located in the city centers and two in suburban areas. The remaining two were located in more remote, rural areas. All schools served pupils of varying income levels and ethnic groups. Initially, 120 pupils were included, but eight did not complete the instruments timely, so the primary sample comprised 112 pupils. A first analysis revealed that eight pupils had not read their novel (it appeared untouched and contained no hyperlinks), and that two additional pupils had recently arrived as refugees and mastered the instructional language insufficiently to match the respondent profile. As a result, the final sample comprised 102 pupils (see Table 1), of whom 41% were boys and 59% girls. Compared to the total Flemish pupil population (Vlaams Ministerie Onderwijs en Vorming, 2018) girls are slightly overrepresented (49.1% in Flemish education). 61 learners are in Grade 7 (60%); 41 are in Grade 8. To verify the sample's representativeness, the pupils' home language (HL) and their mother's education level were equally identified. 74.5% spoke Dutch as HL, 9% were bilingual and 16.6% had a HL other than Dutch. The latter matches the Flemish pupil population of whom 16.4% have another HL. Finally, having a mother with an education level limited to primary education, is considered potentially predictive of lower performance rates. This holds true for 22.6% of Flemish pupils and for 9.2% of the sample, which is therefore considered advantaged.

Along with the seven **teachers** of Dutch of the pupils mentioned above, an additional 50 teachers from 35 secondary schools agreed to participate without pupils. One school's organisational problems led to the withdrawal of their five teachers, and six teachers from four other schools did not complete the instruments on time. The primary sample thus consisted of 46 teachers from 30 secondary schools. Two teachers' novels had not been read and were excluded. The final sample comprised **44 teachers**, of whom 7 men (15.9%) and 37 women (84.1%), all with pupils in grade 7 and 8. Compared to the Flemish teacher population (67,9%), women are overrepresented in the sample. The average teacher's age was 39.2 years (SD = 8.8) (youngest = 26; oldest = 58) and their average teaching experience 13.1 years (SD = 8.5), ranging from 1 to 31 years.

Research Instruments

Data in this study were gathered using three different research instruments: novels, surveys and focus group discussions. The novels were the main data source, while the surveys provided background information on the participants, and the focus group discussions, with a selection of participants, gave a more detailed and deeper understanding of the quantitative data.

Novels

Five novels were used as study material. The participants were required to read those novels which were published in the respondents' mother tongue (Dutch), the language in which pupils at this level are expected to read most. The novels were selected on the basis of the following criteria: acceptable length; sufficient possibilities to add hyperlinks; level of difficulty that matches the target audience; and diversity of topics and genres, resulting in the following list: (a) historical novel (A. Sax, *De hond van Roosevelt*, 65p.); (b) fantasy novel (D. Almond, *De schaduw van Skellig*, 150p.); (c) adaptation (M. De Cock, *Veldslag om een hart*, 126p.); (d) problem novel (C. Cazemier, *#Selfie*, 176p.); (e) realism with unreliable narrator (G. Samson, *Zwarte zwaan*, 207p.). The complete novels were used in their original publication format rather than a (partial) reproduction.

After the publisher's and author's permissions were obtained, the participants received the five blurbs and were invited to rank their top three. We aimed for a balance between the respondents' preferences and an equal representation of each novel in the study. An overall majority of respondents received their first choice, a limited number their second choice, and none received their third choice.

Table 1 shows that the initially perceived balance in the number of respondents per novel was not completely obtained if we consider which participants completed the reading task. The historical novel was read by most respondents (n = 38), whereas the realist novel had the smallest audience (n = 28). However, each novel was read by both learners and teachers, and, despite the underrepresentation of male teachers, in most cases also by both genders. Furthermore, the novels were well distributed over both grades (7 and 8) and over the learners' home languages.

Table 1. Novels and respondent profile

Novel genre		Gender		Home language			Grade	
		Male	**Female**	**Dutch**	**Bilingual**	**Other**	**7**	**8**
Historical novel (n = 38)	Pupils (n= 28)	22	6	18	3	7	13	15
	Teachers (n= 10)	3	7	n.a.			n.a.	
Fantasy novel (n = 29)	Pupils (n= 21)	6	15	19	2	0	13	8
	Teachers (n= 8)	0	8	n.a.			n.a.	
Adaptation (n = 25)	Pupils (n= 16)	1	15	9	1	6	12	4
	Teachers (n= 9)	3	6	n.a.			n.a.	
Problem novel (n = 32)	Pupils (n= 23)	7	16	17	3	3	14	9
	Teachers (n= 9)	1	8	n.a.			n.a.	
Literary realism (n = 22)	Pupils (n= 14)	6	8	13	0	1	9	5
	Teachers (n= 8)	0	8	n.a.			n.a.	
Total (N = 146)	**Pupils (N= 102)**	**42**	**60**	**76**	**9**	**17**	**61**	**41**
	Teachers (N= 44)	**7**	**37**	**n.a.**			**n.a.**	

Surveys

All surveys used in this study were compiled in Dutch and completed on paper (see Appendix 1). They were reviewed by four teachers and piloted with five pupils, none of whom participated in the later stages of the study. The purpose of the questionnaires was to identify the underlying respondent characteristics that might influence hyperlink desirability in literary texts.

Pupils

In order to explain hyperlink desirability, a questionnaire was designed to map the following pupil characteristics: *Gender, Home language, Openness to digital reading, Openness to supporting and enriching information*, and *Reading motivation*. The latter was measured using a validated scale (i.e. MRQ – Motivations for Reading Questionnaire; Wigfield et al., 2004), calibrated on a 5-point Likert scale. The MRQ is composed of 54 items on 11 subscales (e.g. (reading) efficacy, curiosity, recognition). Unrau and Schlackman (2006) observe that intrinsic motivation can be measured using the Efficacy, Challenge, Curiosity and Involvement scales (20 items), and extrinsic motivation using the Recognition and Reading for grades scales (9 items). The EFA on all 29 items (KMO: .82; Bartlett .00) showed that all items loaded on the first factor, with loadings above .30, except for these two items:

- I read about my hobbies to learn more about them.
- Grades are a good way to see how well you are doing in reading.

Consequently, these items were excluded and one composed variable of 27 items, which accounted for pupil reading motivation, was created through sum scores. Internal consistency, measured with Cronbach's alpha, proved very reliable (α= .921).

Teachers

The teacher questionnaire comprised two sections: the first was designed to identify the teachers' media and reading education profile, whereas the second focused on their personal profile and teaching experience. The following teacher characteristics were mapped: *Openness towards digital reading tools, Importance of reading education, Importance of pupil reading motivation, Media self-efficacy, Attitudes towards media in education, Media use, Gender, Age,* and *Teaching experience*. The first two were measured through specifically created items, whereas for the four subsequent characteristics, existing, validated scales were used.

The items on *Openness towards digital reading tools* were created by the researchers and specifically measure teachers' attitudes toward digital reading devices:

- It is important for pupils to get acquainted with different reading devices (e.g. tablets, e-readers).
- I am interested in new reading devices (e.g. tablets, e-readers).
- I regularly use new reading devices (e.g. tablets, e-readers) for my own reading.

The items were analysed through EFA (KMO: .63; Bartlett .00), and the variable composed of all three items showed sufficient reliability (α= .744).

Importance of reading education was measured using eight items also specifically created for this study:

- With regard to literature in my language classes, I go beyond what is required.
- I consider reading education an important part of my language classes.
- I regularly brush up on my knowledge of reading education to prepare for my classes.
- I am an avid reader of children's literature.
- To prepare for my classes, I regularly brush up on my own knowledge of literature.
- I consider myself a good reader.
- I am well informed about current children's literature.
- I believe pupils should regularly visit the library.

EFA-analysis (KMO: .76; Bartlett .00), resulted in the exclusion of items 46 and 49, and the variable composed of the 6 remaining items proved to be sufficiently reliable (α= .798).

Importance of pupil reading motivation was measured using the TBSMRQ-scale (Teacher Beliefs about Student Motivation to Read; Quirk et al., 2010), calibrated on a 5-point Likert scale. Compatible with the MRQ, it consists of similar subscales. The same procedure was applied as mentioned for the MRQ. EFA-analysis of the 23 items (KMO: .65; Bartlett .00), resulted in the exclusion of one item:

- I believe grading is an important way to foster reading development for pupils.

The new composed variable, based on 22 items, proved to be reliable (α= .808).

The *Media self-efficacy (MSE)*; *Attitudes towards media in education (AME)*, and *Media use* scales were adapted from Sang et al. (2010): the Teacher Computer Self-Efficacy scale (CSE) (α = .90); the Attitudes toward Computers in Education scale (ACE) (α = .81); and the Computer use scale (α = .87). These 27 items measure teachers' self-efficacy while using media devices, teachers' general attitudes towards media devices in education and teachers' actual use of media devices in class. The word *computer* from the original scales was extended to *media devices* (i.e. computers, smartphones, tablets, e-readers). Items 4-12 were adapted from the CSE, items 13-20 from the ACE, and 21-30 from the Computer use scale. EFA's on the MSE (KMO: .88; Bartlett .00), AME (KMO: .76; Bartlett .00), and Media use scale (KMO: .75; Bartlett .00) showed that all items loaded on the first factor, with item loadings above .30, except for item 30 from the latter scale, which was excluded:

- I use media devices to communicate with my pupils (e.g., email, online platforms).

The composed variables measuring *Media self-efficacy*, based on 9 items (α= .899); *Attitudes towards media in education*, based on 8 items (α= .847); and *Media use*, based on 9 items (α= .822) proved all to be internally consistent and acceptable compared to the original instruments.

Focus Group Discussions

The final stage of the study comprised focus group discussions with a limited selection of participants to provide a more detailed and deeper understanding of the quantitative data. They were considered an additional supplementary data source (Morgan, 1998), besides the surveys and the main data set, the annotated novels. Focus group discussions ideally have four to ten participants (Liamputtong, 2011), so three focus groups (one with 8 pupils – 6 girls, 2 boys; two with 3 teachers each – 6 women, 0 men) were organized to investigate how both respondent groups experienced adding their desire for hyperlinks across the distinct novel types, and whether differences arose between the hyperlinks each group considered useful and necessary.

Data Collection

All data were collected between February and May 2018. The pupil data were collected in schools and intact classes were probed during their regularly scheduled 50-minute Dutch class. The surveys were administered by the researchers in the presence of the participating language teacher. During the initial 20 minutes, all respondents were first informed about the aim and the hyperlink types detailed below, then annotated one page to practice and received feedback from the researchers. They were told that their responses to the questions depended on their own feelings and experiences in reading and that all answers were acceptable. The pupils were guided to complete three practice items before answering the actual questionnaire, which was completed in the remaining 30 minutes. The teachers were informed and completed their surveys during a similar individual or group session with the researchers, typically during a spare hour in their timetables.

The two hyperlink types that participants could mark, if desirable, were defined as follows:

Type 1: Explanatory Hyperlinks Supporting Information For Pupils, To Increase Comprehension

Explanatory hyperlinks, such as glosses, historical or cultural information, etc. Pupils might observe *"I don't understand this"* or *"I don't know this"* regarding these words or phrases and be willing to know more about them to increase their comprehension.

Examples: "SS insignia" (Historical novel, p.27; P093)
"to blackmail" (Problem novel, p.165; P088)
"Pippi Longstocking" (Literary realism, p.17; T053)

Type 2: Enriching Hyperlinks To Signal Particular Elements Or Beautiful Passages

Enriching hyperlinks with stylistic or literary information, such as original metaphors, citations, references to other stories, paratextual information, etc. Pupils might observe *"This is beautiful or special. I would like to know more"* regarding these words or phrases.

Examples: "How can the bird that is born for joy
Sit in a cage and sing?" (Fantasy novel, p.48; T061)
"and those words burned holes in my head and in my heart"
(Adaptation, p.22; P100)

Afterwards, the respondents had three months to read their novel, and mark the instances for which they would like to have hyperlinks. They were instructed to circle explanatory, and underline enriching hyperlinks. This resulted in 146 annotated novels, 102 by pupils and 44 by teachers (see Table 1). The annotated paper copies were then analysed (hyperlink types and frequency). Finally, three focus groups (one with pupils; two with teachers) were organized to deepen the insights with qualitative data.

Data Analysis

First, the annotated **novels** were analysed in an excel-file where all explanatory and enriching hyperlinks were marked in terms of number and content. In the raw data, all hyperlinks were marked exactly as they had been annotated in the novels. Subsequently, the data was analysed to verify if respondents had not misinterpreted any type 1 hyperlinks as type 2 or vice versa. Only in very obvious cases (e.g. P052 had underlined both types), they were recoded into two types. In case of potential misinterpretation (e.g. explanatory links that might also be interpreted as enriching) the data was left untouched and processed exactly as marked by the respondents. The excel-file was then converted in order to perform further data management and analysis using SPSS software (version 25).

Second, the teacher and learner **surveys** were converted to digital format with *Evasys*-software, and analysed using SPSS software (version 25). As described above, for most variables data reduction was required, as they were measured through a series of items. For EFA, *Principal components analysis* with Varimax-rotation was used (Nunnally, 1978; Kline, 1994) and only items with loadings above .30 were included. Eigenvalues of retained factors had to be ≥ 1 (Kaiser, 1960) and the new scale's reliability (Cronbach's alpha) ≥ .70. Based on these new scales, new composed variables were calculated.

Third, the **novel** and **survey data** files were **merged** to match teacher and learner variables with the annotated hyperlinks. Descriptive statistics, and subsequently multiple linear regression analyses were then performed to verify the type and frequency of hyperlinks, and the impact of the underlying respondent variables. Therefore, two new dependent variables were created: *Hyperlink type desirability* (i.e. n explanatory hyperlinks/total N hyperlinks) and *Hyperlink type frequency* (i.e. total n explanatory hyperlinks/novel pages). Table 2 gives an overview of the pupil variables that were included in the analysis.

The following table gives an overview of the teacher variables that were analysed.

Finally, the three **focus group discussions** were analysed. They were video-taped and transcribed verbatim. During the first coding phase, one researcher conducted an explorative analysis of the transcripts in which the research questions constituted the framework of the initial coding scheme. Data-driven analysis refined and extended the coding scheme (see Appendix 2) and the re-coded data were then analysed with NVivo11 software. The validity structure (Johnson, 1997) of the qualitative data analysis was ensured through *investigator triangulation* and *peer review*: two researchers monitored the data collection and each performed an independent coding of the data (inter-rater reliability). Inconsistencies in coding between both researchers were then discussed. Additionally, *low inference descriptors* were used: the respondents' actual language was described verbatim and reported with direct quotations (see Results section).

RESULTS

The results of the empirical study on hyperlink desirability in literary texts are presented below in two separate sections that each correspond to the underlying research questions. Each section first discusses the overall results, and then focuses on interesting results related to pupil and teacher variables.

Table 2. Overview of dependent and independent variables for pupils

Variable	Name	Number of items	Cronbach's alpha	Categories
Dependent variables				
Hyperlink type Hyperlink frequency	Hyperlink type desirability Hyperlink type frequency			
Independent variables				
1. Gender		1		1 = female 2 = male
2. Home Language		1		1 = language of the region 2 = bilingual 3 = other home language
3. Openness to digital reading		1		0 = no 1 = yes
4. Openness to supporting and enriching information		1		0 = no 1 = yes
5. Reading motivation		27	.921	
6. Literary genre		1		1= historical novel 2= fantasy novel 3= adaptation 4= problem novel 5= literary realism

I. Hyperlink types

RQ1: Which hyperlink types do adolescent readers and their language teachers find desirable in literary texts?

The table below provides an overview of the desired hyperlink types per literary genre.

We hereby distinguish between respondents with no desire for hyperlinks, respondents who marked exclusively explanatory *or* enriching hyperlinks, and respondents who marked both types. The first row shows the results in absolute numbers, the second in relative numbers. If there are no respondents, the box is left blank.

As shown in Table 4, no pupil or teacher chose the 'no hyperlinks at all' option. As mentioned above, novels without hyperlinks that appeared untouched had previously been removed from the sample. The overall majority of pupils (73.5%) and teachers (93%) considered **both hyperlink types** desirable in the novels they read. With regard to the teachers, this holds true across most literary genres, except for one respondent of the problem novel (T010) and two respondents of the realist novel (T026 and T053), who wanted only explanatory hyperlinks. The teachers' educational goals, namely providing a wide range of hyperlinks for the various reader profiles of their pupils, may explain why they mark a combination of both types. As T54 commented in the focus group: "*You ask yourself if the pupils will understand. I specifically targeted our pupils of grade 7*". The pupils show more individually varied patterns of hyperlink desirability. Interestingly, four pupils exclusively marked enriching hyperlinks, which shows that they are attentive to literary style and want to learn more about it.

Table 3. Overview of dependent and independent variables for teachers

Variable	Name	Number of items	Cronbach's alpha	Categories
Dependent variable				
Hyperlink type Hyperlink frequency	Hyperlink type desirability Hyperlink type frequency			
Independent variables				
1. Gender		1		1 = female 2 = male
2. Age		1		
3. Teaching experience		1		
4. Openness towards digital reading tools		3	.744	
5. Importance of reading education		6	.798	
6. Importance of pupil reading motivation		22	.808	
7. Media self-efficacy		9	.899	
8. Attitudes towards media in education		8	.847	
9. Media use		9	.822	
10. Literary genre		1		1= historical novel 2= fantasy novel 3= adaptation 4= problem novel 5= literary realism

As for those pupils who **exclusively** opted for **enriching hyperlinks (type 2)**, interestingly, two out of four read a genre that is commonly considered the least 'literary', namely the problem novel; the other two read the fantasy novel and adaptation, genres where more literary language could be expected. If we examine the former two reader profiles, P036 and P039 (problem novel) are both male, and score respectively low and average on *Reading Motivation* (2.59 and 3.44 out of 5), and negatively on *Openness to digital reading* and *Openness to supporting and enriching information*. Moreover, P036's home language is Turkish, which might imply a limited mastery of the language of instruction. Both equally marked a very limited total number of links (6 and 1). Their result might be explained by a lack of reading motivation and possibly language proficiency (P036), and a lack of interest in additional information in (digital) reading materials (P039). The latter two respondents, P023 (adaptation) and P106 (fantasy novel) are female, respectively highly (4.22) and averagely (3.07) motivated, both with a negative score on *openness to supporting and enriching information*. The fact they only marked enriching links might be explained by their lack of openness to additional information. However, the reasons behind it may differ: highly motivated readers (P023) might see explanatory links (and possibly digital literary reading as such) as an interruption of their immersion, whereas averagely motivated readers (P106) might not like the extra work that explanatory links imply.

Table 4. Hyperlink types

Novel genre	Respondents	No desired hyperlinks	Only Type 1 Explanatory	Only Type 2 Enriching	Type 1 & 2
Historical novel	Pupils (n=28)		14 50%		14 50%
	Teachers (n=10)				10 100%
Fantasy novel	Pupils (n=21)		2 9.5%.	1 5%	18 85.5%
	Teachers (n=8)				8 100%
Adaptation	Pupils (n=16)			1 6%	15 94%
	Teachers (n=9)				9 100%
Problem novel	Pupils (n=23)		4 17%	2 9%	17 74%
	Teachers (n=9)		1 11%		8 89%
Literary realism	Pupils (n=14)		3 21%		11 79%
	Teachers (n=8)		2 25%		6 75%
Total (N=146)	**Pupils (N= 102)**		**23** **22.5%**	**4** **4%**	**75** **73.5%**
	Teachers (N=44)		**3** **7%**		**41** **93%**

Moving on now to consider the respondents that **exclusively** marked **explanatory hyperlinks (type 1),** the pupils' results for the historical novel stand out. Interestingly, half of them (14 out of 28) only annotated explanatory hyperlinks. Moreover, if we examine their profile, it becomes clear that they are all male. Despite a male overrepresentation in this particular pupil subsample (22 out of 28; Table 1), which may be due to the topic (i.e. WWII), it seems therefore possible that gender plays a role in these results. As P048 said: "*I mostly circled* (explanatory hyperlinks) *because of army terminology. I wanted to know which planes or rifles were used.*" Similarly, the only two pupils who exclusively marked explanatory links in the fantasy novel are both also male, whereas the female pupils (15 out of 21) preferred enriching links or both types. The same holds true for three out of four pupils who read the problem novel and two out of three who read the realist novel. Finally, in the adaptation genre no pupils only marked explanatory hyperlinks, while almost all respondents are female (15 out of 16). A possible explanation therefore might be that (at least part of the) male pupils desire more explanatory links to, for instance, technical jargon or vocabulary, whereas female pupils might prefer a combination of explanatory and enriching links, as they seem to be more attentive to literary style, even in the problem novel, a genre that has been criticized for its perceived lack of literary quality (Joosen & Vloeberghs, 2008).

The number of teachers who only marked explanatory hyperlinks is very low (3 out of 44), and concentrated on two literary genres, namely the problem (T010) and the realist novel (T026 and T053). Contrary to the pupils, the three teachers (T010,) are all female, have a similar teaching experience (15 years) and are aged between 30 and 40. They consider *Pupil Reading Motivation* and *Reading education* important, display *Openness towards digital reading tools* except for T053, have high *Media self-efficacy* except for T026, positive *Attitudes towards media in education* and varying scores on *Media use.* The scores of T010 are generally high on all variables, so the fact she marked no enriching hyperlinks might be explained by other variables, such as the novel genre, the author's language use, or the pupils she had in mind while marking the hyperlinks. The profiles of T026 and T053 are quite similar, only differing in terms of *Openness towards digital reading tools* and *Media self-efficacy.* T051 also reported on the realist novel: "*Not much figurative language was used. It was written from the perspective of the main characters, 12-to 14-year-olds,*" which might also explain why some teachers marked only a few enriching links, or none at all. The exact hyperlink frequencies for each novel will be discussed in the next section.

Turning now to the literary genres, the adaptation elicits most hyperlinks from both types in the pupil respondents, whereas the historical novel is the genre with most pupil respondents for explanatory hyperlinks, followed by the realist novel, a genre in which also 25% of the teacher subsample exclusively marked explanatory hyperlinks. The problem and fantasy novel are the genres with the most diversified hyperlink desirability profile. The most remarkable differences between pupils and teachers can be found in the historical and the fantasy novel, where 100% of the teachers prefers both types, whereas respectively 50% and 9.5% of the pupils mark only enriching, and, in the latter novel, 5% only mark explanatory links.

An **independent samples t-test** was used to further analyse the differences in hyperlink types both respondent groups preferred. For the dependent variable *Hyperlink type desirability* (i.e. the preference for explanatory hyperlinks; see Table 2 and 3), homogeneity of variances was violated, as assessed by Levene's test for Equality of Variances ($p= .003$). On average, teachers ($M= 73.77$, $SD= 20.00$) displayed a greater preference for explanatory hyperlinks than pupils ($M= 72.71$, $SD= 31.35$), but this difference was not statistically significant, $t(123.420) = -.244$, $p = .808$.

Finally, the **influence of independent participant variables** on hyperlink type desirability was investigated. For the pupil group, one dependent variable, *Hyperlink type desirability*, and six independent variables were identified (see Table 2 and 3) and the influence of the latter on the former was examined by multiple linear regression analyses. Its significance ($p < 0.05$) was verified and the effect size was examined using the Beta coefficient that indicates which share of the total variance within the dependent variable can be attributed to the independent variable (i.e. contribution on a total of 5). The direction of the effect is marked (positive or negative).

A model composed of all independent variables explains pupil hyperlink type desirability significantly ($F(6,86) = 5.264$, $p < .0005$, adjusted $R^2 = .269$), and accounts for 27% of the variance within the dependent variable. The collinearity between independent variables is correct ($VIF < 1.93$; Tolerance $> .56$). Table 5 shows the results of the analysis.

Hyperlink type desirability in pupils is significantly influenced by two variables, namely literary genre and reading motivation. The variable *literary genre* has the largest Beta coefficient ($\beta =-.359$) and displays the highest significance ($p < .001$). Thus, the hyperlink types pupils prefer depend mostly on the literary genre of the novel, which coincides with the initial observations from Table 4. The problem and fantasy novel have the most diversified - and the historical and realist novel the most specific - hyperlink type desirability profiles, whereas in the adaptation most pupils marked both hyperlink types.

Table 5. The influence of independent variables on the dependent variable Pupil hyperlink type desirability (=p<.01; **=p<.001)*

Independent variables	Pupil Hyperlink type desirability (dependent variable)		
	standardized β	t	p
1. Gender	.120	1.104	.273
2. Home language	.057	.609	.544
3. Openness to digital reading	-.084	-.871	.386
4. Openness to supporting and enriching information	-.168	-1.635	.106
5. Reading motivation	**-.310**	-2.640	.010*
6. Literary genre	**-.359**	-3.396	.001**

A second important variable is *reading motivation*, with a β-value of -.310 and significance at the $p < .01$ level. Its direction indicates that the higher pupils' reading motivation, the lower their preference for explanatory hyperlinks. There are two possible explanations for this result: motivated readers might prefer fewer explanatory hyperlinks, and subsequently more enriching hyperlinks, because they are more attentive to literary style and enjoy the novel more with information about style. However, highly motivated readers might also prefer an overall limited number of hyperlinks, especially explanatory links, because they already fully experience immersion, and do not require (m)any supporting hyperlinks during the reading, as discussed above with regard to P106 and P023. As P049 commented in the focus group discussion: "*Sometimes I was so immersed in the story that I simply forgot to circle* (explanatory hyperlinks)".

For the teachers, ten independent variables and one dependent variable, *Hyperlink type desirability*, were identified (see Table 3), and the same procedure was followed. A model composed of all independent variables explains teacher hyperlink type desirability significantly ($F(10,25) = 2.494$, $p < .05$, adjusted $R^2 = .299$), and accounts for 30% of the variance within the dependent variable. The collinearity between independent variables is correct (VIF < 1.93; Tolerance > .56). Table 6 shows the results of the analysis.

Teacher hyperlink type desirability is significantly influenced by two variables, namely the importance teachers attach to pupil reading motivation, and media use. The variable *importance of pupil reading motivation* has the largest Beta coefficient (β =.474) and displays the highest significance (p=.014). Thus, the more teachers consider their pupils' reading motivation important, the more explanatory hyperlinks (and the less enriching) they will provide for them in the novels they read. The pupil results (Table 5) confirm that this might help to support struggling, less motivated readers, whereas more motivated readers may prefer a smaller percentage (and possibly smaller number) of explanatory hyperlinks. A second important variable is teachers' *media use,* with a Beta coefficient of -.387. Its direction may, however, come as a surprise: the more media tools teachers use, the less explanatory (and the more enriching) hyperlinks they tend to provide for their pupils in the novels.

So far this chapter has focused on the two hyperlink types, namely explanatory and enriching links, that pupils and teachers find desirable in literary texts. The following section will now discuss at what frequency both respondent groups consider these hyperlinks desirable in novels.

Table 6. The influence of independent variables on the dependent variable Teacher hyperlink type desirability (=p<.05)*

Independent variables	Teacher Hyperlink type desirability (dependent variable)		
	standardized β	**T**	**p**
1. Gender	.072	.451	.656
2. Age	-.392	-1.096	.283
3. Teaching experience	.145	.414	.683
4. Openness towards digital reading tools	.102	.672	.508
5. Importance of reading education	.042	.237	.814
6. Importance of pupil reading motivation	**.474**	2.654	.014*
7. Media self-efficacy 8. Attitudes towards media in education 9. Media use	.262 -.242 **-.387**	1.355 -1.330 -2.059	.187 .196 .050*
10. Literary genre	-.006	-.031	.975

II. Hyperlink Frequency

RQ2: At what frequency are the hyperlinks adolescent readers and their teachers find desirable marked in literary texts?

Table 7 shows the summary statistics for the total hyperlink frequencies (both types combined) per novel genre in both respondent groups. The first row shows the results in absolute numbers (i.e. total n hyperlinks), the second row in relative numbers (i.e. total n hyperlinks per page). When displaying the frequency, we consider the number of pages of each novel.

What stands out in the table is that teachers on average mark more hyperlinks than pupils in all novels, except for the problem novel. Pupils mark most hyperlinks per page in the historical novel and the adaptation, whereas teachers do so in the historical and the fantasy novel. The results for the historical novel could be attributed to its length and vocabulary density, which facilitate particularly explanatory hyperlinks. Table 8 shows the frequencies per hyperlink type, which facilitates a more detailed interpretation of the results. The fantasy novel contains not only many unknown concepts, but also many literary and intertextual references (e.g. to William Blake), which explains the highest teacher T2 mean score (30.63). It may be that teachers recognize and mark the latter as enriching hyperlinks, whereas pupils simply overlook them. The very elevated teacher T1 mean score (M=137) for this novel, is additionally due to two particular outliers (242 and 579). Finally, the elevated pupil T1 and T2 mean score for the adaptation is less surprising, as the author uses very descriptive, literary language and many uncommon adjectives, which facilitates both explanatory and enriching hyperlinks. As P051 reported: "*There were many beautiful sentences in the novel*".

As for explanatory hyperlinks, the mean frequency across literary genres is higher for teachers (M=62.64) than pupils (M=38.47). This might be due to the phenomenon described above, where teachers mark links for various pupil profiles, whereas pupils exclusively mark from their own perspective. The pattern for enriching hyperlinks is similar, but differences in mean frequency between teachers

Table 7. Desired hyperlink frequencies per novel

Novel genre	Respondents	Mean	Minimum	Maximum	SD
Historical novel 65p.	Pupils (n=28)	51.75 .79	3 .05	132 2.03	41.40 .64
	Teachers (n=10)	65.90 1.01	24 .37	161 2.48	47.28 .73
Fantasy novel 150p.	Pupils (n=21)	48.14 .32	2 .01	169 1.13	50.43 .34
	Teachers (n=8)	167.62 1.12	10 .07	661 4.41	223.60 1.49
Adaptation 126p.	Pupils (n=16)	75.25 .59	5 .04	340 2.70	89.04 .71
	Teachers (n=9)	96.55 .77	24 .19	268 2.13	83.70 .66
Problem novel 176p.	Pupils (n=23)	39.91 .23	1 .01	174 .99	49.37 .28
	Teachers (n=9)	26.78 .15	9 .05	85 .48	23.82 .13
Literary realism 207p.	Pupils (n=14)	44.14 .21	2 .01	219 1.06	56.94 .27
	Teachers (n=8)	52.75 .25	7 .03	101 .49	37.75 .18
Total (N=146)	**Pupils (N= 102)**	**50.98** **.46**	**1** **.01**	**340** **2.70**	**56.72** **.54**
	Teachers (N=44)	**80.27** **.67**	**7** **.03**	**661** **4.41**	**111.85** **.85**

(17.64) and pupils (12.51) are much smaller. A likely explanation is that teachers' literary knowledge helps them to detect particular passages of interest more easily. A partial explanation also lies in the remarkably elevated pupil mean T2 score on the problem novel (21.78), as opposed to the teachers (5.44). The respective maximum T2 scores (137 and 18) confirm this, which may be explained by the fact that experienced readers evaluate literary value differently than less experienced readers: teachers' literary knowledge might withhold them from marking stereotypical phrases and clichés in problem novels as enriching hyperlinks, as opposed to most pupils.

Turning now to *gender* differences, female pupils mark on average more hyperlinks of both types than male pupils (T1: M=40.40 vs 35.71 and T2: M=15.18 vs. 8.69), and for enriching links the difference is still mitigated by a male T2 outlier of 137. A possible explanation might be that female pupils are on average more motivated readers (Picton, 2014), and therefore pay more attention to literary style. A similar pattern can be observed for the teachers; female teachers' mean scores on explanatory (M=66.32 vs M=43.14) and on enriching hyperlinks (M=18.73 vs. M=11.86) are more elevated than those of their male counterparts.

If we consider *home language* as an additional variable, female bilingual pupils (M= 58.40) and female pupils with another home language (M= 57.67) clearly mark most explanatory hyperlinks, followed by male bilingual pupils (M= 44.75), which might be explained by their mastery of the language, and particularly of specific vocabulary. Surprisingly, the male pupils with another home language (n=8),

Table 8. Desired hyperlink frequencies per hyperlink type

Novel genre	Respondents	Mean	Minimum	Maximum	SD
Historical novel	Pupils (n=28) Teachers (n=10)	T1: 48.68 T2: 3.07	T1: 3 T2: 0	T1: 122 T2: 20	T1: 38.89 T2: 5.43
		T1: 55.70 T2: 10.20	T1: 20 T2: 1	T1: 125 T2: 36	T1: 39.32 T2: 11.23
Fantasy novel	Pupils (n=21) Teachers (n=8)	T1: 38.52 T2: 9.62	T1: 0 T2: 0	T1: 154 T2: 54	T1: 45.68 T2: 12.13
		T1: 137 T2: 30.63	T1: 4 T2: 3	T1: 579 T2: 82	T1: 194.74 T2: 33.23
Adaptation	Pupils (n=16) Teachers (n=9)	T1: 56.13 T2: 19.13	T1: 0 T2: 1	T1: 287 T2: 53	T1: 80.90 T2: 13.87
		T1: 68.22 T2: 28.33	T1: 10 T2: 7	T1: 246 T2: 76	T1: 74.69 T2: 21.00
Problem novel	Pupils (n=23) Teachers (n=9)	T1: 18.13 T2: 21.78	T1: 0 T2: 0	T1: 86 T2: 137	T1: 20.93 T2: 36.91
		T1: 21.33 T2: 5.44	T1: 7 T2: 0	T1: 80 T2: 18	T1: 22.70 T2: 5.57
Literary realism	Pupils (n=14) Teachers (n=8)	T1: 31.21 T2: 12.93	T1: 0 T2: 0	T1: 149 T2: 70	T1: 41.89 T2: 19.47
		T1: 37.13 T2: 15.63	T1: 4 T2: 0	T1: 100 T2: 37	T1: 33.78 T2: 16.01
Total (N=146)	**Pupils (N= 102)** **Teachers (N=44)**	**T1: 38.47** **T2: 12.51**	**T1: 0** **T2: 0**	**T1: 287** **T2: 137**	**T1: 47.86** **T2: 21.54**
		T1: 62.64 **T2: 17.64**	**T1: 4** **T2: 0**	**T1: 579** **T2: 82**	**T1: 96.65** **T2: 20.88**

however, have a T1 score (M= 35.75) very similar to that of female (M=35.07) and male (M=34.50) pupils with the regional home language. As for enriching hyperlinks, the highest scores are also marked by female bilingual pupils (M=30.20), followed by female (M=14.80) and male (M=10.67) pupils with the regional home language. The differences between female (M=8.78) and male (M=4.25) pupils with another home language, and especially female (M=30.20) and male (M=2.75) bilingual pupils are also quite remarkable. Female bilingual pupils clearly seem to desire more hyperlinks of both types than the other pupil subcategories.

Independent samples t-tests were run to further analyse the differences in hyperlink frequencies that pupils and teachers found desirable. For the total number of hyperlinks, homogeneity of variances was violated, as assessed by Levene's Test for Equality of Variances (p= .030). On average, teachers (M= 80.27, SD= 111.85318) marked more hyperlinks than pupils (M= 50.98, SD= 56.72), but this difference was not significant, $t(52.792) = -1.648$, $p = .105$. For the total number of explanatory hyperlinks, Levene's test showed homogeneity of variances was violated (p= .029). On average, teachers (M= 62.64, SD= 96.65) marked more explanatory hyperlinks than pupils (M= 38.47, SD= 47.86), but this difference was not significant, $t(52.330) = -1.577$, $p = .121$. Finally, for the total number of enriching hyperlinks, Levene's test showed homogeneity of variances was not violated (p= .467). On average, teachers (M=

17.64, SD= 20.88) marked more enriching hyperlinks than pupils (M= 12.51, SD= 21.54), but this difference was not significant either, $t(144) = -1.331$, $p = .185$. In conclusion, in terms of hyperlink frequencies, no significant differences between pupils and teachers could be detected.

Using a similar procedure as described for RQ1, through **multiple linear regression analyses** the influence of 6 independent pupil variables and 10 independent teacher variables (see Table 2 and 3) on the dependent variable *hyperlink type 1 frequency* was examined. Analyses on *total hyperlink frequency*, and *type 2 frequency* were also performed for both pupils and teachers, but those models explained total hyperlink frequency and type 2 hyperlink frequency insufficiently. Further research with other variables and bigger sample sizes should shed more light on this issue. The model described below therefore investigates exclusively explanatory hyperlink frequencies in both respondent groups.

For explanatory hyperlinks, a model composed of all independent variables explains **pupil** *hyperlink type 1 frequency* significantly ($F(6,86) = 3.555$, $p < .005$, adjusted $R^2 = .143$), and accounts for 14% of the variance within the dependent variable. The collinearity between independent variables is correct (VIF < 1.93; Tolerance > .56). Table 9 shows the results of the analysis.

Pupil hyperlink type 1 frequency is only significantly influenced by one variable, namely: literary genre. It has a Beta coefficient of -.382 and displays a significance (p=.001). Thus, the number of explanatory hyperlinks pupils mark per page diminishes across the different literary genres, from the historical to the realist novel, which coincides with the results from Table 7. However, *Reading motivation*, which played a role in pupil hyperlink type desirability (Table 5), does not influence hyperlink type 1 frequency in a significant way.

A model composed of all independent variables explains **teacher** *hyperlink type 1 frequency* significantly ($F(10,25) = 2.339$, $p < .05$, adjusted $R^2 = .277$), and accounts for 28% of the variance within the dependent variable. The collinearity between independent variables is correct (VIF < 1.93; Tolerance > .56). Table 10 shows the results of the analysis.

Teacher hyperlink type 1 frequency is only significantly influenced by one variable, namely: importance of pupil reading motivation. It has a Beta coefficient of -.542 and displays a significance (p=.006). Thus, the number of explanatory hyperlinks teachers mark per page diminishes as they consider their pupils' reading motivation more important. Interestingly, its direction is opposite to that of the same variable in Table 7. Consequently, the more teachers consider pupil reading motivation important, the larger the percentage of type 1 in the marked hyperlinks, but the smaller the number of explanatory hyperlinks

Table 9. The influence of independent variables on the dependent variable Pupil HL type 1 frequency (=p<.001)*

Independent variables	Pupil Hyperlink type 1 frequency (dependent variable)		
	standardized β	t	p
1. Gender	-.022	-.190	.850
2. Home language	.174	1.767	.081
3. Openness to digital reading	-.069	-.687	.494
4. Openness to supporting and enriching information	-.014	-.131	.896
5. Reading motivation	.011	.092	.927
6. Literary genre	**-.382**	-3.452	.001*

Table 10. The influence of independent variables on the dependent variable Teacher HL type 1 frequency (=p<.01)*

Independent variables	Teacher Hyperlink type 1 frequency (dependent variable)		
	standardized β	t	p
1. Gender	.064	.395	.696
2. Age	.021	.058	.954
3. Teaching experience	-.229	-.643	.526
4. Openness towards digital reading tools	.109	.704	.488
5. Importance of reading education	-.178	-.981	.336
6. Importance of pupil reading motivation	**-.542**	-2.988	.006*
7. Media self-efficacy	.109	.357	.724
8. Attitudes towards media in education	.070	-1.344	.191
9. Media use	-.248	1.400	.174
10. Literary genre	.178	1.003	.325

per page. A possible explanation might be that teachers opt for quality over quantity, and provide more explanatory than enriching links, but not too many per page, in order to support but yet not overwhelm the struggling reader. As T053 reported in the focus group: "*We need to balance functionality and accessibility on a page*".

Another remarkable finding is that literary genre, as opposed to the results for the pupil group in Table 10, does not play a significant role in the number of explanatory hyperlinks teachers mark per page. However, it might be possible that the size of the teacher sample (N=44) was too small to detect any significant effects for literary genre, as opposed to the large pupil sample (N=102).

DISCUSSION

The results of the study described in this chapter show that most pupils consider explanatory and enriching hyperlinks desirable across five distinct literary genres. However, 20% of the pupils exclusively marked explanatory hyperlinks, and interestingly the overall majority of those 20% were male. A possible explanation therefore might be that male pupils would tend to prefer more explanatory hyperlinks to technical jargon or vocabulary than girls. By contrast, a small number of pupils (4%) exclusively marked enriching hyperlinks. They might not require explanatory hyperlinks in novels for two reasons: already motivated readers might see them as an interruption of immersion, whereas average and low motivated readers might not like the additional work that consulting the hyperlinks entails. Female pupils mark more hyperlinks on average of both types than their male counterparts, and female bilingual pupils seem to desire more hyperlinks of both types than other pupils. Hence, it could conceivably be hypothesised that gender plays a role in hyperlink desirability. However, the observed differences in gender and home language did not become significantly apparent in the subsequent statistical analyses.

As for the teachers, almost all consider both hyperlink types desirable in literary texts, and mark on average more hyperlinks of both types than pupils. One possible explanation is that teachers provide a wide range of hyperlinks for the various reader profiles of their pupils, and thanks to their literary knowledge also more easily detect particular passages of interest. Moreover, female teachers, similar to pupils, mark on average more hyperlinks of both types than male teachers. On average, the percentage of explanatory compared to enriching hyperlinks was larger in the teacher than in the pupil sample, and teachers additionally marked more hyperlinks of both types than pupils, but subsequent analyses did not detect evidence that those differences between both respondent groups were statistically significant.

An explanatory model composed of respectively six variables (pupils), and ten variables (teachers) allowed us to pinpoint those variables that significantly accounted for hyperlink type desirability. For pupils, the independent variables 'literary genre' and 'reading motivation' had the biggest impact, whereas for teachers, the variables 'importance of pupil reading motivation' and 'media use' played the most important role. Additionally, within the explanatory model described above, explanatory hyperlink frequency was significantly accounted for by one pupil variable, literary genre, and by one teacher variable, importance of pupil reading motivation. Combining the results from both models, a possible explanation might be that as teachers consider pupil reading motivation more important, they mark a limited number of explanatory (rather than enriching) links per page to support, but yet not overwhelm struggling readers. One unanticipated finding was that literary genre, as opposed to the results for the pupil group, did not play a significant role in the number of explanatory hyperlinks teachers marked per page. More research is needed to verify whether this holds true for larger teacher samples. With regard to pupils, there is a positive correlation between lack of motivation and the desire for large numbers of explanatory hyperlinks. In contrast, motivated readers may desire more enriching hyperlinks, and possibly a smaller total number of hyperlinks, with a view to maintaining their immersion. The qualitative data seem to underpin this observation.

However, a few limitations of this exploratory study need to be acknowledged. With a small sample size, caution must be applied, as the study might be underpowered to detect small effect sizes of clinical significance. This particularly applies to the teacher sample (N=44). It is therefore possible that the other teacher variables included in the study also influence teacher hyperlink type and frequency desirability, but that their effect can only be measured with a bigger subsample. Furthermore, these findings cannot be extrapolated to all novels of the literary genres included in the study, as novels from the same genre might differ greatly (e.g., topic, length, style), which will equally determine the type and frequency of hyperlinks to be included. Moreover, these data must also be interpreted with caution because each reader is unique, and many unknown variables, such as pupil personality (e.g., curiosity), might also influence their hyperlink preferences. Finally, differences in hyperlink desirability between pupils and teachers might also be interpreted as the latter group's capacity to estimate the former group's reading competencies.

FUTURE RESEARCH DIRECTIONS

Further research is needed to improve our understanding of the role that hyperlinks play in the reading process of digital literary texts in general, and in immersion in particular. It would therefore not only be beneficial to replicate this study with bigger sample sizes for both qualitative and quantitative analyses, and to include additional variables such as pupil and teacher personality, but also to further investigate hyperlink placement in the literary text, and its relation with, for instance, plot development. Further studies can then operationalize these findings by creating hypermedia fiction, digital fiction with multimedia hyperlinks to supporting and enriching information, in order to investigate its influence on immersion, and explore differences between various subsamples of pupils. In the second empirical study of this research project on digital reading, the results of the present study will be implemented in order to further evaluate the potential of multimedia hyperlinks for Mobile-Assisted Language Learning (MALL) and its applications within the classroom environment, particularly in terms of pupils' reading experience of fiction and the impact on immersion. Another fruitful area for further work would be to assess the impact of multimedia hyperlinks on reading comprehension of various text types that are read on different reading devices used for MALL, such as tablets, e-readers and smartphones.

CONCLUSION

As reading on digital screens has become increasingly common in educational and recreational reading, and both scholars and teachers are concerned about pupils' reading motivation and performance, it is important to investigate the potential of technology to support reading among different groups. This chapter presented the results of an experimental study on the desirability of explanatory and enriching hyperlinks and their frequency in novels from five distinct literary genres. A comparative analysis of pupil and teacher perspectives revealed that teachers have a stronger preference for explanatory hyperlinks, and mark on average more explanatory and enriching hyperlinks than pupils. Pupil hyperlink *type* desirability is significantly influenced by respectively literary genre and reading motivation, whereas for teachers, importance of pupil reading motivation and media use have the biggest impact. On the other hand, pupil explanatory hyperlink *frequency* is only significantly influenced by literary genre, whereas importance of pupil reading motivation exerts the biggest influence on teachers.

The implications of these results for the distinct stakeholders related to literary reading and reading education are threefold. First, for schools and policy makers, it is essential to develop well-thought educational policies that include both analogue and various forms of digital reading to foster future pupil reading brains which are still able to carry out the "deep reading" processes at the core of expert reading and essential for critical thinking, personal reflection, imagination and empathy (Wolf, 2018) that are now typically related to reading on paper, while at the same time being flexible enough to process reading experiences on both digital and analogue reading technologies. Explanatory and enriching multimedia hyperlinks in digital texts might play an important role in this process. In that view, there is also a growing need to bridge the gap between the availability of digital (reading) technology at home and at school (OECD, 2015), so that pupils can get acquainted with different technologies (paper books and digital reading tools for MALL) and reading modes (with or without hyperlinks) at school from an early age onwards, in order to develop and train such "biliterate reading brains" (Wolf, 2018).

Second, for L1 and FL teachers and teacher educators, these results can provide new avenues for the development of (teacher training) reading activities, and stimulate the use of mixed (various digital and analogue) reading technologies in the L1 and FL classroom, so that pupils' distinct motivational profiles can be addressed to the maximum while developing deep reading skills.

Finally, for authors and publishing companies, the results of this study show that - even across literary genres - there is a clear interest on behalf of both teachers and pupils in digital reading materials with multimedia hyperlinks. This chapter can provide helpful insights for the development of future novels, but also of FL reading materials that include explanatory and enriching hyperlinks to, for instance, word explanations or cultural elements related to the target culture, to support the reader and facilitate sustained and successful reading experiences.

ACKNOWLEDGMENT

This research was supported by the University of Antwerp.

REFERENCES

Afflerbach, P., & Cho, B. (2009). Determining and describing reading strategies: Internet and traditional forms of reading. In H. S. Waters & W. Schneider (Eds.), *Metacognition, strategy use, and instruction* (pp. 201–225). New York, NY: Guilford.

Baron, N. (2015). *Words onscreen: The fate of reading in a digital world.* Oxford, UK: Oxford University Press.

Barzillai, M., & Thomson, J. M. (2018). Children learning to read in a digital world. *First Monday, 23*(10). doi:10.5210/fm.v23i10.9437

Barzillai, M., Thomson, J. M., Schroeder, S., & Van den Broek, P. W. (2018). Learning to read in a digital world. In M. Barzillai, J. Thomson, S. Schroeder, & P. W. Van den Broek (Eds.), *Learning to Read in a Digital World* (pp. 225–238). Amsterdam, The Netherlands: John Benjamins. doi:10.1075wll.17

Busselle, R., & Bilandzic, H. (2008). Fictionality and perceived realism in experiencing stories: A model of narrative comprehension and engagement. *Communication Theory, 18*(2), 255–280. doi:10.1111/j.1468-2885.2008.00322.x

Carr, N. (2011). *The shallows: What the internet is doing to our brains.* New York, NY: W.W. Norton.

Douglas, Y., & Hargadon, A. (2001). The pleasure of immersion and engagement schemas, scripts and the fifth business. *Digital Creativity, 12*(3), 153–166. doi:10.1076/digc.12.3.153.3231

Fitzsimmons, G. (2016). *The influence of hyperlinks on reading on the web: An empirical approach* (Doctoral thesis). University of Southampton, UK.

Gerrig, R. J. (1993). *Experiencing narrative worlds: On the psychological activities of reading.* New Haven, CT: Yale University Press.

Green, M. C., & Brock, T. C. (2000). The role of transportation in the persuasiveness of public narratives. *Journal of Personality and Social Psychology, 79*(5), 701–721. doi:10.1037/0022-3514.79.5.701 PMID:11079236

Gutiérrez, A. (2014). La ficción progresiva de Leonardo Valencia, o el autor somos todos. *Bulletin of Spanish Studies, 91*(4), 575–594. doi:10.1080/14753820.2014.886900

Hyman, J. A., Moser, M. T., & Segala, L. N. (2014). Electronic reading and digital library technologies: Understanding learner expectation and usage intent for mobile learning. *Educational Technology Research and Development*, *62*(1), 35–52. doi:10.100711423-013-9330-5

Johnson, R. (1997). Examining the validity structure of qualitative research. *Education*, *118*(2), 282–292.

Joosen, V., & Vloeberghs, K. (2012). *Uitgelezen jeugdliteratuur. Een ontmoeting met traditie en vernieuwing*. Leuven, Belgium: Lannoo Campus.

Kaakinen, J. K., Papp-Zipernovsky, O., Werlen, E., Castells, N., Bergamin, P., Baccino, T., & Jacobs, A. M. (2018). Emotional and motivational aspects of digital reading. In M. Barzillai, J. Thomson, S. Schroeder, & P. W. Van den Broek (Eds.), *Learning to Read in a Digital World* (pp. 141–164). Amsterdam, The Netherlands: John Benjamins. doi:10.1075wll.17.06kaa

Kaiser, H. F. (1960). The application of electronic computers to factor analysis. *Educational and Psychological Measurement*, *20*(1), 141–151. doi:10.1177/001316446002000116

Kline, P. (1994). *An easy guide to factor analysis*. Newbury Park, CA: Sage.

Landow, G. (1997). *Hypertext 2.0: The convergence of contemporary critical theory and technology*. Baltimore, MD: Johns Hopkins University Press.

Larson, L. C. (2015). E-books and audiobooks: Extending the digital reading experience. *The Reading Teacher*, *69*(2), 169–177. doi:10.1002/trtr.1371

Liamputtong, P. (2011). *Focus group methodology: Principles and practice*. London, UK: Sage. doi:10.4135/9781473957657

Mangen, A. (2008). Hypertext fiction reading: Haptics and immersion. *Journal of Research in Reading*, *31*(4), 404–419. doi:10.1111/j.1467-9817.2008.00380.x

Mangen, A., & Kuiken, D. (2014). Lost in an iPad: Narrative engagement on paper and tablet. *Scientific Study of Literature*, *4*(2), 150–178. doi:10.1075sol.4.2.02man

Maynard, S. (2010). The impact of e-books on young children's reading habits. *Publishing Research Quarterly*, *26*(4), 236–248. doi:10.100712109-010-9180-5

McGeown, S. P., Osborne, C., Warhurst, A., Norgate, R., & Duncan, L. G. (2016). Understanding children's reading activities: Reading motivation, skill and child characteristics as predictors. *Journal of Research in Reading*, *39*(1), 109–125. doi:10.1111/1467-9817.12060

Miall, D. S. (2004). Reading hypertext – theoretical ambitions and empirical studies. Retrieved from http://computerphilologie.uni-muenchen.de/jg03/miall.html

Miall, D. S., & Dobson, T. (2001). Reading hypertext and the experience of literature. *Journal of Digital Information*, *2*. Retrieved from https://journals.tdl.org/jodi/index.php/jodi/ article/view/35/37

Miall, D. S., & Kuiken, D. (2002). A feeling for fiction: Becoming what we behold. *Poetics*, *30*(4), 221–241. doi:10.1016/S0304-422X(02)00011-6

Mifsud, C. L., & Petrová, Z. (2018). Literacy education in the digital age. In M. Barzillai, J. Thomson, S. Schroeder, & P. W. Van den Broek (Eds.), *Learning to Read in a Digital World* (pp. 165–183). Amsterdam, The Netherlands: John Benjamins. doi:10.1075wll.17.07mif

Miranda, T., Williams-Rossi, D., Johnson, K., & McKenzie, N. (2011). Reluctant readers in middle school: Successful engagement with text using the e-reader. *International Journal of Applied Science and Technology*, *1*(6), 81–91.

Morgan, D. L. (1998). *The focus group guidebook*. London, UK: Sage. doi:10.4135/9781483328164

Nell, V. (1988). *Lost in a book: The psychology of reading for pleasure*. New Haven, CT: Yale University Press. doi:10.2307/j.ctt1ww3vk3

Nunnally, J. (1978). *Psychometric theory* (2nd ed.). New York, NY: McGraw Hill.

OECD. (2015). *Students, computers and learning: Making the connection. PISA*. OECD Publishing.

Picton, I. (2014). *The impact of eBooks on the reading motivation and reading skills of children and young people: A rapid literature review*. London, UK: National Literacy Trust.

Quirk, M., Unrau, N., Ragusa, G., Rueda, R., Lim, H., Velasco, A., ... Loera, G. (2010). Teacher beliefs about reading motivation and their enactment in classrooms: The development of a survey questionnaire. *Reading Psychology*, *31*(2), 93–120. doi:10.1080/02702710902754051

Salmerón, L., Strømsø, H. I., Kammerer, Y., Stadtler, M., & Van den Broek, P. W. (2018). Comprehension processes in digital reading. In M. Barzillai, J. Thomson, S. Schroeder, & P. W. Van den Broek (Eds.), *Learning to Read in a Digital World* (pp. 91–120). Amsterdam, The Netherlands: John Benjamins. doi:10.1075wll.17.04sal

Sang, G., Valcke, M., van Braak, J., & Tondeur, J. (2010). Student teachers' thinking processes and ICT integration: Predictors of prospective teaching behaviors with educational technology. *Computers & Education*, *54*(1), 103–112. doi:10.1016/j.compedu.2009.07.010

Schlochtermeier, L. H., Pehrs, C., Kappelhoff, H., Kuchinke, L., & Jacobs, A. M. (2015). Emotion processing in different media types: Realism, complexity, and immersion. *Journal of Systems and Integrative Neuroscience*, *1*(2), 41–47. doi:10.15761/JSIN.1000109

Simons, M., T'Sas, J., & Mommaerts, M. (2014). *Leesplezier en e-hype: good practices in(ge)beeld en ingezet*. Universiteit Antwerpen.

Sweet, A. P., Guthrie, J. T., & Ng, M. M. (1998). Teacher perceptions and student reading motivation. *Journal of Educational Psychology*, *90*(2), 210–223. doi:10.1037/0022-0663.90.2.210

Tveist, A. K., & Mangen, A. (2014). A joker in the class: Teenage readers' attitudes and preferences to reading on different devices. *Library & Information Science Research*, *36*(3-4), 179–184. doi:10.1016/j.lisr.2014.08.001

Unrau, N., & Schlackman, J. (2006). Motivation and its relationship with reading achievement in an urban middle school. *The Journal of Educational Research*, *100*(2), 81–101. doi:10.3200/JOER.100.2.81-101

Vlaams Ministerie Onderwijs en Vorming. (2018). *Statistisch Jaarboek 2017-2018*. Brussel. Retrieved from https://onderwijs.vlaanderen.be/nl/statistisch-jaarboek-van-het-vlaams-onderwijs-2017-2018

Warschauer, M., & Liaw, M. (2011). Emerging technologies for autonomous language learning. *Studies in Self-Access Learning Journal*, *2*(3), 107–118.

Wigfield, A., Guthrie, J., Tonks, S., & Perencevich, K. (2004). Children's motivation for reading: Domain specificity and instructional influences. *The Journal of Educational Research*, *97*(6), 299–309. doi:10.3200/JOER.97.6.299-310

Wolf, M. (2016). *Tales of literacy for the 21st century*. Oxford, UK: Oxford University Press.

Wolf, M. (2018). *Reader, come home: The reading brain in a digital world*. New York, NY: Harper Collins.

Wolf, M., & Barzillai, M. (2009). The importance of deep reading. *Educational Leadership*, *66*(6), 32–37.

ADDITIONAL READING

Mangen, A., & van der Weel, A. (2016). The evolution of reading in the age of digitisation: An integrative framework for reading research. *Literacy*, *50*(3), 116–124. doi:10.1111/lit.12086

Mangen, A., Walgermo, B., & Brønnick, K. (2013). Reading linear texts on paper vs. computer screens: Effects on reading comprehension. *International Journal of Educational Research*, *58*, 61–68. doi:10.1016/j.ijer.2012.12.002

Takacs, Z., Swart, E., & Bus, A. (2015). Benefits and pitfalls of multimedia and interactive features in technology-enhanced storybooks: A meta-analysis. *Review of Educational Research*, *85*(4), 698–739. doi:10.3102/0034654314566989 PMID:26640299

Wolf, M. (2007). *Proust and the squid: The story and science of the reading brain*. New York: HarperCollins.

KEY TERMS AND DEFINITIONS

Hyperfiction: non-sequential digital literary text, solely composed of written words.

Hypermedia Fiction: digital literary text that offers a linear story with multimedia hyperlinks that meets narrative conventions (continuity in time and space, steady progression of the unfolding plot, social and psychological development of the characters).

Reading Immersion: the cognitive process that unfolds during the act of reading and in which 'the reader dives into the sea, reaches foreign lands (transportation), gets caught (seized by the story), and loses contact with all other realities (to get lost in a book)' (Ryan, 2001, p. 93).

Reading Motivation: the individual's personal goals, values, and beliefs with regard to the topics, processes, and outcomes of reading' (Guthrie & Wigfield, 2000, p. 405).

APPENDIX 1: SURVEYS

Survey Pupils

1. Personal data
2. Gender; Grade; Home language; Education mother
3. Digital reader profile
4. Do you read e-books (electronic, digital) novels? Y/N
5. Do you look up information when you read novels (on paper or digitally) Y/N

- MRQ

Please rate each statement according to how well it describes you: 1 (strongly disagree) – 5 (strongly agree)

1. I don't know that I will do well in reading next year.
2. I am a good reader.
3. I learn more from reading than most students in the class.
4. I like hard, challenging books.
5. If the project is interesting, I can read difficult material.
6. I like it when the questions in books make me think.
7. I usually learn difficult things by reading.
8. If a book is interesting, I don't care how hard it is to read.
9. If the teacher discusses something interesting I might read more about it.
10. I have favourite subjects that I like to read about.
11. I read to learn new information about topics that interest me.
12. I like to read about new things.
13. I enjoy reading books about living things.
14. I read stories about fantasy and make believe.
15. I like mysteries.
16. I make pictures in my mind when I read.
17. I feel like I make friends with people in good books.
18. I read a lot of adventure stories.
19. I enjoy a long, involved story or fiction book.
20. It is very important to me to be a good reader.
21. In comparison to other activities I do, it is very important to me to be a good reader.
22. I don't like vocabulary questions.
23. Complicated stories are no fun to read.
24. I don't like reading something when the words are too difficult.
25. I don't like it when there are too many people in the story.
26. I try to get more answers right than my friends.
27. I like being the best at reading.
28. I read about my hobbies to learn more about them.
29. I like being the best at reading.

30. I like to finish my reading before other students.
31. I like being the only one who knows an answer in something we read.
32. It is important for me to see my name on a list of good readers.
33. I am willing to work hard to read better than my friends.
34. I like having the teacher say I read well.
35. My friends sometimes tell me I am a good reader.
36. I like to get compliments for my reading.
37. I am happy when someone recognizes my reading.
38. Grades are a good way to see how well you are doing in reading.
39. My parents often tell me what a good job I am doing in reading.
40. I look forward to finding out my reading grades.
41. I read to improve my grades.
42. My parents ask me about my reading grade.
43. I visit the library often with my family.
44. I often read to my brother or my sister.
45. My friends and I like to trade things to read.
46. I sometimes read to my parents.
47. I talk to my friends about what I am reading.
48. I like to help my friends with their schoolwork in reading.
49. I like to tell my family about what I am reading.
50. I do as little schoolwork as possible in reading.
51. I read because I have to.
52. I always do my reading work exactly as the teacher wants it.
53. Finishing every reading assignment is very important to me.
54. I always try to finish my reading on time.

Survey Teachers

1. Personal data
2. Gender; Age; Teaching experience
3. TBSMRQ and importance of reading education

Please rate each statement according to how well it describes you: 1 (strongly disagree) – 5 (strongly agree)

1. I believe it is important for students to feel that they can improve as readers while they are in my class.
2. I believe it is important for students to feel that they can learn from reading in class.
3. I believe it is important that students see themselves as a good reader.
4. I believe it is important for students to feel that they can succeed in reading in the classroom.
5. I believe it is important for students to have access to readings that challenge them at their level.
6. I believe it is important to give students questions about their reading that make them think.
7. I believe students will read more difficult material when it is interesting to them.
8. I believe it is important to give students opportunities to learn difficult things through reading.

9. I believe it is important for students to read new information about topics that interest them.
10. I believe it is important to verbally encourage students to find out what interests them.
11. I believe it is important for students to read about new things that interest them.
12. With regard to literature in my language classes, I go beyond what is required.
13. I believe it is important for students to be so interested in what they are reading that they lose track of time.
14. I believe it is important for students to read about a wide variety of topics.
15. When I don't know students' interests, I believe it is important to choose readings on topics that will arouse their interests.
16. When I see that a student has an interest in a topic, I believe it is important to give that student readings that are centrally related to that topic.
17. I believe it is important to select readings that are likely to draw students into a story's narrative.
18. I believe it is important to encourage students to enter the world that the author has created.
19. I consider reading education an important part of my language classes.
20. I believe it is important to encourage students to make pictures in their minds when they read.
21. I believe it is important that students read to improve their grades compared to other reasons for reading
22. I believe it is important that students look forward to finding out their reading grades.
23. I believe it is important that students think that grades are a good way of finding out how they are doing in reading.
24. I believe it is important that students' parents ask about their reading grades.
25. I regularly brush up on my knowledge of reading education to prepare for my classes.
26. I believe that it is important for students to enjoy being the only one who knows an answer in something they read.
27. I believe that it is important for students to strive to get more answers right than their friends.
28. I believe that it is important for students to like finishing their reading before other students in the class.
29. I believe that it is important that students are driven to work hard in order to get better at reading than their friends.
30. I believe that it is important to use competitive activities to promote reading growth.
31. I believe it is important that students read to their brother(s) or sister(s).
32. I am an avid reader of children's literature.
33. I believe it is important that students tell their family about what they are reading.
34. To prepare for my classes, I regularly brush up on my own knowledge of literature.
35. I believe it is important that students visit the library often with their family.
36. I believe it is important that students try to finish their reading on time.
37. I believe it is important that students value finishing every reading assignment.
38. I believe it is necessary for students to think it is important to be good readers.
39. I believe grading is an important way to foster reading development for students.
40. I believe it is necessary for students to view reading as one of the most important activity that they do.
41. I believe it is important to ask students what they want to read.

42. I believe it is important to acknowledge students' perspectives even though they may differ from the teacher's perspective.
43. I believe it is important to provide time for students to read independently.
44. I believe it is important to explain to students why a strategy being taught to them will be useful.
45. I believe it is important to explain to students why they are reading a particular book.
46. I consider myself a good reader.
47. I believe it is important to encourage students to generate questions rather than answer the teacher's questions.
48. I am well informed about current children's literature.
49. I believe pupils should regularly visit the library.

Media Profile

Please rate each statement according to how well it describes you: 1 (strongly disagree) – 5 (strongly agree)

1. It is important for pupils to get acquainted with different reading devices (e.g. tablets, e-readers).
2. I am interested in new reading devices (e.g. tablets, e-readers).
3. I regularly use new reading devices (e.g. tablets, e-readers) for my own reading.
4. When students have difficulty with media devices, I am usually at a loss as to how to help them.
5. I wonder if I have the necessary skills to use media devices for instruction.
6. I generally employ media devices in my classroom ineffectively.
7. Whenever I can, I avoid using media devices in my classroom.
8. I am not very effective in monitoring students' use of media devices in my classroom.
9. Even when I try very hard, I do not use media devices as well as I do other instructional resources.
10. I do not know what to do to turn students onto media devices.
11. I find it difficult to explain to students how to use media devices.
12. Given a choice, I would not invite the principal to evaluate my instruction based on media devices.
13. Media devices provide opportunities for improving the learning performance.
14. The efficiency of the learning process is increased through the use of media devices.
15. Media devices used as learning tools, increase student motivation.
16. Students with learning difficulties can strongly benefit from the didactic possibilities which the use of media devices entail.
17. Media devices increase the level of creativity of students.
18. The use of media devices helps students to achieve better text writing.
19. Knowledge of media devices and practical experience should be more integrated in the curriculum.
20. Media devices can help the teacher to apply differentiation among the students.
21. I would use media devices as a tool for demonstration working with existing presentations, or those someone else has made for me.
22. I would use media devices as a tool to teach new subject knowledge, i.e. the pupils acquire knowledge directly from the media devices.
23. I would encourage pupils in class to search for relevant information on the Internet.
24. I would use educational software with my pupils for learning subject knowledge through drill and practice.

25. I would teach pupils to consider the implications and opportunities of the use of different media devices.
26. I would use media devices as a tool for demonstration working with presentations I have made myself (e.g., Powerpoint).
27. I would ask pupils to undertake tasks or follow up classwork at home on media devices.
28. I would use media devices to assist with differentiation or implementing individual learning plans.
29. I would encourage pupils to work collaboratively when using media devices.
30. I use media devices to communicate with my pupils (e.g., email, online platforms).

APPENDIX 2: CODING SCHEME FOCUS GROUPS

- Hyperlink type
  - Selection process
  - Influences
    - Content
    - Plot
    - Reader profile
    - Literary Genre
    - Other
- Hyperlink frequency
  - Influences
    - Content
    - Plot
    - Reader profile
    - Literary Genre
    - Other

Chapter 5
Social Networking and Language Learning:
Use of Instagram (IG) for Evaluating Oral Communication Skill

Fidel Çakmak
https://orcid.org/0000-0002-3285-7661
Alanya Alaaddin Keykubat University, Turkey

ABSTRACT

This chapter gives a brief overview on the use of social network sites (SNSs) for language learning and presents an empirical study of the use of Instagram (IG), one of the most popular SNSs, to assess learners' oral communication skills in the foreign language classroom. Several studies have mainly revealed the perceptions and preferences of learners in regards to using SNSs for learning grammar, vocabulary, L2 writing, reading, and speaking. In the current study, the performance scores of participants on an oral communication speaking task delivered both on IG and in class, as well as their scores on the Big Five personality traits as measured by the Quick Big Five Personality Test (QBFPT), were examined statistically. The results demonstrate IG facilitated students' performance in oral communication skill significantly and that personality traits do not predict the performance on IG, but the extroverted and conscientious are highly likely to achieve high scores in classroom.

INTRODUCTION

The world has become increasingly interconnected as people use devices and social platforms to form communities, at the same time mobile phones and tablets can also increase isolation as multiple political, ideological and cultural contexts both within and across communities intrude and place demands upon our attention and time. Today's era has marked itself by not only increasing the speed of information exchange dramatically but also increasing interconnectedness of people via mobile technologies. Digital tools and social media (often referred as social networking sites) are premised upon emergent mobile

DOI: 10.4018/978-1-7998-1097-1.ch005

technologies that have made feasible a whole new potential ecology for language learning. When well-configured, these SNSs can provide immersive connections between individuals even when they in fact exist in disparate spaces. In the first decade of the 2000s, Web 2.0 was introduced and enhanced the experience of active social media participation. It enhanced social connections and interaction with one another by introducing liking, posting, connecting, and blogging through the use of various platforms such as blogs, wikis. This enhanced experience increased user participation and social networking sites became commonplace. Over time the idea of mobility has evolved from initially referring to the devices to gradually include the mobility of people and shared exposure to the data underpinning different forms of digital social networks and media. This has allowed users of mobile media to become collaborative and interactive content sharers. Activities such as blogging, wikis, pod/screen casting, multimedia sharing have evolved as ways to stay connected with others (Zeng, 2018). By the end of the first decade of 2000s the social dimension of CALL got the attention of CALL scholars with two publications: The Handbook of Research on Web 2.0 and Second Language Learning by IGI Global, and The next generation: Social networking and online collaboration in foreign language learning by CALICO (Reinhardt, 2019). A volume of articles on the social impact of teaching resources employing Web 2.0 technologies and platforms for digital communication, information sharing and digital collaborations also emerged. This emergence of the social web has created an opportunity for learners to move from data acquisition to data creating and sharing and to connect to other people digitally. Since 2009, social network sites (SNSs) have attracted the attention of researchers and practitioners in the field of language teaching and there has been explosive growth in research publications on social media use in learning (Mao, 2014). This growth has been accelerated by the continued technological advancement in mobile devices, mobile network bandwidth and increasing mobile internet penetration in our lives. The research has explored effectiveness in regards to the learning outcomes, perceptions of learners as users, illuminated design issues and considerations when using social media tools for language learning. However, there has been less emphasis on theoretical and empirical studies on learning in general (Merchant, 2011) and a lack of experimental or longitudinal studies of the use of SNSs specifically (Ellison & Boyd, 2008; Lomicka & Lord, 2016). This chapter aims to provide a brief review on the use of Instagram (IG) as one of the most popular social networking sites for language learning and presents an empirical study of the use of IG to assess learners' oral communication skills in the foreign language classroom.

BACKGROUND

SNSs are online environments where people create a profile on public display and get connected to other people in the same environment for self-representation and data acquisition through personal connections (i.e., followers, friends, etc.). Additionally, the profiles display posts (self-representation data) marking users' connections, and popularity and appreciation data achieved through social activity in the SNS (Zourou, Potolia, & Zourou, 2017). The interactive experience of posters with their audience forms connections and can provide a feeling of attachment in the social world (Merchant, 2011). In another broad definition, SNS refers to "web-based services that allow individuals to (1) construct a public or semi-public profile within a bounded system, (2) articulate a list of other users with whom they share a connection, and (3) view and traverse their list of connections and those made by others within the system. The nature and nomenclature of these connections may vary from site to site." (Ellison & Boyd, 2008, p. 211). Such services encourage individuals to make visible connections among people who may be strangers but who share similar tastes, hobbies, and interests, and could actually be socially connected through mutual acquaintances, and encourages users to communicate with individuals in their social network. The presence of SNSs in our daily life has increased tremendously. From 2010 to 2018 the number of social media users worldwide increased from 0.97 billion to 2.46 billion, and foreseen growth is 2.9 billion by 2020 (Statista, 2017). The popularity of SNSs can be partly attributed to a desire to meet new people, exchange opinions, and to keep in touch with friends and other people that are already in our social circle (Lomicka & Lord, 2016). There can be powerful motives for staying tuned in to online networks for interaction and self-expression (Thorne, 2010). Social interaction and self-expression are two key functions of SNSs that are also necessary for language acquisition; therefore, integration of SNSs into language learning could create a context where language learning can occur (Brick, 2012; Huang & Su, 2018; McBride, 2009; Thorne, 2010) through interactions between digitally connected people and through self-expression (Greenhow, Robelia, & Hughes, 2009; Hwang & Cho, 2018). With this SNS integration, learners can be actively involved in more authentic social and communicative behavior than would be the case in classrooms (Brick, 2012; Chen, Lambert, & Guidry, 2010; Lomicka & Lord, 2009; McBride, 2009; Sun et al., 2017). This puts forward sociocultural approaches advocating for integration of SNSs into learning. Since the sociocultural approach focuses on learning through social interaction within a community of individuals, SNSs are particularly well-suited to the approach. They enable learners to interact with and observe other 'citizens' in the digital community, and to express and exchange opinions through social media sites. This fosters a sense of belonging, confidence in being a member of digital communities, and satisfies the need for self-expression (Doleck, Bazelais, & Lemay, 2017; Greenhow & Robelia, 2009) or presentation of identity (Leier, 2018; Lomicka & Lord, 2016; Reinhardt, 2019). Along similar lines, Haythornthwaite (2019) articulates the kinds of relations built within social networks and interaction where a "learning tie" between actors is established through interaction (p. 29). She remarks that empirical examination of connections between users is of importance since it can form a basic premise for social learning and an input for the design of learning environments. Hamid, Waycott, Kurnia and Chang (2015) clarify in what forms interactions can take place and identified three main interactions in online social network for learning 1) student-student 2) student-teacher 3) student content interactions. This mapping of interaction shows actors in the sites but it is important to bear in mind that SNSs are open to public view unless users have strict privacy, so exceptions to these interactions might also occur, there is always the possibility for unexpected connections with others to occur in addition to these enumerated actors.

INSTAGRAM (IG) AND LANGUAGE LEARNING

Language learning via IG has been explored in several studies. Some highlight the usability of IG in language classrooms due to its specific technical and social features. The majority of existing research is about the perception of the experience when language learners use IG to practice language skills. IG's navigational tools to edit posts, to add location, friends and keywords (hashtags), to tag photos or videos shared, as well as the exchange of opinions with other followers or "friends" through comments/posts (Handayani, 2016) all contribute to improving engagement. These features can help to retain the attention of scholars and practitioners in the educational field, as well as that of visual learners who benefit from the situated contextualized visual data presented in posts and videos. In addition to this, IG can enable users to take part in community building outside the classroom environment (Al-Ali, 2014) where they can enjoy their learning experience (Handayani, 2016). Last but not least, of the SNSs that are popular among the youth such as Facebook, IG and Twitter, IG has gained popularity quickly and is one of the most commonly used among young people (Al-Ali, 2014; Handayani, 2016) ages18 to 29 (Smith, 2015). It is suitable to utilize at most higher education institutes (HEI). Recently several studies have been published which look at the effects of Instagram use as a platform to facilitate language learning.

Yadegarfar and Simin (2016) investigated the effects of IG on grammatical accuracy of word classes among 92 Iranian TEFL students and also explored their perceptions of using IG for learning grammar accuracy for word classes. The experimental group was taught grammatical accuracy for word classes in sessions where English lessons and example sentences focusing on grammar accuracy of the word classes were provided via IG. The control group was exposed to face to face classroom teaching of the same material. Researchers found that the experimental group outperformed the control group in terms of grammatical accuracy of the presented word classes. Moreover, based on the students' responses showing their perceptions on using IG, the researchers noted that the experimental group had positive perceptions towards the use of IG for developing accuracy in relation with context and that IG served well for focusing on both interaction and linguistic content. A study relevant to linguistic structures was carried out by AlGhamdi (2018). In the research design, he set up an account named "i_english_m" which was followed by more than 48000 learners and uploaded English language content comprised of American English expressions and translations from Arabic to English. The posts lasted one minute in duration every day for two weeks. The content had two levels of difficulty: simple and complex. The researcher tracked the responses to each post, and the numbers of likes and views, and then evaluated satisfaction and preference toward the levels of structures. The result showed that the simple posts received the most likes and interactive attention from learners. This indicates that they preferred to engage with simple structures. This preference could be also be explained by the underlying philosophy of an Instagram account which is that users snap posts and connect people "in the moment" to grasp what is going on right now (Abbott, Donaghey, Hare, & Hopkins, 2013, p. 3). The study indicated that daily simple linguistic content might facilitate learners' motivation for language learning. This is a notable observation since the benefits of IG application use in language classes is also mentioned in association with increased motivation for learning echoes observations from other studies (Al-Ali, 2014; Kurniawan and Kastuhandani, 2016; Listinai, 2016; Munday, Delaney, & Bosque, 2016; Wulandari, 2019), supporting the rationale for using IG for classes. What was interesting in this case was that the followers were self-selected. Participation was open to anyone in the Arab world and others living in different countries. Since the account was anonymous, some learners and also family members tagged their friends and other family members to follow the account and take a note of the content. Getting the attention of others and making the learning circle grow highlights the potential for fast growth in an online language learning community.

Morshidi, Embi and Hashim (2019) explored the use of IG to promote reading among 11 young learners (4th graders) in an urban school in Meleka, Malaysia. The pupils were assigned a reading activity mixed with 2 week- reading activities in class followed by a 3-week series of IG video posts. The findings from interview, questionnaire, and observation notes show that IG was fun to integrate when teaching classes for young learners' and that the students were keen on reading in front of the cameras to post the videos on IG. They indicated that they appreciated likes and comments, which boosted their enthusiasm to make more posts on IG. This study's implication is in line with the practical application mentioned in Blattner and Lomicka's (2012) study which claims that SNS can work beyond classroom environment, and enable language progress in a fun environment wherever students feel like connecting.

Although this study shows positive outcomes from the data, there are areas for concern; specifically, the participants' profile of being young pupils. Their age leaves them especially vulnerable to potential online safety threats such as cyberbullying, offensive or vulgar language use in the comments, and exposure to nudge techniques such as likes on IG, allowing the service to track their personal data for marketing aims. These risks must be balanced against the benefits of the platform and ensuring privacy and appropriate moderation could be challenging in such context. Additionally, how reading skills are assessed could be improved during the process. Although it is claimed that respondents' positive perceptions of using mobile SNSs affect performance with an increase in learning experiences (Wong, Tan, Loke, & Ooi, 2015; Yang, 2013), a quantitative metric of increased performance rather than self-reported responses to a questionnaire would provide a more meaningful assessment of the actual effects of integrating IG when teaching reading. Munday, Delaney and Bosque (2016) highlighted the potential of IG to develop language use and cultural awareness by partaking in a challenge of describing a daily prompt (words or grammar phrases in Spanish). Students of Spanish were called to take part in the challenge with photos uploaded, hashtags and quotes added and then tagged with the word given on that day as #InstagramELE. The challenge provided an enhanced opportunity for practicing grammar studied in during those weeks in the month. Students learning Spanish in over 50 classes and in collaboration with over 70 teachers uploaded 32.000 pictures from all around the world. This project was an example of an online community coming together for language learning and provided interpersonal communication opportunities within that community through the posting of comments and/or responding comments. The teachers participated in the challenge by uploading photos for the showcase of culture and provided contextual explanations for words (lexical variations). The researchers concluded from the project that structured IG use in Spanish classes with activities such as the challenge were advantageous in attracting young adults for learning, providing both autonomous learning and cooperative learning opportunities, encouraging intrinsic motivation, and promoting skills growth for strategic and digital competence. Although these advantages of IG were not empirically evaluated in the project, the researchers suggested that the practitioners take privacy issues, the tracking of comments and an improved assessment of the participation in learning into consideration. The taking part in language project shows IG could be a useful digital authentic tool for imbuing students with enthusiasm (Leier, 2018) and promoting cooperation for learning out of the classroom environment and ultimately facilitate the learner becoming an engaged community member actively responsible for learning (Akbari, Naderi, Simons, & Pilot, 2016). Along similar lines, Kessler's (2013) study mentions that new forms of language have been welcomed in a social media context where learners and teachers are given opportunities to partake in meaningful language learning practice:

The digital landscape that we are immersed in not only provides us with access to information, but it also requires that we play an active role in the ongoing exchange of information. After all, it is a participatory culture that is being co-constructed. We are not only surrounded by potential corpora of authentic linguistic input, but we are expected to engage with one another around the construction of new forms of language in the realm of social media. This linguistically dense realm offers English teachers and learners unlimited opportunities to interact around meaningful, authentic, and diverse instances of language practice. (p. 616)

Another study on vocabulary and grammar use by Aloraini (2018) investigated the effect of Instagram posts (vocabulary or grammar) on the amount of EFL learners' production, accuracy of that production and the amount of feedback the user obtained. The findings showed vocabulary posts got more comments and had more errors than grammar posts with a small size effect. As for the accuracy of posts and amount of feedback received, these were not affected by the types of the posts. What this research concluded was that although there might be variation in the amount of posts as output, there was no variation in the quality of data. This would lead us to pay attention to the improvement of the output quality irrespective of the linguistic classification of posts. A study by Sukri, Mustapha, Othman, Aralas, and Ismail (2018) presented an example on the integration of IG in EFL classes. It showcased four students from a Malaysian context with a low proficiency who partook in a classroom activity presenting a short story entitled 'Tanjong Rhu' by Minfong Ho. The students were expected to upload a photo they liked to associate with their favorite character in the story, and then explain why they used that picture to depict their favorite character. The posts made were discussed in successive classroom sessions. This integration, though the procedures and outcome analysis were not refined, highlights how IG and combined language skills: in this case writing and speaking skills could be worked into English classes. In her paper, Handayani (2016) introduced effective simple implementations of IG in skills teaching in a descriptive way. Speaking exercises, posting field trip memories, role plays, review of a famous historical figure and pronunciation practice were all suggested as beneficial IG activities. For reading, sharing reading recommendations was suggested, however, this activity does not target language practice in reading but promotes practical literacy development through likes and suggestions. For writing, descriptive captioning with attention to grammatical structure and word choice based on photo inspiration and writing comments on a post were suggested to improve writing skills, notably creative writing. As for the listening skill, listening to posts of native speakers of English was suggested. This might be questionable since we are well aware of that there are other forms of English such as International English and English as Lingua Franca. In a follow-up study, Handayani (2017) investigated how eight students taking a Paragraph Writing course perceived descriptive captioning and photo inspiration writing activities on IG. When asked how they perceived the use of IG for their writing classes, expectedly, the students reported highly positive attitudes towards the use of Instagram in teaching writing. As a remark, the activities suggested by Handayani (2016; 2017) sound like they would be easily implementable, although their usefulness needs to be quantified and scientifically verified. IG when implemented in a classroom setting should result in an observable improvement in skills which can be verified with quantifiable data. Otherwise, the use of SNSs in the classroom cannot be viewed as anything more than a mere list of additional class activity suggestions. Another paragraph writing activity through IG was investigated by Hopkyns (2017). It had a similar research design, asking for student perceptions of IG both pre class and after two writing tasks on topics of their choice. Topics included observations of fellow university students, siblings, favorite artist and so forth. In association with the pictures posted participants were asked to make a short movie about the subjects and to present a narrative in four paragraphs with several associated pictures. The researcher found that students were well engaged in the activities and felt that these activities enabled the students to improve their writing and speaking skills as well as identify and take joy in the improvement in their writing. Another research focus of IG integration in language classes had to do with the perceptions of students towards use of IG in language classes. Akhiar, Mydin and Kasuma (2017) examined 110 university students' perceptions and attitudes towards the use of Instagram in English language writing classes. The students enrolled in English courses were given a descriptive writing activity based on the photos of people, locations, and various objects posted on IG. They then responded to the questionnaire and took part in group discussions. The findings show that the students perceived that collaboration and cooperation occurred during the writing activity and noted that they had improved communication between their peers and the teachers. Additionally, the students thought they improved their language proficiency. As for the writing activity, the students were very positive about their overall engagement in using IG. What was interesting among the findings is that although the students enjoyed experiencing learning through IG, the majority of them (%67) preferred the face to face lessons over the ones on IG. This indicates that students might consider IG as a supplementary learning tool in support of face to face classrooms, and use it to facilitate teaching and learning processes with fun self-exposure to ideas.

Kurniawan and Kastuhandani (2016) also mentioned effective uses of IG for creative writing activities such as acrostic poem writing and writing captions for past vacation pictures with a brief description of the procedures. Based on the observational notes they claimed that the students could develop creative ideas for their writing tasks and that students were actively involved in the tasks, strengthening their motivation for learning. Additionally and similarly, Shazali, Shamsudin and Yunus (2019) conducted an action research with 25 secondary school students to investigate the use of IG for developing writing ability and the students' attitudes towards IG. Before posting on IG, students were given a self-descriptive writing activity imagining what they would publish on an online website. Students' performance was graded with CEFR grading system and then students were assigned to partake in five writing tasks on IG to elicit feedback from peers in the comment section. Their performance was graded by teachers and when the tasks were completed, students were asked to describe their experience and later filled out a questionnaire about their perception of use of IG in the writing class. The findings showed that students' CEFR level increased and that they adopted positive attitudes towards the use of IG in writing classes. The study's finding in regards to attitudes towards the use of IG is expected, however, the research design lacks generalizable features and the clarification on the writing tasks, and how each of them were to be evaluated could have been more explicit for the practitioners. Another study from a secondary school context was conducted by Sirait and Marlina (2018) in which the researchers explored how students could use IG for peer-review activities when writing descriptive texts. Students were given a list of steps such as opening up an Instagram account, sending the essays to the admin of the group class account and giving feedback and comments. They were then asked to list advantages and disadvantages of IG. Technical features of IG such as captioning, commenting, responding, and editing were mentioned as advantages; while the affordability of connectivity and mobile devices were mentioned as disadvantages. Similar to other research focused on the effect of IG use on descriptive writing skills, yet with a different research design than the ones above mentioned, Soviyah and Etikaningsih (2018) conducted an experimental research designed to investigate the effects of the use of IG on the students' performance in writing descriptive texts. 50 students in a private high school in Yogyakarta, Indonesia were assigned to one of two groups: experimental and control. While the experimental group received the treatment of writing tasks involving using Instagram, the control group received traditional classroom teaching. Both groups received the same material and took a writing test before and after the treatment. Based on the statistical analysis of the pre and post-tests, the researchers found that there was a statistically significant difference between the writing performance of students who used IG and the students who didn't. They concluded that Instagram could serve as an effective educational tool for teaching how to write descriptive texts. Like Soviyah and Etikaningsih (2018), Listiani (2016) also investigated the effect of IG on students' writing summaries of a text. 40 students with high and low motivation were assigned to experimental and control group. The experimental group used Instagram as a teaching technique, whereas the control group received teacher centered writing. Results based on pre and post-test analysis indicated that students using Instagram outperformed their counterparts in writing summary texts, which supports the conclusion that IG can serve as an effective tool in language classrooms. Khalitova and Gimaletdinova (2016) carried out a case study of using IG in teaching English to 50 advanced level students studying at HEI in Russia. They explored students' perception of the mobile- assisted language learning activities and investigated whether using IG in English classes improved listening comprehension. Students were randomly assigned to experimental and control group, where the former were supposed to watch videos (n=34) of 20 seconds each with their mobile phones, and the latter did not receive the treatment. Based on the analysis of results, the findings also indicated that IG improves students' listening comprehension skills especially in the ability to understand opinions and find specific information in speech.

Practicing speaking skills through SNSs has not been examined much in the research of SNSs and language learning. The notable studies were a study by Hitosugi (2011), where Ning, an SNS for the Japanese context, was found to enable students to work in cooperation and provide feedback to each other on their oral communication projects in Japanese. Another study by Sun et al. (2017), in which improvement in speaking skill was observed in the performance of two first grade young learner groups from an urban public elementary in Beijing, China. The experimental group (n=37) who utilized mobile SNSs for speaking practice showed greater improvement when compared to a control group (n=35) who did not use mobile SNSs. The application used by the experimental group was Papa, which is a Chinese audio streaming platform where users can interact for sharing audio files and posting pictures, and also make recordings and comments through their profiles. Oral language improvement was found as an outcome of the combination of classroom activities with the mobile SNS *Papa*. It is suggested that activities promoting social interaction and task-based activities such as information providing or interactive reasoning could be carried out by the integration of a mobile SNS. In line with these studies and their research focus on speaking improvement, Azlan, Zakaria, & Yunus (2019) investigated how integrating Instagram into task-based speaking tasks affected learning and explored pupils' motivation when using IG. Eight low proficiency pupils (four from pre-school and four from primary school) from Negeri Sembilan, Malaysia were selected. The findings from the research showed that IG helped pupils improve their speaking skills and at the same time, motivated them to be involved in task-based learning. Additionally, it made them feel connected with family and friends when sharing their activities. Students also felt excitement about being exposed to the IG integration in speaking classes. Leier (2018) notably indicated that the use of IG helped learners to build online relationships consciously and allowed them to be responsible for curating them. Wulandari (2019) also investigated whether IG Video Blog (Vlog) facilitated EFL's speaking ability. It also examined students' perceptions on the implementation. 28 students taking a Speaking 1 course were assigned an IG Vlog every two weeks, which entailed uploading an Instagram post of their 1-3 minute-long video. They took pretests before the treatment and posttests after the treatment. The findings showed that the integration of IG Vlog into the speaking classes improved the learners' speaking skills in regards to pronunciation, fluency, vocabulary, syntax, the target language use as well as an increase in motivation. With a qualitative research study design, Rahman (2018) investigated students' perspectives on posting photos on IG and how the integration of social media related to boosting their confidence for speaking English. She found that posting on IG did increase their confidence about speaking in that: a) they were less anxious on IG compared to the classroom b) IG boosted reported confidence in speaking c) the students used the platform to collaborate with peers for feedback.

When it comes to the concerns related to the use of IG in the educational environment, the first and major one is related to privacy. Language learning environments on SNSs are open to others publicly and can be archived. Learners (users) should be well aware of the responsibility of their own posts, likes, or comments, which may create risks in future. The online content of any materials should serve primarily as an educational tool for the learning community. Learners should be informed about the code of conduct, and educated as to how to behave in such socially networked platforms (Lomicka & Lord, 2009). Additionally, the terms of service (especially age limitations for opening an account in the country of residence), privacy policies and security of student data should be taken into account when integrating IG use into language learning. Learners' rights and privacy should be protected so that they are inhibited to the least degree when giving their own responses (Aloraini, 2018; Ellison & Boyd, 2008; Erben, Ban, & Castañeda, 2009; Kukulska-Hulme, Lee, & Norris, 2017; Leire, 2018; Lomicka & Lord, 2009; van Dijk, 2006). Additionally, students may not have SNSs accounts or may think participating in SNSs is a waste of time and be unwilling to partake in the use of SNSs as a part of their learning (Abbott, Donaghey, Hare, & Hopkins, 2013; Taşkıran, Gümüşoğlu, & Aydın, 2018). We cannot force students to open an account just for learning, as such could be an imposition. The learners must be willing to use IG for their own learning. Also, assessment of the data for learning outcomes is crucial when using IG for learning. Teachers need to specify how learners can partake in such social network sites for learning and also inform students as to how their participation is going to be assessed beforehand. (Sukri et al., 2018). Learners must be informed about feedback and how feedback can be received both peer to peer, and teacher to learner. (McBride, 2009) Policies should specify the expected outcomes for learning, manage the language data set, and encourage students to avoid any violations of codes of conduct with regards to privacy, shyness, social exclusion, or cyberbullying (Griffith & Liyanage, 2008; Thomas, Reinders, & Warschauer, 2013). Fostering this awareness could be done by training for networking for language learning. The abovementioned studies rarely mentioned the necessity of training for SNS participation and setting expectations for acceptable rules of interaction. If learning is to be facilitated through SNSs, then expectations should be shared before any treatment so that participants know the boundaries in advance and how to appropriately show their presence in the digital community. It is vital to afford time for student to absorb the implications of new technology before they are assigned any learning tasks through (Erben et al., 2009) speaking in the context of SNSs. They should be guided as they negotiate appropriate use of the SNS and make a tangible transition from seeing them as a fun familiar network into a learning zone. This line of thinking, that the training of learners prior to being socially connected on SNS, and giving them time to understand the new treatment, is beneficial has its justification in the research. Laghos and Nisiforou (2018) conclude that students need time to develop success in effectively communicating while learning in web-based groups through planned activities. Students require time to get to know their peers, and then effectively contribute and collaborate with them on the SNS. In other words, students need time to get to know one another and interact with each other in order to build a sense of community. The accumulation of interactions over time leads to on-topic productivity in terms of language learning and discussions being more topic or course related. To conclude, there is a timeline in which we start with a wide open zone, get to know one another, build ties between us and then move on to engage in productive interactions in L2 through sustained network connections.

PERSONALITY

Personality has been viewed in a variety of different ways in social science as a predictor of behavior patterns both socially and academically (Johnson, Krueger, Bouchard, & McGue, 2002). The impact of personality has long been one of the subjects on the language research agenda (Biedroń, 2016). Now that language learners with their individual features are a focal point in designing materials and teaching activities tailored to the learner, cognitive styles and personality can be taken into account as a predictor not only of information processing but also (Messick, 1994) to predict what a learner is "likely to do" (Sharp, 2008, p. 18). In conjunction with predicted behavior, personality is regarded as having an influential impact on second language learning (Lightbrown & Spada, 2006). There are several studies showing the relationship between personality differences as measured by various indicators such as Eysenck Personality Questionnaire (EPQ) (Eysenck & Eysenck, 1985), The Myers Briggs Type Indicator (MBTI) (Myers, McCaulley, Quenk, & Hammer, 1998) and the Five Factor Model (FFM) (e.g., Goldberg, 1992; McCrae & Costa, 2003) and learning strategies and outcomes. Some of the existing research found a direct observable relationship (Griffith, 1991, Li & Quin, 2006) and some did not (Carrell, Prince, & Astika, 1996; Sharp, 2008) between the personality of learners and the strategies they adopt for language learning.

Of the personality indicators mentioned above, The Five Factor Model (FFM) is a more recent and popular personality scale providing "a good representation of the central features of personality" (Dörnyei, 2005, p. 14) in five key categories- *Openness to Experience, Conscientiousness, Extraversion, Agreeableness and Neuroticism.* Dörnyei (2005) discusses personality more within the borders of personality psychology rather than the Big Five model's integration of cross-situational factors. He views personality as a construct involving a more eclectic structure of personality taking situational factors and processes in particular contexts into consideration. Additionally, the research on personality portrays individual differences, though not universally, since the FFM is not valid for all people yet covers a large majority of them (Allik, Hrebícková, & Realo, 2018). However, the investigation of the development of personality with environmental and biological factors taken into consideration could be complementary to explain these traits properly. Dörnyei (2005) describes five broad personality dimensions with a list of adjectives reflecting the high scorers and low scorers on its scales (p. 15).

- Openness (to experience): High scorers are imaginative, curious, flexible, creative, moved by art, novelty seeking, original, and untraditional; low scorers are conservative, conventional, down-to-earth, unartistic, and practical.
- Conscientiousness: High scorers are systematic, meticulous, efficient, organized, reliable, responsible, hard-working, persevering, and self-disciplined; low scorers are unreliable, aimless, careless, disorganized, late, lazy, negligent, and weak-willed.
- Extraversion: High scorers are sociable, gregarious, active, assertive, passionate, and talkative; low scorers are passive, quiet, reserved, withdrawn, sober, aloof, and restrained.
- Agreeableness: High scorers are friendly, good-natured, likeable, kind, forgiving, trusting, cooperative, modest, and generous; low scorers are cold, cynical, rude, unpleasant, critical, antagonistic, suspicious, vengeful, irritable, and uncooperative.
- Neuroticism: High scorers are worrying, anxious, insecure, depressed, self-conscious, moody, emotional, and unstable; low scorers are calm, relaxed, unemotional, hardy, comfortable, content, even tempered, and self-satisfied.

As SNSs provide opportunities to present identity (Leier, 2018; Lomicka & Lord, 2016; Mehdizadeh, 2010; Reinhardt, 2019), a user can utilize this opportunity to express an identity she or he holds for various motives such as seeking a feeling of belonging, popularity or ambition to take part in a particular community. Personality traits in relation to second language learning has garnered research attention. The relationship between personality traits and specific topics of language learning such as speaking confidence (Khany & Ghoreyshi, 2013), the academic achievement of prospective English as a Foreign

Language (EFL) teachers (Kırkağaç & Öz, 2017), communicative competence (Verhoeven & Vermeer, 2002), attitudes towards foreign language learning (Pourfeiz, 2015), and learning strategies (Asmalı, 2014; Sharp, 2008) has been investigated. As for the studies on the relationship between SNSs use and personality traits, differing results as to the direction of the relationship being either positive or negative have been found. De Zúñiga, Diehl, Huber, and Liu's (2017) study demonstrated that extraversion, agreeableness, and conscientiousness are positive predictors of SNS use while emotional stability and openness are negative predictor of the SNS use. Wehrli (2008) investigated the effect of personality traits on online social networking behavior and found a positive effect for neuroticism, a negative influence for conscientiousness, and no effects for openness and agreeableness. (Correa, Hinsley, & Züniga, 2010) found that the use of SNS and extroversion and openness were positively correlated; while, the use of social networks and neuroticism was negatively correlated.

MAIN FOCUS OF THE CHAPTER

Research Context

We surely have been on the move from Social CALL 1.0, where we were just getting to know what digital contexts were available for getting connected which started with opening up accounts on Facebook WhatsApp and Twitter (inclusion literacy), to Social CALL 2.0 where we began to be a part of the online community by using these social sites and actively engaging in those communities (participatory literacy). The more we get active, the more we are inclined to stick to one SNS according to our individual experiences and how much we think a given platform fits our needs. Social CALL 2.0 has moved from introducing people to SNS in order to connect with others, even strangers, with common tastes or interests, to a point where the connecting means also being invested in SNSs because the platform itself is a reflection of certain preferences and identity traits that can place us where we want to be as a user of SNSs. SNSs are very much the subject of recent research under the tag of social CALL. Instagram in particular has been commonly used for online photo or video sharing but has been less empirically researched when compared with Facebook or Twitter (Leier, 2018). The current research has been conducted to contribute to the existing studies, most of which lack quantitative design and evaluation of learning outcomes. The focus of the research is oral communication skill, which is the practice of speaking to communicate in English as a Foreign Language. The research interest in the use of IG has come with the idea of providing a learning zone through a channel with which students are already familiar and already use when using their mobile phones. The author (also a researcher in the context) developed a research question on whether the formal learning environment (classroom) can affect young adult learners' oral performances. Specifically, whether or not IG, which is assumed to be among the mostly frequently used SNSs by young adults, when integrated as a learning tool for temporary period of time affects oral performance. The researcher also explored whether the learners' personality traits would serve as a reliable predictor of their performance on IG and in-classroom. The current study is premised on the quantitative research method and aims to explore whether oral communication performance in two different contexts (classroom and IG) differs significantly, as well as how personality traits affect performance in the two contexts. The study addresses the following research questions:

1. Is there any significant difference between students' oral performance in a classroom context and on Instagram?
2. Do personality traits predict performance in the classroom context and on Instagram?

The hypothesis for the research questions is that students will perform better on IG (Azlan, Zakaria & Yunus, 2019; Rahman, 2018; Sun et. al, 2017; Wulandari, 2019) and that the students who are high scorers in personal trait dimensions such as openness to experience, agreeableness and extraversion, and being open to new experiences will have high scores in their performance in both contexts (de Zúñiga et al., 2017; Raad & Mlačić, 2015; Wehrli, 2008).

Setting and Participants

The research was conducted in a state university during the 2018-2019 winter semester in Turkey. The participants were freshmen students with ages ranging from 18 to 20 who are studying in the program of Teaching English as a Foreign Language. A total of 88 students registered to the course. However, 2 of them were with A1 and C2 proficiency, even though they were involved in the activities, their scores were not counted in this research statistically to avoid any possible outliers. Also 6 of the students did not want to participate in the SNS-natured study. This resulted in 8 participants being excluded from the study. 80 ELT students with B2 level of proficiency did voluntarily participate in the study. 30 of the participants had completed one year of English Language Preparatory Program prior to the treatment. Their level of English was evaluated before the treatment to assess whether or not they had similar language proficiency by employing the Oxford Placement Test (Allen, 2004).

Data Collection Tool

This research consisted of two measurements:

Oral Communication performance was assessed by a grid which has five qualitative aspects of spoken language use of Common European Framework of Reference for Language (CEFR 3.3) (Council of Europe, 2018): Range, accuracy, fluency, interaction and coherence. Each of these aspects is 20 points maximum with a total possible of 100 point aggregate in the five aspects.

The grid was used for assessing students' oral performances in both contexts: classroom and IG. Two raters assessed the performance independently. The scores were compared and measured to ensure reliability. The inter-rater reliability of oral communication performance is 0.88. The discrepancies were resolved in agreement between the raters.

The Quick Big Five Personality Test (QBFPT)

To measure the personality traits, QBFPT, the revised and short form of Goldberg's (1992) 100 descriptions of personal traits were utilized. Each of the five broad personality dimensions is represented with 6 adjectives totaling up a list of 30 adjectives in the test (Verlmuts & Geris, 2005). The test was translated into Turkish and included a 7-point Likert scale from "completely correct" (7) and "completely wrong" (1). The minimum scores of each dimension 1 and maximum is 7. The Cronbach's alphas of the scale were reported. 81 for extraversion, .86 for conscientiousness, .80 for agreeableness, .73 for openness to

Table 1. Assessment grid for spoken language use of the CEFR B2 level (Council of Europe, 2018)

Range: *Has a sufficient range of language to be able to give clear descriptions, express viewpoints on most general topics, without much conspicuous searching for words, using some complex sentence forms to do so.* *---/20pts*
Accuracy: *Shows a relatively high degree of grammatical control. Does not make errors which cause misunderstanding, and can correct most of his/her mistakes.* *---/20pts*
Fluency: *Can produce stretches of language with a fairly even tempo; although he/she can be hesitant as he or she searches for patterns and expressions, there are few noticeably long pauses.* *---/20pts*
Interaction: *Can initiate discourse, take his/her turn when appropriate and end the conversation when he / she needs to, though he /she may not always do this elegantly. Can help the discussion along on familiar ground confirming comprehension, inviting others in, etc.* *---/20pts*
Coherence: *Can use a limited number of cohesive devices to link his/her utterances into clear, coherent discourse, though there may be some "jumpiness" in a long contribution.* *---/20pts*
Total: ---/100

experience and .78 for neuroticism in the validity and reliability study of the test in Turkish by Morsünbül, (2014). Cronbach`s alphas were .87 for extraversion, .84 for conscientiousness, .72 for agreeableness, .72 for openness to experience, and for neuroticism .79.

Procedures

Oral Communication Skills 1 includes a variety of communication-oriented activities such as discussions, debates, individual and group presentations, informative and persuasive presentations, and interviews. It is a compulsory course in the departmental program where students gain speaking practice. Preparation for the integration of IG use for the oral communication classes started with a simple questionnaire to ascertain whether or not students were active on IG at the beginning of the semester. The mean time for experience of use for IG was 3 years 3 months. 92% of participants reported using IG more than Facebook or Twitter and the mean of active time on IG was 2 hours 10 minutes per day. This data indicated that the students were already active on IG and were not required to open up an account just for the assigned task. The students added the researcher (instructor of the course) to their friend list when the course started by voluntarily accepting her friend request. The intervention started in week of 7 of the 14-week-fall-semester and ended in the 9th week. The instructor talked about movie analysis theories and the students did a classroom-activity involving watching parts of the movie *The Man called Ove*. They then discussed it in class as pair work in relation to the types of movie analysis introduced the 6th week. The task assigned on the 7th week was to watch a suggested movie and then work in groups of three to talk about it. In the 8th week, students in group had approximately 1-2 minutes per person to briefly talk about the movie on IG.

The two raters assessed students' oral performances in the class using the grid (See Table 1). For the following week, the instructor elicited the consent forms from students about the IG assignment. Their assignment was to post a brief review of the movie using the principles of analysis introduced in class on their IG. Six of the students did not want to post their video of this assignment on IG but instead submitted the assignment as an audio file, therefore, their data were excluded from the study. 80 students

Figure 1. Student's video on IG for movie analysis
Image used with consent

posted their videos where they talked about the movie they had watched from a critical point of view for 59 seconds in English (See Figure 1). They received likes and comments either in Turkish or English. The raters assessed the posts with the same grid used to assess students' in-class oral performances.

Results

Descriptive statistics for the scores are provided in Table 2. The data indicates that students' performance in IG context was scored higher than their in-classroom performance. In terms of QBFPT scores, they scored highest in the openness dimension while they scored lowest in the extraversion dimension.

A paired-samples t-test was conducted to compare students' oral communication performance in classroom context and on IG. It revealed that there was a significant difference in the scores for students' performance in the classroom context (M=73,12), SD=9,94) and on IG (M=75,43, SD=9,18); t (79) = -2.46, p< .01.

Table 2. Descriptive statistics for scores

	N	M	SD	Min	Max
In-classroom	80	73,12	9,94	50	95
IG	80	73,43	9,18	50	90
Openness	80	32,55	4,58	23	42
Conscientiousness	80	26,86	7,58	9	42
Extraversion	80	26,51	7,45	7	42
Agreeableness	80	24,27	7,05	18	41
Neuroticism	80	24,27	7,05	10	40

As to how well personality traits predict oral communication performance in-classroom and on IG, multiple regression analysis was performed. Regression analysis results indicated that the personality traits significantly predict oral communication performance in the classroom context ($R = .41$, $R2 = .16$, $F = 3.00^*$ $p < .05$). This finding suggests that independent variables accounted for 16% variance of oral communication performance in the classroom context. It was found that extraversion ($\beta = .30$, $p<.05$) and conscientiousness ($\beta = 32$, $p<.05$) significantly predicted oral communication performance in the classroom context. As for the prediction of oral communication performance on IG, the results showed that the personality traits didn't significantly predict oral communication performance on IG ($R = .29$, $R2 = .08$, $F = 1.38$ $p > .05$).

The results indicated that the participant's performance on IG was significantly higher in terms of oral communication skills compared to their performance in the classroom context, suggesting that IG facilitated their performance in oral communication skills. Additionally, the hypothesis was confirmed in terms of increased performance in oral communication skills on IG since the participants performing oral communication skills on IG had significantly higher means than they did in the classroom context. It is not that surprising that students performed more uniformly on the IG assignment. It might be much less stressful and much more fun to record a video, especially if students have the opportunity to only post their best effort, rather than having to make an on the spot presentation in class.

Additionally, the results demonstrated that learners whose personality traits score high for extraversion and conscientious are highly likely to achieve high scores in oral communication skills in a classroom context since they might feel confident in their speaking abilities whereas on IG personality traits do not serve as significant predictors for the oral communication performance on IG. This shows us that the personality traits are not good predictors for the oral communication performance on IG and they cannot reliably predict language performance.

CONCLUSION

This study has reviewed the current developments in SNSs as they relate to language learning, reporting gained language benefits by integrating the use of IG in language teaching. As most previous studies have focused on the perceptions about SNS use in the classroom, and taking note of the few experimental studies on using IG for skill teaching, the author designed the research study to assess whether the use of IG in an Oral Communication Skill courses led to improvements in oral communication skills. The study also examined whether the Big Five Personality Model could predict student oral communication performance. While it found that the model did not work as a predictor for performance on IG, the dimensions such as extraversion and conscientious are significant predictors of the oral communication performance in class. The current study contradicts with the study by de Zúñiga et al. (2017) and Raad & Mlačić (2015). It partially confirms the findings of Öz's (2016) study where he found that openness to experience and extraversion were the strongest predictors of academic motivation-if we regard the term as willingness to take part in learning in an academic environment. It also partially supports the observation in the study by Khany and Ghoreyshi (2013) where extroversion was found to be the most powerful predictor of a students' foreign language speaking confidence and the study by Karadağ and Kaya (2019) which found that there is a significant relationship between willingness to communicate and Big Five personality trait dimensions such as extraversion, openness to experience and conscientiousness. This implies that students' personality does not help us predict significantly whether they would improve language skills or outperform on IG. The study confirms the studies that advocate for the use of IG in language classes but does not support the finding of the studies by Correa, Hinsley and Züniga (2010), de Zúñiga et. al (2017) and Wehrli (2008). The emerging trend shows us that SNSs have pushed the boundaries of what is learning and how learning can occur outside the confines of constructed zones such as classrooms. The popularity of SNSs brings up the question in an opportunistic mind as to whether they could be used for learning too. As long as precautions such as privacy policies and proper security of student data, as well as careful planning to ensure interactions and feedback are appropriate and the expected language use for interactions are specified, it is worth experimenting in integrating SNSs into language learning and skill practice. This will likely empower 21st century learners with competencies such as experiential learning, digitally collaborative and unorthodox practices of learning and help them develop an identity as global citizens and learners in the SNS context. The integration of SNSs might be partial as users might not want to use it exclusively for learning language, but rather partake in it from time to time and as a supplement to take their learning beyond the classroom context. Further research is warranted on how the effects of learner personality characteristics, learners' learning styles, and visual and verbal ability levels come into consideration as possible predictors of performance in language learning through SNSs, as such performance may be more directly related to other factors in the multimedia environment itself.

ACKNOWLEDGMENT

This research received no specific grant funding from any agency in the public, commercial, or non-profit sectors. Dr. Fidel Çakmak gratefully thanks Madlen Ma for always her being there, providing companionship, constant emotional support, and occasionally equipment to conduct research in the field of language learning.

REFERENCES

Abbott, W., Donaghey, J., Hare, J., & Hopkins, P. (2013). An Instagram is worth a thousand words: An industry panel and audience Q&A. *Library Hi Tech News*, *30*(7), 1–6. doi:10.1108/LHTN-08-2013-0047

Akbari, E., Naderi, A., Simons, R. J., & Pilot, A. (2016). Student engagement and foreign language learning through online social networks. *Asian-Pacific Journal of Second and Foreign Language Education, 1*(1), 1–22. doi:10.118640862-016-0006-7

Akhiar, A., Mydin, A. A., & Kasuma, S. A. A. (2017). Students' perceptions and attitudes towards the use of Instagram in English language writing. *Malaysian Journal of Learning and Instruction* [Special Issue], 47–72.

Al-Ali, S. (2014). Embracing the selfie craze: Exploring the possible use of Instagram as a language m-learning tool. *Issues and Trends in Educational Technology, 2*(2), 1–16. doi:10.2458/azu_itet_v2i2_ai-ali

AlGhamdi, M. A. (2018). Arabic learners' preferences for Instagram English lessons. *English Language Teaching, 11*(8), 103–110. doi:10.5539/elt.v11n8p103

Allen, D. (2004). *Oxford placement test*. Oxford, UK: Oxford University Press.

Allik, J., Hrebícková, M., & Realo, A. (2018). Unusual configurations of personality traits indicate multiple patterns of their coalescence. *Frontiers in Psychology*, 1–14. PMID:29515499

Aloraini, N. (2018). Investigating Instagram as an EFL learning tool. *Arab World English Journal, 4*(4), 174–184. doi:10.24093/awej/call4.13

Asmalı, M. (2014). The relationship between the Big Five personality traits and language learning strategies. Balıkesir University Journal of Social Sciences Institute, 17(32), 1–18.

Azlan, N. A. B., Zakaria, S. B., & Yunus, M. M. (2019). Integrative task-based learning: Developing speaking skill and increase motivation via Instagram. *International Journal of Academic Research in Business and Social Sciences, 9*(1), 620–636. doi:10.6007/IJARBSS/v9-i1/5463

Biedroń, A. (2016). Personality factors as predictors of foreign language aptitude. *Studies in Second Language Learning and Teaching, 1*(4), 467–489. doi:10.14746sllt.2011.1.4.2

Blattner, G., & Lomicka, L. (2012). Facebooking and the social generation: A new era of language learning. *Apprentissage Des Langues et Systèmes d'Information et de Communication, 15*, 1–32.

Brick, B. (2012). How effective are web 2.0 language learning sites in facilitating language learning? *Compass: The Journal of Learning and Teaching, 3*, 57–63.

Carrell, P. L., Prince, M. S., & Astika, G. G. (1996). Personality types and language learning in an EFL context. *Language Learning, 46*(1), 75–99. doi:10.1111/j.1467-1770.1996.tb00641.x

Chen, P. S. D., Lambert, A. D., & Guidry, K. R. (2010). Engaging online learners: The impact of web-based learning technology on college student engagement. *Computers & Education, 54*(5), 1222–1232. doi:10.1016/j.compedu.2009.11.008

Correa, T., Hinsley, A. W., & de Zúñiga, H. G. (2010). Who interacts on the web? The intersection of users' personality and social media use. *Computers in Human Behavior, 26*(2), 247–253. doi:10.1016/j.chb.2009.09.003

Council of Europe. (2018). Qualitative aspects of spoken language use - Table 3 (CEFR 3.3): Common reference levels. Retrieved from https://www.coe.int/en/web/common-european-framework-reference-languages/table-3-cefr-3.3-common-reference-levels-qualitative-aspects-of-spoken-language-use

De Zúñiga, H. G., Diehl, T., Huber, B., & Liu, J. (2017). Personality traits and social media use in 20 Countries: How personality relates to frequency of social media use, social media news use, and social media use for social interaction. *Cyberpsychology, Behavior, and Social Networking*, *20*(9), 540–552. doi:10.1089/cyber.2017.0295 PMID:28922034

Doleck, T., Bazelais, P., & Lemay, D. J. (2017). Need for self-expression on Instagram: A technology acceptance perspective. In *3rd International Conference on Computational Intelligence & Communication Technology (CICT)*, 1-3. IEEE. Retrieved from https://ieeexplore.ieee.org/stamp/stamp.jsp?arnumber=7977305

Dörnyei, Z. (2005). *The psychology of the language learner individual differences in second language acquisition*. Mahwar, NJ: Lawrence Erlbaum.

Ellison, N., & Boyd, D. M. (2008). Social network sites: Definition, history, and scholarship. *Journal of Computer-Mediated Communication*, *13*, 210–230.

Erben, T., Ban, R., & Castañeda, M. (2009). *Teaching English language learners through technology*. New York, NY: Routledge.

Eysenck, H. J., & Eysenck, M. W. (1985). *Personality and individual differences*. New York, NY: Plenum. doi:10.1007/978-1-4613-2413-3

Goldberg, L. R. (1992). The development of markers for the Big-Five factor structure. *Psychological Assessment*, *4*(1), 26–42. doi:10.1037/1040-3590.4.1.26

Greenhow, C., Robelia, E., & Hughes, J. (2009). Web 2.0 and classroom research: What path should we take now? *Educational Researcher*, *38*, 246–259. doi:10.3102/0013189X09336671

Griffith, R. (1991). Personality and second language learning: theory, research and practice. In E. Sadtonom (Ed.), *Language acquisition and the second/foreign language classroom*. Singapore: SEAMEO.

Griffith, S., & Liyanage, L. (2008). An introduction to the potential of social networking sites in education. In I. Olney, G. Lefoe, J. Mantei, & J. Herrington (Eds.), *Proceedings of the Second Emerging Technologies Conference 2008* (pp. 76–81). Australia: University of Wollongong. Retrieved from http://ro.uow.edu.au/etc08/9

Hamid, S., Waycott, J., Kurnia, S., & Chang, S. (2015). Understanding students' perceptions of the benefits of online social networking use for teaching and learning. *Internet and Higher Education*, *26*, 1–9. doi:10.1016/j.iheduc.2015.02.004

Handayani, F. (2016). Instagram as a teaching tool? Really? In *Proceedings of the Fourth International Seminar on English Language and Teaching (ISELT-4)*, 320-327. Atlantis Press.

Handayani, F. (2017). Students' attitudes towards using Instagram in teaching writing. *Journal of Educative Studies*, *2*(1), 23–29.

Haythornthwaite, C. (2019). Learning, connectivity and networks. *Information and Learning Science, 120*(1-2), 19–38. doi:10.1108/ILS-06-2018-0052

Hitosugi, C. I. (2011). Using a social networking site in Japanese class. In E. Forsythe, T. Gorham, M. Grogan, D. Jarrell, R. Chartrand, & P. Lewis (Eds.), *CALL: What's your motivation?* (pp. 72–83). Nagoya, Japan: JALT CALL SIG.

Hopkyns, S. (2017). The Instagram identity project: Improving paragraph writing through social media. In *10th International Conference on ICT for Language Learning Conference Proceedings*, 224–228. Libreria Universitaria.

Huang, Y.-T., & Su, S.-F. (2018). Motives for Instagram use and topics of interest among young adults. *Future Internet, 10*(77), 1–12.

Hwang, H. S., & Cho, J. (2018). Why Instagram? Intention to continue using Instagram among Korean college students. *Social Behavior and Personality, 46*(8), 1305–1315. doi:10.2224bp.6961

Johnson, W., Krueger, R. F., Bouchard, T. J. Jr, & McGue, M. (2002). The personalities of twins: Just ordinary folks. *Twin Research, 5*(2), 125–131. doi:10.1375/twin.5.2.125 PMID:11931690

Karadağ, Ş., & Kaya, Ş. D. (2019). The effects of personality traits on willingness to communicate: A study on university students. *Manas Sosyal Araştırmalar Dergisi, 8*(1), 397–410.

Kessler, G. (2013). Teaching ESL/EFL in a world of social media, mash-ups, and hyper-collaboration. *TESOL Journal, 4*(4), 615–632. doi:10.1002/tesj.106

Khalitova, L. & Gimaletdinova, G. (2016). Mobile technologies in teaching English as a foreign language in higher education: A case study of using mobile application Instagram. In *ICERI2016 Proceedings, 1*, 6155-6161. IATED.

Khany, R., & Ghoreyshi, M. (2013). The nexus between Iranian EFL students' Big Five personality traits and foreign language speaking confidence. *European Online Journal of Natural and Social Sciences, 2*(2), 601–611.

Kırkağaç, Ş. & Öz, H. (2017). The role of Big Five personality traits in predicting prospective EFL teachers' academic achievement. *International Online Journal of Education and Teaching (IOJET), 4*(4), 317-328. Retrieved from http://iojet.org/index.php/IOJET/article/view/243/174

Kukulska-Hulme, A., Lee, H., & Norris, L. (2017). Mobile learning revolution: Implications for language pedagogy. In C. A. Chapelle & S. Sauro (Eds.), *The handbook of technology and second language teaching and learning* (pp. 217–233). Oxford, UK: Wiley & Sons. doi:10.1002/9781118914069.ch15

Kurniawan, A. & Kastuhandani, L. A. (2016). Utilizing Instagram for engaging students in their creative writing. In Proceedings of Indonesia Technology Enhanced Language Learning, 4–7. Salatiga, Indonesia: Satya Wacana University Press Universitas Kristen Satya Wacana

Laghos, A., & Nisiforou, E. (2018). Computer assisted language learning social networks: What are they talking about? *Social Networking, 7*(3), 170–180. doi:10.4236n.2018.73014

Leier, V. (2018). Using Instagram in a tertiary German language course: Students' perception and approaches. *The New Zealand Language Teacher*, *44*, 77–90.

Li, J., & Qin, X. (2006). Language learning styles and learning strategies of tertiary level English learners in China. *RECL Journal*, *37*, 67–89.

Lightbrown, M. P., & Spada, N. (2006). *How languages are learned*. Oxford, UK: Oxford University Press.

Listiani, G. (2016). The effectiveness of Instagram writing compared to teacher centered writing to teach recount text to students with high and low motivation. *Journal of English Language Teaching*, *5*(1), 1–8.

Lomicka, L., & Lord, G. (2009). Introduction to social networking, collaboration, and Web 2.0 tools. In L. Lomicka & G. Lord (Eds.), *The next generation: Social networks and online collaboration in foreign language learning*. San Marcos, TX: CALICO.

Lomicka, L., & Lord, G. (2016). Social networking and language learning. In F. Farr & M. Murray (Eds.), *The Routledge handbook of language learning and technology* (pp. 255–268). New York, NY: Routledge.

Mao, J. (2014). Social media for learning: A mixed methods study on high school students' technology affordances and perspectives. *Computers in Human Behavior*, *33*, 213–223. doi:10.1016/j.chb.2014.01.002

McBride, K. (2009). Social networking sites in foreign language classes: Opportunities for re-creation. In L. Lomicka & G. Lord (Eds.), *The next generation: Social networking and online collaboration in foreign language learning (CALICO Monograph Series)* (pp. 35–58). San Marcos, TX: CALICO.

McCrae, R. R., & Costa, P. T. Jr. (2003). *Personality in adulthood: A Five-Factor Theory perspective* (2nd ed.). New York, NY: Guilford. doi:10.4324/9780203428412

Mehdizadeh, S. (2010). Self-presentation 2.0: Narcissism and self-esteem on Facebook. *Cyberpsychology, Behavior, and Social Networking*, *13*(4), 357–364. doi:10.1089/cyber.2009.0257 PMID:20712493

Merchant, G. (2011). Unravelling the social network: Theory and research. *Learning, Media and Technology*, *37*(1), 4–19. doi:10.1080/17439884.2011.567992

Messick, S. (1994). The matter of style: Manifestations of personality in cognitive, learning and teaching. *Educational Psychologist*, *29*(3), 121–136. doi:10.120715326985ep2903_2

Morshidi, A., Embi, M. A., & Hashim, H. (2019). Instagram application: An active tool in cultivating reading behavior. *Journal of Information Systems and Technology Management*, *4*(11), 95–106.

Morsünbül, U. (2014). The validity and reliability study of the Turkish version of Quick Big Five personality test. *Düşünen Adam*, *27*(4), 316–322. doi:10.5350/DAJPN2014270405

Munday, P., Delaney, Y., & Bosque, A. (2016). #InstagramELE: Learning Spanish through a Social Network. *Languages Faculty Publications*. Retrieved from http://digitalcommons.sacredheart.edu/lang_fac/57

Myers-Briggs, I., McCaulley, M. H., Quenk, N. L., & Hammer, A. L. (1998). *MBTI manual: A guide to the development and use of the Myers Briggs type indicator*. Palo Alto, CA: Consulting Psychologists Press.

Öz, H. (2016). The importance of personality traits in students' perceptions of metacognitive awareness. *Procedia: Social and Behavioral Sciences*, *232*, 655–667. doi:10.1016/j.sbspro.2016.10.090

Pourfeiz, J. (2015). Exploring the relationship between global personality traits and attitudes toward foreign language learning. *Procedia: Social and Behavioral Sciences*, *186*, 467–473. doi:10.1016/j.sbspro.2015.04.119

Raad, B., & Mlačić, B. (2015). Big Five factor model, theory and structure. In J. D. Wright (Ed.), *International encyclopedia of the social & behavioral sciences* (pp. 559–566). Oxford, UK: Elsevier. doi:10.1016/B978-0-08-097086-8.25066-6

Rahmah, R. E. (2018). Sharing photograms on Instagram boosts students' self-confidence in speaking English. *Pedagogy Journal of English Language Teaching*, *6*(2), 148–156. doi:10.32332/pedagogy.v6i2.1335

Reinhardt, J. (2019). Social media in second and foreign language teaching and learning: Blogs, wikis, and social networking. *Language Teaching*, *52*(1), 1–39. doi:10.1017/S0261444818000356

Sharp, A. (2008). Personality and second language learning. *Asian Social Science*, *4*(11), 17–25.

Shazali, S. S., Shamsudin, Z. H., & Yunus, M. M. (2019). Instagram: A platform to develop student's writing ability. *International Journal of Academic Research in Business and Social Sciences*, *9*(1), 88–98. doi:10.6007/IJARBSS/v9-i1/5365

Sirait, J. B., & Marlina, L. (2018). Using Instagram as a tool for online peer-review activity in writing descriptive text for senior high school students. *Journal of English Language Teaching*, *7*(1), 291–302. Retrieved from http://ejournal.unp.ac.id/index.php/jelt/article/view/9771/pdf

Smith, A. (2015). *U.S. Smartphone use in 2015*. Retrieved from http://www.pewinternet.org/files/2015/03/PI_Smartphones_ 0401151.pdf

Soviyah, S., & Etikaningsih, D. R. (2018). Instagram use to enhance ability in writing descriptive texts. *Indonesian EFL Journal*, *4*(2), 32–38. doi:10.25134/ieflj.v4i2.1373

Statista. (2017, September). *Number of daily active Instagram users*. Retrieved from https://www.statista.com/statistics/657823/number-of-daily-active-instagram-users/

Sukri, H. I. M., Mustapha, L., Othman, M., Aralas, D., & Ismail, L. (2018). Social media: Engaging language learning. *International Journal of Academic Research in Business and Social Sciences*, *8*(12), 287–294. doi:10.6007/IJARBSS/v8-i12/5013

Sun, Z., Lin, C. H., You, J., Shen, H. J., Qi, S., & Luo, L. (2017). Improving the English-speaking skills of young learners through mobile social networking. *Computer Assisted Language Learning*, *30*(3–4), 304–324. doi:10.1080/09588221.2017.1308384

Taşkıran, A., Koral-Gümüşoğlu, E., & Aydın, B. (2018). Fostering foreign language learning with Twitter: Reflections from English learners. *Turkish Online Journal of Distance Education-TOJDE*, *19*(1), 100–116. doi:10.17718/tojde.382734

Thomas, M., Reinders, H., & Warschauer, M. (Eds.). (2013). *Contemporary Computer-Assisted Language Learning*. New York, NY: Continuum.

Thorne, S. L. (2010). The 'intercultural turn' and language learning in the crucible of new media. In F. Helm & S. Guth (Eds.), *Telecollaboration 2.0 for Language and Intercultural Learning* (pp. 139–164). Bern, Switzerland: Peter Lang.

van Dijk, J. (2006). *The network society: social aspects of new media*. Thousand Oaks, CA: Sage.

Verhoeven, L., & Vermeer, A. (2002). Communicative competence and personality dimensions in first and second language learners. *Applied Psycholinguistics*, *23*(3), 361–374. doi:10.1017/S014271640200303X

Vermulst, A. A., & Gerris, J. R. M. (2005). *QBF: Quick Big Five personality test manual*. Leeuwarden, The Netherlands: LDC Publications.

Wehrli, S. (2008). Personality on social network sites: An application of the Five Factor Model. *ETH Zürich Sociology Working Paper 7*, 1–16.

Wong, C.-H., Tan, G. W.-H., Loke, S.-P., & Ooi, K.-B. (2015). Adoption of mobile social networking sites for learning? *Online Information Review*, *39*(6), 762–778. doi:10.1108/OIR-05-2015-0152

Wulandari, M. (2019). Improving EFL learners' speaking proficiency through IG Vlog. *LLT Journal: A Journal on Language and Language Teaching, 22*(1), 111–125.

Yadegarfar, H., & Simin, S. (2016). Effects of using Instagram on learning grammatical accuracy of word classes among Iranian undergraduate TEFL students. *International Journal of Research Studies in Educational Technology*, *5*(2), 49–60. doi:10.5861/ijrset.2016.1572

Yang, S. (2013). Understanding undergraduate students' adoption of mobile learning model: A perspective of the extended UTAUT2. *Journal of Convergence Information Technology*, *8*(10), 969–979. doi:10.4156/jcit.vol8.issue10.118

Zeng, S. (2018). *English learning in the digital age-Agency, technology, and context*. Singapore: Springer. doi:10.1007/978-981-13-2499-4

Zourou, K., Potolia, A., & Zourou, F. (2017). Informal social networking sites for language learning: Insights into autonomy stances. In T. Lewis, A. Rivens Mompean, & T. Cappellini (Eds.), CALICO monograph "Learner autonomy and Web 2.0", 141–167.

Chapter 6
Examining Oral Performance Characteristics of L2 Learners With the CAF Calculator

Kazumi Matsumoto
Ball State University, USA

Maki Hirotani
Rose-Hulman Institute of Technology, USA

Atsushi Fukada
Purdue University, USA

ABSTRACT

Previous fluency studies typically used small datasets to analyze L2 learners' oral performance with objective fluency measures, and found positive correlation between objective measures and subjective ratings. Findings from a small dataset are difficult to generalize. However, it takes a great deal of time and effort to build a large dataset with various measures. To help facilitate this process, CAF Calculator, which outputs 50 fluency measures, has been developed. In this chapter, CAF Calculator and a workflow to compute fluency measures are introduced along with a study investigating utterance fluency of L2 learners of Japanese at two proficiency levels performing two tasks. The study found significant differences in speed, breakdown, and composite fluency measures between the two groups in both tasks. It also found that task type affects pause locations. It is hoped that the research tools introduced in this chapter will encourage more research on fluency.

INTRODUCTION

Foreign language learners regard speaking fluently as the most important skill to acquire (Harlow & Muyskens, 1994; Houston 2005; Rivera & Matsuzawa, 2007; Tse, 2000). Second Language (L2) instructors have, therefore, focused on improving learners' communicative competence and have increasingly paid attention to measuring oral proficiency. The Oral Proficiency Interview (OPI) of ACTFL (The

DOI: 10.4018/978-1-7998-1097-1.ch006

American Council on the Teaching of Foreign languages) is well-known in the U.S., but the test has practical difficulties. The test requires a time-consuming individual interview process conducted by a certified tester. The results of the test are broad categories, such as novice, intermediate, and advanced with three subcategories, low, mid, and high. It is, therefore, hard to see small amounts of growth in speaking proficiency. Typically, speaking tests administered at many institutions are subjectively rated based on local standards that instructors decide, and their inter-rater reliability may not be consistently high, which would be problematic in terms of fairness. It goes without saying that evaluating L2 learners' oral proficiency on a fair basis as well as tracking their proficiency development in detail are important for both L2 research and language instruction.

Fluency research using objective measures (e.g., speech rate) started in the 80s and it attempted to measure L2 learners' oral proficiency with objective measures of oral fluency. These objective measures are widely used in current second language acquisition research (e.g., de Jong & Mora, 2017; Freed, 1995; Ginther, Dimova, & Yang, 2010; Iwashita, Brown, McNamara, & O'Hagan, 2008; Tavakoli, 2016; Tavakoli, Campbell, & McCormack, 2016; Tavakoli, Nakatsuhara, & Hunter, 2017). Previous research has opened the door to the use of objective measures of oral proficiency as a supplemental tool to subjective rating. Significant positive correlation has been found consistently between objective measures of oral fluency and subjective ratings (Bosker, Pinget, Quene, Sanders, & de Jong, 2012; Freed, 1995; Ginther et al., 2010; Iwashita et al., 2008; Kormos & Denes, 2004; Prefontaine, Kormos & Johnson, 2015; Tavakoli et al., 2017; Towell, Hawkins, and Bazergui, 1996). Thus, objective fluency measures may be useful to include in the assessment of L2 learners' oral production for better understanding of their oral performance and development.

However, such objective measures of fluency have not been widely used for assessments in educational settings due to the cumbersomeness in obtaining the measurements. Computing objective measures manually indeed requires large amounts of time and effort. In order to assist with this process, CAF[1] Calculator[2] (Fukada, Hirotani, & Masumoto, 2019), formerly known as Fluency Calculator (Fukada, Hirotani, Matsumoto, & Huston, 2015a; 2015b) has been developed. In this article, CAF Calculator and a workflow to compute fluency measures will be introduced to demonstrate easy access to the measures for L2 instructors and researchers. Along with the introduction of CAF Calculator, a study which examines oral samples using CAF Calculator will be described as an example. The findings of the study will be discussed to reveal characteristic of L2 learners' oral performance.

BACKGROUND

Fluency has been defined from multiple perspectives. Lennon (1990) provided two senses: broad and narrow senses. The broad sense indicates overall language proficiency, while the narrow sense has to do with smooth and effortless delivery of speech. Segalowitz (2010), moreover, discerned three types of fluency from the cognitive science perspective: cognitive fluency, utterance fluency, and perceived fluency. Cognitive fluency refers to the efficiency of mobilization and integration of the cognitive processes for oral production. Utterance fluency deals with the features of speech which are measurable aspects of fluency, whereas perceived fluency means listener's perception of the speaker's delivery. Of these three aspects, utterance fluency is the one that can objectively be measured. To date, various fluency studies have attempted to find out the features of utterances by mainly measuring three aspects of utterance:

speed fluency, breakdown fluency (i.e., pauses), and repair fluency proposed by Skehan (2003) and Tavakoli and Skehan (2005). In addition to these, recent studies use composite fluency measures (i.e., blend of speed and pause) suggested by Skehan (2014).

Many previous studies found that some of utterance fluency measures were strongly related to perceived fluency (e.g., Bosker, Pinget, Quene, Sanders, & de Jong, 2012; Kormos & Denes, 2004; Prefontaine, Kormos, & Johnson, 2015). These studies found positive correlations between speed as well as composite fluency and subjective rating (Freed 1995; Ginther et al., 2010; Iwashita et al., 2008; Kormos & Denes, 2004; Prefontaine, Kormos, & Johnson, 2015; Tavakoli et al., 2017), and negative correlations between breakdown fluency and oral proficiency (Ginther at al., 2010; Iwashita et al., 2008; Towell et al., 1996).

In addition, pause related fluency measures have a strong connection with perceived fluency (Kahng, 2018) and affected listenability of the speech. Recently, several studies investigated how pauses appeared in oral performance at different proficiency levels. Less fluent speakers tended to have more pauses than fluent speakers (Segalowitz, 2010), and the location of pauses became an important aspect to examine. Fluent speakers took many pauses at boundaries, whereas less fluent speaker tended to take numerous pauses within clauses or units (e.g., Davies, 2003; de Jong, 2016; Freed, 1995; Kahng, 2014; Tavakoli, 2011).

Furthermore, a number of L2 fluency studies have been conducted using different types of tasks in order to find L2 speech characteristics, and the task types mainly divided into two: monologue (Lennon 1990; Nohara & Takamura 2010; Riazantseva 2001; Sakuragi 2011; Tavakoli 2016; Tavakoli & Foster 2008; Tavakoli & Skehan 2005; Towell et al., 1996; Wu & Ortega 2013) and dialogue (Freed 1995; Freed, Segalowitz, & Dewey 2004; Riggenbach 1991; Segalowitz & Freed 2004; Tavakoli 2016). Many researchers agree that task type influences utterance fluency (Derwing 2017; Michel 2011; Riazantseva 2001; Tavakoli 2016; Witton-Davies 2014). Michel (2011) and Tavakoli (2016) found, for example, that L2 learners were more fluent in a dialogue task than in a monologue task. But Tavakoli (2016) draw attention to a reliability issue on fluency measures when analyzing two types of tasks. A dialogic task has different types of pauses compared to a monologue task, such as between-turn pauses, and how those pauses are included to the data affects fluency measures. The previous studies (Mora & Valls-Ferrer 2012; Tavakoli et al., 2016; Witton-Davis, 2014) identified the most reliable measures of utterance fluency in a monologue task (e.g. speech rate, pause length, phonation time ratio, and mean length of run), but they may not apply to a dialogic task (Tavakoli, 2016).

Issues with Computing Fluency Measurements

It is useful to regularly track learners' oral performance. Frequent and equitable evaluation of oral performance using objective fluency measures is cumbersome for instructors and researchers. It is possible to automatically obtain a very small set of objective measurements with Praat and Praat Script Syllable Nuclei (de Jong & Wempe, 2008), but if one wishes to examine additional objective measures for one's own institution as in previous studies (e.g. Bosker, Quene, Sanders, & de Jong, 2014; Kahng, 2018), extra manual measurements and/or computations are necessary. To do anything more than what Praat Script Syllable Nuclei can do would require knowledge of acoustic phonetics and a certain level of computer processing skills.

To help facilitate this issue, CAF Calculator (Fukada, Hirotani & Matsumoto, 2019) was developed. CAF Calculator takes annotated Praat data files as input and automatically calculates and outputs a large set of fluency measurements. By using CAF Calculator, language instructors and researchers without specialized computer knowledge can obtain a large array of objective measurements of the learner's oral performance with less work. They can capture detailed characteristics of the learners' language production at a given point. By accumulating data over multiple data points, they can track the learners' language development as well.

Research Questions

This article will demonstrate how instructors and researchers can examine the characteristics of learners' oral performance by analyzing data from an open-access learner corpus with CAF Calculator. The following are the research questions of this study:

Research Question 1: Were there any significant differences in utterance fluency between novice and intermediate learners of Japanese?
Research Question 2: If so, in which measures were the differences found?
Research Quetsion3: Were there any task differences in utterance fluency?

METHODOLOGY

Corpus Data

The data for this study came from a learner corpus, the International Corpus of Japanese as a Second Language (I-JAS). It was created by the National Institute for Japanese Language and Linguistics and, as of June 2019, houses data collected from 875 learners of Japanese from twelve different language backgrounds and 50 native speakers of Japanese (Sakoda, Konishi, Sasaki, Suga, & Hosoi, 2016). In this learner corpus project, each learner completed six speaking tasks and six written tasks. They also took two Japanese proficiency tests: the Japanese Computerized Adaptive Test[3] and the Simple Proficiency-Oriented Test[4]. The six speaking tasks consist of two storytelling tasks based on a four-frame and a five-frame cartoon, a 30-minute long interview similar to the Oral Proficiency Interview of ACTFL, two role plays, and one picture description task. This study used the oral samples from one of the storytelling tasks and portions of the interviews by English-speaking learners of Japanese.

Participants

The data that we examined came from 42 native English speakers in Australia and the U.S. (15 males and 27 females; two unknown) according to the demographic information provided by I-JAS. All participants were enrolled in Japanese courses at a university in either Australia or the U.S. at the time of the data collection, and their ages ranged between 19 and 24 (One participant's age is unknown). All learners took the Japanese proficiency tests: i.e. J-CAT and SPOT. Since the previous research (Iwasaki 2002) indicated strong correlation between L2 learners' SPOT scores and their oral proficiency (OPI) ratings, we divided the English-speaking learners into two proficiency groups (novice and intermediate) based on their SPOT scores; the score range for the novice group was from 31 to 55, and that for the intermediate group was from 56 to 80 according to the Interpretation of TTBJ [5](Tsukuba Test-Battery of Japanese) score were 13 learners in the novice group and 29 in the intermediate group.

Materials

Oral Samples

Since previous studies showed that task characteristics influence speaking performance (Michel 2011; Riazantseva 2001; Tavakoli 2016; Witton-Davies 2014), the present study examined oral samples from the two tasks separately. As mentioned earlier, two types of oral samples that the present study used were a story telling task and a portion of a 30-minute interview task (discussion parts only). For the storytelling task, the learners were asked to tell a story in as much detail as possible based on a five-frame cartoon. The present study used one of the two storytelling tasks that the learners worked on: the storytelling task based on a five-frame cartoon that was originally used in Tavakoli and Foster's (2008) study.

In the 30-minute interview task, the participants talked about a variety of topics. The present study used only discussion parts of the interview in which they were asked to provide their opinion on the following questions and discuss them with the interviewer: 1). "Which do you prefer, living in the city or in the country?" and 2) "Which is more important to you, time or money?"

The learners' oral samples as well as the transcriptions prepared by the National Institute for Japanese Language and Linguistics were downloaded, and all interviewers' utterances were excluded from both oral samples and transcriptions for data analysis.

Data Analysis

The oral samples were downloaded as sound files from the I-JAS corpus and unnecessary utterances were removed. Then, Praat Script Syllable Nuclei (de Jong & Wempe, 2008) was run on the audio files in Praat and moras, pauses, AS-units, dysfluency, and sentences were manually annotated in Praat[6]. The annotated speech data was saved as TextGrid files and processed by CAF Calculator [7](Fukada, Hirotani, Matsumoto, & Masumoto, 2015a; 2015b), formerly known as Fluency Calculator, to compute objective fluency measures of the oral performance. Descriptions of these tools are given below.

Praat and Praat Script

Praat is acoustic analysis software that can analyze speech signals. Praat Script Syllable Nuclei (de Jong & Wempe, 2008) can automatically detect syllables, sounding parts (i.e., speech sounds), and silent parts. Although manual editing is necessary afterwards, the script saves a lot of time from annotation work. Manual annotation tasks include adding/deleting syllables, identifying and annotating filled pauses, AS-units with/without errors, sentences, repetitions, stutters, self-corrections, and reformulations.

CAF Calculator

The current version of CAF Calculator (v. 1.1.1.) computes 50 complexity, accuracy, and fluency measures of oral performance. By running CAF Calculator on TextGrid files, objective measures of fluency, accuracy and complexity are computed and written to an EXCEL file.

Measurements

Previous studies used a variety of objective fluency measures for comprehensively evaluating oral performance and identified reliable measures to present characteristics of speech fluency (e.g., Ginther. et al., 2010; Segalowitz, French, & Guay, 2018; Tavakoli, 2016; 2018). The present study used some of the fluency objective measures frequently used in previous studies. They can be divided into four categories: speed, breakdown, repair, and composite. The speed category deals with utterance speed such as speech rate (number of moras per minute), whereas the breakdown category looks at mainly filled or unfilled pauses in the utterance. The repair category examines utterances that interrupt speech and slow it down, such as repetition, self-correction, stutter, and reformulation. Finally, the composite fluency measures consider speed and flow in the utterance. Of the fifty fluency objective measures which CAF Calculator outputs, considering the reliable measures which the previous study indicated (e.g., Mora & Valls-Ferrer, 2012; Segalowitz, French, & Guay, 2018; Skehan, 2014; Tavakoli, 2016; 2018; Witton-Davis, 2014), the present study chose two speed fluency measures (speech rate and articulation rate), six breakdown measures (silent pause ratio, silent and filled ratio, silent pause ratio within as units, silent and filled pause ratio within as-units, silent pause ratio between as-units, silent and filled pause ratio between as-units), one repair measure (dysfluency ratio), and two composite measures (mean length of run and phonation time ratio) to examine main characteristics of Japanese learners' oral performance. The following lists the fluency measures used in this study each with a formula and a description.

RESULTS

The present study compared eleven fluency measures of oral performance produced by English speaking novice learners of Japanese to those by intermediate learners. The following two tables show descriptive statistics of the measures and the results of the *t*-tests in the two tasks, the storytelling and the discussion. To avoid running an increased risk of Type 1 error, a Bonferroni-adjusted alpha level of .004 was used.

According to the tables, the novice and the intermediate groups were significantly different in speed fluency measures (i.e., speech rate and articulation rate), breakdown fluency measures (i.e., silent pause ratio and silent and filled pause ratio), and composite measures (i.e., mean length of run and phonation time ratio) in both tasks. Both speed fluency measures showed that the intermediate learners produced a much greater number of moras per minute than the novice learners. The speech rate means showed that the intermediate learners uttered at least twice as many moras per minute as the novice learners did in both tasks. In relation to breakdown fluency, the novice learners took approximately 1.5 times as many silent and filled pauses as the intermediate learners did on average. The mean length run means indicated that the intermediate learners produced approximately 1.6 times as many moras as the novice learners did between pauses in both tasks.

In addition, consistent with the previous studies, the repair fluency measure (i.e., dysfluency ratio) did not significantly differ between the two groups in either task. Regarding the location of pauses, the study found that significant differences in silent pause ratio within as-units or silent pause ratio between as-units appeared depending on the tasks. In the storytelling task, as mentioned above, overall silent and filled pause ratio significantly differed between the groups, but silent pauses within as-units as well as silent pauses between as-units were not significantly different between the two groups. However, silent

Table 1. Fluency objective measures

Fluency aspects	Measures
Speed	1. Speech rate: (Total number of moras) / (Total response time) * 60 Description: Per-minute measure that indicates how many syllables the speaker produced in his/her response.
	2. Articulation rate: (Total number of moras) / (Speech time + Filled pause time +Discourse marker time) * 60 Description: Per-minute measure that indicates how many syllables the speaker produced in his/her response excluding silent pauses, filled pauses, and discourse markers.
Breakdown	3. Silent pause ratio: (silent pause time) / (Total response time) * 100 Description: indicates ratio of silent pauses to total response time
	4. Silent and filled pause ratio: (Silent pause time + Filled pause time) / (Total response time) * 100 Description: indicates ratio of silent and filled pauses to total response time
	5. Silent pause ratio within AS-units: (Silent pause time within AS-units)/(Total response time)*100 Description: indicates ratio of silent pauses within AS-units to total response time
	6. Silent and filled pause ratio within AS-units: (Silent pause time within AS-Units + Filled pause time within AS-Units)/(Total response time)*100 Description: indicates ratio of silent pauses and filled pauses within AS-units to total response time
	7. Silent pause ratio between AS-units: (Silent pause time between AS-unit)/(Total response time)*100 Description: indicates ratio of silent pauses between AS-units to total response time
	8. Silent and filled pause ratio between AS-units: (Silent pause time between AS-unit + Filled pause time between AS-units)/(Total response time)*100 Description: indicates ratio of silent and filled pauses between AS-units to total response time
Repair	9. Dysfluency ratio: (Dysfluency time) / (Total response time) * 100 Description: indicates ratio of dysfluency (repetition, self-correction, stutter, and reformulation) to total response time
Composite	10. Mean length of run: (Total number of moras) / (Number of runs) -*Run* is a sounding interval between pauses Description: indicates numbers of moras between pauses
	11. Phonation time ratio: (Speech time) / (Total response time)*100 Description: indicates ratio of speech time to total response time

Table 2. Two-tailed T-test results on the 11 measures in the storytelling task

Measures	Novice (N = 13)		Intermediate (N = 29)			
	M	SD	M	SD	t	p (2-tailed)
1. Speech rate	82.71	42.90	163.81	47.96	5.23	.000*
2. Articulation rate	194.06	61.30	273.91	52.71	4.32	.000*
3. Silent pause ratio	57.95	9.48	40.19	13.10	-4.96	.000*
4. Silent and filled pause ratio	65.53	10.80	47.25	11.59	-4.82	.000*
5. Silent pause ratio within as-unit	34.29	14.02	23.46	9.90	-2.87	.006
6. Silent and filled pause ratio within as-unit	38.52	15.33	27.66	10.28	-2.33	.032
7. Silent pause ratio between as-unit	23.67	11.83	16.73	8.19	-2.21	.033
8. Silent and filled pause ratio between as-unit	27.01	13.77	19.59	7.98	-1.81	.089
9. Dysfluency ratio	3.31	3.45	2.96	2.02	-.40	.689
10. Mean length of run	4.26	1.28	6.88	1.97	4.39	.000*
11. Phonation time ratio	33.90	11.01	52.63	11.45	4.96	.000*

Note: $*p < .004$ (Bonferroni corrected)

Table 3. Two-tailed T-test results on the 11 measures in the discussion task

Measures	Novice (N = 13)		Intermediate (N = 29)			
	M	SD	M	SD	t	p (2-tailed)
1. Speech rate	66.97	19.29	169.27	59.07	8.38	.000*
2. Articulation rate	187.01	47.63	262.11	55.43	4.23	.000*
3. Silent pause ratio	62.23	7.63	35.39	16.84	-7.11	.000*
4. Silent and filled pause ratio	72.82	5.83	44.38	15.93	-8.44	.000*
5. Silent pause ratio within as-unit	18.81	11.81	16.11	6.86	-.77	.453
6. Silent and filled pause ratio within as-unit	22.06	14.15	20.57	7.64	-.36	.725
7. Silent pause ratio between as-unit	43.42	14.37	19.28	15.51	-4.77	.000*
8. Silent and filled pause ratio between as-unit	50.76	15.90	23.82	16.19	-5.01	.000*
9. Dysfluency ratio	2.91	1.81	3.19	2.43	.376	.709
10. Mean length of run	4.91	1.60	8.23	2.68	4.14	.000*
11. Phonation time ratio	25.76	6.32	54.29	15.30	8.55	.000*

Note: $^*p < .004$ (Bonferroni corrected)

pause within as-units was very close to significant (p=.006). In the discussion task, on the other hand, silent pause ratio between as-units was significantly different between the two groups, but silent pause ratio within as-unit was not.

DISCUSSION

The present study explored characteristics of learners' oral performance using two types of oral production tasks, storytelling (monologue) and discussion (dialogue) from the I-JAS learner corpus. The oral samples produced by forty-two English-speaking learners of Japanese in two proficiency groups, novice and intermediate, were compared using objective fluency measures. The following sections discuss the results by answering the research questions.

RQ1: Were there any Significant Differences Utterance Fluency between Novice and Intermediate Learners of Japanese?

Yes, the present study found significant differences in oral performance between the two groups in both tasks. The results indicated that speech produced by the novice and the intermediate learners of Japanese had different characteristics.

RQ2: If so, in which Measures were the Differences Found?

The present study found significant differences in terms of speech speed, pause, and composite measures between the novice and the intermediate learners of Japanese. This means that learners with higher proficiency spoke faster with more words, were able to speak longer at once, and used fewer pauses than the less fluent learners. The results of the previous studies of English learners (Kahng, 2014; Tavakoli, et al., 2016) suggested that speed fluency, mean length of run, and breakdown fluency measures were consistent measures to distinguish between fluent and dysfluent speech. The present study also found similar results; differences between the two proficiency groups appeared in speech speed, speech flow, and pause ratio measures.

In addition, consistent with the previous studies, dysfluency ratio did not differ between the two proficiency groups in either task. Based on the means, the ratio of dysfluency time to the total response time was very small; only 2.91% of the novice learners' speech and 3.19% of the intermediate learners' speech were dysfluency behaviors in the discussion task, and there was not much difference between the two groups. According to Tavakoli (2018), repair fluency is slow to develop. This insight provides an explanation for the similar repair behaviors observed in the two groups. Compared to the ratio of the silent and filled pause time to the total response time, the dysfluency ratio was much smaller. The learners spent a great deal of time pausing in their speech; 72.82% of the novice learners' speech and 44.38% of the intermediate learners' speech were pauses in the discussion task. Based on these findings, it might be that because they tried to avoid making errors in their utterances by taking many pauses, their repair behavior did not add up too much at all.

Concerning pauses, the results indicated that the novice learners of Japanese used much more pauses than the intermediate learners. According to the previous researchers, pauses are generally known as monitoring tools during the speech production (Kahng, 2014; Kormos, 2006; Skehan, 2014). When the speech production process is automatized and its efficiency improves, pauses are reduced (Lambert & Kormos, 2014). Skehan (2009) suggested that length of run (syllables/moras between pauses) indicates a degree of automatization of speech performance. In light of these notions along with the values of pause related measures and mean length of run (the novice learners: 4.91; the intermediate learners: 8.23), it is reasonable to say that the degree of automatization by the novice learners was low, resulting in inefficient processing.

All in all, the results of the present study clearly indicated that language proficiency influences speech speed, pauses and composite measures.

RQ3: Were there any Task Differences in Utterance Fluency?

Multiple studies of Indo-European language learners have suggested that the nature of the task influenced L2 learners' fluency (Derwing 2017; Michel 2011; Riazantseva 2001; Tavakoli 2016; Witton-Davies 2014). The present study also revealed interesting differences in pauses between the two tasks. As seen in Table 4.1., in the storytelling task, based on the means, the novice learners and the intermediate learners of Japanese had a higher value of silent pause ratio within as-units than that of silent pause ratio between as-units. In the discussion task, on the other hand, the learners in both groups had a higher mean of silent pause ratio between as-units than that of silent pause ratio within as-units.

Both groups changed pause locations depending on the tasks. Kormos (2006) and Levelt (1989) suggested that pauses within clauses were used for *formulating* (i.e. "grammatical encoding" and "phonological encoding" pp.11-12) the speech production and pauses between clauses contributed to *conceptualizing* (i.e. all mental activities related to what and how to express) speech production. In light

Table 4. Means of silent pause ratio within as-units and silent pause ratio between as-units

	Novice (N = 13)		Intermediate (N = 29)	
	Storytelling	Discussion	Storytelling	Discussion
Silent pause ratio within AS-units	34.29	18.81	23.46	16.11
Silent pause ratio between AS-units	23.67	43.42	16.73	15.51

of this theoretical view along with the nature of the tasks, since the learners created a story based on a five-frame cartoon in the storytelling task, they were able to see the overall storyline before they started speaking and did not need to create a story from scratch. This task was also monologue and did not require spontaneous responses to the interlocutor. The discussion task, on the other hand, was a dialogue task that required them to first comprehend what the interlocutor said and then produce responses spontaneously. The learners had to make a response to the interlocutor from scratch, and that response had to be appropriate and effective. Therefore, it seems that the learners tended to take a long time before stating each chunk of speech, contributing to a higher mean for silent pause ratio between as-units than silent pause ratio within as-units. In this way, the present study provides evidence that the task nature affects the location of pauses.

PEDAGOGICAL IMPLICATION

This article analyzed the oral performance of learners of Japanese to show the workflow of the CAF Calculator. The present study found similar results to the previous research on English learners; less fluent speakers had slower speech with more pauses, and their speech length between pauses was shorter than that of fluent speakers. Based on these findings, instructors and researchers would be able to better understand characteristics of fluent and dysfluent speakers. This provides a basis on which to describe fluent and dysfluent behaviors to language learners. It also gives suggestions as to how learners might avoid dysfluent behaviors and improve their speaking ability. For example, the findings of the present and previous studies demonstrate dysfluent L2 speakers frequently had pauses within clauses or units (e.g., de Jong, 2016; Kahng, 2014; Tavakoli, 2011). Language instructors can suggest ways to avoid pauses within clauses by showing appropriate pause locations. During form-focused activities, in particular, language instructors provide a variety of phonetic and/or grammatical feedback to their students. In addition to these corrections, indicating proper pause locations would help make learner's utterances sound more natural. If L2 learners know dysfluent speech characteristics, they would avoid the behavior intentionally, and by practicing, they may eventually become accustomed to having fewer pauses within clauses.

As just described, understanding characteristics of oral performance might lead to understanding language development and help identify efficient language instruction. Instructors can examine learners' language development using this tool and contribute to better instruction through observing the various numerical values of oral performance at their own institutions.

Further Analysis for Effective Instruction Using Caf Calculator

The present study used eleven objective fluency measures to analyze learners' oral performance and displayed the characteristics of oral performance in different proficiency groups. Using CAF Calculator, further analysis can be conducted to understand fluent and dysfluent behaviors. The following describes an idea for further analysis.

Previous studies suggested that pauses are strongly associated with perceived fluency (Kahng, 2018). The present study analyzed pauses in terms of pause ratio as well as pause locations, and found how the learners of Japanese had pauses in the two types of tasks. However, previous studied also indicated that pause frequency and pause length affected perceived fluency (Bosker et al., 2013; Ginther et al., 2010; Tavakoli, 2016). Therefore, further analyses with respect to learners' pause frequency and pause length seem useful for understanding learner's pause pattern and developing instructional strategies. In order to reveal learners' pause pattern in greater detail, the following measures can be used: silent pause length within as-units per minute, silent pause frequency within as-units per minute, silent pause length between as-units, silent pause frequency between as-units, filled pause length within as-units, filled pause frequency within as-units, filled pause length between as-units, and filled pause frequency between as-units. These additional measures are not automatically calculated by CAF Calculator, but are easily calculated by building formulas in EXCEL, combining the base measures CAF Calculator outputs. The following table lists the formula and the description of the measures

Table 5. Breakdown objective measures for further analysis

Fluency aspect	Measures
Breakdown	1. Silent pause length within AS-units per minute: (Silent pause time)/(AS-units time)*60 Description: indicates silent pause length within AS-units per minute.
	2. Silent pause frequency within AS-units per minute: (Silent pause count within AS-units)/(AS-unit time)*60 Description: indicates silent pause frequency within AS-units per minute.
	3. Silent pause length between AS-units per minute: (Silent pause time between AS-units)/(Total response time - AS-unit time)*60 Description: indicates silent pause length between AS-units.
	4. Silent pause frequency between AS-units: (Silent pause count within AS-units)/(Total response time – AS-unit time)*60 Description: indicates silent pause frequency between AS-units per minute.
	5. Filled pause length within AS-units: (Filled pause time)/(AS-units time)*60 Description: indicates filled pause length within AS-units per minute.
	6. Filled pause frequency within AS-units: (Filled pause count within AS-units)/(AS-unit time)*60 Description: indicates filled pause frequency within AS-units.
	7. Filled pause length between AS-units: (Filled pause time between AS-units)/(Total response time - AS-unit time)*60 Description: indicates filled pause length between AS-units per minute.
	8. Filled pause frequency between AS-units: (Filled pause count within AS-units)/(Total response time – AS-unit time)*60 Description: indicates filled pause frequency between AS-units per minute.

CAF Calculator can produce not only fluency objective measures, but also accuracy measures (i.e., number of error-free as-units and error-free as-units ratio) and a complexity measure (i.e., syntactic complexity). CAF Calculator enables language teachers to assess learner's oral performance from multiple perspectives.

CONCLUSION

The present study has shown how an open-access learner corpus was leveraged to acquire a decent-sized dataset and how fluency analysis tools were used to annotate and calculate a large number of objective measures. Although some manual labor is required in data analysis, the amount of cumbersome work can be lessened with the use of Praat and CAF Calculator. The current version of CAF Calculator outputs 50 measures, and additional measures are easily calculated by building formulas in EXCEL, combining the base measures CAF Calculator outputs.

The dataset for this study is not exactly large at 42 records, but this has to do with data availability in the corpus. The I-JAS Corpus is still growing, and there is a plan to boost the number of L1 English learners to 100 by 2020. Needless to say, the methodology will work just as well with 100 records. It is hope that CAF Calculator and a workflow devised for generating objective measures will make oral performance evaluation more easily accessible for language instructors and researchers.

REFERENCES

Bosker, H. R., Pinget, A. F., Quene, H., Sanders, T., & de Jong, N. H. (2012). What makes speech sound fluent? The contribution of pauses, speed and repairs. *Language Testing*, *30*(2), 159–175. doi:10.1177/0265532212455394

Bosker, H. R., Quené, H., Sanders, T., & de Jong, N. H. (2014). The perception of fluency in native and nonnative speech. *Language Learning*, *64*(3), 579–614. doi:10.1111/lang.12067

Davis, A. (2003). *The native speaker: Myth and reality* (2nd ed.). Tonawanda, NY: Multilingual Matters. doi:10.21832/9781853596247

de Jong, N. H. (2016). Predicting pauses in L1 and L2 speech: The effects of utterance boundaries and word frequency. *International Review of Applied Linguistics in Language Teaching*, *54*(2), 113–132. doi:10.1515/iral-2016-9993

de Jong, N. H., & Mora, J. C. (2017). Does having good articulatory skills lead to more fluent speech in first and second language? *Studies in Second Language Acquisition*, 1–13. doi:10.1017/S0272263117000389

de Jong, N. H. & Wempe, T. (2008). Praat script syllable nuclei [Programming script]. Retrieved from https://sites.google.com/site/speechrate/Home/praat-script-syllable-nuclei

Derwing, T. M. (2017). L2 Fluency development. In S. Loewen & M. Masatoshi (Eds.), *The Routledge handbook of instructed second language acquisition* (pp. 246–259). New York, NY: Routledge. doi:10.4324/9781315676968-14

Freed, B. F. (1995). What makes us think that students who study abroad become fluent? In B. Freed (Ed.), *Second language acquisition in a study abroad context* (pp. 123–148). Amsterdam, The Netherlands: John Benjamins. doi:10.1075ibil.9.09fre

Freed, B. F., Segalowitz, N., & Dewey, D. P. (2004). Context of learning and second language fluency in French: Comparing regular classroom, study abroad, and intensive domestic immersion program. *Studies in Second Language Acquisition*, *26*(02), 275–301. doi:10.1017/S0272263104262064

Fukada, A., Hirotani, M., & Matsumoto, K. (2019). *Tools for generating objective measures for oral proficiency research* [Computer software]. Retrieved from http://tell.cla.purdue.edu/CAF-calculator/

Fukada, A., Hirotani, M., Matsumoto, K., & Huston, S. M. (2015a). *Tools for generating objective measures for oral proficiency research.* Retrieved from http://tell.cla.purdue.edu/fluency-calculator/

Fukada, A., Hirotani, M., Matsumoto, K., & Huston, S. M. (2015b). *Fluency Calculator* [Computer software]. Retrieved from http://tell.cla.purdue.edu/fluency-calculator/

Ginther, A., Dimova, S., & Yang, R. (2010). Conceptual and empirical relationships between temporal measures of fluency and oral English proficiency with implications for automated scoring. *Language Testing*, *27*(3), 379–399. doi:10.1177/0265532210364407

Harlow, L. L., & Muyskens, J. A. (1994). Priorities for intermediate-level language instruction. *Modern Language Journal*, *78*(2), 141–154. doi:10.1111/j.1540-4781.1994.tb02025.x

Houston, T. (2005). Outcomes assessment for beginning and intermediate Spanish: One program's process and results. *Foreign Language Annals*, *38*(3), 366–374. doi:10.1111/j.1944-9720.2005.tb02223.x

Ishizaki, A. (2004). Sakubun ondoku ni okeru shokyū gakushūsha no pōzu no tokuchō: eigo bogo washa yon-mei no ōdan-teki shiryō o moto ni (Characteristics of pauses produced by novice learners in an oral reading task: based on the longitudinal study of four native English speakers. *Acquisition of Japanese as a Second Language*, *7*, 26–44.

Ishizaki, A. (2005). How does a learner leave a pause when reading Japanese aloud? A comparison of English, French, Chinese and Korean learners of Japanese and native Japanese speakers. *Japanese-Language Education around the Globe 15*, 75–89.

Iwasaki, N. (2002). Nihongo nōryoku kani shiken (SPOT) no tokuten to ACTFL kōtō nōryoku (OPI) no reberu no kankē ni tsuite [Simple performance-oriented test (SPOT) vs. oral proficiency interview (OPI)]. *Journal of Japanese Language Teaching*, *114*, 100–105.

Iwashita, N., Brown, A., McNamara, T., & O'Hagan, S. (2008). Assessed levels of second language speaking proficiency: How distinct? *Applied Linguistics*, *29*(1), 24–49. doi:10.1093/applin/amm017

J-CAT project team, & International Student Center at University of Tsukuba (n.d.) J-CAT (Japanese computerized adaptive test) Retrieved from http://www.j-cat.org/en/

Ju, I. (2010). Second language proficiency and pausing: Syllable duration and pausing in the Kahng, J. (2014). Exploring utterance and cognitive fluency of L1 and L2 English speakers: Temporal measures and stimulated recall. *Language Learning*, *64*(4), 809–854.

Ju, I. (2010). Second language proficiency and pausing: Syllable duration and pausing in the oral reading of Japanese learners. *Bulletin of the Graduate School of Education. Hiroshima University*, *59*(Part II), 229–238.

Kahng, J. (2018). The effect of pause location on perceived fluency. *Applied Psycholinguistics*, *39*(3), 569–591. doi:10.1017/S0142716417000534

Kormos, J. (2006). *Speech production and second language acquisition*. London, UK: Routledge.

Kormos, J., & Denes, M. (2004). Exploring measures and perceptions of fluency in the speech of second language learners. *System*, *32*(2), 145–164. doi:10.1016/j.system.2004.01.001

Lambert, C., & Kormos, J. (2014). Complexity, accuracy, and fluency in task-based L2 research: Toward more developmentally based measures of second language acquisition. *Applied Linguistics*, *53*(5), 607–614. doi:10.1093/applin/amu047

Lennon, P. (1990). Investigating fluency in EFL: A quantitative approach. *Language Learning*, *3*(3), 387–417. doi:10.1111/j.1467-1770.1990.tb00669.x

Levelt, W. (1989). *Speaking: From intention to articulation*. Cambridge, MA: MIT Press.

Michel, M. C. (2011). Effects of task complexity and interaction in L2 performance. In P. Robinson (Ed.), *Second language task complexity: Researching the cognition hypothesis of language learning and performance* (pp. 141–174). Amsterdam, The Netherlands: John Benjamins. doi:10.1075/tblt.2.12ch6

Mora, J. C., & Valls-Ferrer, M. (2012). Oral fluency, accuracy and complexity in formal instruction and study abroad learning contexts. *TESOL Quarterly*, *46*(4), 610–641.

Nohara, Y., & Takamura, M. (2010). A study on nonfluent utterances that are difficult to understand by Japanese learners: Focusing on corrections and pauses. *Journal of the Graduate School of Humanities and Sciences. Ochanomizu University*, *13*, 117–125.

Prefontaine, Y., Kormos, J., & Johnson, D. E. (2015). How do utterance measures predict raters' perceptions of fluency in French as a second language? *Language Testing*, *33*(1), 53–73. doi:10.1177/0265532215579530

Riazantseva, A. (2001). Second language proficiency and pausing: A study of Russian speakers of English. *Studies in Second Language Acquisition*, *23*(4), 497–526. doi:10.1017/S027226310100403X

Riggenbach, H. (1991). Toward an understanding of fluency: A microanalysis of nonnative speaker conversations. *Discourse Processes*, *14*(4), 423–441. doi:10.1080/01638539109544795

Rivera, G. M., & Matsuzawa, C. (2007). Multiple-language program assessment: Learners' perspectives on first- and second-year college foreign language programs and their implications for program improvement. *Foreign Language Annals*, *40*(4), 569–583. doi:10.1111/j.1944-9720.2007.tb02881.x

Sakoda, K., Konishi, M., Sasaki, A., Suga, W., & Hosoi, Y. (2016). International corpus of Japanese as a second language. *NINJAL Project Review*, *6*(3), 93–110.

Sakuragi, T. (2011). The construct validity of the measures of complexity, accuracy, and fluency: Analyzing the speaking performance of learners of Japanese. *Japanese Association for Language Teaching Journal*, *33*, 157–173.

Segalowitz, N. (2010). *Cognitive bases of second language fluency.* New York, NY: Routledge. doi:10.4324/9780203851357

Segalowitz, N., & Freed, B. F. (2004). Context, contact, and cognition in oral fluency acquisition: Learning Spanish in at home and study abroad contexts. *Studies in Second Language Acquisition, 26*(02), 173–200. doi:10.1017/S0272263104262027

Segalowitz, N., French, L., & Guay, J. D. (2018). What features best characterize adult second language utterance fluency and what do they reveal about fluency gains in short-term immersion? *Canadian Journal of Applied Linguistics, 20*(2), 90–116. doi:10.7202/1050813ar

Skehan, P. (2003). Task-based instruction. *Language Teaching, 36*(1), 1–14. doi:10.1017/S026144480200188X

Skehan, P. (2009). Modelling second language performance: Integrating complexity, accuracy, fluency, and lexis. *Applied Linguistics, 30*(4), 510–532. doi:10.1093/applin/amp047

Skehan, P. (2014). Limited attentional capacity, second language performance, and task-based pedagogy. In P. Skehan (Ed.), *Processing perspectives on task performance* (pp. 211–260). Amsterdam, The Netherlands: John Benjamins. doi:10.1075/tblt.5.08ske

Skehan, P., & Foster, P. (2005). Strategic and on-line planning: The influence of surprise information and task time on second language performance. In E. Rod (Ed.), *Planning and Task Performance in Second Language* (pp. 193–216). Amsterdam, The Netherlands: John Benjamins. doi:10.1075/lllt.11.12ske

SPOT. (n.d.) *About the SPOT (simple performance-oriented test).* Retrieved from http://ttbj-tsukuba.org/mt/p2_e.html

Tavakoli, P. (2011). Pausing patterns: Differences between L2 learners and native speakers. *ELT Journal, 65*(1), 71–79. doi:10.1093/elt/ccq020

Tavakoli, P. (2016). Fluency in monologic and dialogic task performance: Challenges in defining and measuring L2 fluency. *International Review of Applied Linguistics in Language Teaching, 54*(2), 133–150. doi:10.1515/iral-2016-9994

Tavakoli, P. (2018). L2 development in an intensive study abroad EAP context. *System, 72*, 62–74. doi:10.1016/j.system.2017.10.009

Tavakoli, P., Campbell, C., & McCormack, J. (2016). Development of speech fluency over a short period of time: Effects of pedagogic intervention. *TESOL Quarterly, 50*(2), 447–471. doi:10.1002/tesq.244

Tavakoli, P., & Foster, P. (2008). Task design and second language performance: The effect of narrative type on learner output. *Language Learning, 58*(2), 439–473. doi:10.1111/j.1467-9922.2008.00446.x

Tavakoli, P., Nakatsuhara, F., & Hunter, A. M. (2017). Scoring validity of the aptis speaking test: Investigating fluency across tasks and levels of proficiency. *ARAGs Research Reports Online.* Retrieved from http://centaur.reading.ac.uk/73379/

Tavakoli, P., & Skehan, P. (2005). Planning task structure, and performance testing. In R. Ellis (Ed.), *Planning and task performance in a second language* (pp. 239–273). Amsterdam, The Netherlands: John Benjamins. doi:10.1075/lllt.11.15tav

Towell, R., Hawkins, R., & Bazergui, N. (1996). The development of fluency in advanced learners of French. *Applied Linguistics*, *17*(1), 84–119. doi:10.1093/applin/17.1.84

Tse, L. (2000). Student perceptions of foreign language study: A qualitative analysis of foreign language autobiographies. *Modern Language Journal*, *84*(1), 69–84. doi:10.1111/0026-7902.00053

TTBJ. (Tsukuba Test-Battery of Japanese) (n.d.). Interpretation of TTBJ scores. Retrieved from http://ttbj-tsukuba.org/mt/p2_e.html

Witton-Davies, G. (2014). *The Study of Fluency and its Development in Monologue and Dialogue.* (Unpublished doctoral dissertation), Lancaster University, UK.

Wu, S. L., & Ortega, L. (2013). Measuring global oral proficiency in SLA research: A new elicited imitation test of L2 Chinese. *Foreign Language Annals*, *46*(4), 680–704. doi:10.1111/flan.12063

ADDITIONAL READING

Awwad, A., Tavakoli, P., & Wright, C. (2017). "I think that's what he's doing": Effects of intentional reasoning on second language (L2) speech performance. *System*, *67*, 158–169. doi:10.1016/j.system.2017.05.002

De Jong, N., & Wempe, T. (2009). Praat script to detect syllable nuclei and measure speech rate automatically. *Behavior Research Methods*, *41*(2), 385–390. doi:10.3758/BRM.41.2.385 PMID:19363178

Préfontaine, Y., & Kormos, J. (2016). A qualitative analysis of perceptions of fluency in second language French. *International Review of Applied Linguistics in Language Teaching*, *54*(2), 151–169. doi:10.1515/iral-2016-9995

Segalowitz, N. (2016). Second language fluency and its underlying cognitive and social determinants. *International Review of Applied Linguistics in Language Teaching*, *54*(2), 79–95. doi:10.1515/iral-2016-9991

Skehan, P., Foster, P., & Shum, S. (2016). Ladders and snakes in second language fluency. *International Review of Applied Linguistics in Language Teaching*, *54*(2), 97–111. doi:10.1515/iral-2016-9992

Tavakoli, P. (2019). Automaticity, fluency and second language task performance. In Z. E. Wen & M. J. Ahmadian (Eds.), *Researching L2 task performance and pedagogy.* Amsterdam: John Benjamins. doi:10.1075/tblt.13.03tav

Tavakoli, P., & Hunter, A. M. (2018). Is fluency being 'neglected' in the classroom? Teacher understanding of fluency and related classroom practices. *Language Teaching Research*, *22*(3), 330–349. doi:10.1177/1362168817708462

Trentman, E. (2017). Oral fluency, sociolinguistic competence, and language contact: Arabic learners studying abroad in Egypt. *System*, *69*, 54–64. doi:10.1016/j.system.2017.08.007

Wright, C., & Tavakoli, P. (2016). New directions and developments in defining, analyzing and measuring L2 speech fluency. *International Review of Applied Linguistics in Language Teaching*, *54*(2), 73–77. doi:10.1515/iral-2016-9990

KEY TERMS AND DEFINITIONS

AS-Unit: The abbreviation for an Analysis of Speech Unit. It is roughly equivalent to a sentence except that a not very sentence-like utterance such as "Good morning" or "Thank you" qualifies as one.

Between-Turn Pauses: Pauses appearing between conversational turns in a dialogue.

Inter-Rater Reliability: The degree of statistical agreement among raters.

Mora: A small phonological unit that determines timing. Moras coincide with syllables except for a couple of moraic consonants and long vowels.

Syllable: A small phonological unit within a word consisting of onset, nucleus, and coda.

Voiceless Geminate Consonant: A consonant that is held twice as long as a single instance of the same consonant. This is also called a double consonant.

Voiceless High Vowels: High vowels /i/ and /u/ in Japanese become voiceless when they appear in certain environments

ENDNOTES

1 CAF is the abbreviation for Complexity, Accuracy, and Fluency.

2 CAF Calculator is available for free on the official website (http://tell.cla.purdue.edu/CAF-calculator/). Please check there for updates as well.

3 (J-CAT: http://www.j-cat.org/html/en/pages/interpret.html)

4 (SPOT: http://ttbj.jp/mt/p2_e.html).

5 http://ttbj-tsukuba.org/mt/p2_e.html#pageLink03).

6 http://tell.cla.purdue.edu/CAF-calculator/

7 (http://tell.cla.purdue.edu/CAF-calculator/)

Chapter 7
The Potential of Text-to-Speech Synthesis in Computer-Assisted Language Learning:
A Minority Language Perspective

Neasa Ní Chiaráin
Trinity College Dublin, Ireland

Ailbhe Ní Chasaide
Trinity College Dublin, Ireland

ABSTRACT

This chapter describes the potential of text-to-speech synthesis (TTS) as a tool that can transform CALL platforms. Illustrating this point, a specific platform, An Scéalaí, is presented. By incorporating TTS, this platform facilitates the training of literacy skills, writing, and reading, with an emphasis at all times on the spoken language. The platform is described, as is the way in which it functions as a personalised tutor, prompting the learner towards self-correction. The prompts are delivered in both spoken/auditory form (using TTS voices) and in written form. The auditory feedback enables prooflistening, as well as spoken instructions pertaining to specific errors not picked up in the prooflistening process. The learner's progress is monitored throughout and records of the process are harvested for future research. The language in focus is Irish, and the linguistic complexities being targeted in the present implementation are explained, along with the relevant sociolinguistic context.

INTRODUCTION

This chapter deals with the use of text-to-speech (TTS) technology in language learning. The vast potential of TTS technology in language pedagogy has been relatively little explored to date. It is argued here that TTS is a uniquely valuable tool as it allows the spoken language to be placed at the centre of all language learning activities, even activities such as writing and grammatical training. TTS can bring

DOI: 10.4018/978-1-7998-1097-1.ch007

learners' own choice of materials and compositions to life, rather than depending on prerecorded materials. It is a particularly valuable technology in the case of language learning where the learner has limited access to native speaker models of the language. This last is especially true of minority languages, such as Irish, on which the work reported here is based.

An iCALL platform for Irish is described here which illustrates some of the potential advantages that accrue with the integration of TTS. The TTS systems being used are developed as part of the ABAIR initiative (ABAIR, 2019; Ní Chasaide et al., 2017). One feature of the present iCALL development is that it is being done collaboratively with the speech technology building itself, a fact which confers many advantages, as will be discussed below.

Irish is a minority language which is spoken as a community language in pockets, mostly on the western seaboard. However, as an official language of the State, it is widely taught in Ireland. The social and linguistic context is therefore quite different to what typically pertains in the major languages and there are many specific challenges that arise in teaching a minority language where access to native speakers is limited. These issues are discussed in some detail as they provide the context for the present work.

Although developed for the minority language context, the type of educational application and the advantages of the approach would also hold in other language learning contexts. It is therefore presented as a possible model for other languages as the advantages of the incorporation of speech technology have widespread applicability.

The specific iCALL application being described here, *An Scéalaí* (*the Storyteller*), aims to integrate the skills of writing, listening, speaking and reading in a speech-based application. It is intended to promote a holistic approach to the acquisition of different language skills so that the connections are intuitively grasped by learners.

An Scéalaí, currently under development in collaboration with the ABAIR initiative, is a web-based iCALL language learning platform. It is an open-ended platform, modularly constructed which is being added to incrementally. It provides the learner with language learning tools and content as well as feedback prompts for the correction of user-generated content and encourages autonomous language learning. At the same time, it is intended that this platform, by harvesting learner data, will be used by researchers and by platform developers for further refinements of this and other systems. Possibly the greatest strength of the current system derives from the symbiotic relationship between the speech technology and the iCALL researchers and developers and their close collaboration with learners and teachers.

BACKGROUND

Irish: Linguistic and Sociolinguistic Context

The Sociolinguistic Context

Irish is an endangered minority language, which survives as a community language in rural pockets, named '*Gaeltacht*' areas mostly in the West of Ireland. The latest census data shows Gaeltacht areas have a population of 96,090 people (CSO, 2016). Even in the Gaeltacht areas, however, the language is losing ground. Only 66% of those living in the Gaeltacht indicated they could speak Irish and of that group only 21% reported speaking the language daily (CSO, 2016). Although there are many speakers

Figure 1. Gaeltacht areas highlighted in dark green

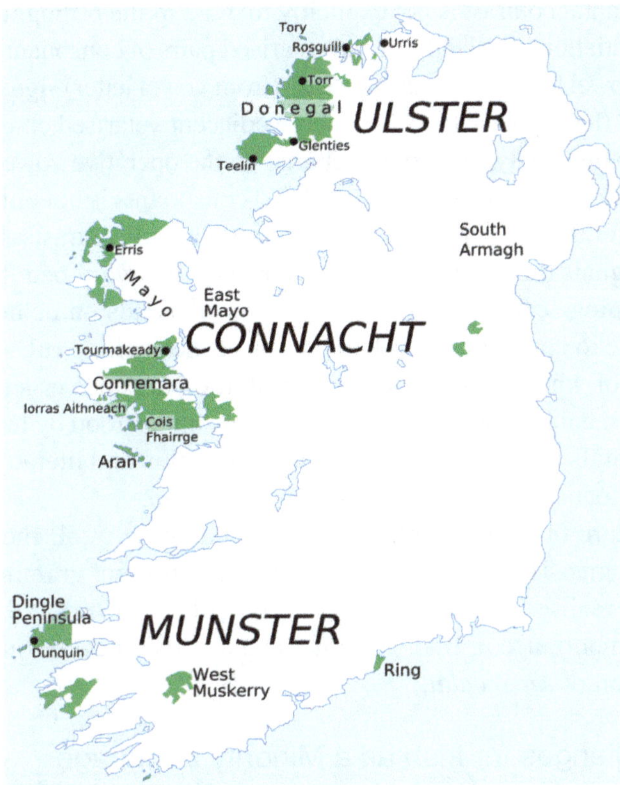

in urban areas it tends to be mostly in individual homes and Irish is relatively rarely overheard on the street. However, it has official status as the first national language of Ireland, and since 2007 is also an official language of the EU.

As common to many minority languages, there is no spoken standard, even though there is a written standard which has been constructed by drawing on the various dialects. This written standard was devised and first published in 1958 (*An Caighdeán Oifigiúil*, 2017). There are three main dialects and a number of subdialects and there are considerable differences among them, particularly at the phonetic/ phonological levels.

Structural Characteristics of Irish

Irish is an Indo-European language but differs structurally in many ways from other western European languages, e.g. in being a VSO language and in certain phonological and morphophonological features. For illustrative purposes here, there is a focus on aspects of the phonology and morphology which can be problematic for learners and which are explicitly addressed in the illustrations of *An Scéalaí* presented below.

The phonological system of Irish contrasts two series of consonants, velarised and palatalised. In simple terms, this means that there are over twice as many consonants in Irish as there are in English. Whereas in English there is a single /b/ phoneme, in Irish there are two e.g., /$\mathbf{b^{j}}$ i:/ bí (*to be*) vs. /$\mathbf{b^{ɣ}}$ i:/ buí (*yellow*). This can be challenging for Irish learners whose first language is English. The situation is further complicated by the fact that the consonantal contrast is not explicitly marked in the orthography. The Roman alphabet doesn't provide for differentiation of palatalised and velarised pairs of consonants. The convention is to use different sets of adjacent vowel letters: an adjacent "i, e" (front vowel letter) signals a palatalised consonant while an adjacent "a, o, u" (back vowel letter) signals an adjacent velarised consonant. One therefore gets unlikely strings of vowel letters, only some of which map to the operative vowel of the syllable.

The phonological contrast of velarised and palatalised consonants is not only fundamental to the pronunciation of Irish but central also to the signaling of grammatical contrasts. For example, the contrast in the final consonants signals the singular vs plural forms of the word bád /bˠ aː **d̪ˠ** / (*boat*) vs báid / bˠ aː **dʲ** / (*boats*). Developing accurate writing in Irish thus depends on being able to hear the phonological contrast, being able to appreciate the phonic basis of the spelling rules that are used to signal it, and having an awareness of where grammatical differentiation depends crucially on this phonological contrast. Despite its fundamental importance, it is not widely understood by learners, or even frequently by teachers, with detrimental consequences for both pronunciation and literacy development. This is an aspect that is specifically focused on in this version of *An Scéalaí.*

A further striking feature of Irish morphophonemics, common to all the Celtic languages, is the *mutation* of initial consonants. This effectively means that in different grammatical contexts the above word '**b**ád' (boat) may be realised as /**b**ˠ aː d̪ˠ /, /**v**ˠ aː d̪ˠ / (**bh**ád) or /**m**ˠ aː d̪ˠ / (**m**bád). These type of mutations are of central importance in Irish grammar and are also focused on in the pedagogical tasks set in the current realisation of *An Scéalaí.*

The Educational Challenges for Irish as a Minority Language

Since the foundation of the Irish State in 1921, it has been recognised that the maintenance of Irish will depend crucially on how successfully it can be transmitted through education to the wider population.

Irish is a compulsory school subject up until school leaving age, yielding a large national cohort of primary and secondary school learners (c. 700,000 learners, Ní Chiaráin, 2014). There is also a substantial number of Irish language learners outside of the Irish education system, both in Ireland and around the world. The latter are partly resultant from the Irish diaspora and are impossible to quantify. However, the traffic on the webpage of the ABAIR initiative (www.abair.ie) alone suggests large global numbers who engage with the language at least to some degree. Irish is a popular extracurricular language in third level institutions in the US and increasingly also in Canada and some other European countries.

The teaching of Irish brings many challenges. As is clear from the examples above, L1 speakers of English need to acquire a very different sound system, orthographic and grammar systems. As mentioned, the phonics of Irish are mostly not understood and not explicitly taught. Parents helping their children with homework are often also thrown by the orthographic conventions of Irish. Increasing levels of feedback from learners and from parents to the ABAIR webpage demonstrates the need for learners to have to access to how written text is pronounced.

A further challenge is that the majority of learners do not have much access to native speaker models. Primary and second level teachers are typically second language learners with differing fluency levels and often feel that they have an incomplete grasp of the language and its structures. For adult learners, particularly those abroad, the tendency is to focus on the written language. The opaque orthography and the complex sound system can often present a major barrier to their acquisition of the language as a spoken entity.

Lack of materials are often decried and especially the lack of digitally-based interactive materials that harness modern technology, as available for 'major' world languages. The lack of engaging materials, allied to the compulsory nature of the subject in school, can impact on learners' motivation.

This is where a speech-based iCALL platform, which inherently integrates the spoken language with the written, and which exposes learners continuously to native speaker models, can offer a technology-based solution to many of the difficulties that Irish language education currently encounters.

TEXT-TO-SPEECH SYNTHESIS DEVELOPMENT: THE ABAIR INITIATIVE

Text-to-speech systems take any written text and output the spoken version. TTS is being developed for Irish in conjunction with underpinning linguistic resources, which are generally not available for minority languages (ABAIR, 2019; Ní Chasaide et al., 2019). Furthermore, along with the core technology, user applications are being developed that respond to the critical needs of the Irish language community, and particularly the need for applications in education, hence the present iCALL platform.

The ABAIR group is a multidisciplinary team of speech scientists, engineers, phoneticians, linguists, computer scientists, programmers and CALL experts. This is an ideal environment for speech-based CALL development and the advantages that accrue are discussed in later sections. To appreciate this, it is worth considering how TTS systems are built, and this is illustrated in Figure 2.

The synthetic voice is typically built on the voice of a single representative speaker, from whom a corpus of high-quality recordings are made. In these corpora, the individual sounds are identified, transcribed and aligned to the acoustic signal. As each sound is heavily influenced by neighbouring sounds, the corpus must contain instances of each sound in every possible context. The very extensive consonantal system of Irish demands a very large corpus, but the precise size and structure of the corpus depends on the speech engine to be used. In the case of the Irish synthesis, multidialect coverage was envisaged from the outset and corpora were required for each dialect. Currently, synthetic voices for the three main dialects are publicly available on the www.abair.ie webpage and these include male and female voices. Ongoing work is targeting childrens' voices and further dialects.

At the heart of a TTS system lies the elaboration of linguistic-phonetic components, all of which for Irish must be provided for the individual dialects. Specific components include: ***text normalisation*** to allow numerals, acronyms, symbols, etc. to be rewritten as text; ***letter-to-sound*** rules to map written forms to the sequence of sounds they represent (e.g. mapping 'mhaoigh' → /wiː/, *Ulster dialect*). Where

Figure 2. Some linguistic components at the heart of a text-to-speech synthesis system

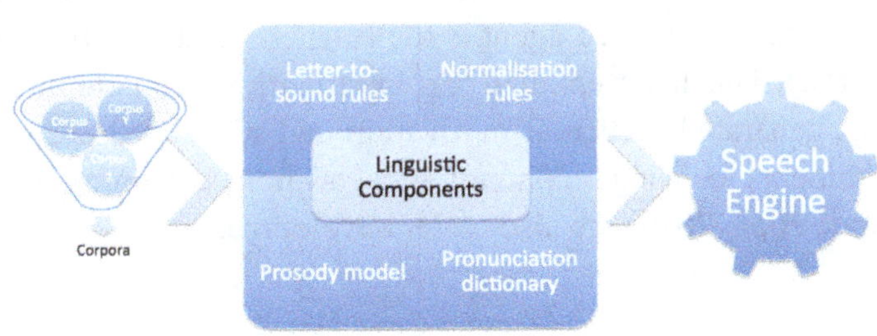

written forms do not conform to the letter-to-sound rules of the language, they are stored in a ***pronunciation dictionary*** (e.g., 'bhfaighidh' → /w:i/, *Ulster dialect*). The ***voice prosody*** module formulates prosodic patterns of voice, melody and rhythm. The iCALL platform described below exploits not only the ABAIR synthetic voices but also these linguistic-phonetic components that lie under the hood.

There are numerous types of speech engines that can be deployed and this aspect of the technology is advancing rapidly, particularly with the improvement in the performance of deep neural networks (DNNs). In the case of ABAIR, there are currently a variety of technologies in use, and some approaches are better suited to particular applications. For example, the requirements of a screenreader for the visually impaired where very high speeds of output are needed would be very different from those that might be most appropriate in an educational game for children.

Due to public demand, the TTS systems are increasingly being deployed to develop applications in the following areas:

- **Applications for the wider Irish language community:** from its first hosting on the web, which allows free and easy access to the synthetic voices, ABAIR has been used as a public facility across the globe. To facilitate users, a web-reader addon can be downloaded from the site which speech-enables any Irish language online text.
- **Disability and access:** in the case of the minority languages, the lack of speech technologies results in exclusion of those with disabilities. For example, those with a visual impairment do not have access to the written word. In the case of Irish a screenreader for the visually impaired has been developed (McGuirk, 2015) which outputs speech and Braille simultaneously. Work is also ongoing on applications for literacy development with a particular focus on those with dyslexia (Barnes et al., 2017).
- **Documentation and preservation:** a synthetic voice of a dialect preserves a *virtual speaker* for posterity and, of itself, provides a 'live' form of documentation (Ní Chasaide et al., 2015). Even for the most endangered dialect, this provides tangible support for local communities concerned with its maintenance.
- **Language learning:** this is an area that is central to the future of the language, as explained above. The simple direct use of the ABAIR webpage is growing for all kinds of tasks, e.g. figuring out how to pronounce Irish names; helping children with their reading/writing homework, etc. In the period January – March 2019, 155,305 hits were recorded.

More complex speech-enabled CALL platforms are being piloted for Irish that exploit the TTS voices and linguistic underpinnings. Pilot developments so far include interactive educational games (Ní Chiaráin and Ní Chasaide, 2016a), virtual scenarios where the avatars are Irish-speaking (Ní Chiaráin and Ní Chasaide, 2015). Particularly popular with school-going children has been an initial dialogue-based CALL interface, where an animated agent is the learner's dialogue partner (Ní Chiaráin and Ní Chasaide, 2016b and 2016c). This latter entailed text input with speech output. Trials with 228 school children showed that this generated particular motivation and stimulated engagement for learners (Ní Chiaráin, 2014). Developments focusing on phonological awareness training (Ó Broin et al., 2018) and early literacy (Ní Chasaide et al., 2019) are also under development.

The iCALL platform, which forms the application discussed in detail in the present chapter, is a currently ongoing development, where the synthetic voices and the system-internal linguistic-phonetic components e.g. letter-to-sound rules are being used to develop writing skills, while reinforcing other language skills. It allows for open-ended extension to address a full range of phonological, morphophonological and grammatical aspects of language learning.

TTS IN CALL

TTS to date has not gained widespread acceptance in CALL (Gupta & Schulze, 2012), although this situation is beginning to change. To date synthetic speech has been used, for example, to help learners with individual language skills such as aids to pronunciation (Liakin, Cardoso and Liakina, 2017; Wik, 2011; Wik & Hjalmarsson, 2009), reading assistants (Lopes et al., 2010), facilitators of dictation exercises and of listening comprehension (Huang et al., 2005; Pellegrini et al., 2012; Coniam, 2013) or talking dictionaries (Handley, 2009).

One reason why TTS was not used more widely may be that the quality of some of the TTS systems was sometimes fairly poor with an unnatural voice quality (Sha, 2010). However, TTS technology has evolved rapidly so that the better systems today provide very intelligible and natural speech output, something that was generally not the case some years ago. For the major world languages, therefore, TTS is now a mature technology which is ripe for deployment in educational settings. In evaluations of the intelligibility of the Irish TTS synthetic voices (carried out in the context of the Chatbot application), 73% of 288 students rated them at points 4 or 5 on a Likert scale – agree/agree strongly – for "the synthesised voices were sufficiently clear to make the speech intelligible". Similar results were found for naturalness ratings (Ní Chiaráin, 2014).

TTS allows open-ended speech output as opposed to prerecorded speech prompts. This allows for creative tasks where the learner controls the materials as opposed to the more tightly constrained, predetermined materials that would be required if prerecorded prompts were used. Note that for many CALL applications that are currently being piloted as part of the ABAIR initiative, such as *An Scéalaí,* prerecorded natural voices would not be suitable since the user is composing in a creative, open-ended way. Furthermore, program developers or teachers would be very constrained if obliged to prerecord prompts.

Following on the above, the inclusion of TTS means there can be a far greater degree of interactivity between learner and machine where at least a semblance of a genuine communicative experience can be achieved, i.e. the negotiation of meaning between an avatar of some type and a learner. This concept of negotiation of meaning is a central theoretical underpinning for communicative approaches to language teaching/learning (Foster & Snyder Ohta, 2005; Pica, 1994). In terms of CALL, the negotiation of meaning would involve an avatar being capable of reacting to a learner's input in a meaningful way so that the learner can make the appropriate adjustments to his/her input in order to make grammatically and contextually appropriate inputs. TTS synthesis will be central to the development of such learning/teaching platforms and this would open up an entirely new vista in the direction of CALL development.

PEDAGOGICAL CONSIDERATIONS IN CALL AND ICALL

Pedagogical considerations are paramount to all CALL applications. A criticism often levelled is that applications are technology-led, insofar as a piece of technology is developed and gets deployed in an educational context regardless of its educational value. The functionalities built into the *An Scéalaí* platform are guided by relevant second language acquisition (SLA) theories as the technology extends learning opportunities in ways that would be difficult if not impossible to achieve through traditional classroom methods (Sykes et al., 2008). Some particularly important considerations are:

- **Experiential learning:** a key guiding principle in all learning is that the process must involve an element of *action* on the part of the learner (Dewey, 1997). This 'experiential' or 'hands-on' learning is the most fundamental guiding principle of *An Scéalaí* and is detailed below.
- **Communicative Approaches:** the predominant approach to language learning in recent decades has been the 'communicative approach' where the emphasis was overwhelmingly on spoken communication. This approach emerged in reaction to the grammar-translation method, which led to a focus on writing skills and writing accuracy but where learners achieved little spoken or communicative competence.
- **Task-based Language Teaching/Learning:** within the communicative approach there has been considerable emphasis on task-based language teaching (TBLT), based in turn on the idea of "learning by doing" or "experiential learning" (González-Lloret, 2016). Many different types of 'tasks' fall under this methodology (see Ellis, 2003 and Skehan, 1998) but common characteristics of 'tasks' are that they are meaning oriented, communicative in nature, and focus on the content of the message. In order to emulate the real-life situation in which the target language is likely to be used, a CALL platform should ideally have tasks which have a real-world significance, have a focus on meaning, and be able to engage the learners' interests (Dickey, 2011).

In line with TBLT theory the *An Scéalaí* platform has been constructed so that learners would have the opportunity to interact with the computer in a meaningful way in order to achieve some given aim. It allows the setting of different kinds of tasks and the one described in the present trials entails the writing of an open-ended reflective piece – a personal diary about a learner's own language learning journey. This speaks to Chapelle's proposed framework (2001) whereby the task should be authentic, practical, focused on meaning, and appropriate to the learners' level and learning goals.

In reaction to the criticisms of the communicative approach, authors such as Chapelle have emphasised the need to provide opportunities within TBLT to focus on form (Long, 1991). In keeping with this approach, the primary focus of *An Scéalaí* is to increase learners' awareness and acquisition of form and this is actualised through a series of learning steps detailed under 'TTS speech-enhanced platform: An Scéalaí' below.

An Scéalaí sets out to create an environment where learners' attention is directed to language form, use and meaning. There are various learning steps included, and the manner in which feedback given is designed to promote **noticing** of linguistic features (Schmidt, 1990, 2001). Self-correction is used as a way of developing meta-awareness of linguistic structure, helping the learner acquire and consolidate their own production of these features.

A rather different point has been raised about CALL platforms, i.e. that the **cognitive load**, i.e. the mental effort required for successful completion of a task, should not be too great (Pillay, 1994). As will be discussed in the following section, the numbers of linguistic features to be trained at one time are kept to a maximum of three, helping to **scaffold** learners attempts in writing. Screen layout and material presentation are also important factors to bear in mind relative to cognitive load.

- **Integrated Skills Development:** a common criticism of the communicative approach is that it sometimes fails to bring along those writing skills that create the "all-important interface between explicit processes of language learning and implicit processes of spontaneous language use" (Little, Dam & Legenhausen, 2017, p.210). *An Scéalaí* is aligned with a communicative philosophy for language learning and while this current instantiation is initially directed particularly at writing skills, integration with the other three language skills is integral to the process. Linkage between the written and the spoken word is pervasive and learners are simultaneously training the phonics basis and their pronunciation skills. The goal is that the learner is gaining a holistic competence, where advances in one skill is simultaneously bringing on the others.
- **Autonomous Language Learning:** autonomy of the learner is an important consideration, particularly in the Irish context where many learners may not live near Irish-speaking populations and some may not have much access to a human teacher.

An Scéalaí is primarily an autonomous tool which can be used in two different ways. On the one hand it can be used under the guidance of a teacher. In this scenario, the teacher has an organisational role in that he/she can coordinate the selection of a task, have oversight of the learners' progress, and access learner data. Such data may provide the teacher with material relevant to the linguistic progress of that particular group and he/she can in turn design further lessons which give highly relevant feedback. However, the learner remains central and controls the language production and selection of self-correction mechanisms to be used. The teachers' role is as 'facilitator' rather than 'director' of learning.

On the other hand, the autonomous distance learner is equally catered for by the platform and differs from the group learner only in lacking one of the many sources of input, i.e. that of a formal teacher.

- **Intelligent CALL and one-to-one tutoring:** Intelligent CALL allows us to mimic the intelligent one-to-one interaction that happens between a teacher and a learner by applying concepts, techniques, algorithms, and technologies from artificial intelligence (Gamper and Knapp 2002; Heift and Schulze 2007; Nerbonne 2003; Schulze 2008). As pointed out by Luckin et al. (2016)

One-to-one human tutoring has long been thought to be the most effective approach to teaching and learning. … Unfortunately, one-to-one tutoring is untenable for all students. Not only will there never be enough human tutors, it would also never be affordable. All of this begs the question: how can we make the positive impact of one-to-one tutoring available to all learners across all subjects?

In its simplest terms, *An Scéalaí* is an answer to Luckin et al. (2016) in that it sets out to make one-to-one tutoring available to all learners at all levels. It integrates 'intelligent' prompting in the platform that functions as a system-internal tutor for learners as they progress through the steps.

In the current version of *An Scéalaí*, its suitability as a teacher-led activity for the classroom is emphasised, as at this stage in its development, a suitable environment is needed to provide teacher and learner evaluations. Nonetheless, the tool itself is designed so that it does not in fact require a teacher, but rather provides one-to-one tutoring to the autonomous learner.

TTS SPEECH-ENHANCED PLATFORM: *AN SCÉALAÍ*

A fundamental approach in the design of *An Scéalaí* is that it should be a platform whereby the four skills of writing, listening, reading, and speaking are integrated in a holistic learning experience and where the spoken language remains constantly at the heart of every activity. At its current stage of development, *An Scéalaí* begins with an exercise in productive creative writing and the example provided here is geared towards upper second level learners. The emphasis is on developing writing skills but in a way that ensures auditory linkage of the spoken and written language. Thus, the written is at all times associated with the spoken form.

Feedback allows the learner to identify errors in the written forms and to correct those errors. *An Scéalaí* is designed to permit the autonomous learner to write and correct/improve his/her productions. This process is facilitated in two ways:

1. Audio proofing using TTS voices allows learners to listen to their written text. This is a potent way for them to identify where the output doesn't match the intended output and to correct accordingly.
2. Natural Language Processing (NLP) prompts draw attention to specific errors. These are based on prior available linguistic components, embedded in the TTS system (see Figure 2 above) or on freely available tools such as spelling/grammar checkers (Scannell, 2013). The prompts provide feedback on errors made and in certain cases colour coding is used to highlight the type of error involved.

While the student is learning 'grammatical rules', the auditory channel provides simultaneous training in the phonological and morphophonological rules, which lie at the heart of many aspects of Irish grammar. The learner's experience and the types of feedback facilities for correction are detailed below.

An Scéalaí is, however, more than a language learning tool. While it can be used for autonomous learning, it can also be deployed as an extension of the teacher's classroom work and the platform includes features that allow teachers to intervene, provide feedback and manage the process if that is desired. It is also intended as a research platform insofar as all learner interactions with the system and all spoken and written records are harvested to be used in future analysis and research. These latter aspects are described in the 'Beyond the Learner' section below.

How it Works: The Steps Involved

For the user *An Scéalaí* involves a six step process, where Steps 3 and 4 are the critical language learning modules. The six steps are illustrated in Figure 3 and briefly explained here. A more detailed discussion of Steps 3 and 4 follow below.

Step 1. Register and Complete Learner Profile: On first registering, learners create their individual accounts and fill out a profile with some background information, e.g. their language level, the amount of contact they have with native speakers, etc. This is a preliminary step which is only required on the first occasion of use. It is needed for the management, feedback and data collection processes.

Step 2. Write: Learners begin by writing their own composition (the current default being tested involves the user writing a reflective diary on their language learning journey). The choice of writing task is, of course, open to the teacher and learner.

Step 3. Listen and Correct (prooflisten): Learners listen to the spoken version of their text read aloud by the ABAIR voice. The user chooses the dialect they wish and the gender of the speaker. This provides not only exposure to native-like speech, but provides essentially a proof-listening tool which is a powerful aid for self-correction.

Step 4. Read and Correct: The system prompts the user to correct other specific errors. These errors are 'detected' through the use of the linguistic components of the TTS system (ABAIR, 2019) and NLP tools that have been developed elsewhere for Irish (Scannell, 2013).

Step 5. Speak: A recording facility is built into this application and in Step 5 learners are encouraged to read their stories aloud. They then listen to their own recordings and compare them to the ABAIR synthetic native-speaker version. They are encouraged to do this iteratively and if they wish, they can play particular words/phrases on a loop. This can greatly help in consolidating native-pronunciation memory.

Step 6. Save: Learners can save their work in their own personal account online and access it at any time, taking up where they left off. If so desired, teachers can also have access to the learners' input. This would permit teachers as well as learners to monitor progress over time. A simple voice memo feedback mechanism is included, which allows the teacher to provide feedback and communicate specific guidance to individual learners in a quick way.

It should be noted, however, that all user interactions are saved along with text and speech files. These are needed for a variety of research on language acquisition and platform development. These aspects are elaborated below.

Elaboration on Process, Steps 3 and 4

Step 3. Listen and Correct: Having corrected the written text, learners then listen to the spoken version of their text read aloud by an ABAIR voice in the dialect of their choice.

There are a number of dividends to this exercise. Firstly, it helps establish the phonics basis for the Irish written forms. The complexities of the sound system and the opaque orthography result in many learners (and parents helping children with homework) being unsure as to how text should be pronounced. The fact that the text is spoken aloud provides repeated habituation in establishing the sound-to-letter mappings of the language, reinforcing the spelling regularities that are often not explicitly taught.

Furthermore, having the text read aloud provides a proof-listening step in the correction process. The synthetic voice reads out exactly what has been written and this makes audible those errors that may not have been spotted hitherto. In this way learners are guided by ear to correct their errors. It turns out that errors that would not be spotted by eye tend to jump out when the text is accordingly mispronounced. An example concerns the use of the acute accent to identify long vowels (in Irish *fada* meaning 'long').

Figure 3. An Scéalaí functionality summarised in 6 key steps

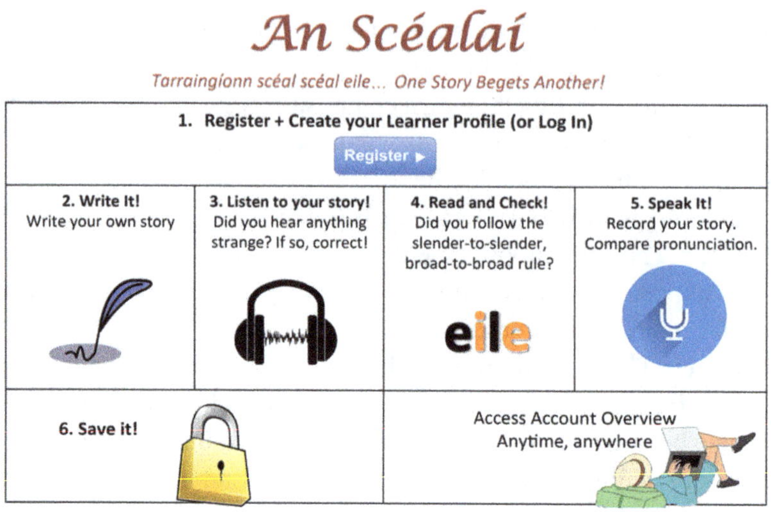

Figure 4. Illustration of the importance of the acute accent in Irish

sean	/ʃ **a** n̪ˠ/	*old*	ait	/**a** tʲ/	*strange*
séan	/ʃ **e**: n̪ˠ/	*to deny*	áit	/**a**: tʲ/	*place*
Seán	/ʃ **a**: n̪ˠ/	*name*			
solas	/sˠ **o** l̪ˠ **ə** sˠ /	*light*	cead	/cʲ **a** d̪ˠ /	*permission*
sólás	/sˠ **o**: l̪ˠ **a**: sˠ /	*comfort*	céad	/cʲ **e**: d̪ˠ /	a hundred

This is a very frequent error which can arise either as a simple typo or from a poor grasp of how the acute accent affects the vowel pronunciation. In the words in Figure 4 below, the presence and location of the accent determines the pronunciation and potentially also the meaning of the word.

Initial trials with *An Scéalaí* have shown how very frequent this type of error is and how amenable it is to correction by ear.

Linking the spoken language to the written text is useful in every language. Putting the spoken language at the centre of the development of writing skills increases learners' exposure to the spoken language. This is all the more critical in the case of a minority language, such as Irish, where many learners have limited access to native speaker models of the language.

Step 4. Read and Correct: as mentioned above, linguistic knowledge is used to prompt the learner towards self-correction.

This version of *An Scéalaí* accesses the letter-to-sound rules of the synthesiser to signal where the spelling rule that signals consonant quality in Irish is violated, e.g. the use of front vowel letters with a velarised consonant (see description in 'Structural Characteristics of Irish' above). Where this is the case, the colour of the font and background highlighting is used to draw the learners' attention to the probable spelling error, and they are explicitly encouraged to check/correct their own text accordingly.

Figure 5. Example from An Scéalaí where slender-to-slender, broad-to-broad rule is violated

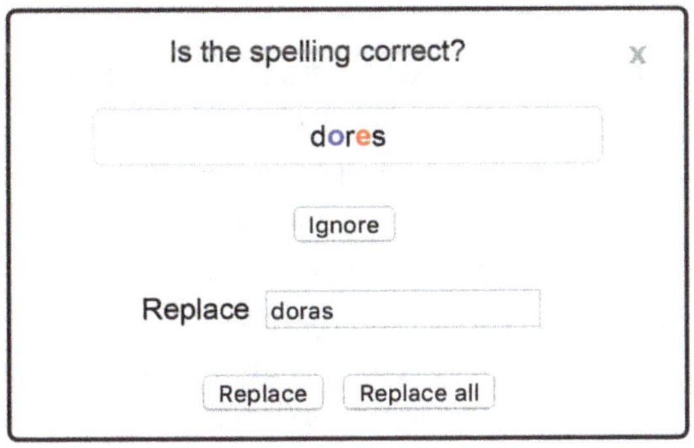

Note that although the spelling rule alluded to here holds across most cases, there are instances where it is not adhered to (e.g. the noun 'Gaeltacht'). However, although it would be possible to include a pronunciation lexicon of the synthesiser to allow these irregular forms, the decision was made not to do so at this point. This obliges the learner to be actively engaged in the self-correction process, using their judgment and further raising their awareness of irregular forms.

A further NLP module allows the signaling of errors involving the morphophonemic rules pertaining to mutation of initial consonants (also described in 'Structural Characteristics of Irish' above).

The NLP modules are included on a phased basis so that the learner focuses on one specific aspect at a time. Having too many different items being corrected simultaneously could be quite confusing and demotivating. Precisely how to phase the introduction of specific items is something that will require considerable testing and evaluation for learners at different levels.

The examples here focus on specific items that are particularly challenging for learners of Irish. Each language presents specific challenges and therefore the content planning for NLP inclusion requires expertise in the structure of the language as well as availability of such rules in digitally accessible form.

THE LEARNER PERSPECTIVE: KEY FEATURES OF *AN SCÉALAÍ*

Autonomous learning is promoted by this application as the combination of auditory and explicit visual corrective feedback affords learners the opportunity to iteratively improve their own work. It can also serve as an adjunct to ongoing classroom activities. For individual adult learners and those located outside of Ireland, a tool such as this is badly needed to develop a broad range of language skills. It has therefore been made available online and is openly available to the public.

Figure 6. Example where mutation of initial consonant is signaled for correction after the possessive pronoun 'mo' (my): "I am living in Dublin with my **mother** *and my* **brothers.***"*

Holistic learning is at the heart of the system where the spoken language and literacy skills are simultaneously trained.

Learner Profiles Used to Present Appropriate Content

The profile information provided by the learner upon registration will be used to select the most appropriate initial content that matches the learners' level and the precise corrective feedback needed.

Adaptive to different levels of proficiency: *An Scéalaí* is intended for all levels. It is currently being used with second and third level students who already have a reasonable competence in the language. However, as the learner data is harvested, the research team can, in the longer term, aspire to implementing learner models in order to be able to generate personalised feedback and adaptive learning paths individual to each learner (see e.g., Amaral and Meurers 2007; Bull 2000).

Overview of progress over time: currently, as the data is logged and timestamped, learners and teachers can both review material and learners can self-evaluate their progress. A future feature will allow the learners' progress to be tracked and visually presented through their individual accounts. This should help give positive feedback that will in turn increase motivation.

BEYOND THE LEARNER: IMPLICATIONS FOR THE TEACHER, THE RESEARCHER, THE DEVELOPER

Teaching

An Scéalaí, although essentially an autonomous learning tool, is suitable for extending classroom work, where the teacher acts as a facilitator of learning. The learners' data can be viewed by the teacher who can use it to track the progress of individuals in the group and to provide them with tailored feedback. Teachers might also use this information to plan where their own teaching input might best be directed. Although not yet implemented, the intention is that the teacher can incorporate content so that the materials become increasingly streamlined with the curriculum for various learning cohorts.

Research

Language Acquisition Research: as learners' written and spoken interactions are recorded, a corpus of L2 Irish is collected that will be used for various types of research. Firstly, the learner corpora, both written and spoken, will provide many insights into the process of language learning, into the most frequently occurring types of errors for learners at different stages. They will also provide insights into the effectiveness of the corrective mechanisms that are being used.

Over time, it is hope that this will provide normative data. This would allow information on a particular learner's level and rate of progress relative to a wider population, which could be made available to learners/teachers as well as curriculum developers.

Speech Technology Research: there is an inherent symbiotic relationship between the speech technology and iCALL development. Whereas this iCALL platform is possible due to the prior TTS development of ABAIR, it in turn offers an ideal testbed for evaluation of that same technology. The network of learners using the system can be used to garner evaluations of the quality of synthetic voices, etc. as the technology evolves. Furthermore a feedback mechanism allows them to flag or report unsatisfactory aspects of the speech technology.

When speech recognition becomes available, the recordings of the spoken data will be pooled and utilised in building targeted recognition systems for learners at different levels. The fact that spoken data is available with matching text and the fact that information on the linguistic background of the individual learners is available means that this corpus will eventually be a very useful resource for recognition development. The inclusion of recognition will open up a more extensive role in pronunciation training, dictation and ultimately interactive dialogue, with many promises for the future.

Platform Development

An Scéalaí is designed as a modular system. The linguistic NLP components that have been incorporated at this stage are only the beginning of what is intended to become a more comprehensive package which can be adapted to early learners right through to more proficient learners, who may simply need to improve on specific aspects.

The recorded interactions by the learner will guide the further development of NLP modules and will prompt refinements as to how the platform can be made more effective. A cyclical programme of iCALL development driven by the learner data is envisaged. For example, once normative data is available showing where a particular learner is and how they're progressing in relation to a larger population of learners, the platform can be made more adaptive so that the learner is directed to the most appropriate content for them.

CONCLUSION

Text-to-speech synthesis technology is ripe for exploitation in CALL. For many languages it has reached a point where the intelligibility and quality of the speech is very high. The present iCALL platform is but one possible way in which TTS can be deployed. It serves as an illustration of how classroom activities can be extended through autonomous learning with immediate personalised feedback in an interactive setting.

Although *An Scéalaí* focuses on writing skills, it presents a **holistic** language learning experience where the **spoken language is centre stage** and writing and reading skills are closely integrated. This ready access to native speaker speech is particularly important in the minority language setting, and as such it addresses some of the sociolinguistic challenges discussed earlier in this chapter.

By constantly linking the written and spoken word, the platform provides subliminal training on Irish phonics. This specific implementation also provides explicit phonics training in that the spelling prompts draw attention to the written correlates of the phonological system of Irish. In this and the other examples mentioned, the platform is geared to the specific linguistic issues of Irish, which tend to be resistant to correction. Each language presents specific challenges and the platform is ideally suited to allow tailoring of this kind.

The present work benefitted from a multidisciplinary environment where the iCALL platform building was carried out hand in hand with the speech technology development. This may not always be possible, but the greater the contact between CALL developers and the technology developers, the richer the pedagogical offerings that can be delivered. For example, the linguistic components of the

TTS systems were here deployed in the iCALL platform. As discussed above, the advantages flow in both directions – the iCALL application motivates the technology development and provides an ideal context for its evaluation.

In comparison to major world languages, the provision of technology for Irish is still very limited, and this is something that will hopefully progress rapidly with the Digital Plan for the Irish Language currently under preparation (Rialtas na hÉireann, *forthcoming*). With the addition of further technologies, the possibilities of what can be achieved grow logarithmically. For example, speech recognition would allow the incorporation of pronunciation correction tools (Cucchiarini, Nejjari and Strik, 2012) to aid learners acquire native-like speech patterns. Once speech recognition is available the inclusion of dialogue systems and conversational agents open up even more possibilities. This would transform iCALL platforms in allowing a communicative context where the learner negotiates meaning with a virtual interlocutor. Even the very primitive dialogue system developed for Irish and described in Ní Chiaráin and Ní Chasaide (2019) was found to be highly popular with second level school learners, motivating them to push their boundaries in language production. For Irish, although open-ended recognition and dialogue systems are some time away, given the multidisciplinary environment of the team, even initial limited domain versions will be built with CALL objectives in mind and deployed and tested in that context.

Most of this chapter has implications for language teaching in a broad sense, regardless of what language is involved. However, as indicated in the title, it is particularly pertinent for the minority language context, as the needs for such applications are greater, as are the potential dividends. The ABAIR experience to date provides a roadmap for other minority languages and highlights how speech and language technologies can stimulate joint applied research in the educational sphere, that potentially can transform the learning and teaching of these languages.

REFERENCES

ABAIR. (2019). *ABAIR: An Sintéiseoir Gaeilge – The Irish Language Synthesiser.* Retrieved from www.abair.ie

Amaral, L., & Meurers, D. (2007). Conceptualizing student models for ICALL. In C. Conati, K. McCoy, & G. Paliouras (Eds.), *User modeling 2007* (pp. 340–344). Heidelberg, Germany: SpringerVerlag. doi:10.1007/978-3-540-73078-1_41

An Caighdeán Oifigiúil. (2017). Gramadach na Gaeilge – An Caighdeán Oifigiúil. (The official standard for writing in the Irish language). Tithe an Oireachtais. Retrieved from https://data.oireachtas.ie/ie/oireachtas/caighdeanOifigiul/2017/2017-08-03_an-caighdean-oifigiuil-2017_en.pdf

Barnes, E., Ní Chiaráin, N., & Ní Chasaide, A. (2017). Departures from the "norm": how the phonology, morphology and orthography of the Irish language impact on literacy instruction and acquisition. Exploring the literacy landscape: Celebrating 40 years of research and practice, *Literacy Association of Ireland 40th Annual Conference, Marino Institute of Education,* Dublin, Sept. 29-30, 2016, Literacy Association of Ireland, pp. 22–32.

Broin, Ó., D., Barnes, E., Ní Chiaráin, N., & Ní Chasaide, A. (2018). A story based game for teaching phonics for the Irish language. In *Proceedings of the 12th European Conference on Games Based Learning*, Sophia Antipolis, France, pp. 495-503.

Bull, S. (2000, June). Individualized recommendations for learning strategy use. In G. Gauthier, C. Frasson, & K. VanLehn (Eds.), *Intelligent tutoring systems.* In *Proceedings 5th International Conference, ITS 2000,* Montréal, Canada, (pp. 594–603). Berlin, Germany: Springer Verlag 10.1007/3-540-45108-0_63

Chapelle, C. (2001). *Computer applications in second language acquisition: Foundations for teaching, testing, and research.* Cambridge, UK: Cambridge University Press. doi:10.1017/CBO9781139524681

Coniam, D. (2013). Computerized dictation for assessing listening proficiency. *CALICO Journal, 13*(2&3).

CSO. (2016) Census of Population 2016 – Profile 10 Education, Skills and the Irish Language. Retrieved from https://www.cso.ie/en/releasesandpublications/ep/p-cp10esil/p10esil/ilg/

Cucchiarini, C., Nejjari, W., & Strik, H. (2012). My pronunciation coach: Improving English pronunciation with an automatic coach that listens. *Language Learning in Higher Education, 1*(2), 365–376. doi:10.1515/cercles-2011-0024

Dewey, J. (1997). *Experience and education.* New York, NY: Simon & Schuster.

Dickey, M. D. (2011). Murder on Grimm Isle: The impact of game narrative design in an educational game-based learning environment. *British Journal of Educational Technology, 42*(3), 456–469. doi:10.1111/j.1467-8535.2009.01032.x

Ellis, R. (2003). *Task-based language learning and teaching.* Oxford, U.K.: Oxford University Press.

Foster, P., & Snyder Ohta, A. (2005). Negotiation for meaning and peer assistance in second language classrooms. *Applied Linguistics, 26*(3), 402–430. doi:10.1093/applin/ami014

Gamper, J., & Knapp, J. (2002). A review of intelligent CALL systems. *Computer Assisted Language Learning, 15*(4), 329–342. doi:10.1076/call.15.4.329.8270

Gass, S. M. (2017). *Input, interaction, and the second language learner* (2nd ed.). Routledge Linguistics Classics Edition. doi:10.4324/9781315173252

González-Lloret, M. (2016). Technology and task-based language teaching. In S. Thorne & S. May (Eds.), *Language, education and technology. Encyclopedia of language and education* (3rd ed.). Cham, Switzerland: Springer. doi:10.1007/978-3-319-02328-1_16-1

Gupta, P. & Schulze, M. (2012). Human language technologies (HLT). Module 3.5. In G. Davies (Ed.), *Information and Communications Technology for Language Teachers (ICT4LT),* Slough, Thames Valley University [Online]. Retrieved from http://www.ict4lt.org/en/en_mod3-5.htm

Handley, Z. (2009). Is text-to-speech synthesis ready for use in computer-assisted language learning? *Speech Communication, 51*(10), 906–919. doi:10.1016/j.specom.2008.12.004

Heift, T., & Schulze, M. (2007). *Errors and intelligence in CALL: Parsers and pedagogues*. New York, NY: Routledge.

Huang, S., Liu, C., & Gao, Z. (2005). Computer-assisted item generation for listening cloze tests and dictation practice in English. In *Proceedings of ICWL*, 197–208. 10.1007/11528043_19

Liakin, D., Cardoso, W., & Liakina, N. (2017). The pedagogical use of mobile speech synthesis (TTS): Focus on French liaison. *Computer Assisted Language Learning*, *30*(3-4), 3–4, 325–342. doi:10.1080/09588221.2017.1312463

Little, D., Dam, L., & Legenhausen, L. (2017). *Language Learner Autonomy: Theory, Practice and Research*. UK: Multilingual Matters.

Long, M. H. (1991). Focus on form: A design feature in language teaching methodology. In K. de Bot, D. Coste, R. Ginsberg, & C. Kramsch (Eds.), *Foreign language research in crosscultural perspectives* (pp. 39–52). Amsterdam, The Netherlands: John Benjamins. doi:10.1075ibil.2.07lon

Lopes, J., Trancoso, I., Correia, R., Pellegrini, T., Meinedo, H., Mamede, N., & Eskenazi, M. (2010). Multimedia Learning Materials. In *Proceedings IEEE Workshop on Spoken Language Technology SLT*, Berkeley, CA. pp. 109–114.

Luckin, R., Holmes, W., Griffiths, M., & Forcier, L. B. (2016). *Intelligence Unleashed. An argument for AI in Education*. London, UK: Pearson.

McGuirk, R. (2015). *Exploration of the use of Irish language synthesis with a screen reader in the teaching of Irish to pupils with vision impairment*. (Unpublished MPhil dissertation). Trinity College, Ireland.

Nerbonne, J. A. (2003). Natural language processing in computer-assisted language learning. In R. Mitkov (Ed.), *The Oxford handbook of computational linguistics* (pp. 670–698). Oxford.

Ní Chasaide, A., Ní Chiaráin, N., Berthelsen, H., Wendler, C., & Murphy, A. (2015). Speech technology as documentation for endangered language preservation: The case of Irish, *Proceedings of the 18th International Congress of Phonetic Sciences*, Glasgow, UK. pp. 1-5.

Ní Chasaide, A., Ní Chiaráin, N., Wendler, C., Berthelsen, H., Murphy, A., Barnes, E., & Gobl, C. (2019). Leveraging Phonetic and Speech Research for Irish Language Revitalisation and Maintenance. Ifn S. Calhoun, P. Escudero, M. Tabain & P. Warren (Eds.) Proceedings of the International Congress of Phonetic Sciences, Melbourne, Australia 2019 (pp. 994-998). Canberra, Australia: Australasian Speech Science and Technology Association Inc.

Ní Chasaide, A., Ní Chiaráin, N., Wendler, C., Berthelsen, H., Murphy, A., & Gobl, C. (2017). *The ABAIR initiative: Bringing spoken Irish into the digital gpace* (pp. 2113–2117). Stockholm, Sweden: INTERSPEECH.

Ní Chiaráin, N. (2014). Text-to-speech synthesis in computer-assisted language learning for Irish: Development and evaluation. [Doctoral thesis] CLCS, Trinity College, Ireland.

Ní Chiaráin, N. & Ní Chasaide, A. (2015). Evaluating synthetic speech in an Irish CALL application: Influences of predisposition and of the holistic environment. In *Proceedings of SLaTE 2015: 6th Workshop on Speech and Language Technologies in Education*, pp. 149–154.

Ní Chiaráin, N. & Ní Chasaide, A. (2016a). The Digichaint interactive game as a virtual learning environment for Irish. In *Proceedings of CALL communities and culture - short papers from EUROCALL 2016,* Limassol, Cyprus, Research-publishing.net, pp. 330–336. 10.14705/rpnet.2016.eurocall2016.584

Ní Chiaráin, N. & Ní Chasaide, A. (2016b, November) Faking intelligent CALL: the Irish context and the road ahead. *Proceedings of the Joint Workshop on NLP for Computer Assisted Language Learning and NLP for Language Acquisition at SLTC*, Umeå, Sweden, (130), Linköping University Electronic Press, Linköpings universitet, pp. 60–65.

Ní Chiaráin, N. & Ní Chasaide, A. (2016c). Chatbot technology with synthetic voices in the acquisition of an endangered language: Motivation, development and evaluation of a platform for Irish. In *Proceedings of the Tenth International Conference on Language Resources and Evaluation (LREC 2016)*, Portoro, Slovenia, May 23-28, 2016, European Language Resources Association (ELRA), pp. 3429-3435.

Ní Chiaráin, N. & Ní Chasaide, A. (2019). Integrating dialogue within an iCALL platform for Irish. *Proceedings of DiGo 2019: Dialog for Good - Workshop on Speech and Language Technology Serving Society*. Stockholm, Sweden (forthcoming).

Pellegrini, T., Costa, Â., & Trancoso, I. (2012). Less errors with TTS? A dictation experiment with foreign language learners. In *Proceedings Thirteenth Annual Conference of the International Speech Communication Association.*

Pica, T. (1994). Research on negotiation: What does it reveal about second-language learning conditions, processes, and outcomes? *Language Learning*, *44*(3), 493–527. doi:10.1111/j.1467-1770.1994.tb01115.x

Pillay, H. K. (1994). Cognitive load and mental rotation: Structuring orthographic projection for learning and problem solving. *Instructional Science*, *22*(2), 91–113. doi:10.1007/BF00892159

(Forthcoming). Rialtas na hÉireann (2019). Plean Digiteach don Ghaeilge 2019-2029. *Digital Plan for the Irish Language.*

Scannell, K. (2013). An Gramadóir: An open source grammar checking engine, Version 0.70. Retrieved from https://cadhan.com/gramadoir/

Schmidt, R. (1990). The role of consciousness in second language learning. *Applied Linguistics*, *11*(2), 129–158. doi:10.1093/applin/11.2.129

Schmidt, R. (2001). Attention. In P. Robinson (Ed.), *Cognition and second language instruction* (pp. 3–32). Cambridge, UK: Cambridge University Press. doi:10.1017/CBO9781139524780.003

Schulze, M. (2008). AI in CALL – Artificially inflated or almost imminent? *CALICO Journal*, *25*(3), 510–527. doi:10.1558/cj.v25i3.510-527

Sha, G. (2010). Using TTS voices to develop audio materials for listening comprehension: A digital approach. *British Journal of Educational Technology*, *41*(4), 632–641. doi:10.1111/j.1467-8535.2009.01025.x

Skehan, P. (1998). *A cognitive approach to language learning*. Oxford, UK: Oxford University Press.

Sykes, J. M., Oskoz, A., & Thorne, S. L. (2008). Web 2.0, synthetic immersive environments, and mobile resources for language education. *CALICO Journal*, *25*(3), 528–246. doi:10.1558/cj.v25i3.528-546

Wik, P. (2011). The virtual language teacher: Models and applications for language learning using embodied conversational agents. [PhD Thesis]. KTH Royal Institute of Technology.

Wik, P., & Hjalmarsson, A. (2009). Embodied conversational agents in computer assisted language learning. *Speech Communication*, *51*(10), 1024–1037. doi:10.1016/j.specom.2009.05.006

ADDITIONAL READING

Arnold, N., & Ducate, L. (2019). *Engaging Language Learners through CALL: From Theory and Research to Informed Practice*. Sheffield, UK: Equinox Publishing.

Eskenazi, M. (2009). An overview of spoken language technology for education. *Speech Communication*, *51*(10), 832–844.

Hamp, E. (1951). Morphophonemes of the Keltic Mutations. *Language*, 27.

Heift, T., & Schulze, M. (2012). *Errors and Intelligence in Computer-Assisted Language Learning: Parsers and Pedagogues*. London, UK: Taylor & Francis Ltd.

Thomas, M., Reinders, H., & Warschauer, M. (Eds.). (2014). *Contemporary Computer-Assisted Language Learning*. London, UK: Bloomsbury Publishing PLC.

Thorne, S. L., & May, S. (Eds.). (2017). *Language, Education and Technology, Encyclopedia of language and education*. Cham: Springer International.

KEY TERMS AND DEFINITIONS

Gaeltacht Areas: Regions in which the Irish language is the predominant vernacular. These areas are mainly located on the western seaboard of Ireland. These zones have official recognition from the Irish Government as particular cultural and socioeconomic zones.

Initial Mutation: A feature common to all Celtic languages where initial consonants undergo 'mutations' (consonantal alternations, e.g. of stops and fricatives). Specific grammatical contexts trigger this process e.g. the formation of a question or a negative statement. Take, for example, the case of the verb 'dún' (*to close*) in the present tense: '**Dún**ann sí' *(statement: 'she closes');* 'An **ndún**ann sí?' *(question: 'does she close?');* 'Ní **dhún**ann sí' *(negative statement 'she does not close').* Verbal forms in Irish can have as many as 42 inflected forms.

Intelligent Computer-Assisted Language Learning (iCALL): The intersection of Speech Technology and Natural Language Processing with Computer-Assisted Language Learning. iCALL uses techniques from the field of artificial intelligence in order to bring 'understanding' of language to the development of CALL tools resulting in the semblance of 'genuine' interactivity between the learner and the system.

Morphophonemics: The branch of linguistics that studies the relationship between written (morphological) and aural (phonological or phonetic) processes.

Natural Language Processing: A field of study, which brings together different disciplines such as linguistics, computer science and artificial intelligence in order to computationally process human language for specific purposes, e.g. generating 'intelligent' feedback in language learning activities based on unpredictable learner input.

Personalisation in CALL: The process by which a language learning environment tailors itself to meet an individual learner's needs or preferences, e.g. providing individualised feedback on a particular learner's written input.

Text-to-Speech Synthesis: Refers to artificially generated speech produced by computer, usually on the basis of text input, i.e., a text-to-speech system (TTS).

Chapter 8
There Is Nothing to See. Or Is There?
Visualizing Language Through Augmented Reality

Josef Buchner
St. Gallen University of Teacher Education, Switzerland & University of Duisburg-Essen, Germany

Julia Weißenböck
University of Salzburg, Austria

ABSTRACT

Augmented Reality has the potential of transforming teaching. By applying this technology themselves, students progress from consumers of technology to producers of their own digital content, and they can make it available to a large audience and create an immersive learning experience. After a language trip to Scotland, teachers and two classes of 17-year-old students from a school in Salzburg, Austria, decided to present the outcome of their project work from their stay abroad in a more interactive and innovative way. Instead of inviting parents and friends to an evening of PowerPoint, the students designed AR posters and visitors were invited to an interactive gallery walk during which they had the chance to explore thematic posters related to Scotland. While at first glance the posters seem rather bland, digital content, which the students produced, opens up by scanning the pictures on the poster with a special AR app. This project successfully combines language and digital skills and has shown to motivate students to engage even deeper with their topic.

INTRODUCTION

There is a long tradition in research on how to integrate and/or use technology for teaching and learning. Nowadays the debate is more current than ever due to the constant availability of multimedia learning opportunities via smartphone and the Internet. Recent studies show that almost all young people are in possession of a mobile phone, have access to the world wide web and use online-platforms' like

DOI: 10.4018/978-1-7998-1097-1.ch008

YouTube for self-organized learning activities (European Union, 2017; Ikrath & Speckmayr, 2016; Medienpädagogischer Forschungsverbund Südwest (mpfs), 2017; Rat für Kulturelle Bildung, 2019). So, for students, the use of technology has already become part of their everyday lives. At school or university, this awareness is usually still lacking. Although numerous meta-analyses have repeatedly confirmed that the use of computer technologies can lead to learning success, satisfaction and increased motivation taking into account pedagogical and didactic requirements, their use is the exception rather than the rule (Bangert-Drowns, Kulik, & Kulik, 1985; Fazal, 1996; Hattie, 2008; Kulik & Kulik, 1991). This is also because digital media are mostly understood as pure instructional media. The potential for creativity, communication or collaboration then falls short (*Partnership for 21. Century Learning*, 2015). A relatively new approach is that of digital skills, i.e. those skills that young people need to grow up in the increasingly digital world. Here, learning is no longer only with media, but also about media. If several perspectives such as technological, social and application perspectives are taken into account, young people can mature from mere consumers of digital media content to critically-reflected personalities. One way to promote this is to let learners slip into the role of designers of their own media products (EU Science Hub, 2018; European Union, 2018; Gesellschaft für Informatik, 2016). In the literature this approach is referred to as *Learning by Design* or design-based learning. This teaching method is strongly oriented towards problem-based learning or approaches to research-based learning (Kolodner, Camp, et al., 2003). Initially applied strongly in science teaching, today there are also many studies on learning by design in the field of foreign language teaching. The Erasmus+ project PALM (Promoting Authentic Language Acquisition in Multilingual Contexts, (Mewald & Wallner, 2018)) is particularly worth mentioning here. As part of this project, children, teenagers and teachers from eight different countries created digital learning materials for the respective first languages, but also for the languages to be learned at school. The main focus was on texts, videos and quizzes, which can now be accessed via a homepage. The aspect of authentic learning and language use deserves special mention here. Thus, no artificial situations were created, instead authentic situations of language use were used for the design process. (Bajner, 2018; Mewald & Wallner, 2018; Mewald, Wallner, & Buchner, 2018). Authenticity and the feeling of working on relevant tasks with the help of learning by design are only two of the advantages of this approach which are also repeatedly mentioned in literature (Kearney & Schuck, 2006; Schuck & Kearny, 2006).

Most of the implementations of learning by design concentrate on the production of videos, texts, images and, in rare cases, digital games. In this chapter we would like to present a project that, using a relatively new technology, enabled learners to produce linguistic artifacts: Augmented Reality.

First, we will provide an overview of the research results of learning by design implications, including existing projects with augmented reality. The next section presents the AR Scotland project in detail. The curriculum, competences to be acquired and the learning by design process are described. Subsequently, we present results from the accompanying research of this learning experience, the focus is on the motivation of the learners as well as the acceptance of Augmented Reality as (learning) technology. The outlook on open research questions and a conclusion will complete this chapter.

BACKGROUND

Designing Traditional Media

The Learning by Design approach goes back to constructivist approaches that put learners in the focus of schooling and teaching. Students should be actively involved in the learning process and, above all, enabled to make their own decisions and, if necessary, to revise them. Kolodner et al. (2003) use problem-based learning, case-based reasoning and research-based learning for their developed model. The design process is understood as an iterative process that is started, adapted and completed by several students working together on the basis of a problem. The method is independent of analog or digital media, the learning process should be the focus. With the advent of the Internet and the massive spread of digital devices, the production of digital artefacts as part of this teaching method has increasingly been added. If media products are created in the context of education, they are also referred to as learner-generated content. Meanwhile, there are many implementations of this form of teaching in the most diverse subjects. The authenticity of the learning experience, working on real problems, the promotion of interdisciplinary competences such as digital abilities and skills, motivational aspects, satisfaction, teamwork and the possibility of personalized learning are regarded as valuable potentials (Frydenberg & Andone, 2016; Kearney, 2011; Kolodner, Gray, & Fasse, 2003; Orús et al., 2016). These potentials were primarily verified by the authors mentioned by creating videos. Videos are of course very good because they are close to the life of the students. YouTube is also a platform that can make the products available to a broad public. For example, Kearney and Schuck (2006) have had students create videos for partner schools and infomercials in English and French lessons. In their research, they refer above all to the naturalness of this task, the promotion of communication and the expansion of literacy skills. Orús et al. (2016) also used YouTube to make the created media products visible. Students created their own videos for the field of marketing. The results of this study lead to the conclusion that the creation of videos not only had a positive effect on the acquisition of knowledge, but also on learner satisfaction. Students in higher education also benefit from this method. Zahn et al. (2014), for example, have found a positive influence on learning success and Stevenson, Länsitie, Kogler and Bauer (2015) have given prospective teachers a collaborative video challenge as a task. It was found that the teachers entered into a strong dialogue with each other, worked together internationally and also exchanged views on how videos can enrich teaching. Henderson et al. (2010) showed in their study that even young learners in primary education are able to create short movies. Outcomes of this research in the English classroom are pedagogical changes on the side of the teachers, higher learner autonomy and learning outcomes as well as a positive influence on reflectivity. A few implementations make use of the possibilities of game design. Gershenfeld (2011) argues that players should also become designers of computer games. An implementation for the promotion of media and information literacy skills can be found by Costa, Tyner, Henriques and Sousa (2017). Learners between the age of 9-14 developed their own games on the mentioned topics, the promotion of critical thinking and participation are named as the most important outcomes by the authors.

Designing Immersive and Augmented Media

Not only the proliferation of devices has changed teaching and learning in recent years, but also the technological evolution of software. This is especially true for Augmented and Virtual Reality (VR). While a few years ago bulky devices were needed for the display of such immersive or augmented content, today this is already possible with inexpensive cardboard glasses or for augmented reality even via smartphone or tablet. Wössner (2019) recommends the *CoSpacesEdu* tool to put learners in the role of designers of their own immersive virtual worlds. The vocabulary work as well as the grammar can be worked on. Other learners can then explore these virtual spaces and continue learning on their own. The potentials of VR for the learning of language are given, for an overview we recommend Andujar and Buchner (2019). We now focus on Augmented Reality (AR), the computer-aided extension of reality. Real objects or images always remain part of the learning medium. With an app, analogue materials can then be extended by digital multimedia content. A static image then becomes, for example, an animation or a graphic in a book becomes a simulation that visualizes the course of a content (Azuma, 1997; Billinghurst Mark, Kato Hirakazu, & Poupyrev Ivan, 2001; Klopfer & Sheldon, 2010). Many studies have already shown that AR can have positive effects on motivation, learning success, learner satisfaction and the understanding of learning content. Above all, AR can be used as a supportive medium that makes the invisible visible (Akçayır & Akçayır, 2017; Radu, 2014; Wu, Wen-Yu Lee, Chang, & Liang, 2013).

Most studies focus on how AR can be used for teaching and learning. Only a few focalise on students' own production of AR content. Mathews (2010) can be mentioned here as an example. In his study, learners had the task of redesigning the school's immediate neighborhood. To do this, they first looked at the squares and places, took photos and videos, and developed ideas on how the appearance of the places could be changed. After this phase, the students attended an AR-game-design course and acquired the skills necessary to create an AR-game. All contents of the AR-game were created by the students themselves. Location-based technology was used. This means that the AR elements are displayed when you open the game at a certain location. Mathews (2010) sums up that this project not only promoted literacy skills among the students, but also created a sense of social responsibility. The students felt responsible for their immediate neighborhood and wanted to make a lasting difference. Here, too, the author refers to the naturalness of the learning experience. For the language classroom the study of Ho, Nelson and Müeller-Wittig (2011) can be mentioned. In this implementation, the students designed a virtual AR museum and independently created the necessary artefacts. The authors refer in particular to the possibility of collaborative work and the presentation of multimodal content. The use of AR has also not constricted the realization of the language curriculum, but has been able to promote not only technical but also interdisciplinary competences. This study also investigated the extent to which AR was accepted as a technology and whether AR or the creation of AR was perceived as complicated or relatively simple. The feedback of the participants was very positive to a large extent. Buchner and Höfler (2018) and Di Serio, Ibanez and Kloos (2013) use a similar approach to teach grammatical rules and acquire new vocabulary. In addition, AR can simulate an authentic language exchange between well-known figures from the children and youth world and the pupils.

This study also inspired our project that will be described in the next section. We will therefore also address the question of the acceptance of AR in our research project. Furthermore, we will present results that reflect the motivational effect of learning by design implementation with AR.

THE AR SCOTLAND PROJECT

Legal Background and Frameworks

Austrian Curriculum

In the general part of the Austrian curriculum the use of technology and digital media is explicitly called for, irrespective of the subject. The didactic potential of digital media should be used for teaching, digital competence should be fostered and ideally students create their own digital content using technology. Furthermore, the curriculum specifically mentions product-oriented approaches. Project work is to be conducted by using computers, research to be conducted online, results to be presented using digital media and communication as well as cooperation is to be documented by means of ICT technology, also in a foreign language. Additionally, individualisation and differentiation are mentioned as important steps for students to become independent learners, all of which is covered in the Scotland AR project (BMBWF (Bundesministerium für Bildung, Wissenschaft und Forschung), 2018)

CEFR

While the Austrian curriculum naturally presents the primary legal framework, the Common European Framework of References for Languages (CEFR) is equally important and also serves as the basis for the Austrian curriculum in terms of language competence. In year 3 of the upper forms, 17-year-old students in Austria work at a lower B2 level in terms of language competence. As outlined in the description of the project, depending on what each group (or each student) decided to include in their AR poster, various language skills were trained. Since it was a prerequisite for each student to include at least 2 minutes of speaking, everyone trained their spoken production, such as sustained monologue and addressing audience. In terms of overall spoken production students at lower B2 level can do the following:

Can give clear, detailed descriptions and presentations on a wide range of subjects related to his/her field of interest, expanding and supporting ideas with subsidiary points and relevant examples (Council of Europe, 2018, p. 69).

More specifically, when addressing audiences students can "give a clear, systematically developed presentation, with highlighting of significant points, and relevant supporting detail" (ibid, p. 74), which, in the course of this project, every student had to do. Additionally, linguistic features play a role when language is used, skills such as general linguistic and vocabulary range, grammatical accuracy and vocabulary control were equally practiced, just as phonological features, e.g. overall phonological control, sound articulation and prosodic features. Spoken fluency, part of the pragmatic features outlined in the CEFR, is a noticeable feature in both audio and video included in the AR posters. Some students also did an interview with each other or with people in Scotland for which they needed their spoken interaction skills, especially interviewing and being interviewed. It can already be seen that in terms of speaking alone, this project offers a myriad of potential.

Moreover, many students also included text in their AR posters and thus further honed their writing skills. The B2 overall written production descriptor in the CEFR outlines that students *[c]an write clear, detailed texts on a variety of subjects related to his/her field of interest, synthesising and evaluating information and arguments from a number of sources (ibid, p. 75),*

providing another link to the AR project by emphasising the importance of using information from a number of sources for the final product.

Finally, as part of the research that the students conducted during the project, they also needed to apply their receptive skills, reading and listening, in order to extract information from various sources.

digi.komp12

Autumn 2018 saw a significant change in the Austrian education system: a new curriculum for basic digital competence was introduced. This curriculum outlines what students should be able to do after 8 years of school (14 years of age). While there is no curriculum for the upper forms, there is, however, a competence model called digi.komp12, which contains descriptors of ICT competence after 12 years of school. Especially area 3 "Angewandte Informatik" [applied computer science] („digi.komp: Informatik 5. Klasse", n. d.) includes descriptors of competence that students need for this project, such as:

3.1. Produktion digitaler Medien
[production of digital media]

Ich kann Standardsoftware zur schriftlichen Korrespondenz, zur Dokumentation, zur Publikation von Arbeiten, zur multimedialen Präsentation sowie zur Kommunikation sicher anwenden.

[I can safely use standard software for written correspondence, documentation, publication and multi-modal presentation, as well as communication]

Ich kann Arbeitsergebnisse zusammenstellen und multimedial präsentieren.

[I can summarise results and present them in a multimodal way]

Ich kann digitale Medien in Form von Text, Ton, Bildern und Filmen sachgerecht bearbeiten, produzieren und publizieren.

[I can edit, produce and publish digital media as text, audio, pictures and film] (ibid)

digi.kompP

Along with the implementation of a curriculum for basic digital skills and accompanying competence models for students came a competence models for teachers as well. The so-called digi.komP outlines eight areas of competence for professional development. Teachers wanting to work on the Scotland AR project (or create a similar one) need competence in the following areas:

D1: Digital Lehren und Lernen [Teaching and Learning Digitally]

Ich kann im Unterricht passende und unpassende soziale Settings bei der Nutzung digitaler Technologien bedenken.

[I can consider suitable and unsuitable settings for the use of digital media in my lessons.]

Ich kann unter Berücksichtigung unterschiedlicher technischer und organisatorischer Voraussetzungen mediengestützte Unterrichtsszenarien durchführen und evaluieren.

[I can apply and reflect on e-learning/e-teaching scenarios, considering technical and organizational pre-requisites.]

Ich kann digitale Unterstützungssysteme zur Unterrichtsgestaltung verwenden.

[I can use digital support for designing my lessons]

Ich kann mediengestützte Projektarbeiten initiieren, begleiten und evaluieren.

[I can initiate, guide and evaluate digital project work.]

E1: Digital Lehren und Lernen im Fach [teaching and learning digitally within one's subject]

Ich kann mein didaktisches Wissen zum Umgang mit digitalen Medien mit den Spezifika meines Faches vereinen.

[I can combine my didactic knowledge about the usage of digital media with subject-related specifics.]

Ich kann mit Hilfe von digitalen Medien Lernsettings gestalten und dabei auf die fachspezifischen Besonderheiten Rücksicht nehmen.

[I can design e-learning settings and keep in mind subject-related specifics.]

Ich kann aktivierenden, differenzierten und kompetenzorientierten Fachunterricht mit digitalen Medien realisieren.

[I can carry out active, differentiated and competence-based lessons using digital media.]

Ich kann interaktive und soziale Lernformen im Fach mit Hilfe digitaler Medien fördern.

[I can foster interactive and social learning scenarios with the help of digital media.]

(Brandhofer et al. 2016)

As can be seen from the information presented, the AR Scotland project is firmly based on legal frameworks relating to the two main content areas it encompasses: language and digital skills.

Project Description

History

It is part of the school's profile that students in the 3rd year of the upper forms at BG Zaunergasse, Salzburg, go on a language week to an English-speaking country. Prior to this week, the students are grouped and assigned topics related to the country and culture they are visiting. It is the students' task then to gather as much information as possible before, during and after the language week and then present the outcomes of their findings to family and friends. In the past the preferred mode of presentation was PowerPoint. Family and friends were invited to a themed evening at school during which all of the groups, one after the other, presented their outcomes. While this procedure always worked flawlessly and doubtlessly fulfilled its purpose, students and teachers alike started to raise criticism as to whether this procedure was really necessary since these evenings could become rather drawn out and the presentations were mostly the same, the topic being the only thing that set them apart from each other.

The New Project

Based on personal experiences and the feedback received, the English teachers of this year's year 3 classes decided to take a different turn on things and use augmented reality for the project about their language trip to Scotland. It was hoped to re-create more interest in the outcomes of and more motivation for working on the project by applying an innovative and interactive form of presentation. Instead of creating PowerPoint presentations, the students' task was to design augmented reality posters, where the augmented content holds the outcomes of the students' research on the respective topic. These multimodal posters contained "two or more ways, or mode[s], of communication, so that their combined meanings as a whole is greater than either mode separately or their simple combination" (Dressman, 2010, p. 71). This was made possible through a cooperation with the St. Gallen University of Teacher Education, which generously provided the students with access to the augmented reality programme XTend. The designed posters included pictures and text related to the topic the students had been assigned, and while at first glance these posters might not hold a lot of information, each picture was augmented with digital content. Once the respective pictures were scanned with XTend, an augmented reality app, digital content appeared on the screen "provid[ing] a composite view for the user with a combination of the real scene viewed by the user and computer generated virtual scenes" (Kesim & Ozarslan, 2012, p. 298). This digital content was created by the students themselves and ranged from picture galleries, as well as audio and video files to weblinks providing additional information. In doing so the posters became a "dynamic source of information" (Kesim & Ozarslan, 2012, p. 301). Figures 1-3 are examples of such posters.

Procedure

As a first step students had to decide which pictures they were going to use as so-called "markers" (i.e. the pictures on the posters that could later be scanned). Secondly, using the AR programme, these markers were then augmented with digital content for which the programme offers various possibilities, e.g. weblinks, audio, video, picture, picture gallery, AR areas (more than one content per marker) as well as 3D objects and panorama pictures. What they wanted to include as augmented content was completely

Figure 1. Crimes & Criminals (© 2019, Julia Weissenboeck, used with permission)

Figure 3. Mary Queen of Scots (© 2019, Julia Weissenboeck, used with permission)

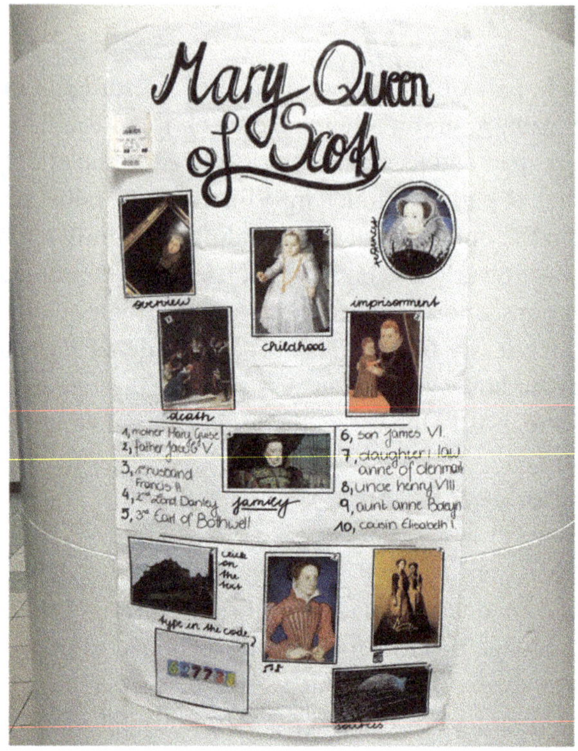

Figure 2. Scottish Music and Dance (© 2019, Julia Weissenboeck, used with permission)

up to the students, which is one of the main differences to PowerPoint. While PowerPoint provides a very linear, sequential structure with relevant information mainly presented in text and picture only, the augmented reality project emphasized a more design-based approach which means that students are better able to gain understanding of the world around them if they themselves have the opportunity to create real objects, the ‘real objects’ being the posters and augmented content in various forms (Papert & Harel, 1991). Through actively engaging the students in the creation of process, they “become designers with Augmented Reality in order to develop higher order thinking” (Bower, Howe, McCredie, Robinson, & Grover, 2014, p. 7) with the “teacher [mainly] act[ing) as a facilitator” and students working on their own, using the respective technology (Kamarainen et al., 2013, p. 554). Students actively using technology is also fully in line with the Austrian curriculum that specifically calls for the integration of new technology into teaching (see above). As van’t Hooft points out:

digital technologies should be integrated in ways that provide students with learning experiences that go beyond what can be done without technology. One way to do this is by turning students into producers of information and having them create content rather than simply consume (van’t Hooft, 2016, p. 99).

Outcome

The final posters were presented to the public as part of an interactive gallery walk with the title “Scotland? There is nothing to see. Or is there?”. Visitors were asked to download the free app Xtend prior to their visit to the exhibition. Then the visitors could walk from poster to poster, scanning the marker pictures that seemed interesting to them and then viewing the augmented content on their phones or tablets, while the students acted as guides and ushers. This interactive exhibition well shows one of the big advantages of using technology and mobile devices. As digital devices are part of today’s everyday life, they “offer substantial power in taking learning opportunities outside the four walls of the classroom“ (McQuiggan, Kosturko, McQuiggan, & Sabourin, 2015, p. 11) and the used technology “breaks down classroom walls. [...] easily connecting teachers, students, families” (Saumell, 2016, p. 17). This project thus successfully and meaningfully integrated technology in teaching by actively involving the students in the creation of content by using their language and digital skills.

Figures 4-9 offer an insight into the results of the project. All of these pictures have been augmented using the programme Xtend. In order to view the digital content, follow these steps:

1. download the free app “Xtend” (Xtend interactive) from your app store
2. install it on your mobile device and when prompted, allow the app access to your camera
3. open the app, finish the brief tutorial and click on “AR scan”
4. hold your mobile device over the picture, make sure that it fills the whole frame
5. digital content will appear (might take a while)

Figure 10 provides a link to a video about the whole project.

Figure 4. Bernt Rostad (https://www.flickr.com/photos/brostad/6260703154) "Haggis", https://creative-commons.org/licenses/by/2.0/legalcode

Figure 5. howardmizuki (https://pixabay.com/photos/library-college-back-to-school-948965/) "library 948965_1920", https://pixabay.com/service/license/

EVALUATION OF THE AR SCOTLAND PROJECT

For the evaluation of our project we follow already existing research results on Learning by Design and Augmented Reality. First, we present the results from the area of motivation, then we check the acceptance of AR in general, the learning possibilities with AR as well as the user-friendliness of AR for the design of own learning artefacts. A total of 40 learners took part in the project, 27 of whom were female (67.5%) and 13 males (32.5%). The average age of our volunteers is 16.8 years with two abstentions.

Figure 6. Paul Moreelse (https://commons.wikimedia.org/wiki/File:Paulus_Moreelse_-_Girl_-_1623.jpg) "Girl wearing a Gold Chain" (https://creativecommons.org/publicdomain/zero/1.0/legalcode)

Figure 7. Ungry Young Man (https://www.flickr.com/photos/markusunger/41677929370) "Sir William Wallace Memorial", https://creativecommons.org/licenses/by/2.0/legalcode

Figure 8. Introduction (© 2019, Nadine Kurz, Elena Schwarz, Amna Syed, used with permission)

Figure 9. SJ van Schaik (https://commons.wikimedia.org/wiki/File:Whisky_distilleries_and_regions_in_Scotland.svg) "Whisky distilleries and regions in Scotland", https://creativecommons.org/licenses/by-sa/4.0/legalcode

Figure 10. Project video. Link to a video of the whole project, visualising how the scanning works and how digital content is made accessible.

Results

Motivation

As outlined above, previous studies on learning by design have shown that learners have found such design-based learning experiences extremely motivating. We have now used an adapted version of the *Intrinsic Motivation Inventory* (Ryan, 1982), the *Short Scale Intrinsic Motivation* (KIM, Wilde, Bätz, Kovaleva, & Urhahne, 2009). The Intrinsic Motivation Inventory has been used in several studies examining the motivational perception of participants during and after tasks. It has been revised and complemented by researchers all over the world (Center for Self-Determination Theory, 2019). One version is the Short Scale Intrinsic Motivation (KIM) by Wilde et al. (2009). They designed the questionnaire for the evaluation of extracurricular learning venues with the hint that the questionnaire can be adapted for the evaluation of open and action-oriented teaching scenarios (Wilde et al., 2009, p. 42). Intrinsic motivation is a qualitative expression of motivation that goes back to the self-determination theory of motivation (Deci & Ryan, 2000; Ryan & Deci, 2000). Interest and enjoyment, perceived competence and perceiving freedom of choice are characterized as factors for experiencing intrinsic motivation. Pressure and fear are inhibiting factors. We collected the data for these scales via a seven-point Likert-scale (1 = not at all, 7 = fully agree). For example, the scale Interest/Enjoyment was raised through the question “Creating content with Augmented Reality was fun for me” and the scale Perceived Choice with “I could choose how I did it when creating the Augmented Reality content” (for the whole questionnaire see Appendix 1). Table 1 shows Cronbach’s Alpha, the mean values with standard deviation and the number of items collected for each scale.

As the values in Table 1 show, the approach of Learning by Design with AR has led to high approval for the areas of interest/enjoyment, perceived competence and perceived freedom of choice. These values suggest, with reference to the self-determination theory of motivation (Deci & Ryan, 2000; Ryan & Deci, 2000), that students found this form of learning highly motivating. Only the scale pressure/tension narrows the result a little, since the average value is higher here compared to other studies (e.g. Buchner & Zumbach, 2018). This can certainly be explained by the fact that the creation of the learning artefacts was a must-do-task and served as a basis for assessment.

Table 1. Results of the short scale intrinsic motivation (N = 40; higher scores show agreement)

Scale	Mean (SD)	Cronbachs' Alpha	Items
Interest/Enjoyment	4.40 (1.62)	.94	3
Perceived Competence	4.76 (1.25)	.85	3
Perceived Choice	5.04 (1.42)	.89	3
Pressure/Tension	3.34 (1.66)	.75	3

Technology Acceptance

The use of video, text or games in the context of Learning by Design has to do with the fact that these technologies have already been anchored in the public domain for many years. With newer technologies in particular, it is important to ask to what extent they are accepted by actors in society. Research on technology acceptance is mainly based on the work of Davis (1989) and the further development of his TAM model by Venkatesh and Davis (2000), Venkatesh, Morris, Davis and Davis (2003) and Venkatesh and Bala (2008). On the basis of this work, Kothgassner et al. (2013) have developed the *Technology Usage Inventory* (TUI), which now collects essential psychological influencing factors of technology use and acceptance, allows a prediction of the use independent of the technology as well as an evaluation of new technologies after their use. A central question for this project is to what extent the presentation of linguistic and written artifacts with AR is accepted, AR as technology is perceived as useful or accessible, and above all whether creating your own AR content is complicated or easy. To answer these questions, we used an adapted version of the *Technology Usage Inventory* (TUI, Kothgassner et al., 2013). Examples of the questionnaire are "I want to know more about AR technology", "The use of Augmented Reality was complicated" and "I think, Augmented Reality is dangerous". The participants answered these and the other 13 questions on a seven-point Likert-scale (for the whole questionnaire see Appendix 2). Table 2 shows the mean values with standard deviation for all scales, Cronbach's alpha and the number of items.

The results in Table 2 show us that the participating students are interested in technology and rate the accessibility of AR as good. Interesting is their assessment of usability. Here the values show that creating AR content was perceived as relatively easy. The skepticism towards new technologies is low among these test subjects.

Table 2. Results of the adapted version of the technology usage inventory, TUI (N = 40; higher scores show agreement)

Scale	Mean (SD)	Cronbachs' Alpha	Items
Interest in Technology	4.40 (1.25)	.85	3
Usability	4.98 (1.34)	.90	3
Usefulness	2.47 (1.36)	.88	4
Skeptical of Technology	2.13 (0.76)	.42	3
Accessibility	3.56 (1.20)	.54	3

Limiting it must be noted that Cronbach's alpha for the scale skeptical of technology and the scale accessibility is smaller than .7. This means that the internal consistency of these two scales must be questioned. Further use of the questionnaire will be necessary here in order to check Cronbach's alpha again.

In order to be able to say more about the relationship between motivation and technology acceptance, we calculated a Spearman correlation.

The interest in technology and the perception of usefulness correlate significantly ($r = 0.318$, $p = 0.045$) with interest/enjoyment. Usability also correlates significantly ($r = 0.315$, $p = 0.048$) with interest/enjoyment and further highly significantly ($r = 0.562$, $p = 0.000$) with perceived competence. A negative correlation was also found for usability for the scale pressure/tension ($r = -0.405$, $p = 0.010$). Skepticism towards new technologies correlates negatively with the scale interest/enjoyment ($r = -0.315$, $p = 0.048$).

Discussion

Our evaluation confirms existing research results on the approach of Learning by Design. The active involvement in designing content is perceived as extremely motivating by the learners. Self-determination theory served as the basis for our study. The creation of linguistic and written products as AR elements was perceived by the students as an interesting and joyful activity. They felt competent and were allowed to act autonomously. Both factors are important characteristics for intrinsic motivation. At the same time, the students certainly also felt a certain pressure, as it was not possible to put together a PowerPoint presentation or similar for the presentation of their results at short notice. Rather, they had to get involved from the very beginning, create ideas and finally put their ideas into practice. This can simply overburden some learners, especially if they have not yet gained experience with project-based forms of learning.

AR as a technology is accepted by the students and is largely appreciated positively. The usefulness does not seem to have completely convinced this group, but the interest in technology in general and the ease of use for their own creation have met with high approval.

The calculation of the correlations shows that when using AR in the context of design-based tasks, usability must be emphasized. This correlates strongly with the perceived pleasure. If learners are interested in technology, working with it also makes them happy. It has to be emphasized that usability also contributes to the experience of competence. If the learners often fail due to the operation of a certain technology, this could also have a negative effect on the experience of competence. Learners who adopt a very critical or skeptical attitude towards modern technologies also experience the design of AR as less joyful.

FUTURE RESEARCH DIRECTIONS

Future research projects can test different AR applications for their suitability for the Learning by Design approach. In addition, 3D objects should also be used for the AR elements. In our context, videos, texts or images were primarily designed. Modeling 3D visualizations is certainly a competence that qualifies for future professions.

In addition to collecting data using quantitative survey methods, future studies should also use qualitative research methods. This would certainly provide an even more accurate picture of the individual learning process. In the field of foreign language learning, comparative studies can still produce interesting results. How much do students learn in design-based learning scenarios compared to traditional approaches? Is producing AR content for a later exhibition more motivating or effective than designing such an exhibition with videos, pictures or classic presentations? Investigations with comparison groups are absolutely necessary here and must be taken into consideration for future research projects.

If one already thinks ahead about the possibilities of Virtual Reality, one has to explore to what extent the development of language-based virtual worlds can promote the acquisition of abilities and skills. It is also important to investigate the acceptance of VR and also the acceptance of VR glasses, which could perhaps be perceived as disruptive in the learning process.

CONCLUSION

In summary it can be said that through the application of modern technology, in this case AR, a traditional language learning scenario was transformed into a design-based learning experience. This learning experience was highly motivating for the participating learners and AR as a technology was also accepted by the majority and considered valuable for the learning and design process. Future projects that wish to combine AR and learning by design can draw on our results. The usability of the AR application is particularly important. If this is perceived positively, the pupils experience joy, interest and competence in the course of the learning by design process. The results have also shown once again that problem-based, project-based and case-based learning scenarios of this kind give pupils many degrees of freedom. This autonomy in their own learning process is one reason why such scenarios are perceived as intrinsically motivating.

The AR Scotland project was also oriented towards the requirements of the curriculum and took into account new demands on teachers and pupils. The promotion of digital competences can also be realized in subject teaching, here foreign language teaching. To this end, teachers also need appropriate competences, e.g. to be able to present the AR application. Pupils acquire important skills for the future, such as the presentation of content in the form of text and sound or the production of media products, which are integrative in the subject lessons. The teaching of language and its rules thus succeeds hand in hand with innovative technologies and promotes central competences for the age of digitization.

REFERENCES

Akçayır, M., & Akçayır, G. (2017). Advantages and challenges associated with augmented reality for education: A systematic review of the literature. *Educational Research Review*, *20*, 1–11. doi:10.1016/j.edurev.2016.11.002

Andujar, A. & Buchner, J. (2019). The potential of 3D virtual reality (VR) for language learning: an overview. In I. A. Sanchez, P. Isaias, P. Ravesteijn, & G. Ongena (Hrsg.), *Proceedings of the 15th International Conference Mobile Learning 2019* (S. 153–156). Utrecht, The Netherlands: iadis. 10.33965/ml2019_201903R002

Azuma, R. (1997). A survey of augmented reality. *Presence (Cambridge, Mass.)*, *6*(6), 355–385. doi:10.1162/pres.1997.6.4.355

Bajner, M. (2018). Digital competence through authentic, self-made texts. *R&E-SOURCE - Open Online Journal for Research and Education*, (12). Retrieved from https://journal.ph-noe.ac.at/index.php/resource/article/view/583

Bangert-Drowns, R. L., Kulik, J. A., & Kulik, C.-L. C. (1985). Effectiveness of computer-based education in secondary schools. *Journal of Computer-Based Instruction*, *12*, 59–68.

Billinghurst, M., Kato, H., & Poupyrev, I. (2001). The MagicBook-Moving seamlessly between reality and virtuality. *IEEE Computer Graphics and Applications*, *21*(1), 6–9. doi: 10.1109/38.920621

BMBWF (Bundesministerium für Bildung, Wissenschaft und Forschung). (2018). *Verordnung des Bundesministers für Unterricht und Kunst vom 14. November 1984 über die Lehrpläne der allgemeinbildenden höheren Schulen* [Regulation by the Austrian Federal ministry of Education, Science and Research from Nov. 14, 1984 about the curricula in secondary forms].

Bower, M., Howe, C., McCredie, N., Robinson, A., & Grover, D. (2014). Augmented reality in education - Cases, places and potentials. *Educational Media International*, *15*(2), 1–15. doi:10.1080/09523987.2014.889400

Brandhofer, G., ... (2016). *digi.kompP - Kompetenzmodell* [dig.kompP - Competence Model]. Retrieved from http://www.virtuelle-ph.at/wp-content/uploads/2016/09/digi.kompP-Grafik-und-Deskriptoren-1.pdf

Buchner, J., & Höfler, E. (2018). Flipped learning mit augmented reality. In J. Buchner, C. Freisleben-Teutscher, J. Haag, & E. Rauscher (Eds.), *Inverted Classroom: Vielfältiges Lernen* (pp. 61–66). Retrieved from http://skill.fhstp.ac.at/wp-content/uploads/2017/09/23489_TdL_sh_270218_final.pdf#page=63

Buchner, J. & Zumbach, J. (2018). Promoting intrinsic motivation with a mobile augmented reality learning environment. In I. A. Sanchez & P. Isaias (Hrsg.), *Proceedings of the 14th International Conference Mobile Learning 2018* (S. 55–61). Lisbon, Portugal: iadis.

Center for Self-Determination Theory. (2019). Intrinsic Motivation Inventory (IMI). Retrieved from https://selfdeterminationtheory.org/intrinsic-motivation-inventory/

Costa, C., Tyner, K., Henriques, S., & Sousa, C. (2017). *Games for media and information literacy - developing media and information literacy skills in children through digital games creation*. 2997–3004. doi: 10.21125/edulearn.2017.1627

Council of Europe. (2018). *Common European framework of reference for languages: Learning, teaching, assessment: Companion volume with new descriptors*. Retrieved from https://rm.coe.int/cefr-companion-volume-with-new-descriptors-2018/1680787989

Davis, F. D. (1989). Perceived usefulness, perceived ease of use and user acceptance of information technology. *Management Information Systems Quarterly*, *13*(3), 319–339. doi:10.2307/249008

Deci, E. L., & Ryan, R. M. (2000). The "what" and "why" of goal pursuits: Human needs and the self-determination of behavior. *Psychological Inquiry*, *11*(4), 227–268. doi:10.1207/S15327965PLI1104_01

Di Serio, Á., Ibáñez, M. B., & Kloos, C. D. (2013). Impact of an augmented reality system on students' motivation for a visual art course. *Computers & Education*, *68*, 586–596. doi:10.1016/j.compedu.2012.03.002

digi.komp: Informatik 5. Klasse [digi.komp: ICT 5th form]. (n.d.). Retrieved from https://digikomp.at/?id=585

Dressman, M. (2010). *Let's poem: The essential guide to teaching poetry in a high-stakes, multimodal world.* New York, NY: Teachers College Press.

European Union. (2017). *Being young in Europe today - digital world.* Retrieved from http://ec.europa.eu/eurostat/statistics-explained/index.php?title=Being_young_in_Europe_today_-_digital_world

European Union. (2018). *DigiComp: The European Digital Competence Framework.*

Fazal, M. B. (1996). *Effectiveness and costs of computer-based instruction in higher education and defense training: A meta-analysis* (Doctoral dissertation). University of Southern California, CA.

Frydenberg, M., & Andone, D. (2016). Creating micro-videos to demonstrate technology learning and digital literacy. *Interactive Technology and Smart Education, 13*(4), 261–273. doi:10.1108/ITSE-09-2016-0030

Gershenfeld, A. (2011). From player to designer: Engaging and empowering youth through making video games. *Knowledge Quest, 40*(1), 7.

Gesellschaft für Informatik. (2016). *Dagstuhl-Erklärung: Bildung in der digitalen vernetzten Welt* [Dagstuhl Declaration: Education in a digital and interconnected society]. Retrieved from https://www.gi.de/aktuelles/meldungen/detailansicht/article/dagstuhl-erklaerung-bildung-in-der-digitalen-vernetzten-welt.html

Hattie, J. (2008). *Visible learning: A synthesis over 800 meta-analyses relating to achievement.* London, UK: Routledge. doi:10.4324/9780203887332

Henderson, M., Auld, G., Holkner, B., Russell, G., Seah, W. T., Fernando, A., & Romeo, G. (2010). *Students creating digital video in the primary classroom: Student autonomy, learning outcomes, and professional learning communities. 24*(2), 9.

Ho, C. M. L., Nelson, M. E., & Müeller-Wittig, W. (2011). Design and implementation of a student-generated virtual museum in a language curriculum to enhance collaborative multimodal meaning-making. *Computers & Education, 57*(1), 1083–1097. doi:10.1016/j.compedu.2010.12.003

Ikrath, P. & Speckmayr, A. (2016). Digitale Kompetenzen für eine digitalisierte Lebenswelt. Eine Jugendstudie der AK Wien, durchgeführt vom Institut für Jugendkulturforschung [Digital skills for a digital world. A youth survey by the AK Vienna, conducted by the Institute for youth culture research]. Vienna, Austria.

Kamarainen, A., Metcalf, S., Grotzer, T., Mazzzuca, D., Tutwiler, M., & Dede, C. (2013). EcoMOBILE: Integrating augmented reality and porbeware with environmental education field trips. *Computer Education, 68*, 545–556. doi:10.1016/j.compedu.2013.02.018

Kearney, M. (2011). A learning design for student-generated digital storytelling. *Learning, Media and Technology, 36*(2), 169–188. doi:10.1080/17439884.2011.553623

Kearney, M., & Schuck, S. (2006). Spotlight on authentic learning: Student developed digital video projects. *Australasian Journal of Educational Technology*, *22*(2). doi:10.14742/ajet.1298

Kesim, M., & Ozarslan, Y. (2012). Augmented reality in education: Current technologies and the potential for education. *Procedia: Social and Behavioral Sciences*, *47*, 297–302. doi:10.1016/j.sbspro.2012.06.654

Klopfer, E., & Sheldon, J. (2010). Augmenting your own reality: Student authoring of science-based augmented reality games. *New Directions for Youth Development*, *128*(128), 85–94. doi:10.1002/yd.378 PMID:21240956

Kolodner, J. L., Camp, P. J., Crismond, D., Fasse, B., Gray, J., Holbrook, J., ... Ryan, M. (2003). Problem-based learning meets case-based reasoning in the middle-school science classroom: Putting learning by design into practice. *Journal of the Learning Sciences*, *12*(4), 495–547. doi:10.1207/S15327809JLS1204_2

Kolodner, J. L., Gray, J., & Fasse, B. B. (2003). Promoting transfer through case-based reasoning: Rituals and practices in learning by DesignTM classrooms. *Cognitive Science Quarterly*, 119–170.

Kothgassner, O. D., Felnhofer, A., Hauk, N., Kastenhofer, E., Gomm, J., & Kryspin-Exner, I. (2013). *TUI: Technology Usage Inventory*. Retrieved from https://www.ffg.at/sites/default/files/allgemeine_downloads/thematische%20programme/programmdokumente/tui_manual.pdf

Kulik, C.-L. C., & Kulik, J. A. (1991). Effectiveness of computer-based instruction: An updated analysis. *Computers in Human Behavior*, *7*(1-2), 75–94. doi:10.1016/0747-5632(91)90030-5

Mathews, J. M. (2010). Using a studio-based pedagogy to engage students in the design of mobile-based media. *English Teaching*, *9*(1), 87–102.

McQuiggan, S., Kosturko, L., McQuiggan, J., & Sabourin, J. (2015). *Mobile learning: A handbook for developers, educators, and learners*. Hoboken, NJ: John Wiley & Sons.

Medienpädagogischer Forschungsverbund Südwest (mpfs). (2017). *JIM-Studie 2017. Jugend, Information, (Multi-) Media. Basisstudie zum Medienumgang 12- bis 19-Jähriger* [JIM Survey 2017. Youth, information (multi-)media. Basic study on media usage]. Retrieved from https://www.mpfs.de/fileadmin/files/Studien/JIM/2016/JIM_Studie_2016.pdf

Mewald, C., & Wallner, S. (2018). Digitalised materials for young foreign language learners on the interactive website PALM. In M. Bouckaert, M. Konings, & M. van Winkelhof (Eds.), *Meaning-focused materials for language learning* (pp. 189–204). Newcastle-upon-Tyne, UK: Cambridge Scholars.

Mewald, C., Wallner, S., & Buchner, J. (2018). *How-to: Lexical notebooks 1 & 2*. Video. Retrieved from https://www.youtube.com/watch?v=etjnHfKWzN4

Orús, C., Barlés, M. J., Belanche, D., Casaló, L., Fraj, E., & Gurrea, R. (2016). The effects of learner-generated videos for YouTube on learning outcomes and satisfaction. *Computers & Education*, *95*, 254–269. doi:10.1016/j.compedu.2016.01.007

Papert, S., & Harel, I. (1991). *Constructionism*. Westport, CT: Ablex Publishing.

Partnership for 21. Century Learning. (2015). Retrieved from https://www.p21.org//our-work/p21-framework

Radu, I. (2014). Augmented reality in education: A meta-review and cross-media analysis. *Personal and Ubiquitous Computing*, *18*(6), 1533–1543. doi:10.100700779-013-0747-y

Rat für Kulturelle Bildung. (2019). *Jugend/Youtube/Kulturelle Bildung. Horizont 2019* [Youth/Youtube/Cultural Education. Horizont 2019]. Retrieved from https://www.rat-kulturelle-bildung.de/fileadmin/user_upload/pdf/Studie_YouTube_Webversion_final.pdf

Ryan, R. M. (1982). Control and information in the intrapersonal sphere: An extension of cognitive evaluation theory. *Journal of Personality and Social Psychology*, *43*(3), 450–461. doi:10.1037/0022-3514.43.3.450

Ryan, R. M., & Deci, E. L. (2000). Intrinsic and extrinsic motivations: Classic definitions and new directions. *Contemporary Educational Psychology*, *25*(1), 54–67. doi:10.1006/ceps.1999.1020 PMID:10620381

Saumell, V. (2016). Principle for meaningful technology integration. *Modern English Teacher*, *25*(2), 15–17.

Schuck, S., & Kearny, M. (2006). Capturing learning through student-generated digital video. *Australian Educational Computing*, *21*(1), 15–20.

E. U. Science Hub. (2018). *Digital competence framework for educators (DigCompEdu)*. Retrieved from https://ec.europa.eu/jrc/en/digcompedu

Stevenson, B., Länsitie, J., Kogler, C., & Bauer, P. (2015). *Exploring co-creation of educational videos in an international collaborative context*, *11*(2), 10.

van't Hooft, M. (2016, März 30). Taking it to the street: Using QR codes to tell student-created (hi) stories on location. Retrieved from https://www.socialstudies.org/publications/socialeducation/march-april2013/taking_it_to_street

Venkatesh, V., & Bala, H. (2008). Technology acceptance model 3 and a research agenda on interventions. *Decision Sciences*, *39*(2), 273–315. doi:10.1111/j.1540-5915.2008.00192.x

Venkatesh, V., & Davis, F. D. (2000). A theoretical extension of the technology acceptance model: Four longitudinal field studies. *Management Science*, *46*(2), 186–204. doi:10.1287/mnsc.46.2.186.11926

Venkatesh, V., Morris, M. G., Davis, G. B., & Davis, F. D. (2003). User acceptance of information technology: Toward a unified. *Management Information Systems Quarterly*, *27*(3), 425–478. doi:10.2307/30036540

Wilde, M., Bätz, K., Kovaleva, A., & Urhahne, D. (2009). Überprüfung einer Kurzskala intrinsischer Motivation (KIM) [Testing a short scale of intrinsic motivation]. *Zeitschrift für Didaktik der Naturwissenschaften*, *15*, 31–45.

Wössner, S. (2019). Immersives Fremdsprachenlernen [Immersive language learning]. *Computer+Unterricht*, *114*, 28–31.

Wu, H.-K., Wen-Yu Lee, S., Chang, H.-Y., & Liang, J.-C. (2013). Current status, opportunities and challenges of augmented reality in education. *Computers & Education*, *62*, 41–49. doi:10.1016/j.compedu.2012.10.024

Zahn, C., Schaeffeler, N., Giel, K. E., Wessel, D., Thiel, A., Zipfel, S., & Hesse, F. W. (2014). Video clips for YouTube: Collaborative video creation as an educational concept for knowledge acquisition and attitude change related to obesity stigmatization. *Education and Information Technologies*, *19*(3), 603–621. doi:10.100710639-013-9277-5

ADDITIONAL READING

Buchner, J. (2017). Impact of an augmented reality learning environment on teacher motivation in an in-service training on creative multimedia technology. *R&E-Source*. Retrieved from https://journal.ph-noe.ac.at/index.php/resource/article/view/501

Chen, C.-H., Wu, F. G., Su, C. C., & Lee, P.-Y. (2007). Augmented interface for children Chinese learning. *Seventh IEEE International Conference on Advanced Learning Technologies (ICALT 2007). IEEE Computer Society*, 268–270. 10.1109/ICALT.2007.76

Chen, Y. L. (2016). The Effects of Virtual Reality Learning Environment on Student Cognitive and Linguistic Development. *The Asia-Pacific Education Researcher*, *25*(4), 637–646. doi:10.100740299-016-0293-2

Dunleavy, M., & Dede, C. (2014). Augmented Reality Teaching and Learning. In J. M. Spector, M. D. Merrill, J. Elen, & M. J. Bishop (Eds.), *Handbook of Research on Educational Communications and Technology* (pp. 735–745)., doi:10.1007/978-1-4614-3185-5_59

Freitas, R., & Campos, P. (2008). SMART: A System of augmented reality for teaching 2nd grade students. *Proceedings of the 22nd British Computer Society Conference on Human-Computer-Interaction (HCI 2008)*, 27–30. Liverpool John Moores University, UK.

Klopfer, E. (2008). *Augmented learning: Research and design of mobile educational games*. Cambridge: MIT Press. doi:10.7551/mitpress/9780262113151.001.0001

Mayer, R. E. (Ed.). (2014). *The Cambridge Handbook of Multimedia Learning* (2nd ed.). Cambridge, UK: Cambridge University Press. doi:10.1017/CBO9781139547369

Milgram, P., & Kishino, F. (1994). A Taxonomy of Mixed Reality Visual Displays. *IEICE Transactions on Information and Systems*, *77*(12), 1321–1329.

KEY TERMS AND DEFINITIONS

Augmented Reality: Computer-aided enhancement of reality.

Digital Skills: ICT competences that are necessary to participate in the digital society in the 21st century.

Intrinsic Motivation: Highest form of experiential motivation. Willingness to carry out tasks and actions on the basis of inner conviction.

Learner-Generated Content: Learning artifacts created by students themselves.

Learning by Design: Constructivist approach based on problem-based learning and case-based reasoning. Learners become producers.

Technology Acceptance Model (TAM): Theory about the conditions necessary for users to accept and use technologies.

Technology Usage Inventory (TUI): Questionnaire containing essential psychological factors influencing technology acceptance and use, allowing predictions to be made about technology use regardless of the type of technology, and which can be used to evaluate new technologies.

APPENDIX 1

Table 3. Adapted Version of the Short Scale Intrinsic Motivation (KIM, Wilde et al., 2009)

	not at all					fully agree
Creating content with Augmented Reality was fun for me.						
I found creating content with Augmented Reality interesting.						
Creating augmented reality content was fun.						
I am satisfied with my performance while creating the augmented reality elements.						
When creating the Augmented Reality content, I was very skilled.						
I think I was pretty good at creating the augmented reality content.						
I was able to control the activity myself while creating the augmented reality content.						
I could choose how I did it when creating the augmented reality content.						
I could do it the way I wanted.						
I felt under pressure.						
I felt tense.						
I was worried that everything was going to be fine.						

APPENDIX 2

Table 4. Adapted Version of the Technology Usage Inventory (TUI, Kothgassner et al., 2013)

	not at all					fully agree
I already know a lot about technology.						
I am interested in new technical equipment.						
I would like to know more about new technical equipment.						
The application of Augmented Reality was easy for me to understand.						
The application of Augmented Reality was easy overall.						
The application of Augmented Reality was complicated.						
Using Augmented Reality can help me to learn.						
Augmented Reality could help me to do learning tasks more conveniently.						
If I could afford this technology, I would buy it.						
Augmented Reality can help me learn.						
I think using augmented reality can be bad for some children.						
I think augmented reality is dangerous.						
Augmented reality would bring me more disadvantages than advantages.						
I think that almost everyone can afford augmented reality.						
I think that anyone can experience augmented reality.						
It is easy to experience augmented reality at home.						

Chapter 9
MOOCs in the Language Classroom:
Using MOOCs as Complementary Materials to Support Self-Regulated Language Learning

Barbara Conde Gafaro
The Open University, UK

ABSTRACT

In this chapter, the author discusses the role of Massive Open Online Courses (MOOCs) in the field of foreign language education. It examines the relationship between MOOCs and self-regulated learning, and goes on to explore the repurposing of content-based MOOCs as complementary materials to foster students' self-regulated behaviour and practice of the target language inside and out the language classroom. A research project that has explored self-regulated language learning in a blended context with content-based MOOCs is also discussed here, with particular reference to the self-regulatory strategies employed by a group of language students at University level. A series of recommendations that supports the use of content-based MOOCs as part of face-to-face language courses are provided at the end of this chapter.

INTRODUCTION

Learning a foreign language in this digital era has been engendered by the increasing number of open educational resources and technological developments available online. Massive Open Online Courses (henceforth, MOOCs) represent the development of online learning at a massive scale (Daniel, 2012). They are designed for large numbers of participants who can access the courses anytime, anywhere as long as they have an internet connection (Mcauley, Stewart, Siemens, & Cormier, 2010). The enrolment process in MOOCs tends to differ from other types of online learning programmes offered by academic

DOI: 10.4018/978-1-7998-1097-1.ch009

institutions, since anyone can enrol in MOOCs regardless of their prior academic achievement (Wiley, 2015), which means that no formal qualifications or entry tests are required to study in these courses.

The term MOOC was coined by Dave Cormier to explain a way of learning in the networked world (Cormier, 2010), which started in 2008 with an online course called *Connectivism and Connective Knowledge (CCK08)* that George Siemens, Stephen Downes and Dave Cormier facilitated at the University of Manitoba in Canada (Cormier, 2008). The open distance course attracted a total of 2,200 enrolled and non-enrolled students worldwide (Downes, 2009) who instead of receiving organisational scaffolding, they were provided with a digital network to enable them to share ideas, connect with people and co-create knowledge with the groups of people they were interested in within the course. This innovative course reinforced the principles of connectivist pedagogy, in which many people and their ideas are connected in an online network to learn and collaboratively construct knowledge (Sharples, 2019). This pedagogy of connectivist learning (Siemens, 2005) subsequently underpinned the development of connectivist MOOCs.

Although the pedagogical approaches of MOOCs have varied since 2008 (McLoughlin & Magnoni, 2017), the format of these courses usually requires people to regulate their own learning to successfully engage with them (Littlejohn & Hood 2018). As explained by Downes (2012), "one big difference between a MOOC and a traditional course is that a MOOC is completely voluntary. You decide that you want to participate, you choose how to participate, then you participate" (para. 9, emphasis in original). Although Downes (2012) was referring specifically to participation in connectivist MOOCs, participants in content-based MOOCs can also choose the content and activities they engage with, decide how to interact with the course and when to do so (Milligan & Littlejohn, 2016). However, students who have not dealt with the flexible instructional design of these courses before can be challenged by what appears to be a free choice in the way they approach their learning.

Within the field of foreign language education, content-based MOOCs can potentially provide students with flexible learning opportunities to engender their self-regulated behaviour and practise their target language while accessing knowledge that may be relevant to their personal or academic interests. These type of courses often present well-structured and engaging audio-visual content (Margaryan, Bianco, & Littlejohn, 2015), which can become useful resources for language students to practise or improve their language skills. Nevertheless, there is little research in this field that examines the self-regulated learning behaviour of foreign language students while engaging with content-based MOOCs as part of their instructional language courses. Accordingly, the focus of this chapter is on the integration of content-based MOOCs as blended elements in classroom-based language courses, with particular reference to the affordances of such integration in regards to students' self-regulated learning and the practice of the target language inside and out the classroom.

This chapter therefore addresses four main questions:

- How MOOCs can contribute to students' self-regulated learning?
- What features of MOOCs are useful for language education?
- How language educators have integrated MOOCs as part of their instructional language courses?
- What self-regulatory strategies do university students employ when working with MOOCs as part of an academic English course?

In order to do this, it first identifies the main features of content-based MOOCs that engender self-regulated learning, before going on to examine the affordances of content-based MOOCs to language learning and the approaches to the integration of theses into classroom-based language courses. The chapter ends by reporting some findings of a case study that investigated self-regulatory language learning strategies in an academic English course using content-based MOOCs, and by providing some recommendations when working with content-based MOOCs as part of an instructional language course.

BACKGROUND

MOOCs

Open educational resources, distance learning technologies and connectivist pedagogy have witnessed the origins of MOOCs (Yuan & Powell, 2013; Siemens, 2005), which represent the evolution of online learning at a large scale. These courses should not be deemed as a new phenomenon when referring to learning at scale (McAndrew & Scanlon, 2013), yet, one of the main features of MOOCs seems to be related to what Ferguson and Shum (2012) describe as the dimension of "Open Communities" (p. 311), since these courses welcome thousands of people all over the world who have access to internet connection as well as a desire for learning any subject regardless of their academic background.

After registering in these online courses, MOOC students are anticipated to manage resources as well as their time when engaging with the content of the courses (Milligan & Littlejohn, 2016; de Waard, Kukulska-Hulme, & Sharples, 2015). Unlike in face-to-face courses where teachers guide the students along the course, within MOOCs students are meant to navigate the course material on their own and evaluate their progress based on individual or peer feedback (Milligan, Littlejohn, & Hood, 2016). Taking these aspects in turn, MOOC students are often anticipated to self-regulate their learning while interacting with the content and activities of such courses.

Self-regulated learning is explained as "the degree to which students are metacognitively, motivationally, and behaviorally active participants in their own learning processes" (Zimmerman (1989, p. 329). This notion of self-regulation is strongly influenced by Bandura's (1986) socio-cognitive theory, which states that peoples' actions are the result of the interplay of personal, environmental, and behavioural factors. Bandura and Zimmerman's theoretical interests are compatible and both agree that the scope of self-regulated learning involves a dynamic interaction between students, their behaviour, motivation and their social and physical context. In other words, it is an ongoing process in which students develop a conscious sense of responsibility for their learning through the use of self-regulatory strategies that help them to guide, regulate, and inform their learning, which is influenced by personal, environmental, and behavioural factors. Students in MOOCs are challenged to adopt an active role in their learning, as discussed next.

The Learning Design of MOOCs

The flexible instructional design of MOOCs is a characteristic feature that often requires students to regulate their learning (Littlejohn & Hood, 2018). The flexibility of some MOOC platforms tends to create an environment where students have a sense of control over their learning. For instance, in the MOOC platform called edX, students can choose courses based on skill level, language and learning pathway, i.e. self-paced courses or instructor-paced. The instructor-paced pathway follows a planned schedule for assignments and students can complete the course within a specific time, whereas the self-paced path does not have due dates and students can complete the course at their own pace (edX Help Center, 2018a). In both types of courses, pre-recorded lectures are available 24 hours a day, allowing flexibility for students, who can access course material whenever is more convenient for them (edX Help Center, 2018b). This flexibility allows students to "create their pathway to learning" (Ronaghi, Saberi, & Trumbore, 2015, p.102), which facilitates the regulation of "their own time, resources and learning path" (McLoughlin & Magnoni, 2017, p. 61).

Most MOOCs provide an overview and display the syllabus of their content and have the same set of weekly activities for all students (Margaryan et al., 2015). Nonetheless, they allow students to engage with the material by alternative routes (Bates, 2012). Students can work with the material in different ways based on their individual interests, learning needs or prior knowledge (Motzo & Proudfoot, 2017). On the other hand, unlike in face-to-face courses where students have more possibilities to engage in dialogue with their teachers, students have less interaction with MOOC educators and low chances of receiving personalised feedback from them (Littlejohn & Milligan, 2015; Israel, 2015; Orsini-Jones, Pibworth-Dolinski, Cribb, Brick, Gazeley-Eke, Leinster, & Lloyd, 2015; Decker, 2014). Hence, the learning design of MOOCs often challenges students' capacity to learn on their own and take responsibility for their learning.

MOOC Pedagogies

The pedagogical approaches of MOOC platforms also influence such capability to learn independently as well as the patterns of interaction between students and MOOC educators. The Open University, for example, developed a MOOC platform named FutureLearn, which is underpinned by a social-constructivist pedagogical approach to support social learning (Ferguson, Clow, Beale, Cooper, Morris, Bayne, & Woodgate 2015). Conversely, Sharples argues that MOOC platforms such as Coursera and edX were developed following an instructivist approach (as cited in Bayne & Ross, 2015, p. 36), in which the learning goals are predetermined by an instructor and students have limited interactions with their online peers (Littlejohn, 2013). MOOC educators in FutureLearn often expect that students interact with other participants, since the latter platform is underpinned by the Conversational Framework (Papathoma, 2019), which means that each teaching section should be linked to discussions where students are able to share ideas and information with others (Ferguson & Sharples, 2014). This type of MOOC educators' beliefs alongside the pedagogical underpinnings of those platforms may, therefore, influence the engagement patterns adopted by participants in MOOCs and the way they assume responsibility for their learning.

Overall, MOOCs are often designed for people who have the capacity to learn independently. Students in MOOCs can choose what course content they engage with and how and when to do so (Hood, Littlejohn & Milligan, 2015; de Waard et al., 2015). This flexibility in learning combined with the option of navigating through the course in a non-linear way, as well as the lack of personalised learning and feedback, or even group tuition from an instructor, bring their own challenges. Instead of offering scaffolds that support people who are not able to assume responsibility for their learning (Littlejohn & Hood 2018), MOOCs often challenge students' capacity to self-regulate their learning. Hence, students are anticipated to use strategies that maintain self-discipline in progressing through MOOCs (Bowden, 2018). The following section reviews previous studies that have investigated the self-regulatory strategies that MOOC students have employed as a way to take responsibility for their learning.

Researching Self-Regulated Learning in MOOCs

Using self-regulated strategies is crucial to plan, monitor and evaluate one's performance and hence facilitate the learning process (Zimmerman, 2000). The strategies employed by students tend to reflect their self-regulated behaviour in MOOCs (Littlejohn, Hood, Milligan, & Mustain, 2016). For instance, Kizilcec, Pérez-Sanagustín and Maldonado (2017) examined the self-regulatory strategies of 4,831 students across six MOOCs in different subjects offered by the same provider in Coursera. Kizilcec et al., (2017) administered a survey that measured participants' self-regulated learning and gathered user interactions with course material and records of course achievement. Findings show that goal setting and strategic planning, such as revising a lecture two weeks before an assessment, were the most effective self-regulatory strategies that helped participants to achieve their learning goals in MOOCs.

Participants in Kizilcec et al.'s (2017) study also reported using self-evaluation and task strategies when revisiting assessments and lectures. Individual differences also emerged based on participants' backgrounds. University students with a master's degree or PhD reported higher levels of goal-setting, strategic planning and task strategies than those with a bachelor's degree, despite lower levels of help-seeking and self-evaluation. By contrast, working professionals reported higher levels of self-regulated learning in goal setting, strategic planning, and help-seeking than the two previous groups of students, "despite lower levels of self-evaluation" (Kizilcec et al., p. 26). Highly self-regulated students use diverse learning strategies to support their learning within different situations (Zimmerman, 2000), though a standard feature of all three groups of students was lower levels of self-evaluation.

Likewise, some research projects have carried out some interventions to scaffold students and prompt them to use effective self-regulatory strategies in MOOCs. For example, Gutiérrez-Rojas, Alario-Hoyos, Pérez-Sanagustín, Leony, and Delgado-Kloos (2014) designed a mobile application to help less experienced MOOC users to self-regulate their learning. The digital tool called MyLearningMentor assists the students with strategic planning and providing tips to make the most of MOOCs. However, according to Pérez-Sanagustín and Maldonado (2016), the effectiveness of the application has yet to be assessed.

Moreover, a case study conducted by Chen and Chen (2015) reported positive outcomes regarding self-regulated learning in MOOCs when working with a small sample of students. In that case-study, a small group composed of four MOOC students gathered during the six weeks of a chosen MOOC to share and reflect upon the strategies of (1) Goal setting, (2) Environment structuring, (3) Task strategies, (4) Time management, (5) Help-seeking and (6) Self-evaluation. Findings from discussion recordings, observation notes, interviews, goal setting sheets and weekly reflection journals showed that face-to-face study groups might help MOOC students to share and develop self-regulatory strategies and strengthen their learning in the MOOC. Despite being a small-scale study, this study is one of the few retrievable case studies that have offered a support system outside the MOOC environment for the practical use of self-regulatory strategies.

Interacting with content-based MOOCs may not only stimulate the employment of self-regulated learning strategies, but also support students in the practice of a foreign language. The next section, therefore, comments on the benefits that the content and activities of these online courses can have when learning a language for academic or professional purposes. The section starts by discussing how working with MOOCs can help students to improve their language skills, before moving on to look at examples of integration of content-based MOOCs into classroom-based language courses.

INTEGRATING CONTENT-BASED MOOCS INTO CLASSROOM-BASED LANGUAGE COURSES

MOOCs and their Contribution to Language Learning

Generally, MOOC content is delivered in the language of the country where the provider is located (e.g. FutureLearn in The United Kingdom, Coursera, Kadenze and Udacity in the United States, XuetangX in China, MiríadaX in Spain, and FUN in France). The wide range of content-based MOOCs offered in different languages may represent a learning opportunity for language students who seek to learn or improve their target language for specific purposes while accessing knowledge, which is not necessarily for language learning purposes, but it may be relevant to the students' profession or academic studies.

The quality of presentation of MOOC resources can stimulate students to practise their target language, at their own pace (Read, Bárcena, & Kukulska-Hulme, 2016). Findings from a large-scale questionnaire study, which examined and compared the course environment of 76 MOOCs from different platforms, indicated that nearly all the MOOCs studied, from Coursera, edX, Udacity, Canvas and FutureLearn, were "well-packaged, comprising well-organised, clearly articulated and attractively presented course material" (Margaryan et al., 2015, p. 82). The structure of MOOC materials is also easy to follow on a weekly basis (Beaven, 2013), which tend to facilitate the interaction with subject-specific information in the target language.

Furthermore, the instructional design of MOOCs tends to include video lectures, subject-specific articles, discussion forums and other social networking tools (Sokolik, 2016), which can facilitate the practice of the skills of listening, reading and writing of language students (Beaven, 2013). Some videos, for example, expose students to competent speakers of the target language, including non-native speakers (Read et al., 2016). Hence, language students may find opportunities to practise or improve, much precisely, their language skills within MOOC activities and resources.

Altogether, content-based MOOCs often present well organised and engaging audio-visual content in the target language that may be related to students' academic or professional disciplines. Likewise, students can engage, according to their learning needs, with different activities and multimedia resources to work on the receptive and productive language skills; mainly writing instead of synchronous oral interactions due to the massive scale of MOOCs (de Waard et al., 2015; McLoughlin & Magnoni, 2017; Motzo & Proudfoot, 2017; Read et al., 2016; Sokolik, 2016). Lastly, these courses usually offer social learning spaces where students can interact with participants across the globe, using the target language, which exceeds the affordances of language learning in classroom-based courses (Rubio, Fuchs, & Dixon, 2016, p. 179). Next section analyses the way MOOCs have been used as part of classroom-based language courses.

Using MOOCs as Complementary Materials to Support Self-Regulated Learning

Although MOOCs as stand-alone courses are likely to offer resources and activities that support the practice of a target language, the use of MOOCs in conjunction with instructional language courses may stimulate students in the process of taking responsibility for their learning, outside and inside the language classroom. Some blended practices with MOOCs tend to follow a flipped approach to learning (Israel, 2015; Orsini-Jones, Conde Gafaro, & Altamimi, 2017a; Orsini-Jones, Zou, Borthwick, & Conde, 2017b). This approach is often used to take advantage of the MOOC content (Kloos, Muñoz-Merino,

Alario-Hoyos, Ayres, & Fernández-Panadero, 2015; Orsini-Jones, 2015) and to carry out interactive activities in the classroom such as problem-solving and discussions in order to confirm students' understanding of the material (Muñoz-Merino et al., 2016; Orsini-Jones, Conde, Borthwick, Zou, & Ma, 2018). Similarly, Israel (2015) classifies the common blended practices with MOOCs into (1) "Single MOOC Adoption", i.e. students enroll and complete the online assignments as part of their traditional program and (2) "Multiple MOOC Adoption", i.e. the integration of MOOC content and the use of other online technologies in the classroom (p. 105). The "MOOC as an added value" is another blended approach described by Pérez-Sanagustín, Hilliger, Alario-Hoyos, Kloos, & Rayyan (2017) in which the MOOC is considered useful to "acquire some extra knowledge or develop cross-curricular skills" (p. 52). With this in mind, a review of some studies that have integrated MOOCs into face-to-face language learning settings and the type of blended practice adopted is reported below.

MOOCs have mainly been integrated into instructional courses of computer science and engineering at university where teachers and students have repurposed the content of these courses to complement their teaching and learning concepts. Early studies examined the impact on both teachers and students of the integration of MOOCs into undergraduate courses in computer science such as Machine Learning and Databases (Bruff, Fisher, McEwen, & Smith, 2013; Caulfield, Collier, & Halawa, 2013; Holotescu, Grosseck, Cretu, & Naaji 2014). Taking into account that the first MOOCs focused on the disciplines of computer science and engineering (Sandeen, 2013, p. 34), it was likely that first MOOCs were blended within university degree programs in these disciplines, rather than in the field of language learning. Research on the integration of MOOCs into traditional degree programs also took some time to move from embedding customised MOOCs into social sciences campus-based courses to exploring language education in face-to-face and online learning, including MOOCs for language learning purposes as stand-alone courses (Rubio, 2015; Dixon & Fuchs, 2015). However, the former tended to focus only on language student interaction patterns studying Spanish and German respectively without considering similar blended MOOC formats.

A year after being proclaimed the year of the MOOC (Pappano, 2012), a case study in the area of English for Academic Purposes (EAP) was conducted by Beaven (2013) in which 20 university students followed a content-based MOOC as an additional learning resource for their face-to-face EAP course. Each student chose a MOOC, from the Coursera platform, related to their degree programs at the University of Ferrara (Italy). Following that, they discussed their experiences in a forum created in the Virtual Learning Environment (VLE) of the EAP course and completed two questionnaires to obtain their feedback about the project. Based on the data gathered from the qualitative questionnaires and students' posts in the online discussion forum, Beaven's (2013) findings highlight the positive impact a blended MOOC as an added value has on EAP students' motivation, the practice of English and learning skills. The study encouraged students to take responsibility for their language learning by allowing them to choose a course and to work with the MOOC resources that were most significant to their language learning needs. However, the study did not demonstrate how content-based MOOCs could foster students' self-regulated learning and the use of self-regulatory strategies to monitor and assess their learning process.

The production of language MOOCs (LMOOCs) also burst into the distance education field at that time. According to Martín-Monje and Bárcena (2014), "LMOOCs are dedicated Web-based online courses for second languages with unrestricted access and potentially unlimited participation." (p. 1). By way of illustration, a group of university lecturers at The National Distance Education University (UNED) designed an English language course for adult absolute beginners as "part of the first large-scale experience with MOOCs as a tool for English language learning in Spain" (Chacón-Beltrán, 2014, p. 242). This short MOOC of English as a foreign language, offered on the UNED Abierta platform, was designed for students to acquire and practice the thousand most frequent words in English that would help them read short texts afterwards. A total of 44,410 people joined the course and more than 10% of them finished it, which may be due to the short length of the course, 24 hours of study in that case. Findings, based on 364 participants who completed more than 80% of the course and a post-questionnaire, suggest that middle-aged Spanish adults are interested in using LMOOCs as a way to continue learning English for personal or professional reasons. Unfortunately, there is insufficient data from the post-questionnaire that shows the self-regulatory strategies used by the middle-aged Spanish adults in order to successfully complete more than 80% of the course. Yet, Chacón-Beltrán (2014) claims that such positive results should encourage researchers to explore new ways of facilitating large-scale language education, inside or outside the university system.

The linguistic target of LMOOCs seemed to expand from vocabulary practice to speaking tasks. The latter was observed in the tandem MOOC: English-Español project designed by Appel and Pujolà (2015) to offer an online space for speaking interaction in which native speakers of English collaborated synchronously with speakers of Spanish while both learned each other's languages. The tandem project reached 1284 Spanish and English proficient speakers who had opportunities to "reflect on their skills, and keep record of their own progress from the feedback given by different native speakers", through a portfolio that included the recordings of the task interactions and the ratings provided by the tandem partner (p. 1699). Nonetheless, no evidence is presented to illustrate how MOOC students employed any learning strategies they had acquired in subsequent conversations with their tandem partner after reflecting on their performance using the portfolios.

Another important intervention study in foreign language learning was carried out by de Waard & Demeulenaere (2017) at the high-school level in Belgium. They designed a MOOC-CLIL project to explore the possible motivational and self-efficacy benefits that 42 students could derive from integrating content-based MOOCs into their English and French instructional classes. The research project was divided into three stages to enhance students' language, digital skills and their online self-regulated learning. In the first stage, a MOOC description was presented to familiarise students with different MOOC platforms. Following a flipped classroom approach, they worked with a given MOOC at home and completed activities in class based on selected MOOC content. Then, in the second stage, they selected a course related to their personal interests and worked with the help of the teacher, who provided online resources to help them understand specific terms. In the final phase, students made a short video to share their learning MOOC experience with next year's cohort.

The online self-regulated learning of the 42 participants was monitored through semi-structured focus group interviews limited to 9 students per group and an adapted Skills and Attitudes Measuring (SAM) scale used by the teacher. The researchers also administered a pre- and post- online survey, adapted from the Motivated Strategies for Learning Questionnaire (MSLQ) designed by Pintrich, Smith, Garcia and McKeachie (1991), which has a combined focus on motivation and self-regulated learning. There is not much information from the findings that accounts for other sub-processes of self-regulated learning, except for strategic planning which increased 9% according to the post-survey results and help-seeking from peers which rose from 61% to 83%.

So far, this chapter has reviewed previous research projects that have incorporated MOOCs into academic language courses, including a tandem MOOC project and a MOOC-CLIL project at a secondary school. The projects reviewed above studied the benefits of using the audio-visual content and materials of content-based MOOCs in face-to-face courses with regards to students' motivation, collaborative learning skills and English proficiency. Nevertheless, there are not many studies that have investigated self-regulated learning in blended MOOC formats, particularly in the field of language learning, which is limiting our understanding about the domain of self-regulated learning within the integration of MOOCs into the language classroom. This gap in the literature prompts the need to explore how foreign language students self-regulate their learning when working with MOOCs as part of their classroom-based language courses. Accordingly, the next section goes on to discuss some of the findings of a case study that investigated self-regulated language learning strategies in an academic face-to-face English course using content-based MOOCs.

Repurposing MOOCs to Explore Self-Regulated Language Learning Strategies: Case Study

Self-regulated learning refers to the process in which students deploy metacognitive, cognitive and social strategies, such as setting learning goals, relying on memory to learn vocabulary and asking for help, respectively, in order to take responsibility for their language learning (Read, Bárcena, & Rodrigo, 2010). Within MOOCs, the ultimate responsibility rests on the students (Littlejohn & Milligan, 2015), who are encouraged to regulate their learning using self-regulatory strategies (de Waard, 2015). Taking those principles into consideration, Conde Gafaro (2019) carried out a case study to investigate the self-regulatory learning strategies that university students adopt while engaging with content-based MOOCs as part of an English for Academic Purposes (EAP) course.

The research design of the case study was segmented in three stages in order to identify the self-regulatory strategies that thirteen participants used before and after engaging with MOOCs during eight weeks of their EAP course offered at an Italian university: (1) Pre- questionnaire (2) MOOC engagement (3) Post-questionnaire. Both online questionnaires were adapted from the Motivated Strategies for Learning Questionnaire (Pintrich et al., 1991). In the second stage, participants chose a MOOC related to their study programme using Class Central, an online website that classifies MOOCs by subjects, providers and universities. They decided how and when to work with the course content that was considered relevant to their language learning needs. Voluntary semi-structured interviews were consequently carried out to complement the quantitative data from the two questionnaires.

Data collection and analysis were based on Zimmerman's (2000) cyclical model of self-regulated learning, which illustrates the self-regulatory strategies that are used before, during, and after efforts to learn. Findings show that participants were more likely to review the course description of the MOOC to see how it was organised. This focus on strategic planning might be associated with a lack of familiarity with MOOCs on the part of participants; hence the need to have an overview of the structure and topics of the online course in advanced (Conde Gafaro, 2019). Conversely, help-seeking strategies were employed more in face-to-face EAP lessons.

Some of the findings also relate to the results of Kizilcec et al.'s (2017) study regarding the lower levels of help-seeking and self-evaluation reported by the participants. They rarely asked for help to their online peers or the MOOC educator nor they kept a track of what they were doing to assess their own performance after working with the MOOC. However, participants were more likely to employ strategic planning strategies before enrolling in the MOOC than in the language course, which based on Kizilcec et al.'s (2017) findings, is one of the most effective self-regulatory strategies that help students to accomplish their goals in the MOOC. Similarly, metacognitive monitoring strategies, as well as strategies that were considered effective for participants while learning English, such as memorisation, were also used during the EAP course and the MOOC. Findings also highlight participants' positive attitudes towards the use of MOOCs to supplement classroom activity in an EAP course at university level. Conde Gafaro's (2019) case study built upon previous empirical studies of engagements with content-based MOOCs as part of face-to-face language courses at university and high-school level, and provided a deeper understanding of students' self-regulated language learning strategies in more depth than previous studies.

RECOMMENDATIONS

Integrating content-based MOOCs into face-to-face academic language courses, for instance, offers students a self-regulated language learning experience. This blended learning practice requires students to interact with academic online content and a flexible learning structure, characteristic of MOOCs, in order to support their self-regulated behaviour while accessing knowledge in the target language that may be relevant to their academic disciplines. Likewise, it provides students with a direct instruction approach to scaffold their self-regulated learning processes, since teachers "can try to increase the range of learning strategies at students' disposal and to assist students to achieve self-regulation" (Ellis & Shintani, 2014, p.26). Thus, some recommendations are outlined below for language educators interested in using content-based MOOCs as part of an academic classroom-based language course:

- Students should have a B1 or B2 level of proficiency in the target language, according to the Common European Framework of Reference for Languages (CEFR), in order to engage with the academic content of the MOOCs, since at these level students can understand the main ideas of complex texts about their academic field (Council of Europe, 2018).
- Identify suitable MOOCs that are released at a time that matches the dates of the academic classroom-based language course.
- Get students familiar with the MOOC platforms by providing a clear step-by-step guide with instructions on how to choose and access the courses.
- Students should have previous knowledge of the main topic addressed in the MOOC to compensate for the possible uncertainty that doing a MOOC for the first time involved.
- Suggest a minimum amount of time to weekly work with the MOOC.
- Define the aspects students should focus on during the blended learning experience, i.e. linguistic features, academic content or both.
- Last but not least, an understanding of self-regulated learning as an ongoing process is required to encourage students to develop awareness and the appropriate use of metacognitive, cognitive and social strategies that are essential to monitor and adjust their language learning throughout the MOOC engagement.

FUTURE RESEARCH DIRECTIONS

Previous studies on blended learning practices revealed that well-designed material and activities found in content-based MOOCs have helped high-school and university students to enhance their target language, collaborative and digital learning skills and their online self-regulated learning. Future research on this area of educational technology and language learning should investigate the role(s) of language teachers who decide to implement this blended MOOC practice, which is focused on scaffolding students' self-regulated learning. It would be interesting to identify the strategies that teachers may use to encourage students to participate in these online learning environments that challenge their self-regulated behaviour while studying at their own pace and connecting with other students around the world, so they can con-construct knowledge and practise their language skills. Language educators alongside researchers should also explore the design and implementation of classroom activities, involving MOOC content or resources, which can help students to assess their learning, more precisely reflect on and adapt their performance while practising their target language. This type of research may contribute to close the gap between classroom learning and online learning in order to enhance students' self-regulated language learning process.

CONCLUSION

This chapter has focused on the integration of content-based MOOCs as blended elements in classroom-based language courses, considering, in particular, the affordances of such integration regarding students' self-regulated learning and the practice of the target language inside and out the classroom. It has commented on the affordances of using these courses for self-regulated learning purposes by outlining first the features of MOOCs that contribute to engender students' self-regulated learning, discussing research on self-regulated learning in MOOCs, and then reviewing a series of research projects that have blended MOOCs into face-to-face language courses at a university and school level. Lastly, a case study that has examined the self-regulatory strategies of university students taking a content-based MOOC as part of an academic English course was also discussed, followed by some recommendations and discussion of future research directions.

In summary, there is a variety of research on the use of MOOCs within contexts different from the ones for which they are developed. MOOCs are often used in face-to-face education to support learning experiences by taking advantage of MOOCs' technology and content. In the language learning field, however, there is little research that shows the potential benefits and challenges of these blended MOOC practices with regards to the self-regulated learning behaviour of language students at different educational levels. Adopting these blended learning practices may help language educators to understand how students regulate their learning and how MOOCs may challenge the use of self-regulatory strategies in studying the target language for academic or professional purposes. Lastly, understanding the advantages and challenges of working with content-based MOOCs should encourage language educators to explore alternative learning opportunities that support students in the process of taking responsibility for their language learning, outside and inside the classroom.

ACKNOWLEDGMENT

I would like to thank my supervisors from The Open University, Dr. Tim Lewis, Dr. Tita Beaven and Dr. Rebecca Ferguson who inspire me to make the most of this PhD journey. And, a special thank you to Dr. Rebecca Ferguson for taking the time to discuss with me the first ideas of this book chapter.

REFERENCES

Appel, C. & Pujolà, J. T. (2015). Pedagogical and technological issues in the instructional design of a tandem MOOC. In *Proceedings of E-Learn: World Conference on E-Learning in Corporate, Government, Healthcare, and Higher Education* (pp. 1696-1705). Kona, Hawaii: Association for the Advancement of Computing in Education (AACE).

Bandura, A. (1986). *Social foundations of thought and action: A social cognitive theory*. Englewood Cliffs, NJ: Prentice-Hall.

Bates, T. (2012, Aug. 5). What's right and what's wrong about Coursera-style MOOCs? Retrieved from https://www.tonybates.ca/2012/08/05/whats-right-and-whats-wrong-about-coursera-style-moocs/

Bayne, S., & Ross, R. (2015). MOOC pedagogy. In P. Kim (Ed.), *Massive open online courses* (pp. 23–45). New York, NY: Routledge.

Beaven, A. (2013). Using MOOCs in an academic English course at university level. In A. Beaven, A. Comas-Quinn, & B. Sawhill (Eds.), Case studies of openness in the language classroom (pp. 217-227). © Research-publishing.net. doi:10.14705/rpnet.2013.000122

Bowden, P. (2018, Jan. 7). Five years of MOOC trends: Pat Bowden recounts her experiences of taking 89 MOOCs, and analyzes the trends she has noticed in that time. [Web log post]. Retrieved from https://www.class-central.com/report/five-years-mooc-trends/

Bruff, D. O., Fisher, D. H., McEwen, K. E., & Smith, B. E. (2013). Wrapping a MOOC: Student perceptions of an experiment in blended learning. *Journal of Online Learning and Teaching / MERLOT*, *9*(2), 187–199.

Caulfield, M., Collier, A., & Halawa, S. (2013, Oct. 6). Rethinking online community in MOOCs used for blended learning [Web log post]. Retrieved from http://www.educause.edu/ero/article/rethinking-online-community-moocs-used-blended-learning

Chacón-Beltrán, R. (2014). Massive online open courses and language learning: The case for a beginners' English course. *Procedia: Social and Behavioral Sciences*, *141*, 242–246. doi:10.1016/j.sbspro.2014.05.042

Chen, Y. H., & Chen, P. J. (2015). MOOC study group: Facilitation strategies, influential factors, and student perceived gains. *Computers & Education*, *86*, 55–70. doi:10.1016/j.compedu.2015.03.008

Conde Gafaro, B. (2019). Repurposing MOOCs for self-regulated language learning in an English for academic purposes course. In A. Beaven, A. Comas-Quinn, & B. Sawhill (Eds.), *Case studies of openness in and beyond the language classroom*. © *Research-publishing.net*. doi:10.14705/rpnet.2019.37.970

Cormier, D. (2008, Oct. 2). *The CCK08 MOOC – Connectivism course, 1/4 way*. [Web log post]. Retrieved from http://davecormier.com/edblog/2008/10/02/the-cck08-mooc-connectivism-course-14-way/

Cormier, D. (2010, Dec. 8). *What is a MOOC?* Retrieved from https://www.youtube.com/watch?v=eW3gMGqcZQc

Council of Europe. (2018). *Global scale - Table 1 (CEFR 3.3): Common reference levels.* Retrieved from https://www.coe.int/en/web/common-european-framework-reference-languages/table-1-cefr-3.3-common-reference-levels-global-scale

Daniel, J. (2012). Making sense of MOOCs: Musings in a maze of myth, paradox and possibility. *Journal of Interactive Media in Education, 18*, 1–20. Retrieved from http://wwwjime.open.ac.uk/jime/article/view/2012-18

de Waard, I. (2015). MOOC factors influencing teachers in formal education. *Revista Mexicana de Bachillerato a distancia, 7*(13).

de Waard, I. & Demeulenaere, K. (2017). The MOOC-CLIL project: Using MOOCs to increase language, and social and online learning skills for 5th grade K-12 students. In Q. Kan & S. Bax (Eds.), Beyond the language classroom: Researching MOOCs and other innovations (pp. 29-42). Research-publishing.net. doi:10.14705/ rpnet.2017.mooc2016.669

de Waard, I., Kukulska-Hulme, A., & Sharples, M. (2015). Self-directed learning in trial FutureLearn courses. [University Catholique de Louvain, Belgium.]. In *Proceedings of the European Stakeholder Summit EMOOCS, 2015*, 234–243.

Decker, G. L. (2014). MOOCology 1.0. In S. D. Krause & C. Lowe (Eds.), *Invasion of the MOOCs: The promises and perils of massive open online courses* (pp. 3–13). Anderson, SC: Parlor Press.

Dixon, E. & Fuchs, C. (2015). Face to face, online, or MOOC–How the format impacts content, objectives, assignments, and assessments. In E. Dixon & M. Thomas (Eds.), Researching language student interaction online: From social media to MOOCs. (pp. 89–105). San Marcos, TX: CALICO monograph.

Downes, S. (2009, April 25). New technology supporting informal learning [Web log post]. Retrieved from https://halfanhour.blogspot.co.uk/2009/04/new-technology-supporting-informal.html

edX Help Center. (2018a). What is self-paced or instructor-paced? Which is my course? Retrieved from https://support.edx.org/hc/en-us/articles/115011202847-What-is-self-paced-or-instructor-paced-Which-is-my-course-

edX Help Center. (2018b). What time is the class? Retrieved from https://support.edx.org/hc/en-us/articles/206503538-What-time-is-the-class-

Ellis, R., & Shintani, N. (2014). *Exploring language pedagogy through second language acquisition research.* London, UK: Routledge.

Ferguson, R., Clow, D., Beale, R., Cooper, A. J., Morris, N., Bayne, S., & Woodgate, A. (2015). Moving through MOOCS: Pedagogy, learning design and patterns of engagement. In G. Conole, T. Klobucar, C. Rensing, J. Konert, & E. Lavoué (Eds.), *Design for teaching and learning in a networked world* (pp. 70–84). Lecture Notes in Computer Science Springer International. doi:10.1007/978-3-319-24258-3_6

Ferguson, R., & Sharples, M. (2014). Innovative pedagogy at massive scale: Teaching and learning in MOOCs. In Proceedings of 9th European Conference on Technology Enhanced Learning - EC-TEL 2014 (pp. 98–111). Graz, Austria: Springer.

Ferguson, R., & Shum, S. B. (2012). Towards a social learning space for open educational resources. In A. Okada, T. Connolly, & P. Scott (Eds.), Collaborative Learning 2.0: Open educational resources (pp. 309–327). Hershey, PA: IGI Global. doi:10.4018/978-1-4666-0300-4.ch017

Gutiérrez-Rojas, I., Alario-Hoyos, C., Pérez-Sanagustín, M., Leony, D., & Delgado-Kloos, C. (2014). Scaffolding self-learning in MOOCs. [Lausanne, Switzerland: Ecole Polytechnique Federale de Lausanne.]. *Proceedings of the European MOOC Stakeholder Summit, 2014*, 43–49.

Holotescu, C., Grosseck, G., Cretu, V., & Naaji, A. (2014). Integrating MOOCs in blended courses. In *Proceedings of the International Scientific Conference of eLearning and Software for Education*, 1, pp. 243-250. Bucharest, Romania: eLSE. doi:10.12753/2066-026X-14-034

Hood, N., Littlejohn, A., & Milligan, C. (2015). Context counts: How students' contexts influence learning in a MOOC. *Computers & Education*, *91*, 83–91. doi:10.1016/j.compedu.2015.10.019

Israel, M. (2015). Effectiveness of integrating MOOCs in traditional classrooms for undergraduate students. *International Review of Research in Open and Distance Learning*, *16*(5), 102–118. doi:10.19173/irrodl.v16i5.2222

Kizilcec, R. F., Pérez-Sanagustín, M., & Maldonado, J. J. (2017). Self-regulated learning strategies predict student behavior and goal attainment in massive open online courses. *Computers & Education*, *104*, 18–33. doi:10.1016/j.compedu.2016.10.001

Kloos, C. D., Muñoz-Merino, P. J., Alario-Hoyos, C., Ayres, I. E., & Fernández-Panadero, C. (2015). Mixing and blending MOOC technologies with face-to-face pedagogies. In *Proceedings of Global Engineering Education Conference (EDUCON*), 2015 IEEE (pp. 967-971). New York, NY: Institute of Electrical and Electronics Engineers (IEEE).

Littlejohn, A. (2013). Understanding massive open online courses. *CEMCA EdTech Notes*, 1–12.

Littlejohn, A. & Hood, N. (2018). Reconceptualising learning in the digital age: The [un]democratizing potential of MOOCs. Gateway East, Singapore: Springer.

Littlejohn, A., Hood, N., Milligan, C., & Mustain, P. (2016). Learning in MOOCs: Motivations and self-regulated learning in MOOCs. *The Internet and Higher Education*, *29*, 40–48. doi:10.1016/j.iheduc.2015.12.003

Littlejohn, A. & Milligan, C. (2015). Designing MOOCs for professional students: Tools and patterns to encourage self-regulated learning. *eLearning Papers, 42*, 38–45.

Margaryan, A., Bianco, M., & Littlejohn, A. (2015). Instructional quality of massive open online courses (MOOCs). *Computers & Education*, *80*, 77–83. doi:10.1016/j.compedu.2014.08.005

Martín-Monje, E., & Bárcena, E. (Eds.). (2014). *Language MOOCs: Providing learning, transcending boundaries*. Berlin, Germany: De Gruyter Open.

McAndrew, P., & Scanlon, E. (2013). Open learning at a distance: Lessons for struggling MOOCs. *Science*, *342*(6165), 1450–1451. doi:10.1126cience.1239686 PMID:24357302

McAuley, A., Stewart, B., Siemens, G., & Cormier, D. (2010). *The MOOC model for digital practice.* Retrieved from https://oerknowledgecloud.org/sites/oerknowledgecloud.org/files/MOOC_Final.pdf

McLoughlin, L. & Magnoni, F. (2017). The Move-Me project: Reflecting on xMOOC and cMOOC structure and pedagogical implementation. In Q. Kan & S. Bax (Eds.), Beyond the language classroom: researching MOOCs and other innovations (pp. 59-69). Research-publishing.net. .mooc2016.671 doi:10.14705/rpnet.2017.mooc2016.671

Milligan, C., & Littlejohn, A. (2016). How health professionals regulate their learning in massive open online courses. *The Internet and higher education, 31*, 113–121. doi:10.1016/j.iheduc.2016.07.005 PMID:27928198

Milligan, C., Littlejohn, A., & Hood, N. (2016). Learning in MOOCs: A comparison study. In *Proceedings of the 4th European MOOCs Stakeholder Summit on experiences and best practices in and around MOOCs (EMOOCS 2016).* (pp. 15-26). University of Graz, Austria.

Motzo, A. & Proudfoot, A. (2017). MOOCs for language learning–opportunities and challenges: The case of the open university Italian beginners' MOOCs. In Q. Kan & S. Bax (Eds.), Beyond the language classroom: Researching MOOCs and other innovations (pp. 85-97). Research-publishing.net. doi:. mooc2016.673 doi:10.14705/rpnet.2017

Muñoz-Merino, P. J., Ruipérez-Valiente, J. A., Delgado Kloos, C., Auger, M. A., Briz, S., de Castro, V., & Santalla, S. N. (2016). Flipping the classroom to improve learning with MOOCs technology. *Computer Applications in Engineering Education, 25*(1), 15–25. doi:10.1002/cae.21774

Orsini-Jones, M. (2015). *Innovative pedagogies series: Integrating a MOOC into the MA in English language teaching at Coventry University-Innovation in blended learning practice.* Retrieved from https://www.heacademy.ac.uk/system/files/marina_orsini_jones_final_1.pdf

Orsini-Jones, M., Conde, B., Borthwick, K., Zou, B., & Ma, W. (2018). B-MELTT blending MOOCS for English language teacher training. *ELT Research Papers 18.02.* 1-41. Retrieved from https://www.teachingenglish.org.uk/sites/teacheng/files/Pub_J121_Blending%20MOOCs%20for%20English%20language%20teacher%20training_FINAL_Web.pdf

Orsini-Jones, M., Conde Gafaro, B., & Altamimi, S. (2017a). Integrating a MOOC into the postgraduate ELT curriculum: reflecting on students' beliefs with a MOOC blend. In Q. Kan & S. Bax (Eds.), Beyond the language classroom: Researching MOOCs and other innovations (pp. 71-83). Research-publishing.net. doi:10.14705/rpnet.2017.mooc2016.672

Orsini-Jones, M., Pibworth-Dolinski, L., Cribb, M., Brick, B., Gazeley-Eke, Z., Leinster, H., & Lloyd, E. (2015). Learning about language learning on a MOOC: How massive, open, online and "course"? In *Proceedings of the 2015 EUROCALL Conference* (pp. 450-457). Research-publishing.net. doi:10.14705/rpnet.2015.000374

Orsini-Jones, M., Zou, B., Borthwick, K., & Conde, B. (2017b). B-MELTT (blending MOOCs for English language teacher training): A distributed MOOC flip to explore local and global ELT contexts and beliefs. In *Proceedings of 18th International CALL Conference: CALL in Context* (pp. 608–625). Berkeley, CA: University of California. Retrieved from http://call2017.language.berkeley.edu/wp-content/uploads/2017/07/CALL2017_proceedings.pdf

Papathoma, T. (2019). *MOOC educators: who they are and how they learn.* (Doctoral dissertation). Retrieved from http://oro.open.ac.uk/59302/7/Thesis%20Final%20Papathoma_03.03.2019.pdf

Pappano, L. (2012, Nov. 2). The year of the MOOC. Education life. *The New York Times.* Retrieved from https://www.nytimes.com/2012/11/04/education/edlife/massive-open-online-courses-are-multiplying-at-a-rapid-pace.html

Pérez-Sanagustín, M., Hilliger, I., Alario-Hoyos, C., Kloos, C. D., & Rayyan, S. (2017). H-MOOC framework: Reusing MOOCs for hybrid education. *Journal of Computing in Higher Education*, *29*(1), 47–64. doi:10.100712528-017-9133-5

Pérez-Sanagustín, R. & Maldonado, M. J. J. (2016). How to design tools for supporting self-regulated learning in MOOCs? Lessons learned from a literature review from 2008 to 2016. In *Proceedings of Computing Conference (CLEI), 2016 XLII Latin American* (pp. 1-12). New York, NY: Institute of Electrical and Electronics Engineers (IEEE).

Pintrich, P. R., Smith, D. A. F., Garcia, T., & Mckeachie, W. J. (1991). A manual for the use of the motivated strategies for learning questionnaire (MSLQ). Retrieved from https://files.eric.ed.gov/fulltext/ED338122.pdf

Read, T., Bárcena, E., & Kukulska-Hulme, A. (2016). Mobile and massive language learning. In E. Martín-Monje, I. Elorza, & B. G. Riaza (Eds.), *Technology-enhanced language learning for specialized domains: Practical applications and mobility* (pp. 151–161). New York, NY: Routledge.

Read, T., Bárcena, E., & Rodrigo, C. (2010). Modelling ubiquity for second language learning. *International Journal of Mobile Learning and Organisation*, *4*(2), 130–149. doi:10.1504/IJMLO.2010.032633

Ronaghi, F., Saberi, A., & Trumbore, A. (2015). NovoEd, a social learning environment. In P. Kim (Ed.), *Massive Open Online Courses* (pp. 96–105). New York, NY: Routledge.

Rubio, F. (2015). The role of interaction in MOOCs and traditional technology-enhanced language courses. In E. Dixon & M. Thomas (Eds.), Researching language student interaction online: From social media to MOOCs. (pp. 63–88). San Marcos, TX: CALICO monograph.

Rubio, F., Fuchs, C., & Dixon, E. (2016). Language MOOCs: Better by design. In E. Martín-Monje, I. Elorza, & B. G. Riaza (Eds.), *Technology-enhanced language learning for specialized domains: Practical applications and mobility* (pp. 177–188). New York, NY: Routledge.

Sandeen, C. (2013). Integrating MOOCS into traditional higher education: The emerging “MOOC 3.0” era. *Change: The Magazine of Higher Learning*, *45*(6), 34–39. doi:10.1080/00091383.2013.842103

Sharples, M. (2019). *Practical pedagogy: 40 new ways to teach and learn.* London, UK: Routledge.

Siemens, G. (2005). Connectivism: A learning theory for the digital age. *International journal of instructional technology and distance learning, 2*(1), 3-10.

Sokolik, M. (2016). Academic writing in MOOC environments. In E. Martín-Monje, I. Elorza, & B. G. Riaza (Eds.), *Technology-enhanced language learning for specialized domains: Practical applications and mobility* (pp. 165–176). New York, NY: Routledge.

Wiley, D. (2015). The MOOC misstep and the open education infrastructure. In C. J. Bonk, M. M. Lee, T. C. Reeves, & T. H. Reynolds (Eds.), MOOCs and Open Education around the world (pp. 43–51). New York, NY: Routledge. doi:10.4324/9781315751108-2

Yuan, L. & Powell, S. (2013). MOOCs and open education: Implications for higher education- A white paper. *JISC CETIS* 1-21. Retrieved from https://publications.cetis.org.uk/wp-content/uploads/2013/03/MOOCs-and-Open-Education.pdf

Zimmerman, B. J. (1989). A social cognitive view of self-regulated academic learning. *J. Educ. Psychol., 81*(3), 329–339. doi:10.1037/0022-0663.81.3.329

Zimmerman, B. J. (2000). Attaining self-regulation: A social cognitive perspective. In M. Boekaerts, M. Zeidner, & P. Pintrich (Eds.), *Handbook of self-regulation* (pp. 13–39). San Diego, CA: Academic Press. doi:10.1016/B978-012109890-2/50031-7

ADDITIONAL READING

Martín-Monje, E., Castrillo, M. D., & Mañana-Rodríguez, J. (2018). Understanding online interaction in language MOOCs through learning analytics. *Computer Assisted Language Learning, 31*(3), 251–272. doi:10.1080/09588221.2017.1378237

Matthies, S. (2015). MOOCs and Adult language students. In *Proceedings of 8th ICT for Language Learning Conference* (pp. 54-57). Florence, Italy: Libreria Universitaria Edizioni.

O'Flaherty, J., & Phillips, C. (2015). The use of flipped classrooms in higher education: A scoping review. *The Internet and Higher Education, 25*, 85–95. doi:10.1016/j.iheduc.2015.02.002

Orsini-Jones, M., & Smith, S. (Eds.). (2018). Flipping the blend through MOOCs, MALL and OIL – new directions in CALL. Research-publishing.net.

Wong, J., Baars, M., Davis, D., Van der Zee, T., Houben, G. J., & Paas, F. (2019). Supporting self-regulated learning in online learning environments and MOOCs: A systematic review. *International Journal of Human-Computer Interaction, 35*(4-5), 356–373. doi:10.1080/10447318.2018.1543084

Zhang, X. (2017). Researching into a MOOC embedded flipped classroom model for college English Reading and Writing course. In Q. Kan & S. Bax (Eds.), Beyond the language classroom: researching MOOCs and other innovations (pp. 15-27). Research-publishing.net. doi:10.14705/rpnet.2017.mooc2016.668

Zimmerman, B. J. (2011). Motivational sources and outcomes of self-regulated learning and performance. In B. J. Zimmerman & D. H. Schunk (Eds.), *Handbook of self-regulation of learning and performance* (pp. 49–64). New York, NY: Routledge.

Zimmerman, B. J., & Moylan, A. R. (2009). Self-regulation: where metacognition and motivation intersect. In D. J. Hacker, J. Dunlosky, & A. C. Graesser (Eds.), *Handbook of Metacognition in Education* (pp. 299–315). New York, NY: Routledge.

KEY TERMS AND DEFINITIONS

Blended Learning: Usually combines face-to-face classroom instruction with the delivery of online material, which provides a new type of language learning environment different from the traditional language classroom. Blended learning practices also tend to integrate Online Intercultural Exchange projects and MOOCs into established curricula.

Class Central: An online search engine tool that classifies MOOCs into specific subjects and languages. It also displays the courses' availability, the platform where these are delivered and some reviews and rates for each course.

Connectivist Pedagogy: It emphasises learning as a process of sharing specialised information sets with other people to construct knowledge through interaction and collaboration. In other words, it encourages social networking among people to learn together in this digital era.

Flipped Learning: A learning approach that consists in working with a specific (online) material at home and using class time for interactive activities related to the topic that students are supposed to work on outside the language classroom.

LMOOC: Massive Open Online Courses where content, activities and resources are specifically related to foreign language learning. Unlike content-based MOOCs, LMOOCs materials are designed for language learning purposes.

MOOC Providers: Academic institutions that offer courses, most of them for little or no cost, in a digital education platform such as Coursera, edX, FutureLearn, etc.

Open Educational Resources: Freely accessible educational materials used by anyone who has access to the internet connection in order to fulfil different teaching or learning purposes.

Self-regulatory Strategies: Self-initiated actions employed to improve learning and performance such as setting daily or weekly learning goals to complete a specific task.

Chapter 10
The Role of Technology in Audio Text Comprehension for English as a Second Language

Pablo Armando Alcaraz-Valencia
Universidad de Colima, Mexico

Laura Sanely Gaytán-Lugo
Universidad de Colima, Mexico

Sara Catalina Hernández Gallardo
Universidad de Guadalajara, Mexico

Rocio Maciel Arellano
Universidad de Guadalajara, Mexico

ABSTRACT

This chapter proposes to establish a starting point in the design of technology specialized in the development of listening comprehension skills from a theoretical perspective, when learning English as a second language. Therefore, an exploration about how technologies have evolved in the practice of learning English as a second language was required. Likewise, authors look at fundamental aspects of technology literacy and how this is rooted in users' contexts. From this, they conceptualize the symbolic competence through the ecological theory in order to design a computer-assisted language learning practice. Findings led to the conclusion that a guide on how to build specialized technology in English learning as a second language does not exist. What indeed exists is the application of recycled technology created for other purposes but used for English learning. Authors propose a practice where the meaning is explored through the understanding of what happens in the context by using 3D holograms as an optical illusion.

DOI: 10.4018/978-1-7998-1097-1.ch010

INTRODUCTION

The research connected to the acquisition of English as a second language is a key component, due to the social and economic implications that lead to the proficiency of this language (Calderon, 2015). Therefore, a variety of learning theoretical approaches have emerged, which mainly focus on how the student builds knowledge (Buck, 2001; Flowerdew & Miller, 2005; Brown, 2006; Richards, 2008; Nation, & Newton, 2009; Abdelhamid, 2012). Proficiency of this language acts in response to a multifactorial structure that is determined by the experiences, knowledge, and the context in which the student is currently immersed in. (Lambert, & Wolfgang, 1973; Levy, 1997; Lam, 2000; Egbert, Paulus, & Nakamichi, 2002; Richards & Renandya, 2002; Kormos, & Csizér, 2008).

A variety of supporting tools are utilized in the learning process of this language. From audios/stereo, to virtual reality applications. (Slattery, & Willis, 2001; Dunleavy, Dede, & Mitchell, 2009; Liu, & Chu, 2010; Liu, & Tsai, 2013; Merchant, Goetz, Cifuentes, Keeney-Kennicutt, & Davis, 2014). However, the search to improve any approach that relates to the aforementioned learning process continues. Unfortunately, while in the field, we tend to assume that the user is the one who needs to adapt to the applied technology, and not the other way around (Norman, 2013)

Creating change in the cognitive process of a student to stimulate or boost the improvement of a certain skill, is more related to the type and quality of the interaction with the technology than the actual tool per see. (Ybarra, & Green, 2003.) In this sense, the success in the use of technology depends of a process which involves the design of a practice that is sustained in a theoretical way, built from a socio-cultural perspective where it must involve the knowledge regarding to the user, and its own context. For that reason, this research aims to provide answers to the following questions: What kind of frameworks are currently available in the construction of technology when learning English as a second language? And more specifically, how to design technology as a support tool that allows listening comprehension of texts in English as a foreign language?

THE EVOLUTION OF TECHNOLOGIES IN THE ENGLISH LEARNING PROCESS AS A SECOND LANGUAGE

The learning process through the use of technologies is directly influenced by the industry that develops hardware and software (Warschauer, 1996). Thus, the type of teaching materials that are created, are mainly a consequence of the physical and processing abilities of modern technology. Therefore, due to the advent of personal computers and color monitors, the development of teaching materials in the language learning process was made possible.

Warschauer & Healey (1998) describe in three stages (CALL) the assistance computers provide in the learning process of languages. The first stage, where they integrated computer tools with language learning strategies, consisted in making possible the filling of blank spaces during the reconstruction of ideas within a text. The issue with this stage, was that there was a lack of contextualization between what the student needed to fill out the empty space, and what the computer was capable of providing. This enabled a feedback that only included "correct" or "incorrect" (Bax, 2003).

Warschauer (1996) and Bax (2003) agree that the second stage began when computers developed email and multimedia material management abilities, which were later constituted as everyday life elements for those that acquired a new language. The idea consisted in providing the student with a context that, combined with a tutor or virtual guide, would enable directing the student in the linguistic production and recognition of the target language. This meant developing abilities, both oral and written, to enrich visual and auditive comprehension. Consequently, practices that favored memorization were starting to be a thing of the past, and a new comprehension proposal emerged. Because of this, audio cassettes were beginning to be substituted with computer rooms.

Throughout the third stage, the potential of audiovisual materials was fused with the Internet's capacity to establish remote communication within students that could simultaneously display communication abilities in real time. The latter meant placing all bets in the development of teaching materials based on an approach that would focus on social constructivism (Reinking, 2005)

Literacy Technology in the English Learning Process as a Second Language

Fraillon, Ainley, Schulz, Friedman & Gebhardt (2013) define literacy technology as an individual's ability to utilize computers and research, create and communicate. Gillen (2014) considers that literacy technology is not subjected to a group of immovable abilities, but are in constant change, which fluctuate according to the emergence of new technological-cultural interactions.

Dudeney, Hockly & Pegrum (2013) and Pegrum (2009) explain that digital literacy corresponds to the conjunction and implementation of four areas: (1) the use of language, (2) the use of the information, (3) the ability to create connections based on the flow of information that is received, and (4) the redesign of metadata. In this sense, Dunedey et al. (2013) comments that literacy based on the language when learning English as a second language, consists in more than recognizing grammatical structures surrounding oral and written texts. An image, a sound, a video, just to mention a few, are also ways of communication, and in some instances, substitute or improve the comprehension of a written or spoken word.

On the other hand, Pedgrum (2015) mentions that literacy implies: (1) the filtering of data that is compiled when browsing through the internet, and (2) the launch of an alert system that "turns on" and "turns off" when dismissing what results to be more, or less relevant. In regard to this research, the objective of digital literacy consists in establishing a practice principle, through which the student's ability is exercised in the production of abilities that will enable them to communicate.

Technology Adaptability in the Teaching of English as a Second Language

The characteristics of any group of students varies according to age, nationality, region, education's effectiveness, frequency in the use of technology outside the classroom, teachers role in the classroom, students education and socioeconomic level, motivation level in regards to acquiring a language, contact frequency with target language, students' preferred way of communication, efficiency regarding prevailing production abilities in each of the students, the role of technology in the classroom, among other things (Arslan & Şahin-Kizil, 2010; Hwang, Chen, Shadiev, Huang, & Chen, 2014; Butler, & Zeng, 2015; Oliver, Philp, & Duchesne, 2017).

Listed below are three examples regarding the use of technology, and the variability in the interaction objectives that can be established with it:

1. In Japan, English students use computers, and through educative games, develop the necessary skills to complete a standardized multiple-choice national test. That specific game presents the user with six game options, which are chosen based on the student's predilection, whose objectives (specifically the improvement of listening skills) vary according to the activity (Butler, Someya & Fukuhara, 2014).
2. In some rural areas of Uruguay, through videoconferences, a qualified Argentinian professor leads a group of students which are individually provisioned with a low-cost laptop, and a screen that can be seen throughout the entire classroom. Out of the many different activities that can be implemented through a long-distance course, the professor displays a group of images to the students that are directly related to specific new vocabulary. The activity awakens the curiosity to identify the correct pronunciation of the word, helps develop questions related to the vocabulary, ask students to jot down those questions on the computer and allows students to work with other classmates. In the classroom, the long-distance teacher is helped by an assistant that is not an expert but has a functional level of understanding of the English teaching process (Plan Ceibal, 2017).
3. A case study that took place throughout eight months, where a professor in the United States reads a chapter from a book to a group of students, out of which two thirds of them are not English native speakers. All of the students have a laptop, in which they write down all the ideas that are generated in real time while listening to the reading presented by the teacher. As the reading (which is presented out loud) unfolds, the professor makes pauses for those that are listening, allowing them to write down the ideas that emerge. The written data created by the students is saved to the class' general blog. This activity takes place two times a week in segments of twenty minutes. Furthermore, they dynamic of reading a text out loud can be substituted with a video that can be seen collectively by everyone in the classroom, and it requires all of the students to periodically write down the ideas that are developing as it relates to the topic (Zheng & Warschauer, 2015).

As we can see, educational technology can adapt to the social context where the learning process takes place. The latter can be achieved regardless of the level of sophistication of the implemented tools, the repercussions or the extent of the subjects covered. Also, the teacher's creativity is an important role since it will determine what, and how to do more with the technological resources that are available. The learning process and technology enables the production of knowledge based on the type of interaction the user has with it.

The knowledge generation is underlying to the nature of technology, which requires a specific behavior from the user, and based off the interaction, it facilitates the development of specific linguistic abilities. Communication abilities such as listening and reading comprehension, become both receptive and production skills, which require an expansion and support of other communication channels (visuals, sounds, tactile or odorants) to emphasize the possibilities of information comprehension that is presented to the user.

Technology Implementation in the Learning Process of English as a Second Language

The technology for learning purposes can be used in different ways and each of them affects (1) the acquisition of a language, and (2) the quality of the linguistic production that is generated. Examples of this can be interpreted through the following questions: How to work best when teaching listening comprehension (team or individually)?, in what way does communication should occur: (a) person to person, (b) only with the teacher – instructor - guide, or (c) with the computer (digital system)?, the implementation of technology is dedicated to understand or to produce?, is the use of technology part of a long term project or it is an isolated activity?, what is the role of technology in the activities carried out?, what are the theories behind the digital materials and the technology implemented?, how much freedom does the student have over the use of technological material?, how is the interaction process generated? and how does technology support or improve learning?

So that, there are three dimensions through which it is viable to implement technology to increment the linguistic production capacity in those that learn English as a second language: (1) Meskill & Quah (2003) propose the development of online digital contexts, where students' motivation develops through socio-affective educational processes; (2) Levy & Stockwell (2006) and Hubbard (2009a) propose to implement a socio-historic perspective of technology, where a specific set of tools are used as a working mechanism through which culture takes place; (3) Ducate & Arnold (2006) and Hubbard (2009a) propose implementing technology according to the linguistic production as it relates to the specific abilities that will be developed: listening, oral, reading and writing.

In this sense, the objectives within the technological research in the acquisition of a second language can greatly vary. Ware & Hellmich (2014) present a hierarchy in regard to research objectives: (1) the research focused in the learning process results, which consists of identifying what and how technology is able to produce measurable knowledge, through quantitative methods; and (2) research focused on the opportunities, which are known to be of exploratory nature, where the objective consists of how to implement and calibrate emerging technology in the acquisition of a second language. These research approaches do not cancel each other out, which means that it is common to find research papers that implement both approaches.

Furthermore, Reiser & Dempsey (2002) mention that knowledge takes place after undertaking an interaction process, which requires a reflection upon the given information, to internalize it, store it, and connect it with previous experiences, and consequently enhancing the ability of prediction underlying future events that with similar characteristics, are approached (in terms of comprehension) with effectiveness.

THEORY WITHIN COMPUTER ASSISTANCE LANGUAGE LEARNING (CALL) AS A FRAME OF REFERENCE

Levy & Stockwell (2006) explain that in the scope of CALL it does not exist a definitive theory that explains any distinctive nature of learning. The authors refer that CALL turns out to be a consumer of theories from other sources, with the purpose of covering three areas: (1) the teaching process, (2) the development of technology in the language learning process, and (3) the way it is developed a research from that specific point.

In this sense, Hubbard (2008) presents a review of 113 theories in regard to CALL where it is confirmed the absence of a framework that justify a theoretical approach in the development of technology, when learning a second language.

So, how to validate the integration of hardware and software that focuses on a theoretical approach in the learning process of a second language? On Table 1 it is presented a framework designed by Hubbard & Levy (2016), where theoretical perspectives are classified in order to integrate hardware and software in CALL.

In this sense, Hubbard & Levy (2016)'s framework offers a way to validate the integration of technology, based on the nature of certain skills to be improved.

The Ecological Approach *instantiation* in the CALL Theory

According to Kramsch (2002, 2006), Van Lier (2004) and Kramsch & Vork Steffensen (2008), the ecological theory in the learning process of a second language consists on interacting with a comprehensive environment or context, which expand beyond the classroom. This requires that linguistics get to be seen from a perspective where the context resembles a panacea, which in terms of knowledge, it stands firmly above socio-cultural and ethnographic factors. This theoretical perspective of the learning process requires a holistic approach towards the language. In other words, language does not constitute an isolated entity, contained within a system of linguistic structures with the purpose of being transferred and applied as if it were a formula. For these authors, language is something experiential, filled with multi-sensorial experiences, which help to generate the interaction awareness necessary for the development of knowledge based on personal, situational and cultural relationships; and the reasoning regarding different factors, which gradually continue modifying the production and evolution of the target language. (See Figure 1).

Table 1. Hubbard & Levy (2016)'s framework dedicated to the integration of hardware and software in CALL

Theorical Perspective	Description of Theorical Approach
Provision	The objective of this theoretical approach is to showcase that an ideal communication climate moderated through computers, can be examined as a group of characteristics.
Instantiation	It was developed to categorize the learning process theories, such as the theory of the assignment or ***ecological theory***, which are placed in a learning environment where language and technology can be identified as explicit elements for analysis.
Adaptation	It is referred to a process that begins with a borrowed theory. Subsequently, the researcher is led to propose a change in the structure of the base theory, an then justify the information that emerges as a result of the research in CALL.
Assembly	Emerges due to the needs of multiple theories within a single investigation. Furthermore, these specific theoretical sources are combined, while simultaneously maintaining their individuality.
Synthesis	Implements a variety of theories to be combined as a single entity. The results of this theory involve an object that can be categorized as: a new theory, a frame of reference or an approach.
Construction	Focuses in the production of a new native theory in CALL. It is a rarely implemented theory. Unlike the rest of the categorized theories, which build upon one or more theoretical structures, this one feeds off of former theories, while avoiding its disadvantages.
Refinement	This theory operates in a parallel manner with all of the categorized theories on this table. This, because as time goes by, some theories can improve, fail, or simply disappear.

Figure 1. The ecological theory approach refers to one topic by promoting the understanding of other subject. Source: personal elaboration.

Thorne (2010) explains that both the internet and the development of mobile devices and other technologies, provide an ideal support for the language learning process as it's transferred outside classrooms, where those that are acquiring English as a second language, can develop their own identity within the learning process of non-academic social contexts, under their own terms, and at a young age.

On the other hand, Mitchell, Myles & Marsden (2013) consider that the development of technology in the language learning process, influenced by disciplines such as sociolinguistic, psycholinguistic and computer linguistic, has led the development of target software; virtual environments, online videogames and social media.

Olivera & Schleppegrell (2015) and Fu & Matoush (2015) mention that from a socio-cultural approach, the key to success in the development of technological tools for language acquisition, relies on the emphasis of contexts underlying the progressive teaching interaction, both with the environment and with other individuals. The latter, achieved through the meditation of how that specific interaction takes place, where the understanding of a certain type of mental or behavioral mechanism, depends on the interaction with the development of stories within significative contexts in terms of experiences, which result in the transformation of human activity.

The Symbolic Competence

No matter where a speech travels through or where it takes place, the only aspect that stays with it, is its structure; however, it's value, significance, and function, depends on the speaker's rendition, which underlies the different ways of personal indexing of information, and it is subject to the market value as a cultural product (Blommaert, 2005).

Hult (2014), Kramsch (2006, 2009, 2011), Kramsch & Whiteside (2008) and Vinall (2012), mention that social actors within a multilingual environment are capable of activating more than one communicative competence, which symbolically speaking, is used for the purpose of comprehending a second language (See Table 2). Authors agree that the word "symbolic" as it pertains to this context, not only refers to the representation of people and objects within physical or imaginary settings, but also the development of perceptions, attitudes, beliefs, aspirations and values; through entities that represent and symbolize them. This way, the symbolic competence corresponds to the acute display of a specific skill, from which we are able to maneuver the linguistic codes of the target language, and the spatial and temporary resonances of the proper or consolidated language.

According to MLA (2007), a symbolic competence multiplies the possibilities of practicing, consequently increasing the possibilities of finding meanings through various codes of presence. Thus, the distinction in the language learning process could possibly not come from managing one or many linguistic systems, but from a person's multilingual ability in operating between languages. Based on these terms, Kramsch & Whiteside (2008) highlight that the symbolic competence is a relational construct that operates through four intertwined phases: (1) in the positioning of the topic, (2) the historicity, (3) the creation of alternative realities and (4), the reassessment of thoughts and actions.

However, according to Vinall (2016), symbolic competence is in reality perceived in a complex way: (1) How to identify a symbolic competence? (2) Where can a symbolic competence be found? (3) Specifically, for the language teaching and learning process, what new meanings emerge from a symbolic competence? (See Table 3).

Consequently, the author proposes a frame of reference from which it is possible to generate the analysis of information within a text, with the use of a symbolic competence: (a) correlation, (b) transgression and (3) potential.

TECHNOLOGY INTEGRATION IN THE EDUCATION CONTEXT OF A SECOND LANGUAGE

In the introduction of this work, two questions were presented: What kind of frameworks are currently available in the construction of technology when learning English as a second language? The short answer

Table 2. The manifestation of the symbolic competence (Hult, 2014)

Sub constructs of the symbolic competence	Definition
Positioning of the topic	Refers to the understanding of identities, suppositions, emotions, and values that are indexed in a local manner in different languages.
Historicity	Consist of the implementation of semiotic resources, to evoke a shared understanding of a social group regarding the past.
Creation of alternative realities	It's a discursive calling that relates to another scale of social organization, different in time and space.
Reassessment	Corresponds to the modification of expectations and social norms, to influence how speakers perceive interaction contexts.

Table 3. A layered interpretation of a symbolic competence (Vinall, 2016)

Characteristics of a symbolic competence	Textual analysis	Self-reflective dimension
1. Correlation	How can we identify (the analyzed topic)?	How can we identify ourselves in (the analyzed topic)?
2. Transgression	Where can we find it in (the analyzed topic)?	Where do we find ourselves in regard to the (analyzed topic)?
3. Potential	How can we reassess (the analyzed topic)?	How can we reassess to ourselves (the analyzed topic)?

is: a guide on how to develop technology that focuses on the language learning process does not exist. The closest to it consists in implementing recycled technology; non-specialized technology, initially developed for solving other issues, not for academic purposes, and which has been adapted to attempt developing knowledge in the language learning process. There also isn't any clarity as to which development method should be implemented, through which (with some level of efficiency and effectiveness) accomplishes the integration or construction of technological tools that stimulate specific abilities in the learning process of a second language. According to what was previously stated, the integration of a technology tool in this subject, follows the theoretical perspectives of the learning process, linked to expected work competences, which in reality could derive in the generation of knowledge.

Meanwhile, how can you design a technology support tool that enables the comprehension of English texts as a second language? For an academic activity where it is required the audio text comprehension in the English language learning process as a second language, we must explore the meaning by understanding what is happening from the internalization of the information through the Vinalls (2016)' interpretation of the symbolic competence: (1) the correlational level, which consists in the student identifying itself in the text. The proposed levels of recognition revolve around the dimensions: history, historical topic, or a cultural representation. For each level of recognition, a graphic and auditive environment must exist, and from the interaction it is possible that the user can get contextualized within the topic at hand. (2) Transgression, which adds on to the process of generating meaning based on the analysis of the text. It consists on stimulating the student with the interactive support (auditive and graphic material), so they can infer or suppose in a relational manner, all of this based on their own perspective of the analyzed topic. And (3) the potential stage of the symbolic competence, which consists of the resulting conclusion of the previous interactions with the audio-visual material. The idea of this last stage encourages a reflection of the established conjectures, pertaining to the type of recognition established between the student, the text, and the graphic-auditive material (See Figure 2).

Figure 2. Representation of the global theoretical approach that this work proposes in order to design a practice and technology that explores the significance of English audio texts as a second language. Source: personal elaboration

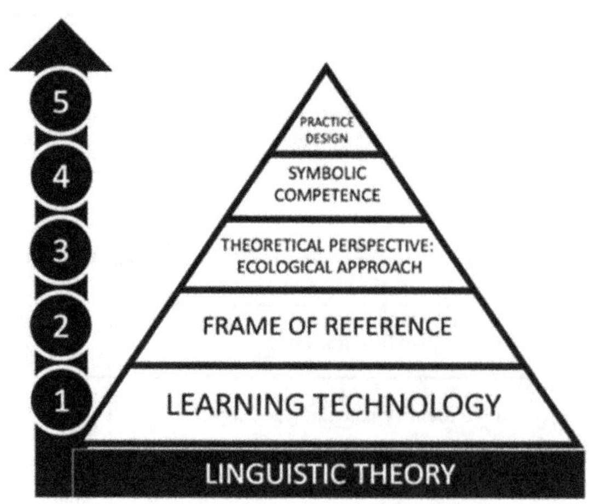

In this way and based on the analysis of information proposed in the previous table, we propose a conceptual idea of the 3D hologram interaction tool dedicated to stimulating people's cultural baggage. The goal of this process is to instance a starting point from which it is possible to analyze the information that lies in the graphic and auditory environment. In this order it would be possible to conclude with meaningful information by interacting with the context as a story, encouraging the students to use of their own cultural baggage by making inferences and assumptions through the holographic tool.

Thus, according to the three immersion levels proposed for the practice to be developed in Table 4, the initial stage of the practice design (the correlational level) consisted in make the student recognize himself within the auditory text. At this level, the phases of recognition suggested by the practice revolve around three dimensions: (1) as a story, (2) as a historical theme, or (3) as a cultural representation. Consequently, these levels of recognition at a practical level offer a graphic and auditory environment that allows from the interaction, to contextualize the user by using their own cultural baggage. This last in order to make the selection of an option when exploring the subject.

Table 4. Goals in the analysis of information in audio texts through the interaction with 3D holograms. Source: personal elaboration

Symbolic compoetence levels	Conceptual goals	Objectives in the skills	Objectives in practice
Correlation	1 - As a story 2 - As a historical subject 3 - As a cultural representation	To stimulate the cultural baggage to position the user in the subject	To provide a graphic and auditory environment that allows the user to contextualize through the selection of options
Transgression	1 - Through the texts 2 - Through time and space 3 - Through modalities 4 - Through cultures 5 - Through languages	To stimulate the ability to infer or assume logically – relationally, in order to rethinking about new ideas from the contextualization	The graphical and auditory environment is poured into a system that evaluates relational connections through the selection of options
Potential	1 - In the cultural and historical relationship of the students. 2 - In relation to the broader social, political and economic realities and power structures.	To provide feedback regarding the meaning of the produced auditory text that the user has generated	The graphic environment links the conceptual nodes to show an approximation of the meaning of the auditory text

Figure 3. The users make a correlation of information based on previous (personal) experiences to infer / or assume the subject presented as a graphic and auditory holographic interface. The goal is to stimulate the cultural baggage, as a mechanism of contextualization while it is presented the theme that will developed. Source: Personal elaboration.

The previous figure rises to the second stage in the process of generating meaning from the analysis of audio texts, at the transgression level of the symbolic competence. That is, under the graphic and auditory support that the tool provides, the student - user is encouraged to infer or suppose relationally and identify where the user is with the respect of the subject that is analyzed. The idea is to present consecutive menus of four selectable options, which are linked to each other to lead the user to a conclusion, whose value can vary according to the level of transgression generated.

Finally, in the third stage (the potentiality) of de symbolic competence, the system presents the user with a possible conclusion because of the previous interactions with the holographic multimedia three-dimensional material. The idea of this last stage is to promote a reflection of the decisions that led the user to the selection of options carried out, based on the inferences or assumptions about the subject in hand. Decisions on selecting options are made concerning on the depth of the dive with the information, that is, the type of recognition established between the user, the audio-text and the graphic-auditory material.

Thus, the work documented up to this point involved the development of theoretical considerations that derived into: (1) the integration of an unprecedented global theoretical approach in the design of practices to explore the meaning of auditory texts (see Table 4), as well as in (2) the establishment of guidelines for the design, development and evaluation of specialized technology through an interaction tool with 3D holograms as optical illusions.

Figure 4. Transgression means to transit in the immersion process by concluding information while the comprehension of the auditory text happens. This last by selecting options with a different value of accuracy in the inferred or supposed information. Source: Personal elaboration.

Figure 5. The potentiality of immersion is presented at the end of the interaction process. This last stage provides feedback on the comprehension of auditory texts in the learning of the English language as a second language. Source: Personal elaboration.

CONCLUSION

Technology in the language learning process awakens in the user communication abilities that otherwise in a traditional class would occur in one dimension. Therefore, different technologies have been used throughout the years to develop abilities connected to this learning process. However, where they are sophisticated or not, these technologies have been created for other objectives; developing a recycling of these tools when adapting them for the use of the previously mentioned learning process; and thanks to the ingenuity of the instructor, which considers the context characteristics, is how these technologies come into play in the learning process.

Given the above, a guide on how to design specialized technology in the development of listening comprehension skills when acquiring English as a second language, does not exist. Therefore, this study addresses (from a theoretical basis) a starting point for the creation of such said specialized tools. A fundamental point relies on the symbolic competence, defined as the skill that shapes the context in which one acquires a language. Thus, this paper uses the given interpretation of such said competence, represented in the correlational, transgression and potential stages, which allow the instructor or designer to develop these technological tools that are focused solely in the comprehension of auditive texts.

REFERENCES

Abdalhamid, F. (2012). *Listening comprehension strategies of Arabic-speaking ESL learners* (Doctoral dissertation), Colorado State University, Colorado.

Arslan, R., & Şahin-Kızıl, A. (2010). How can the use of blog software facilitate the writing process of English language learners? *Computer Assisted Language Learning*, *23*(3), 183–197. doi:10.1080/09588221.2010.486575

Bax, S. (2003). CALL—Past, present and future. *System*, *31*(1), 13–28. doi:10.1016/S0346-251X(02)00071-4

Blommaert, J. (2005). *Discourse: A critical introduction*. Cambridge, UK: Cambridge University Press. doi:10.1017/CBO9780511610295

Brown, S. (2006). *Teaching Listening*. New York, NY: Cambridge University Press.

Bruegge, B., & Dutoit, A. H. (2002). *Object-Oriented software engineering. Conquering complex and changing systems*. DF, Mexico: Prentice Hall.

Buck, G. (2001). *Assessing Listening*. New York, NY: Cambridge University Press. doi:10.1017/CBO9780511732959

Butler, Y., Someya, Y., & Fukuhara, E. (2014). Online games for young learners' foreign language learning. *ELT Journal*, *68*(3), 265–275. doi:10.1093/elt/ccu008

Butler, Y. & Zeng, W. (2015). Young learners' interactional development in task-based paired-assessment in their first and foreign languages: A case of English learners in China. *Education 3-13, 43*(3), 292–321. doi:10.1080/03004279.2013.813955

Calderon, D. (2015). Introducción: El derecho a aprender inglés. *Sorry, el aprendizaje del inglés en México*, Mexicanos Primero, 28. Recuperado desde http://www.mexicanosprimero.org/

Ducate, L., & Arnold, N. (2006). *Calling on CALL: From theory and research to new directions in foreign language teaching*. TX: CALICO.

Dudeney, G., Hockly, N., & Pegrum, M. (2013). *Digital literacies: Research and resources in language teaching*. Harlow, UK: Pearson.

Dunleavy, M., Dede, C., & Mitchell, R. (2009). Affordances and limitations of immersive participatory augmented reality simulations for teaching and learning. *Journal of Science Education and Technology, 18*(1), 7–22. doi:10.100710956-008-9119-1

Egbert, J., Paulus, T., & Nakamichi, Y. (2002). The impact of CALL instruction on classroom computer use: A foundation for rethinking technology in teacher education. Language Learning & Technology, *6*(3), 108-126. Retrieved from https://eric.ed.gov/?id=EJ653846

Flowerdew, J., & Miller, L. (2005). *Second language listening theory and practice*. New York, NY: Cambridge University Press. doi:10.1017/CBO9780511667244

Fraillon, J., Ainley, J., Schulz, W., Friedman, T., & Gebhardt, E. (2014). *Preparing for life in a digital age*. Cham, Switzerland: Springer International Publishing. doi:10.1007/978-3-319-14222-7

Fu, D., & Matoush, M. M. (2015). *Focus on literacy*. Oxford, UK: Oxford University Press.

Gillen, J. (2014). *Digital literacies*. London, UK: Routledge. doi:10.4324/9781315813530

Hubbard, P. (2008). Twenty-five years of theory in the CALICO Journal. *CALICO Journal, 25*(3), 387–399. Retrieved from https://www.learntechlib.org/p/74286/

Hubbard, P. (2009a). *Computer assisted language learning. Critical Concepts in Linguistics series*. New York, NY: Routledge.

Hubbard, P. (2009b). Developing CALL theory: A new frontier. In *Proceedings of the JALTCALL Conference*. Nagoya, Japón.

Hubbard, P., & Levy, M. (2016). Theory in computer-assisted language learning research and practice. In F. Farr & L. Murray (Eds.), *The Routledge handbook of language learning and technology* (pp. 24–38). Routledge Handbooks Online.

Hult, F. M. (2014). Covert bilingualism and symbolic competence: Analytical reflections on negotiating insider/outsider positionality in Swedish speech situations. *Applied Linguistics*, *35*(1), 63–81. doi:10.1093/applin/amt003

Hwang, W.-Y., Chen, H. S. L., Shadiev, R., Huang, R. Y.-M., & Chen, C.-Y. (2014). Improving English as a foreign language writing in elementary schools using mobile devices in familiar situational contexts. *Computer Assisted Language Learning*, *27*(5), 359–378. doi:10.1080/09588221.2012.733711

Kormos, J., & Csizér, K. (2008). Age-related differences in the motivation of learning English as a foreign language: Attitudes, selves, and motivated learning behavior. *Language Learning*, *58*(2), 327–355. doi:10.1111/j.1467-9922.2008.00443.x

Kramsch, C. (2002). *Language acquisition and socialization. Ecological perspectives*. London, UK: Continuum.

Kramsch, C. (2006). From communicative competence to symbolic competence. *Modern Language Journal*, *90*(2), 249–252. doi:10.1111/j.1540-4781.2006.00395_3.x

Kramsch, C. (2009). Discourse, the symbolic dimension of intercultural competence. In A. Hu & M. Byram (Eds.), *Interkulturelle kompetenz und fremdsprachliches lerneen. Modelle, empirie, evaluation*. Tübingen: Gunter Narr.

Kramsch, C. (2011). The symbolic dimensions of the intercultural. *Language Teaching*, *44*(3), 354–367. doi:10.1017/S0261444810000431

Kramsch, C., & Vork Steffensen, S. (2008). Ecological perspectives on second language acquisition and socialization. In *Encyclopedia of Language and Education* (pp. 2595–2606). Boston, MA: Springer US; doi:10.1007/978-0-387-30424-3_194

Kramsch, C., & Whiteside, A. (2008). Language ecology in multilingual settings. Towards a theory of symbolic competence. *Applied Linguistics*, *29*(4), 645–671. doi:10.1093/applin/amn022

Lam, Y. (2000). Technophilia vs. technophobia: A preliminary look at why second-language teachers do or do not use technology in their classrooms. *Canadian Modern Language Review*, *56*(3), 389–420. doi:10.3138/cmlr.56.3.389

Lambert, W. & Wolfgang, A. (1973). Culture and language as factors in learning and education. En Annual Simp on Cultural Factors in Learning, Washington.

Levy, M. (1997). *Computer-assisted language learning: Context and conceptualization*. Oxford, UK: Oxford University Press.

Levy, M. Y., & Stockwell, G. (2006). *CALL dimensions: Options and issues in computer assisted language learning*. Mahwah, NJ: Lawrence Erlbaum Associates.

Liu, P.-H., & Tsai, M. (2013). Using augmented-reality-based mobile learning material in EFL English composition: An exploratory case study. *British Journal of Educational Technology*, *44*(1), E1–E4. doi:10.1111/j.1467-8535.2012.01302.x

Liu, T., & Chu, Y. (2010). Using ubiquitous games in an English listening and speaking course: Impact on learning outcomes and motivation. *Computers & Education*, *55*(2), 630–643. doi:10.1016/j.compedu.2010.02.023

Mantle, M. W., & Lichty, R. (2012). *Managing the unmanageable: Rules, tools, and insights for managing software people and teams*. Addison-Wesley.

Merchant, Z., Goetz, E., Cifuentes, L., Keeney-Kennicutt, W., & Davis, T. J. (2014). Effectiveness of virtual reality-based instruction on students' learning outcomes in K-12 and higher education: A meta-analysis. *Computers & Education*, *70*, 29–40. doi:10.1016/j.compedu.2013.07.033

Meskill, C. & Quah, J. (2013). Researching language learning in the age of social media. En T. Michael, H. Reinders, & W. Mark (Eds.), Contemporary computer-assisted language learning (pp. 39–54). London, UK: Bloomsbury.

Mitchell, R., Myles, F., & Marsden, E. (2013). *Second language learning theories*. New York, NY: Routledge.

MLA. (2007). Foreign languages and higher education: New structures for a changed world. *Profession*, (1), 234–245. doi:10.1632/prof.2007.2007.1.234

Nation, I., & Newton, J. (2009). *Teaching ESL/EFL Listening and Speaking*. New York, NY: Routledge Taylor and Francis Groups.

Norman, D. (2013). *The design of everyday things: Revised and expanded edition*. AZ: Basic Books.

Oliveira, L., & Schleppegrell, J. (2015). *Focus on grammar and meaning*. Oxford, UK: Oxford University Press.

Oliver, R., Philp, J., & Duchesne, S. (2017). Children working it out together: A comparison of younger and older learners collaborating in task-based interaction. *System*, *69*, 1–14. doi:10.1016/j.system.2017.08.001

Pegrum, M. (2009). *From blogs to bombs: The future of digital technologies in education*. Crawley, Australia: UWA Publishing.

Pegrum, M. (2015). Language and literacies for digital lives. In E. Martín-Monje, I. Elorza, & B. García Riaza (Eds.), *Technological advances in specialized linguistic domains: Learning on the move* (pp. 16–36). Londres, UK: Routledge.

Plan Ceibal. (2017). *10 años Plan Ceibal: Hicimos historia haciendo futuro*.

Reinking, D. (2005). Multimedia learning of reading. In R. E. Mayer (Ed.), Cambridge handbook of multimedia learning (pp. 355-74). Cambridge, UK: Cambridge University Press. doi:10.1017/CBO9780511816819.024

Reiser, R., & Dempsey, J. (2002). *Trends and issues in instructional design and technology*. Boston, MA: Pearson.

Richards, J. (2008). *Teaching listening and speaking*. Cambridge, UK: Cambridge University Press.

Richards, J., & Renandya, W. (2002). *Methodology in language teaching: An anthology of current practice*. Cambridge University Press. doi:10.1017/CBO9780511667190

Slattery, M., & Willis, J. (2001). *English for primary teachers: A handbook of activities and classroom language* (Vol. 1). Oxford, UK: Oxford University Press.

Thorne, S. (2010). The 'intercultural turn' and language learning in the crucible of new media. In S. Guth & F. Helm (Eds.), *Telecollaboration 2.0 for Language and Intercultural Learning* (pp. 139–164). UK: Peter Lang.

Van Lier, L. (2004). *The ecology and semiotics of language learning. A sociocultural perspective*. Dordrecht, The Netherlands: Kluwer Academic. doi:10.1007/1-4020-7912-5

Vinall, K. (2012). ¿Un legado histórico?: Symbolic competence and the construction of multiple histories. *Journal of Linguistics and Language Teaching*, *4*(1), 102–123.

Vinall, K. (2016). "Got Llorona?": Teaching for the development of symbolic competence. *Journal of Linguistics and Language Teaching*, *8*(1), 1–16.

Ware, P., & Hellmich, E. (2014). CALL in the K-12 context: Language learning outcomes and opportunities. *CALICO Journal*, *31*(2), 140–157. doi:10.11139/cj.31.2.140-157

Warschauer, M. (1996). Computer-assisted language learning: An introduction. In S. Fotos (Ed.), *Multimedia language teaching* (pp. 3–20). Tokyo, Japan: Logos International.

Warschauer, M., & Healey, D. (1998). Computers and language learning: An overview. *Language Teaching*, *31*(2), 57–71. doi:10.1017/S0261444800012970

Ybarra, R. & Green T. (2003). Using technology to help ESL/EFL students develop language skills. The Internet TESL Journal, *9*(3).

Zheng, B., & Warschauer, M. (2015). Participation, interaction, and academic achievement in an online discussion environment. *Computers & Education*, *84*, 78–89. doi:10.1016/j.compedu.2015.01.008

ADDITIONAL READING

Alcaraz-Valencia, P. A., Gaytán-Lugo, L. S., & Gallardo, S. C. H. (2018). An Exploratory Study on the Interaction Beyond Virtual Environments to Improve Listening Ability When Learning English as a Second Language. In *Optimizing Human-Computer Interaction with Emerging Technologies* (pp. 306–331). IGI Global. doi:10.4018/978-1-5225-2616-2.ch013

Andujar, A. (2016). Benefits of mobile instant messaging to develop ESL writing. *System*, *62*, 63–76. doi:10.1016/j.system.2016.07.004

Andújar, A., & Cruz-Martínez, M.-S. (2017). Mobile Instant Messaging: Whatsapp and its Potential to Develop Oral Skills. *Comunicar*, *25*(50), 43–52. doi:10.3916/C50-2017-04

Arnó–Maciá, E. (2012). The Role of Technology in Teaching Languages for Specific Purposes Courses. *Modern Language Journal*, *96*, 89–104. doi:10.1111/j.1540-4781.2012.01299.x

Chambers, A., & Bax, S. (2006). Making CALL work: Towards normalisation. *System*, *34*(4), 465–479. doi:10.1016/j.system.2006.08.001

Chapelle, C. A. (2003). *English language learning and technology*. John Benjamins. doi:10.1075/lllt.7

Hsueh-Jui, L. (2008). A Study of the Interrelationship between Listening Strategy Use, Listening Proficiency Levels, and Learning Style. *Arecls*, *5*, 84–104.

Klímová, B. (2018). Mobile phones and/or smartphones and their apps for teaching English as a foreign language. *Education and Information Technologies*, *23*(3), 1091–1099. doi:10.100710639-017-9655-5

Kukulska-Hulme, A. (2009). Will mobile learning change language learning? *ReCALL*, *21*(2), 157–165. doi:10.1017/S0958344009000202

Shadiev, R., Liu, T., & Hwang, W.-Y. (2019). Review of research on mobile-assisted language learning in familiar, authentic environments. *British Journal of Educational Technology*. doi:10.1111/bjet.12839

Tang, Y., & Hew, K. F. (2017). Is mobile instant messaging (MIM) useful in education? Examining its technological, pedagogical, and social affordances. *Educational Research Review*, *21*, 85–104. doi:10.1016/j.edurev.2017.05.001

Tomlinson, B. (Ed.). (2011). *Materials development in language teaching*. Cambridge University Press.

Warburton, S. (2009). Second Life in higher education: Assessing the potential for and the barriers to deploying virtual worlds in learning and teaching. *British Journal of Educational Technology*, *40*(3), 414–426. doi:10.1111/j.1467-8535.2009.00952.x

Winn, W. (1993). *A conceptual basis for educational applications of virtual reality. Technical Publication R-93-9, Human Interface Technology Laboratory of the Washington Technology Center*. Seattle: University of Washington.

KEY TERMS AND DEFINITIONS

CALL: An approach to teach and learn language through the support of computers.

Context: Underlying information to an object, person, or situation scenario, which allows serve as a starting point to generate inferences or assumptions about a subject, all of this based on our past experiences.

Ecological Theory: Environmental system based on the development of persons through different scenarios which influence their cognitive, moral and relational development.

Hologram: An image that cheats the brain's visual perception in order to recreate three dimensional objects in our own dimension.

Instructional Design: The practice of creating instructional experiences that make acquiring knowledge and skills more efficient, effective and attractive.

Listening for Comprehend: To pay attention when listening to undertake strategies that lead a person to understand the perceived audio input.

Symbolic Competence: The ability to manipulate symbolic representation, symbolic action and symbolic power.

Virtual Environment: Recreation of reality (or an alternative reality) encapsulated into a software program.

Chapter 11
Student Binge Studying, Recall, and Success in a Blended Korean Class

Julie Damron
Brigham Young University, USA

Jennifer Dobberfuhl Quinlan
Brigham Young University, USA

ABSTRACT

With more than 5 million college students taking at least one online or blended college course, (Seaman et al, 2018) and 5 percent of K-12 students taking an online or blended course (The Evergreen Education Group), research regarding student success in blended and online classes has begun to emerge and shed light on potential trouble spots. This chapter uses an action research model to further examine binge studying in two, beginning university Korean classes in relation to success on exams, overall course grades, language recall after a four-week break, and continued enrollment in Korean. Findings indicate a correlation between overall page views and course grade, between overall page views in 101 to retention in 102 and a correlation was also evident between binge studying and 102 assessment. A negative correlation was seen between binge studying and 101 final exam scores. The strongest correlation appeared to be between non-binge studiers, or those which dispersed their study throughout the semester, and language retention in Korean 102.

INTRODUCTION

Significant attention has been given in the literature to the trend of increasing rates of online education. Because of its widespread availability, convenience to students and instructors alike, and accommodation for diverse groups of students, online classes are offered more widely and more freely, with online education available at a majority of degree-granting institutions in the United States (Allen & Seaman 2016; Ryan, Kaufman, Greenhouse, She, & Shi, 2014). Similarly, blended learning, which integrates

DOI: 10.4018/978-1-7998-1097-1.ch011

online and face-to-face components, is also on the rise in the United States and is frequently offered as an option for students to integrate into their learning experience (Bonk & Graham, 2006). As of 2014, 70.7% of institutions providing higher education offered distance courses in some form, whether that was fully online or blended (Allen & Seaman, 2017).

Not only are online classes widely and readily available, but increasingly more students are taking advantage of this opportunity: Allen and Seaman (2016) indicate that while only 9.6% of undergraduate students were taking at least one distance course in 2002, by 2016, this number had risen to 29.7%. The range of courses offered online and in blended formats is expansive: almost any course can be found offered online, including world languages. Enrollments in online and blended world language courses are also on the rise, though the number is still modest in comparison to the number of online courses offered across the world (Martin-Monje, 2017).

This chapter starts by exploring the distance education landscape broadly, including important elements that may contribute to success in online *courses*. Next, typical online student profiles, including characteristics which may impact *student* success in online courses, are discussed in order to give contextual meaning to the research questions and conclusions in this study. The methods section details the setting, sample, and hypothesis/research questions for the study. Finally, findings are evaluated and implications for action are considered in the results and discussion sections of the chapter.

Distance Education Landscape

Who is offering online/blended coursework, and who is taking the courses? A handful of universities are significant contributors to online/blended course offerings in higher ed. For instance, Southern New Hampshire University, Western Governors University, Brigham Young University-Idaho, University of Central Florida, and Grand Canyon University were among the leaders of universities offering distance education, earning the top five spots of distance enrollments from 2012-2015 (Allen & Seaman, 2017). In general, however, public universities lead out, both in in terms of number of students enrolled in distance courses as well as number of distance courses offered. Additionally, larger institutions (i.e. those with over 15,000 enrollments) consist of the most students participating in online classes (Allen & Seaman, 2016).

With the continual development of new technology, online teaching techniques have progressed rapidly, allowing students to participate in the learning process through recording audio, editing photos, completing multimedia projects, working on collaborative documents, and utilizing other online resources (Jabeen & Thomas, 2015). These applications and tools have revolutionized the course of online and blended teaching and allow students and instructors to participate in their education more actively and regularly. Despite the access to many tools and applications, some researchers question whether these resources used in online and blended settings produce significantly different results than those provided in the traditional classroom setting. Previous attempts to understand the effectiveness of online courses and their teaching tools and methods have involved examining course learning outcomes, student satisfaction, and student perception of learning (Weldy, 2018). Fendler, Ruff & Shrikhande found that some studies reveal online learning produces inferior outcomes, some show online produce equivalent outcomes, and yet others show online produces outcomes superior to those of traditional classroom counterparts (2016). Whatever these outcomes and opinions indicate about the effectiveness of online courses, the fact remains that an effective course, whether online or face to face, is influenced by a wide range of factors.

FACTORS IN COURSE SUCCESS

In any course, there are two important roles that contribute to a student's success in the class: the instructor and the learner. The role of the instructor is an important one in online classes. Recent research indicates that the instructor's presence in on online course is one of several factors contributing to effective online teaching (Fendler, Ruff, & Shrikhande, 2016). Beyond teaching the material, the instructor must also facilitate online participation and gauge when students are struggling with content, all without the physical presence of the students to help the instructor discern which students are succeeding in the class. In order to produce successful students, an online instructor must place the responsibility of learning on the students and promote interaction among them to best facilitate learning, as an instructor would do in a traditional class, but to an even larger extent.

Instructor Role

A study of English students in the United Arab Emirates underscores the importance of the instructor. Students were surveyed following their completion of an entry-level language course. Results indicated reading and listening skills were better for students who interacted with instructors than for those who had online interactive sessions with peers in place of instructor interactions. Students found learning with an instructor to be more effective than simply studying by themselves or with peers (Jabeen & Thomas, 2015). In online and blended language classes, this notion is particularly important; students consistently report that face to face interaction with their instructor is of a great value to them (King, 2016; Damron & Dobberfuhl Quinlan, 2018).

Opportunities for Interaction

Like traditional courses, in an online or blended course, students need opportunities for frequent formative assessment in addition to methods of communication with instructors in order to inform both the instructor and themselves about their progress and to receive feedback (Goertler & Gacs, 2018). Without these essential practices--practices that an instructor can promote and facilitate--student success in online classroom may be impeded. Thus, in an online or blended course, the instructor must take on a wide range of roles: not only must they instruct students in course material, but they also must act as a facilitator, content creator, assessment evaluator, and course organizer, among other roles they must take on to benefit the students and their learning process (Martin-Monje & Castrillo, 2016).

Student Characteristics

Despite the significant attention it has received, various arguments exist as to what a student must do or be to succeed academically online: some have investigated demographics and their influence on student performance in the online and blended classroom setting; others have investigated factors that may influence a student's performance that may go beyond basic demographics, such as participation, motivation, and study habits or learning style (Li, 2019; Yukselturk & Bulut, 2007; Breslow et al., 2013; Martin-Monje, Castrillo, & Manana-Rodriguez, 2013). Each of these is discussed further in Online Student Profiles.

ONLINE STUDENT PROFILES

Understanding the learner profile of online students may shed light on these success factors. Many students who take online courses are younger, educated men of developed countries seeking formal education at the higher level (Shrader, Wu, Owens, & Santa Ana, 2016). This demographic is well-represented at universities, and universities are the primary offerers of online courses. Therefore, this relationship is not surprising.

The profile of students enrolled in online or blended language courses, on the other hand, is slightly different. Like students enrolled in other online courses, most online language learners are pursuing higher education and have university qualifications; however, they differ in the fact that online or blended language students are primarily female and are in a slightly higher age bracket; they tend to be in their 30s, as opposed to the majority of online students in general, who are typically in their 20s (Martin-Monje, 2017).

Additionally, students taking online courses are typically undergraduates – undergraduate enrollments outnumber graduate students 5 to 1 (Allen & Seaman, 2017) – but enrollment rates among both groups are rising. From 2014 and 2016, the percentage of undergraduate students enrolled in online education courses rose by 3%, and graduate enrollment rates increased by 4% (Lederman, 2018).

Many students who favor online learning are otherwise limited by family and work responsibilities (Beqiri, Chase, & Bishka, 2010; Bocchi, Eastman, & Swift, 2004). While this particular information does not indicate what demographic of student is represented by those who may be limited by work and family responsibilities, it does provide insight into the background of many students who take courses online.

Impact of Demographics

Before examining the effect of student demographic on their ability to succeed in an online or blended course, it is important to first define what constitutes success in a non-traditional course. One measure of success in an online course can be found in a student's ability to complete the course, as online courses have notoriously low completion rates. One study showed that reported rates of completion fell between 5% and 12% (Morris, Hotchkiss, & Swinnerton, 2015). However, this definition is just one way of defining success in an online or blended course. Success in an online or blended course setting can also be defined more generally by not just student course completion rates, but also by final exam grade or overall course (Conijn, Van den Beemt, & Cuijpers, 2018).

Traditionally, in studies of demographics of students taking online courses--and more expansive studies that investigate how demographics affect student success in the course-- three main demographics have been studied: age, gender, and educational background (Li, 2019). A 2007 study investigating the relationship between demographic information--age, gender, and so on--and student success in an online computer programming course indicates no relationship between a student's basic demographic information and their success in a course (Yukselturk & Bulut, 2007). This finding, at least in regard to age, was confirmed by Breslow et al., who stated that no correlation was found between age and success in a course, where success was defined as grade awarded rather than completion to determine a student's achievement (2013).

However, contradictory evidence surrounding the effect of age on student performance in an online course was explored in Martin-Monje, Castrillo, and Manana-Rodriguez's research that found statistically significant differences between age groups and their study habits, which may in turn influence success in the course. Students in the age groups 21-30 and 41-50 were found to have had statistically significant differences between their frequency of downloaded materials; that is, students from the 41-50 age group downloaded course materials significantly more (Martin-Monje, Castrillo, & Manana-Rodriguez, 2017). Whether there is a correlative relationship between frequency of course material download and individual student outcomes was not determined, but it seems that older students, at least outwardly, expressed better study habits.

In one study, students in both online and traditional classroom settings for an undergraduate finance class were compared based on their final exam score. Several predictors for their success on the final exam were also evaluated, including declared major, gender, and learning style. While the correlation between gender and grade on the final exam and major and grade on the final exam did not exhibit a statistically significant relationship, learning style and success in the class demonstrated a correlative relationship. Results indicated that active, balanced, and reflective learners were distributed more in the top quarter than in the bottom (Fendler, Ruff, & Shrikhande, 2016). Within the online class, however, males were more likely to be in the top quarter than females, though the cause of this trend is unclear. Thus, high performance in an online or blended course cannot be due simply to basic demographic differences like gender or age differences alone.

Beyond the basic demographic differences of age and gender, other more narrow demographic separations did seem to predict student success in the online classroom, that is, whether they stayed in the class or dropped out before the course ended. Among these factors were educational background, relevant experience and skills, and psychological attributes (Lee & Choi, 2011).

Academic background, or a student's aptitude for academic skills and previous academic experience, can predict a student's success in a course. Students with lower GPAs, SAT scores, and generally lower markers for academic success experienced higher dropout rates in online courses, whereas students with higher scores in each of these areas were more likely to be among the students who did their online courses to completion (Lee & Choi, 2011).

Previous experiences and skills can also indicate student success in an online or blended course; that is, students with previous knowledge of the subject matter being taught, or skills necessary for a university level course, overall experienced more success in online courses (Lee & Choi, 2011). Additionally, students with a background in online courses specifically, meaning those who have previously taken an online course, receive higher scores on exams and assignments in the class, presumably because they have learned the skills necessary to succeed in online classes, and like the format of online classes and continue to pursue their education in that way (Levy, 2007).

For example, Morris, Hotchkiss, and Swinnerton (2015) identified that learners' age, previous experience with online learning, educational attainment, and job status predicted their learning outcomes. Kennedy, Coffrin, De Barba, and Corrin (2015), on the other hand, reported that the best predictor for success is instead the learners' prior knowledge of the topics covered in the course. Breslow et al.(2013), conversely, found that prior knowledge of the course's topics did not, in fact, predict success; rather, students' educational attainment could significantly predict success.

In addition, many assume that because today's undergraduate students are digital natives, meaning they grew up with access to the internet and the tools it offers, they will succeed in online learning. However, success in online learning goes beyond simple familiarity with technology although often, students may grow overwhelmed or disoriented by the technological aspect of the online classroom (Knox, 2014). Despite this challenge that may present itself to some students, technologically adept students do not necessarily have a head start in online classes, simply because of the fact that they are more familiar with using technology (Comer, Lenaghan, & Sengupta, 2015).

Factors such as gender, prior knowledge of course content, technological adeptness, and previous academic skill sets are often beyond a student's control, and though they may influence a student's success in the course, failure because of one of these factors--or success, for that matter--is through no fault of the student on their own. However, there is a wide variety of other factors that may preclude a student's success.

The most general predictor for success for students in an online course that lies within the individual student's own control is simply their activity in the course: students who are more active in the class and in their learning process are more likely to succeed (Conijn, Van den Beemt, & Cuijpers, 2018).

Active Participation and Motivation

Being an active participant in their own learning can manifest itself in many ways among students, including the student's attitude towards learning; their motivation and desire for taking an online course; their self-control and focus; and so on (Lee & Choi, 2011). For example, students with an internal locus of control--that is, students with the mindset that their own behavior causes outcomes, rather than external circumstances, such as luck, the quality of an instructor, or the design of a course--tend to perform better in online courses (Lee & Choi, 2011). Students with an internal locus of control are those who tend to be self-motivated and have a good measure of self-control, and predictably, experience more success than students who have an external locus of control and blame their surroundings and environment for their circumstances. Students with an internal locus of control are able to maintain their work ethic and motivation despite setbacks, which may be numerous in an online or blended course, particularly language courses: from reduced interaction with instructors to frustration with a new language to distraction and lack of self-motivation (Lee & Choi, 2011).

Student motivation is also an important predictor for success in online courses. In general, online and blended courses experience a fair amount of diversity in student desire to take the course, with some students taking the course for academic preparation or requirement, while others take online classes to simply examine the process of online education (Li, 2019). The array of reasons that students take online courses can also be helpful in indicating whether a student will excel or struggle in a particular course; students who are motivated by a desire to study the particular topic offered by their course or a desire to learn and gain something out of their online course will predictably experience different outcomes than students who are unmotivated or who lack a desire to work or interest in the course (Li, 2019).

The varying degrees of effort and motivation expressed among students enrolled in online courses can often produce different grade outcomes for students. Like traditional classes, online and blended classes also have certain traits, practices, and behaviors that can predict success among the students in the class.

One such behavior is active participation, which often reflects a student's motivation and interest in the class and can predict their success in the class. Students who participate both frequently and actively in the class tend to experience more success in the course. Student participation manifests itself in various forms, and Martin-Monje, Castrillo, and Manana-Rodriguez (2018) identify five different types of participants in online classes, categorizing them as (1) "no-shows," or students who register but never log in, (2) "observers," who log in, browse content, but do not take assessments in any form, (3) "drop-ins," who engage somewhat in online classroom activity, (4) "passive participants," who consume content available through the course but do not actively participate with the material, and (5) "active participants," who fully participate in the online course.

Additionally, Anderson and Elloumi (2014) identified five different styles of student engagement used to predict student learning outcomes in the class: (1) viewers, who watch lectures but do not hand in assignments, (2) solvers, who hand in assignments but generally do not view lectures, (3) all-rounders, who strike a balance between viewing lectures and turning in assignments, (4) collectors, who download lectures and hand in few assignments, (5) bystanders, who perform little activity in the course.

While these categories are perhaps not definitive, Martin-Monje, Castrillo, and Manana-Rodriguez argue that online language learners--particularly those who are initially classified as no-shows or observers--should be encouraged to become active participants, through participating regularly, submitting assignments, and viewing teaching material (Martin-Monje, Castrillo, and Manana-Rodriguez, 2017).

Participation allows students to solidify concepts they have learned, teach other students, clarify new information, and discover areas where they are unsure (Dallimore, Hertenstein, & Platt, 2017). However, because of the online setting, participating, for students, and facilitating participation, for instructors, may be a difficult task.

In order to facilitate participation online, many courses employ online discussion formats such as forums. Successful students more frequently participated in online discussions--making comments, engaging with other students, and so on--than did non-successful students, as concluded in a study by Martin-Monje, Castrillo, and Manana-Rodriguez (2018) examining students enrolled in a "How to succeed in the English B1 Level Exam" course.

Impact of Study Habits

Student success in an online course cannot be limited to participation only; students who hope to achieve high marks in an online or blended course must also engage in proper study habits outside of the online classroom to help them do so. Students who, on the other hand, engage in poor study habits often experience a lack of success in the online classroom. David Glenn (2007) suggests one poor study habit that may predict a student's failure in their courses: cramming. This poor study habit may also be extended to students' in the online classroom.

A traditional classroom example of cramming is exemplified in the practice of studying flashcards--a common study method of students, particularly in language classroom settings where memorization of vocabulary is a required skill. When studying flashcards, unfamiliar facts are repeated continuously in immediate succession and students convince themselves of the false notion that they have memorized facts better than they actually have. David Glenn (2007) suggests that an improved approach involves intervals of time, where studying is spaced out and cycled.

Extending this suggestion to the online classroom could involve spacing out studying by extending the time between log-ins, and logging in consistently over a longer period of time rather than logging in infrequently to view content, but occasionally engaging in an intense burst of studying, perhaps immediately before an exam (Brinthaupt & Shin, 2001).

A poor study habit similar to cramming that may predict a student's lack of success in an online course is procrastination. Like students who cram, students who procrastinate leave much of their coursework and study until the last minute. Not only does this practice often lead to stress for students, but can also predict lower grades for students; in one study investigating undergraduates enrolled in an introductory psychology course, on both term papers and exams, procrastinators received lower grades than their non-procrastinator counterparts (Brinthaupt & Shin, 2001).

Thus, in order to avoid this failure in online academic courses, successful students must not practice these poor study habits. Rather, students must actively engage in proper study habits, including participation in the classroom setting, timely completion of assignments, and studying course material frequently and consistently.

Students who completed the course, which perhaps may not be indicative of a high grade in the course but at least indicated more success than the students who dropped out a point during the semester, were found to submit assessments with greater frequency than subjects who did not complete the course (Martin-Monje, Castrillo, & Manana-Rodriguez, 2018). Thus, student success in an online or blended can be related to their completion of assessments. Students who more frequently complete assessments are more likely to stay enrolled in the course.

Additionally, increased exposure to online learning materials can predict student success (Conijin, Van den Beemt, & Cuijpers, 2018). For example, in one study of an online programming class, students who frequently replayed videos were 33% less likely to drop the course than students who rarely replayed videos (Sinha, Jermann, Li, & Dillenbourg, 2014). Moreover, students who watched significantly more videos than average were 37% less likely to drop the course compared with students who watched average or less than average proportions (Sinha, Jermann, Li, & Dillenbourg, 2014).

Thus, students who are exposed more to learning materials, and exposed frequently, experience more success. However, number of exposures may not directly predict course outcomes, such as final exam grade, quiz grades, overall grade in the course, and so on. As stated previously, time spent between exposures to material is often important in a student's learning outcome (David Glenn, 2007).

Re-exposure to course topics is especially important leading up to major assessments, such as the final exam, typical of many traditional and online courses, as it helps students retain and reinforce information previously learned in the course (Glenn, 2007). In one study, researchers demonstrated this principle as they investigated student usage of course items--in this case, videos and modules related to the topics covered--in a graduate level Quantitative Modeling and Analysis class (Conijn, Van den Beemt, & Cuijpers, 2018).

Throughout the semester, several modules were available to students to present learning outcomes of the course and provide instruction in a specified topic. Researchers investigated trends among students who started accessing--and re-accessing--module material earlier on in the term to see what relationship this had with whether or not they passed the final exam (Conijin, Van den Beemt, & Cuijpers, 2018). Their findings suggested that students who passed the exam, on average, started reviewing the modules earlier on in the semester than the students who failed the exam. In addition, their findings indicated that students who passed the course final exam invariably showed more constant activity in the course's modules throughout the semester than the students who failed the final exam; in general, these students almost exclusively neglected to access the modules throughout several weeks of the semester (Conijn, Van den Beemt, & Cuijpers, 2018).

The general finding of this study can be summarized by the idea that students who spread out their learning, which includes exposure to new material, review of material for upcoming tests, and frequency of revisiting material, succeeded more on the final exam (Conijn, Van den Beemt, & Cuijpers, 2018). It can therefore be concluded that in order to experience success in online courses, students must view and engage in course material not only frequently, but also regularly and consistently throughout the course of the term. Spacing out the learning process online can mean that students log in consistently throughout the term, rather than leading up to the final exam, and expend time and effort doing so, rather than just passively viewing material (Glenn, 2007).

Within the online and blended language classroom specifically, similar factors can predict success for students enrolled in the course. When used appropriately by students and instructors, the online and blended classrooms, learners of the language can actively engage in their learning process and experience success in the course (King, 2016). For example, time spent viewing course material in an online language class is positively correlated with success in the course (Erçetin, Lee, Kadel, Madden, & Gazi, 2018)

Language courses in particular additionally require a level of interaction with peers, classmates, and instructors in order to facilitate language learning and solidify principles. In online and blended courses, then, it is important that students engage with their classmates and receive feedback from them; in doing so, they are able to correct misunderstood principles, engage in their own learning process, and solidify concepts covered in the course (Martin-Monje, Castrillo, & Manana-Rodriguez, 2017).

However, it is not clear that engagement with peers will necessarily correspond to a high score achieved in the course, that is, submission of peer feedback and participation in activities that involve classmates do not necessarily predict success in an online language course. This may be attributed to the fact that students are not communicating in the target language of the course when interacting with their peers, so they are not actually improving their language skills (Martin-Monje, Castrillo, & Manana-Rodriguez, 2017).

Thus, the brunt of the work in a course must come down to the students themselves and their engagement with course material and work ethic: students must be frequently accessing materials, reviewing previously learned concepts, logging in to view content regularly rather than sporadically, and avoiding binge studying in the weeks leading up to major assessments, in order to truly succeed in an online or blended language course (Erçetin et al., 2018; Martin-Monje, Castrillo, & Manana-Rodriguez, 2017).

In light of online and blended language learning research, both broad and more specific, questions concerning binge studying, regular and consistent page views within a course and success in that course, seems in its infancy. In an effort to better understand the relationship between interaction in the online portion of a language class with success in that class, researchers carried out the following research.

METHODS

Research for this project took place at a large university in the western United States during the fall 2018 and winter 2019 semesters. Researchers used an action research model to examine student grades on final exams in two, sequential blended learning model Korean classes. Data were gathered from the Canvas (the learning management system) on minutes online, page views, and student scores. Students took a written final exam in class at the end of Korean 101 and the same exam ('assessment test') at the beginning of Korean 102 in the next semester. Data from these results were also gathered from the Canvas Learning Management System. Details about the setting, participants, and research questions for this study are detailed below.

Setting

This study took place at a university where a blended model of Korean 101 and 102 was taught on campus. It is relevant to understand the difference between the setting for the blended courses as compared to the traditional face-to-face (F2F) counterparts.

The F2F version of the Korean courses involved meeting together five days a week in a classroom on campus. Teaching assistants (TAs) led class sessions three times a week, and the professor led class sessions the other two days a week. In the blended model, the number of instructional days was decreased to four days per week, in which one was led by the professor and three were led by the TAs.

In the blended version, specific content and activities took place online, and F2F meetings were reserved for conversation, applying principles learned and practiced online, and providing live peer/TA feedback on written and oral assignments.

Participants

Students who participated in this study were enrolled in Korean 101 in fall 2018 and Korean 102 in winter 2019. Fifty-four students who were all enrolled in Korean 101 were included at the beginning of the study. Ten of the participants were male and forty-four were female students. Sixteen students were freshmen, eight were sophomores, twelve were juniors, fourteen were seniors, three were post baccalaureate students and one was a master's student.

Information regarding students' declared majors was used to identify whether students were enrolled in Korean 101 as an elective or as a required class to fulfill major course requisites. At the time of the study, two students had declared a Korean major, two students had declared a linguistics major (Korean fulfills a graduation requirement for linguistics majors), nineteen students were either open major or undeclared and all other students had declared majors unrelated to Korean language such as economics, microbiology, nursing, exercise science, illustration, etc.

Hypotheses and Research Questions

Action research is designed to inform the researcher's practice with the intent of being able to improve his/her practice. For this study, the researchers hypothesized that students in blended Korean 101 were binge studying and that this practice had a negative impact on language retention in Korean 102 as well as on likelihood to choose to continue Korean studies into 102. This study was designed to explore the following research questions:

1. Do overall 'page views' correlate with success in Korean 101?
2. For students who continued Korean studies into 102, did the overall page views in 101 look different than those of the students who took 101 only?
3. Do page views specifically devoted to studying the final exam review correlate with success on the written final exam?
4. Do students whose page views online are evenly dispersed throughout the semester perform better on the final exam than students whose page views come primarily immediately before the final exam?
5. After a four-week break, do students who had more overall page views during the course have better recall on the same assessment?

To analyze collected data, researchers used a Pearson correlation to examine possible relationships. Additionally, a paired samples t-test was used to evaluate student scores on the final exam from Korean 101 and the assessment test from the start of Korean 102 to explore the amount of recall after a lapse in time.

It is important to note that data collected and analyzed was tied to page views rather than minutes spent online. Because minutes spent online may include time where a student was logged in but stepped away from the computer, page views was determined to be a measures less subject to variability. While a student could also click on a page and do nothing on that page or spend quite a bit of time working on activities and consuming content on that page, by not measuring actual time spent, the researchers felt the measure was more specific than measuring minutes. Rather than trying to quantify how much time students spent with material, researchers tried to capture how many times specific pages were visited.

RESULTS

This study was designed to assist the researchers in refining their practice related to Korean 101 and 102. Data was collected to try to capture whether students were binge studying in the blended Korean 101 courses and, if so, the impact of the practice on student grades in 101 as well as their retention upon entering 102. Results for each of the research questions is detailed below, followed by potential implications for practice (in the Discussion portion of this chapter).

Relationship Between Overall Page Views Online and Success

Mean score on the Korean 101 final exam was 78.8; upon exempting zero scores, the mean score was 85.0. Mean number of total course 106. There was a very weak positive correlation of page views to final exam grade (r= 0.10). See Figure 1. Note, student final exam scores reflect some grades over 100. There were up to five extra credit points available, thus totaling 105 points possible on the exam.

The researchers also considered overall page views and overall course grade (OCG) in Korean 101. Results revealed a slightly stronger but still weak positive correlation ($r=0.16$). See Figure 2.

Page Views for Continuers into 102

Of 54 total students, 29 continued on to take Korean 102 after completing 101. Twenty-five of the 54 did not continue into102. For students that did not continue to 102, the mean number of overall page views and final exam scores were lower than those that did continue to 102 (see Table 1).

Binge Exam Preparation Relationship to Final Exam Scores

Page views devoted specifically to final exam study showed a weak positive correlation to final exam score in Korean 101 (r= 0.11). Pages devoted to final exam preparation included preparing for the final exam, final review page/final review slides, and practice final exam. See Figure 3. Considering final exam preparation may consist of reviewing content from anywhere in the course and not just the final exam prep materials, a correlation was also run between course page views during the month of December (prior to the final exam) and final exam score, which also revealed a weak positive correlation (see Figure 4).

Figure 1. Correlation of total course page views to final exam grade for Korean 101 (N=54; r=0.1)

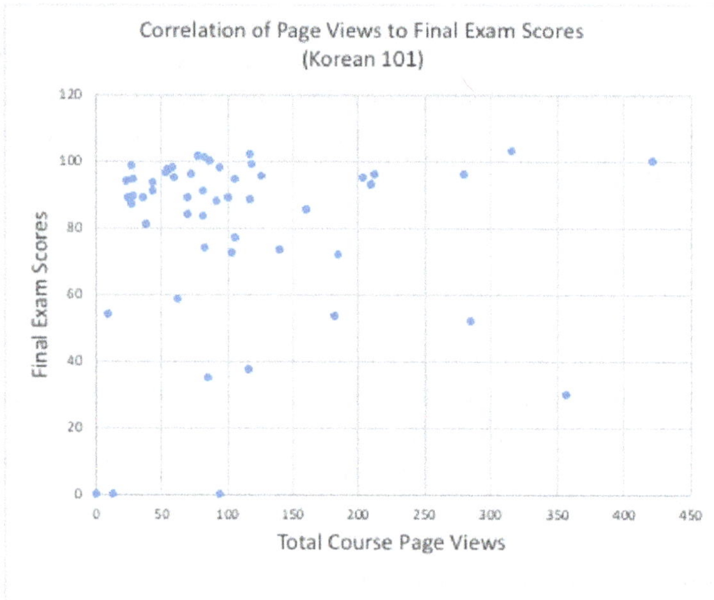

Figure 2. Correlation of total course page views to overall course grade for Korean 101 (N=54; r=0.16); grade based on 4.0 grade scale

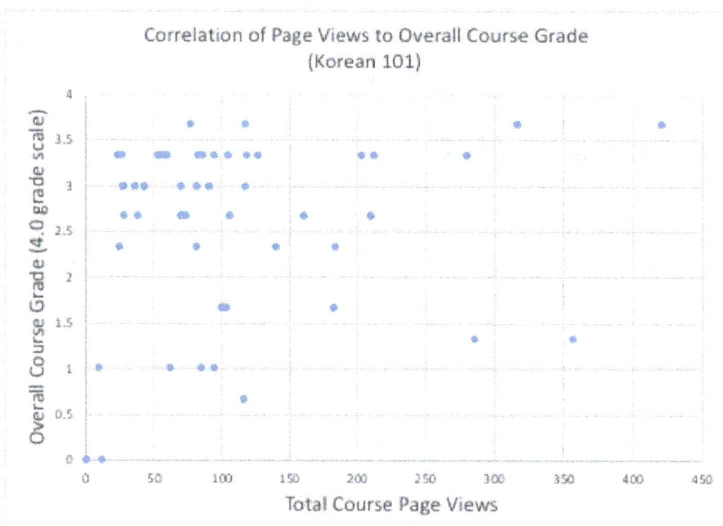

Table 1. Mean final exam scores and page views in Korean 101

	Students who did not take 102 (N=25)	Continued to 102 (N=29)
Mean score on final exam (Korean 101)	72.5	86.5
Mean number of page views (in Korean 101)	70.7	137

Figure 3. Correlation of page views for final exam preparation content to final exam scores in Korean 101 (N=54; r=0.11)

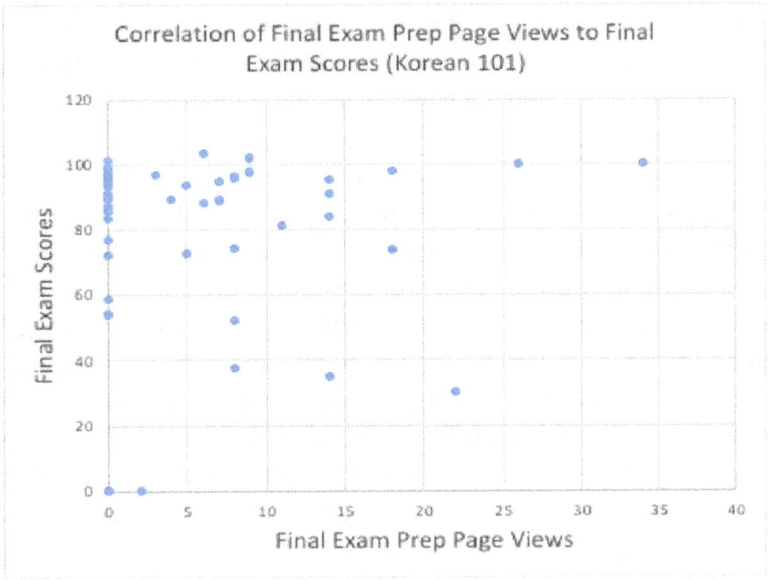

Figure 4. Correlation of page views during the month of December to final exam scores in Korean 101 (N=54, r=0.13)

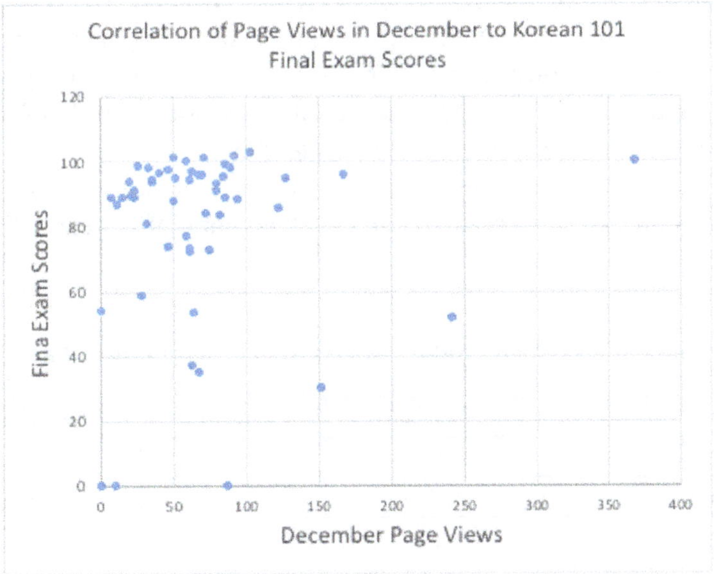

Binge Studying Relationship to Final Exam Grades

The final exam for Korean 101 was administered during the third week of December. Researchers evaluated page views throughout the semester (see Table 2). Researchers expected page views to be high in the first month of the semester, based on departmental historical observations. Researchers also theorized that students were binge studying by viewing pages many more times in the month where the final exam takes place. These hypotheses proved out in the data compiled for this study.

Researchers hypothesized that binge studying would have a negative impact on final exam scores in 101; however, students who showed evidence of binge studying had a mean final exam score of 82.6. Mean final exam score for students with even dispersion of page views was 66.4. The total number of page views was divided by four (total number of months the course was taught) to determine mean number of page views per month; adding 20% to this mean was the criteria for determining "higher page views" in the month of December than the mean. Data revealed that on average students accessed course pages more than twice as much in the month of December than in any other month of the semester.

Researchers also used a Pearson Correlation to evaluate the relationship of mean page views per month to final exam grades. Binge studiers had a 0.12 correlation, whereas non-binge studiers showed a correlation of 0.22 (see Table 3). Although non-binge studiers had a lower mean final exam score, the correlation between evenly dispersed page views in the course and final exam score was stronger than for binge studiers.

After a four-week break, do students who had more overall page views during the course have better recall on the same assessment?

Two weeks into students' winter semester in Korean 102, which represented a four-week break from when they had their final exam in Korean 101, they were given the same assessment as they had for their final exam in Korean 101. A correlation of total page views in Korean 101 to Korean 102 assessment scores revealed nearly no correlation (r= -0.02). Evidence of binge studying via exam prep pages in 101 and December page visits in 101 showed varied correlations (r= 0.3 for prep pages views and -0.11 for December views). See Figures 5, 6, and 7; note that results were not controlled for outliers.

Finally, a paired samples t-test was run to evaluate retention from Korean 101 final exam (pre-test) to Korean 102 assessment (post-test). There was no statistical difference in student scores on the same assessment when taken at the end of Korean 101 and the beginning of Korean 102 (p=0.680), thus accepting the null hypothesis.

Table 2. Sum of page views per month and mean page views per month per student

	September	October	November	December
Total page views (N=54)	1316	845	650	2992
Mean number of page views per student	24	16	12	55

Table 3. Impact of binge studying on final exam scores; correlation of binge studying to final exam score

Evidence of binge studying	Yes (N=41)	No (N=13)
Mean final exam score (Korean 101)	82.6	66.4
Correlation (binge studying to 101 final exam score)	.12	.22

Figure 5. Correlation of overall course page views in Korean 101 to assessment score in Korean 102

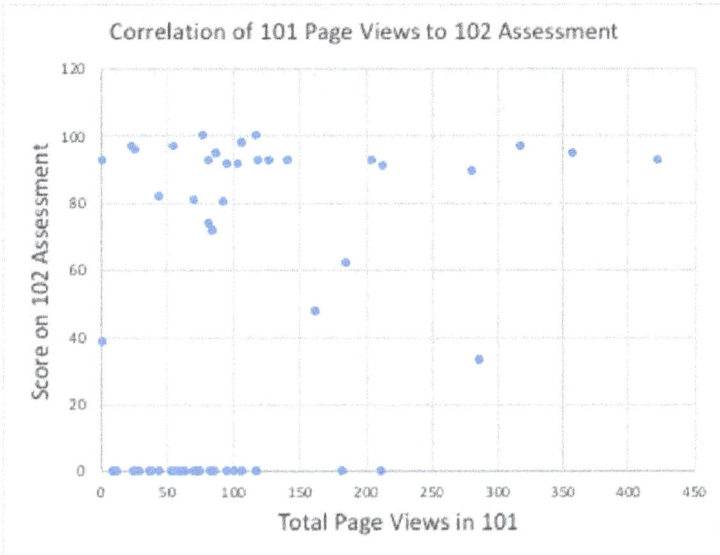

Figure 6. Correlation of page views in Korean 101 during the month of December to assessment score in Korean 102

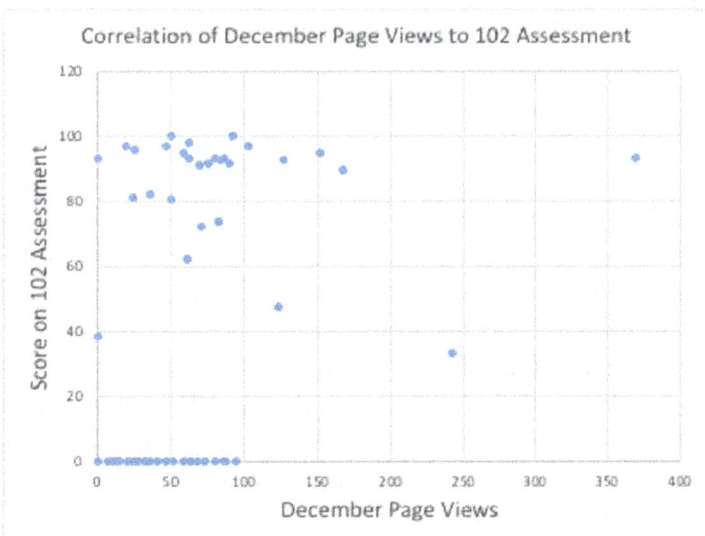

Figure 7. Correlation of final exam preparation page views in Korean 101 to score on assessment in Korean 102

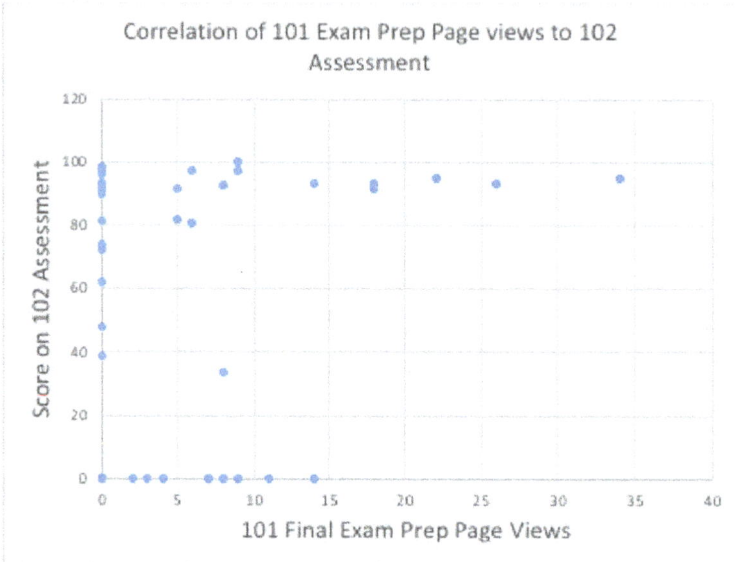

Note that 25 students did not continue from Korean 101 to 102. As a result, researchers excluded the data from those students to evaluate this question. Thus, sample size to answer this research question was 29 rather than the sample of 54 for the other research questions.

DISCUSSION

Researchers discovered only a very weak positive correlation between overall page views and success in Korean 101 and an even weaker correlation between overall page views in 101 to retention in 102. As researchers hypothesized, students in fact were accessing course content more frequent, by 20% or more, in the month of December than the rest of the months in the semester, which suggests evidence of binge studying. However, despite the evidence that over 75% of students were binge studying, those students appeared to score higher on their final exam in Korean 101, and there appeared to be little correlation between having binge studied and the score on the 102 assessment (indicator of retention). The strongest correlation appeared to be between non-binge studiers, or those which dispersed their study throughout the semester, and language retention in 102.

Implications for Action

First, due to the low correlation of page views online and student success, researchers determined that learning experiences which take place in the classroom contribute to the success of students as well as the online elements and that online elements alone do not seem to have the likelihood of deeply impacting student success. This is an important consideration in ensuring that classroom activities are well-planned and build upon what has been done online. Likewise, it was identified that these findings may help ensure that faculty and TAs do not mistakenly assume certain results from online participation alone.

Second, the stronger correlation between not binge studying and higher language retention (or score on the Korean 102 assessment) led researchers to consider possible ways to introduce pacing into the online elements of the course. Because retention is a significant value in Korean 101-102 courses (and into upper levels), faculty and TAs could be more sensitized to ways to prevent binge studying and ways to carefully plan curriculum design to support evenly dispersed work. Likewise, for students intending to pursue Korean beyond the 102 course, researchers identified the value in emphasizing to students the relationship between evenly dispersed work in the course and retention.

Implications for Future Research

After evaluating this data, the researchers intend to pursue further study to consider factors impacting how much time students spend with online materials in the blended courses (Korean 101 and 102). What variables might play into the frequency and regularity with which students access online content? The researchers also identified potential research tied to this same question in the face-to-face environment. Are students binge studying in the F2F classes as well? What affects the frequency and regularity with which they attend class and study materials outside of class? Likewise, is there a correlation between binge studying and student success in fully F2F courses? Researchers could also explore the traits of students who continued into 102 those who took 101 only, especially noting the higher final exam grades and higher average page views for students who continued into 102. Finally, researchers have considered evaluating these same questions in fully online versions of Korean 101 and 102; is there evidence of binge studying? And if so, what relationship does it appear to have with student success? Because the researchers are involved with teaching face-to-face, blended and fully online versions of Korean 101 and 102, a comparative study exploring the impact of binge studying on student scores and on retention may be valuable.

REFERENCES

Anderson, T., & Elloumi, F. (2004). *Theory and practice of online learning* (2nd ed.). Edmonton, Canada: AU Press.

Beqiri, M. S., Chase, N. M., & Bishka, A. (2010). Online course delivery: An empirical investigation of factors affecting student satisfaction. *Journal of Education for Business*, *85*(2), 95–100. doi:10.1080/08832320903258527

Bocchi, J., Eastman, J. K., & Swift, C. O. (2004). Retaining the online learner: Profile of students in an online MBA program and implications for teaching them. *Journal of Education for Business*, *79*(4), 245–253. doi:10.3200/JOEB.79.4.245-253

Bonk, C. J., & Graham, C. R. (2005). *The handbook of blended learning: Global perspectives, local designs*. Hoboken, NJ: John Wiley & Sons.

Breslow, L., Pritchard, D., DeBoer, J., Stump, G., Ho, A.,D., & Seaton, D. (2013). Studying learning in the worldwide classroom: Research into edX's first MOOC.

Brinthaupt, T. M., & Shin, C. M. (2001). The relationship of academic cramming to flow experience. *College Student Journal*, *35*(3), 457–471.

Comer, D. R., Lenaghan, J. A., & Sengupta, K. (2015). Factors that affect students' capacity to fulfill the role of online learner. *Journal of Education for Business*, *90*(3), 145–155. doi:10.1080/08832323.2015.1007906

Conijn, R., Van, B., & Cuijpers, P. (2018). Predicting student performance in a blended MOOC. *Journal of Computer Assisted Learning*, *34*(5), 615–628. doi:10.1111/jcal.12270

Dallimore, E. J., Hertenstein, J. H., & Platt, M. B. (2017, March 27). How do students learn from participation in class discussion? *Faculty Focus: Higher Ed Teaching & Learning.*

Damron, J. & Dobberfuhl-Quinlan, J. (2018, July 4-6). An exploratory study of student learning. In *CALL Proceedings.* Universiteit Antwerpen.

Erçetin, G., Lee, S., Kadel, R., Madden, A., & Gazi, Y. (2018, July 4). How can learner analytics data inform language MOOC design? In *Proceedings of the 19th International CALL Conference*, Bruges, Belgium. Academic Press.

Fendler, R. J., Ruff, C., & Shrikhande, M. (2016). Evaluating characteristics of top and bottom performance: Online versus in-class. *American Journal of Distance Education*, *30*(2), 109–120. doi:10.1080/08923647.2016.1153350

Glenn, D. (2007). Why cramming doesn't work [1]. *The Chronicle of Higher Education*, *53*(40).

Goertler, S., & Gacs, A. (2018). Assessment in online German: Assessment methods and results, linguistics and language behavior abstracts (LLBA). *Die Unterrichtspraxis*, *51*(2), 1–21. doi:10.1111/tger.12071

Jabeen, S. & Jesse Thomas, A. (2015). Effectiveness of online language learning [2]. In *Proceedings of the World Congress on Engineering and Computer Science*, 1. Academic Press.

King, A. (2016). *Blended language learning: Part of the Cambridge Papers in ELT series*. Cambridge, UK: Cambridge University Press.

Knox, J. (2014). Digital culture clash: "Massive" education in the E-learning and digital cultures MOOC. *Distance Education*, *35*(2), 164–177. doi:10.1080/01587919.2014.917704

Lederman, D. (2018). Who is studying online (and where). Inside Higher Ed.

Lee, Y., & Choi, J. (2011). A review of online course dropout research: Implications for practice and future research. *Educational Technology Research and Development*, *59*(5), 593–618. doi:10.100711423-010-9177-y

Levy, Y. (2007). Comparing dropouts and persistence in e-learning courses. *Computers & Education*, *48*(2), 185–204. doi:10.1016/j.compedu.2004.12.004

Li, K. (2019). MOOC learners' demographics, self-regulated learning strategy, perceived learning and satisfaction: A structural equation modeling approach. *Computers & Education*, *132*, 16–30. doi:10.1016/j.compedu.2019.01.003

Martin-Monje, E. (2017). *Mind the L in LMOOCs: The importance of language learning in massive courses.* Venice, Italy: Conference: IV Congresso della Societa di Didattica delle Lingue e Linguistica Educativa (DILLE) 2017.

Martin-Monje, E. & Castrillo, M. D. (2016). A redefinition of the teacher and student roles in language MOOCs: The example of 'how to succeed in the English B1 level exam'. In *LLAS 11th Annual e-learning Symposium,* Southampton, UK.

Martín-Monje, E., Castrillo, M. D., & Mañana-Rodríguez, J. (2018). Understanding online interaction in language MOOCs through learning analytics. *Computer Assisted Language Learning*, *31*(3), 251–272. doi:10.1080/09588221.2017.1378237

Morris, N., Hotchkiss, S., & Swinnerton, B. (2015a). Can demographic information predict MOOC learner outcomes? In Proceedings of the European MOOC Stakeholder Summit, 199-207, Mons, Belgium.

Ryan, S., Kaufman, J., Greenhouse, J., She, R., & Shi, J. (2016). The effectiveness of blended online learning courses at the community college level. *Community College Journal of Research and Practice*, *40*(4), 285–298. doi:10.1080/10668926.2015.1044584

Seaman, J. E., Allen, I. E., Seaman, J., & Babson Survey Research Group. (2018). *Grade increase: Tracking distance education in the United States.* Place of publication not identified: distributed by ERIC Clearinghouse.

Shrader, S., Wu, M., Owens, D., & Santa Ana, K. (2016). Massive open online courses: Participant activity, demographics, and satisfaction. *Online Learning Journal*, *20*(2), 199.

Sinha, T., Jermann, P., Li, N., & Dillenbourg, P. (2014). Your click decides your fate: Inferring information processing and attrition behavior from MOOC video clickstream interactions. In *Proceedings of the EMNLP 2014 Workshop on Analysis of Large Scale Social Interaction in MOOCS.* 10.3115/v1/W14-4102

Watson, J., & Murin, A. (2010). *Keeping pace with K-12 online learning. An Annual Review of Policy and Practice*. Evergreen Education Group.

Weldy, T. G. (2018). Traditional, blended, or online: Business student preferences and experience with different course formats. *E-Journal of Business Education & Scholarship of Teaching*, *12*(2), 55–62.

Chapter 12
Gamification as a Teaching Resource for English-Medium Instruction and Multilingual Education at University

Maria Del Mar Sanchez Perez
https://orcid.org/0000-0002-6550-9030
University of Almeria, Spain

Alicia Galera Masegosa
University of Almería, Spain

ABSTRACT

Gamification has emerged in recent years as a resource that incorporates game-related elements and mechanics into the classroom to foster students' motivation, engagement, and further competences. With the proliferation of English-medium instruction (EMI) and multilingual degree programs at university nowadays, new and innovative teaching tools are desirable to help students cope with the double-challenging task of acquiring new and complex disciplinary content through a foreign language. This chapter provides an overview of some recent computer-based gamification tools that may be applied in EMI and multilingual university settings. More specifically, the main features and advantages of these tools for these specific educational contexts are explored. This chapter may be useful for researchers and practitioners in the field of EMI and multilingual teaching in higher education.

INTRODUCTION

English-medium instruction (EMI) and multilingual education is gaining momentum in higher education institutions worldwide. The increasing need to foster internationalization at universities has resulted in an exponential growth of courses and degree programs taught through a foreign or second language (e.g. English) (Doiz et al., 2012; Macaro et al., 2017; Smit & Dafouz, 2012). Within these settings,

DOI: 10.4018/978-1-7998-1097-1.ch012

students find an extra degree of difficulty. The challenging task of acquiring new and complex knowledge is reinforced by the fact that it is taught in a language different from the native one, which may lead to increasing student inhibition, content understanding issues and low in-class participation rates. Consequently, a strong demand for new teaching methods and techniques to cope with these issues in EMI and multilingual environments has arisen among the university teaching community (Sánchez-Pérez & Salaberri, 2017). Faced with this need, new and innovative teaching resources to facilitate the teaching and learning of subject content through a non-native language are increasingly emerging. This is the case of gamification, a teaching technique defined as "using game-based mechanics, aesthetics, and game thinking to engage people, motivate action, promote learning, and solve problems" (Kapp, 2012, p. 10). Gamification incorporates game-related elements and mechanics into the classroom with the aim of fostering students' motivation, engagement and further competences. The recent interest in gamification as a tool in educational contexts has given rise to an ever-growing number of studies aiming to provide evidence that demonstrates that gamification is a powerful and effective teaching resource in computing sciences (Caton & Greenhill, 2014; Ibáñez et al., 2014). Given the connection between motivation and language learning (Brown, 2007), gamification has also gained increasing attention in second language education (cf. Al-Hadithy & Ali, 2018; Figueroa, 2015; Garland, 2015, among many others). In addition, university students have been found to be particularly sensitive to gamified courses in different fields and disciplines (Barata et al., 2013; Caton & Greenhill, 2014; Fanfarelli & McDaniel, 2017; Wiggins, 2016).

Within this context, our main contention is that EMI and multilingual settings may benefit from game elements and mechanics at university levels. In order to support our claims, we first present an outline of the recent challenges faced by tertiary level teachers in multilingual contexts, and the reasons that point to the suitability of computer-based elements in these settings. Then, we provide an overview of some of the most relevant aspects of gamification, including insights from some of the most recent studies engaged in assessing the advantages and drawbacks of gamification as a pedagogical technique. Lastly, we address some recent computer-based gamification tools that may be applied in EMI and multilingual university settings. In particular, we explore the main features and advantages of these tools for these specific educational contexts.

MULTILINGUAL EDUCATION AND COMPUTER-MEDIATED TECHNIQUES AT TERTIARY LEVEL

The approaches to multilingual education at university level have been reconsidered since the implementation of the European Higher Education Area (EHEA). This involves a complete restructuring of the traditional concepts ranging from the use of new technological resources to the understanding of new roles of both students and lecturers. This new academic system needs to adapt to the emerging demands of the higher education arena and, for such purposes, some changes are necessary, which means certain separation from traditional educational concepts and methods.

A new aspect supported by the EHEA is the implementation of activities aimed to develop critical and creative thinking among students. Accordingly, new learning ways are constantly appearing, along with new methods and resources to improve the teaching-learning process (Pérez et al., 2009). Within the EHEA, the cognitive needs and the level of the students have a crucial role, especially when the teaching is conducted in a foreign language. This educational context demands a dynamic position of the lecturers, who have to adopt an attitude of continuous quest to find valid solutions in their daily practice. In this context, university lecturers work as "content learning instructors" and "language learning advisors" (Dudley-Evans, 1997, p. 60) at the same time.

The introduction of multilingual education in Europe and Spain has materialized through different teaching and learning approaches, e.g. CLIL (Content and Language Integrated Learning) and EMI (English-Medium Instruction), among others. However, CLIL has been more associated with earlier educational stages, such as primary or secondary education, while EMI is the most widespread term used to refer to multilingualism in higher education. Additionally, unlike CLIL, which involves a dual focus on content and foreign language learning (FLL), EMI is characterized by the exclusive use of English as the language of instruction and it primarily focuses on content learning (Dearden, 2015). Despite these differences regarding their conceptual and methodological features, both approaches share some common principles underlining multilingual education, such as the use of student-centered, communicative-oriented, task-based and problem-solving methods.

In this sense, some recent studies have revealed the potential of combining such bilingual approaches with computer-based methods and techniques due to the common pedagogical principles shared by them. According to some authors (Pérez-Cañado, 2018; Pérez-Cañado & Ráez Padilla, 2015; Scott & Beadle, 2014), both EMI and computer-based techniques promote among students a more active, participatory, relevant and authentic learning; encourage a more collaborative, communicative and interactive teaching process; foster the use of task-based and problem-solving methods; promote critical thinking, creativity, risk-taking, autonomy and intercultural communicative competence; and encourage motivation, self-confidence and positive attitudes towards learning, among others.

Within this scenario, the use of innovative and computer-based teaching techniques such as gamification appear as ideal resources to foster the benefits of these approaches in multilingual and EMI contexts at university in order to provide students with opportunities to engage in a more active, dynamic and motivating learning process.

GAMIFICATION

This section is devoted to the revision of the notion of gamification as a resource that may be (and has in fact been) used in a wide variety of contexts, ranging from selling strategies and promotional campaigns to motivational techniques in working and teaching environments. We provide an overview of some the most relevant pieces of research on the topic, placing special emphasis on the field of education, particularly in second language and tertiary education.

Concept and Evolution

Finding a unified definition of gamification is a very difficult enterprise. Over the last decades, many researchers have devoted their work to the analysis and assessment of gamification and its impact on a vast variety of disciplines and fields.

A broadly accepted definition of gamification is the use of "game-based mechanics, aesthetics, and game thinking to engage people, motivate action, promote learning, and solve problems" (Kapp, 2012, p. 10). In line with this author, and also following pioneering work in the gamification of business and work environments by Werbach & Hunter (2012), we regard gamification as a strategy that applies game elements, techniques and mechanics commonly found in games to non-game contexts in order to promote desired behaviors in individuals.

One of the first precursors of gamification as we know it today can be traced back to the badge system of rewards used by Boy Scouts at the beginning of the 20th century. Boy Scouts are awarded badges as they achieve their goals in several subject areas, so adding badges to one's uniform is synonym of pride and accomplishment. This symbolic expression of achievement clearly parallels video game elements (Rosewell, 2012; Silvers, 2011). In his reflection about precursors of gamification, Nelson (2012) claims that gamification was preceded by some of the motivational techniques in the factories of the Soviet Union to increase productivity in the early-mid 20th century, and then later in American management.

Gamification has been used to increase sales, improve learning, encourage production, promote health and sustainability, etc., which means that clients, customers, learners, employees, or people in general can be addressed through gamification techniques. The use of game elements to encourage desired behaviours has been present in different contexts, such as business management, advertising and marketing campaigns among others. Kumar and Herger (2013) point to loyalty programs as a straight forward instantiation of how the dynamics of games can be used as a selling strategy. In fact, the use of game elements is nowadays widespread in our society. Competition between two or more participants, points to measure progression, badges that represent accomplishments or levels of difficulty are only some of the components of games that have been incorporated to many activities of our everyday lives (see Werbach & Hunter, 2012 for an account of game components, mechanics and dynamics). In sum, gamification is a technique rooted in psychology that fosters motivation and engagement, taking advantage of our competitive spirit and eagerness to achieve goals and attain success (Burke, 2014; Christians, 2018). Work, learning processes, and many other activities can be redesigned in order to turn them into enjoyable experiences that involve fun and commitment. Additionally, gamification strengthens the link between organizations, their employees and the users of their products (Wood & Reiner, 2015).

Although gamification has been generally regarded as a successful resource, especially in business and marketing settings, it has also received criticism and raised controversy (Deterding et al. 2011). It has been noted that misuses of gamification can go beyond motivation and engagement, taking steps towards excessive influence on behaviour and manipulation (Christians, 2018; Pixie, 2010), or even exploitation (Bogost, 2015). However, the potential effectiveness of gamification when used properly keeps engaging the interest of researchers in many fields, including teaching and learning environments.

Gamification in Educational Settings

Prensky (2001) refers to students of today as *digital natives*. This means that today's learners have grown up surrounded by digital technology: the Internet, emails, video games, mobile phones, etc. They have been born and raised in the digital era, and are thus "native speakers of the digital language of computers, video games and the Internet" (p.1). It is only logical to think that students in the digital generations have had their minds shaped by the continuous stimuli of this new technology. This posits a problem as far as education is concerned, since "today's students are no longer the people our educational system was designed to teach" (Prensky, 2001, p. 1).

Lack of engagement and motivation are difficult challenges that teachers often struggle to overcome. Given the tight connection between learning and motivation (Brown, 2007), approaching education from a perspective that includes game elements and technology, as the language spoken fluently by today's students, seems then the logical thing to do.

Some scholars have pointed to the parallelism between game elements and the traditional school system, as they seem to nicely fit the process of learning (de Byl, 2013). Rewards and punishments (final grades), points for task completion, levels (academic years), etc. can guide learners towards objectives and help them follow the path to success (Glover, 2015; Hung et al, 2017; Lee & Hammer, 2011). But despite these similarities, the educational system seems to fail in engaging and motivating students, so we agree that "the existence of game-like elements does not translate directly to engagement. Understanding the role of gamification in education, therefore, means understanding under what circumstances game elements can drive learning behavior" (Lee & Hammer, 2011, p. 2). Therefore, gamification as an effective teaching strategy needs to be carefully designed in principled ways. Systematic approaches to gamification are likely to increase the effectiveness of the gamified course, as it enhances the customization of the program and optimizes resources (Dicheva et al. 2015; Hung, 2018). Huang & Soman's (2013) Educational Gamification Model includes five steps to be taken in the implementation of gamification in education: (i) identifying and understanding the profile of the target audience and the context in which the education program will take place; (ii) stating what the learning objectives are; (iii) staging the experience by designing the structure of the program; (iv) determining what resources will be needed; and (v) selecting the gamification elements that will be applied.

We may thus assume that the proper design and successful implementation of gamified activities in learning environments implies high cost in terms of time and effort on the part of the teacher. In this respect, Lee & Hammer (2011) suggest that teachers may find the gamification process very absorbing. In addition, these authors claim that students may associate learning exclusively to external rewards. However, much of the literature on the topic points to a positive impact of gamification on educational contexts.

One of the main strengths of gamification relates to its motivational power. In their investigation of the psychological dimension of motivation and gamification, Sailer et al. (2013) analyze different motivational perspectives and advocate for the connection between motivational mechanisms and game elements. In later research, Sailer et al. (2017) provide empirical results based on psychological theories that show how different game elements relate to the improvement of different aspects of motivation, lending further support to the gamification-motivation link.

Gamification has been found to be a powerful tool that fosters motivation and engagement also within educational environments (Landers & Landers, 2014, Reiners et al., 2012), specially through competition and the obtention of immediate rewards (Lee & Hammer, 2011).

Apart from improving motivation and engagement, recent studies suggest that gamification promotes effective learning in that "the retention rate on information that is displayed in a fun, more interactive environment is generally much higher than traditional textbook reading" (Christian, 2018, p. 8). Lee & Hammer (2011) underscore other advantages of gamifying the educational system: games provide guidance through the learning process; multiple ways to achieve goals allow students to experiment and deal with failure, short-term feedback, among many others.

Even if gamification in education has not yet been implemented as successfully as it has in the business and marketing worlds, there is a growing number of research projects that aim at the effective incorporation of gamification in learning environments, and second language education is no exception.

Gamification in Second Language Education

In the light of our previous discussion, motivation seems to be essential in educational settings. There is evidence that points to the strong correlation between motivation and successful learning processes in general, and second language acquisition in particular (Clement et al. 1994; Pintrich & Schunk, 1996; Schmidt and Watanabe, 2001; Lasagabaster et al., 2014).

This connection has generated a growing number of studies that explore the potential benefits of creating a playful atmosphere that involves challenges and rewards in the language classroom (Figueroa, 2015). For example, Perry (2015) presents a case study in which gamification is used in teaching French as a second language. His research shows positive results both in student's perception, motivation and learning. Along similar lines, Iaremenko (2017) shows how online games enhance motivation in learners of English as a second language.

The effects of online games in second language education have been largely explored in the last few years. Al-Hadithy & Ali (2018), for instance, conduct an experiment that involves the use of an online game ('Kahoot') in an English course for undergraduate Law students. The experiment shows that the incorporation of the game to the course improves engagement and motivation, and also increases learning. The implications of using Kahoot in the language classroom have also been addressed by Medina & Hurtado (2017).

A growing number of the recent research on gamification as a tool in second language education has been engaged in the design, testing and assessment of mobile phone applications in the context of second language learning (Rosell-Aguilar, 2017). In this context, Hasegawa et al. (2015) designed a mobile phone application making use of gamifying elements with the aim of improving English vocabulary. Their study shows that the platform increased student's motivation and improved vocabulary memorization. Alvarado et al. (2016) also explored and highlighted the positive implications of mobile apps for students of English. However, some authors argue that language learning mobile phone applications lack an appropriate pedagogical component and tend to focus too much on translation (Burston, 2014).

In spite of being a controversial topic (Deterding et al., 2011) and the mixed results obtained in some empirical studies (Domínguez et al., 2013), the potential benefits of gamification on motivation and performance have encouraged researchers and teachers to engage themselves in the exploration and incorporation of this resource in second language teaching. However, even if the amount of online games and mobile apps designed to aid second language learning is impressive, it is important to notice that the successful implementation of gamification in second language teaching is not as simple as incorporating game elements into the dynamics of the classroom. As we already remarked with regard to education in general, the age of students, time on task or length of instruction are some of the aspects that need to be taken into consideration (see Garland, 2015 for a detailed account of the different variables that should be considered in the design of a gamified course in second language education, and the implications for language teaching). In addition, the desired outcome should be clear during the design process of the gamified course, to make sure that the appropriate game elements, and not others, are included (Sailer et al., 2017).

The suitability of some of the most popular online games and mobile applications for language learning has been extensively discussed in recent gamification literature (see Lam, 2014 for 'Fling the teacher' and 'Jeopardy'; Murar, 2015 for 'Edmodo'; Plump & LaRosa, 2017, Pede, 2017 and Al-Hadithy & Ali (2018) for 'Kahoot!'; Majdoub, 2016 for 'Classcraft'; Ahmed, 2016, Munday, 2016 and Savvani, 2019 for 'Duolingo'; Hwang, 2018 for 'Man bites the dog'; among many others).

A detailed account of the features of each of these games and apps is beyond the scope of this study. For our current purposes, we have selected the ones that, in our view, may be most beneficial for university students engaged in EMI and multilingual education (see 'Analysis of Gamification Tools').

Gamification in Higher Education

In the last years, a growing amount of research has been conducted in order to assess the impact of gamification in university settings. According to Garland (2015), studies involving gamified applications with university students display better results than the implementation of gamification in lower levels of education. Also, the availability of the internet for university students facilitates the incorporation of gamification tools in this context (Bicen & Kocakoyun, 2018).

Most research conducted with university or older students have obtained positive results. Barata et al. (2013) gamified two years of a five-year college course. The comparison of data between the gamified and non-gamified years showed that gamification had a positive impact on student's participation, proactivity and learning. O'Donovan et al. (2013) also obtained successful results in the gamification of a university course, in terms of problem solving, attendance, learning and general engagement. Apart from the implications regarding performance, attendance and learning achievements, the perceptions of university students after taking a gamified course have also been assessed with positive results (Alexander, 2017; Bicen & Kocakoyun, 2018).

In some cases, the exploration of the impact of some game elements on university courses has given rise to mixed results. In the case of badges as a gamification strategy, for instance, Haaranen (2014) obtained a negative response from students, while Fanfarelli & McDaniel (2017) show improvement both in engagement and performance. Domínguez et al. (2013) gamified a university course on technology by using badges and leaderboards, obtaining mixed results between the control and the gamified groups.

All in all, most empirical studies indicate that gamification strategies in higher education tend to improve motivation, engagement and learning achievements. Therefore, and having also found evidence about the suitability of gamification strategies in the language classroom, we postulate that the use of computer-based gamification tools may be beneficial in EMI and multilingual university settings.

ANALYISIS OF GAMIFICATION TOOLS

In this section, three computer-based gamification tools that may be applied in EMI and multilingual university settings will be explored. After careful examination, we have found that Classcraft, Kahoot and the gamified environments created through the TeCola Project present features and advantages that may be particularly useful in these specific educational contexts.

Classcraft

ClassCraft is an online platform which allows transforming the classroom into a role-playing game. It was initially created by Shawn Young, a high-school science teacher, with the purpose of engaging his students (Sant, 2014). This resource is available online, which can be accessed from any computer, and as a smartphone application. Therefore, both students and teachers can access the platform from any device with an internet connection. The teacher can use the platform either in the computer lab or in a conventional classroom with internet access and a video projector. Additionally, as university students usually bring their mobile phones to their lessons, they can access the platform either inside or outside the classroom. Classcraft has a Premium version, which requires a monthly payment, but also a free version with similar functionalities. The service only requires installation if used as a mobile app – which is not necessary if used through computer. The game operates like a usual real-time video game, therefore, events occur in real time to other users' devices (Sánchez, Young & Jouneau-Sion, 2017).

The purpose of using ClassCraft is to turn the class into the setting for a role-playing scenario, in which the teacher can promote certain behaviors related to the management of the class. In Papadakis & Kalogiannakis's (2018) words, "it can be considered to metaphorize the functioning of a classroom as a battlefield combining collaboration and competition" (p. 368). It offers a system of rewards or sanctions to help manage the behavior and performance of students throughout the course. As students earn points they rise in level and progress through the game, allowing them to gain powers and customize their avatars (e.g. acquiring accessories, costumes, pets, etc.). The duration of the game is determined by the teacher and students can access the platform both during class hours and out of class, as each student has a unique account that can be accessed through the mobile application or the website (Sánchez, Young, & Jouneau-Sion, 2017).

Classcraft allows the teacher to create each student's avatar and assign them a role or it may be the student who creates his/her own character. Students can be grouped into four to five or six members and they need to work together to succeed. Therefore, the performance of each group member will affect the rest of the group. Each group has, at least, three defined characters: a Warrior, a Healer and a Mage. Warriors are in charge of protecting the team; they are strong characters, but they cannot use their powers as often as other characters. Healers are the responsible for healing their group mates when they lose health points (HP) to prevent them from falling in battle and their entire group from losing points. Finally, Mages can use their powers very often, which gives certain advantages to their teams.

According to the students' performance, they can get a series of different points:

- *Experience Points (XP):* they can be acquired by demonstrating positive behavior or by completing positive actions (e.g. completing activities in class, active participation, etc.). These points will allow the students to raise their level and gain new powers which have benefits for the whole group.
- *Action points (AP):* they are necessary to use the powers and they regenerate every day.
- *Health points (HP):* these points are lost when students engage in inappropriate behaviors (e.g. inappropriate language, task failure, breaking class rules, etc.). Should any student lose all his/her HP points, he/she will fall in battle and the whole group will lose HP points.

Both rewarding and penalising actions can be changed and adapted to the needs of the classroom. Classcraft offers default options but these can be modified, both the numerical value of points earned or lost and the behaviour by which those points are lost or earned.

As students earn points and increase their level, they can obtain different powers that give them privileges both for themselves and for the whole group. Thus, each time a student increases his/her level, he/she obtains a *power point* which can be used to get one of the available powers. Some of these powers are determined by Classcraft and cannot be modified. However, there are other powers that can be customized. Powers are divided into:

- *Basic powers*: only need one power point to be acquired;
- *Intermediate powers*: requires two power points to be acquired;
- *Advanced powers*: requires three power points to be acquired.

ClassCraft has been proven to be a useful tool to promote dynamism, collaboration and solidarity in the classroom as students are organized in groups in which each student is assigned a role which entails some responsibilities within the entire group (Sanchez, Young, & Jouneau-Sion, 2017). Additionally, it fosters the sense of responsibility, as each student's performance has an impact on the rest of the group. Particularly, there are recent studies which evidence the positive impact on students' engagement (Papadakis & Kalogiannakis, 2018) and students' motivation in the foreign language classroom (Rivera-Trigueros, 2018). In this sense, ClassCraft seems to be an appropriate potential gamification tool to be implemented in EMI and multilingual university lectures, due to different factors: easy access, low cost, possibility of use inside and outside the classroom, easy handling and understanding of its operation and the benefits reported on students' engagement and motivation, which are aspects to be developed in university lessons where the instruction is given in a non-native language.

Kahoot

Kahoot is a free online learning platform initially launched in 2013. It allows the creation of questionnaires, discussions or quizzes or the selection of some available from a repository. It is a free online tool currently used by more than 30 million people worldwide. It is inspired by the behavioral learning methodology and the student-centered approach (Plump & laRosa, 2017).

Kahoot can be used in the classroom to learn or reinforce a particular learning aspect where students can compete and collaborate in the resolution of certain questions proposed by the lecturer. Students can log in with their own username or "alias" and they can answer the questions available in the platform through the use of their smartphone - without the need to download the application. Students can play either individually or in group. Questionnaires and quizzes are accessible to all participants and they can be reused and modified to adapt to particular needs or demands that may subsequently arise. Questions can include 95 characters maximum, while answers can include up to 60 characters (Göksün & Gürsoy, 2019).

Questions are formulated to all participants and answers are displayed on the participant screens. Response to the questions has a time limit, which can be also modified or extended for more challenging questions. Students cannot answer the following questions until the rest of participants have answered the previous one or when the time limit is over. Additionally, the software allows incorporating extra linguistic elements such as pictures, sounds or videos.

Kahoot is a very easy and simple tool. The usual functioning in the classroom starts by the signing up of the lecturer, who creates the classroom game environment and displays it to the class - with the help of a computer, a projector and internet access. Students do not need to register and can participate through their smartphones hones after logging in. Questions can be customized (e.g. randomizing the order of the questions or the answers in multiple-choice options). Students can answer the questions by pressing a button on the app and they get points for each correct answer.

At the end of the questionnaire or quiz students can know their score which appears in a rank with the results and statistics of the rest of their classmates, so feedback is easily and rapidly obtained, either individually or for the whole class. Scores will depend of the number of right answers, as well as the answer speed. The lecturer can set up a rewarding system for higher-scored students and he/she can save the results – either exporting them to drive or excel – to subsequently revise them for assessment purposes.

Kahoot is perceived by students as a useful, engaging and motivating tool, as well as an effective resource for foreign language improvement (Bicen & Kocakoyun, 2018), academic achievement and formative assessment (Göksün & Gürsoy, 2019). Within this environment, students become the protagonists of their own learning experience and the role of the lecturer takes a back seat, being a mere informant or monitor of the teaching and learning process. This is precisely one of the advantages for considering the use of such tools in EMI and multilingual educational settings, where a more active and participatory role of the students is pursued in contrast to traditional university teaching methods.

TeCoLa Project

The TeCoLa Project (*Pedagogical Differentiation through Telecollaboration and Gaming for Intercultural and Content Integrated Language Teaching*) is an Erasmus+ project funded by the European Commission. It arose as a consortium of experts in the field of foreign language teaching and learning, intercultural communication, teacher training and technology-based pedagogy from six different European countries (Belgium, Germany, Netherlands, France, UK and Spain). As a successor of a previous project (*Telecollaboration for Intercultural Language Acquisition, TILA*), it started in 2016 and is expected to finish in 2019 with the purpose of developing and evaluating innovative gamified telecollaboration approaches in foreign language and subject-matter educational environments (Pennock-Speck & Clavel-Arroitia, 2018).

The TeCoLa project is inspired by recent communicative teaching approaches such as task-based teaching, with a special focus on the socio-cultural dimension in language learning (Colpaert & Spruyt, 2018). The task-based approach in foreign language teaching is considered among scholars as one of the most complete and advanced approaches to language teaching (Ellis, 2003; Van den Branden, 2006), and especially, in technology-based teaching environments (Ortega & González, 2014). Language tasks, understood as activities "in which a person engages in order to attain an objective, and which necessitates the use of language" (Van den Branden, 2006, p. 4), are the minimal units of language lesson planning and syllabus design and they are the heart of the language learning process. Within the task-based approach, students are considered as language users, and not only as language learners (Ellis, 2003).

Users within this project have the opportunity to participate in online gamified scenarios to engage in authentic meaningful communication with students from other countries. Such involvement has, therefore, "the potential to facilitate interaction and collaboration, enhance language learning and intercultural communicative competences (ICC) in a technology-mediated setting" (Colpaert & Spruyt, 2018, p. 140). These are in fact some of the principles shared by computer-based teaching methods and bilingual educational approaches such as EMI, which makes this project suitable for the context dealt with in the present study, even though it was initially addressed to secondary education.

The use of telecollaboration in virtual gamified environments to communicate and collaborate through authentic task-based interactions among people from different language, socio-cultural and educational backgrounds is, therefore, one of the milestones of TeCoLa (Jauregi & Melchor-Couto, 2017; Melchor-Couto, 2019). Particularly, the main objectives of this project are summarized as follows (TeCoLa, 2019):

- To provide pre- and in-service instructors with the necessary skills and abilities to use gamification and telecollaboration in order to foster intercultural communicative competence as the main goal in FLL, create authentic communication conditions, promote autonomous collaboration and authenticity in task-based situations, and foster learner agency and personalised learning.
- To promote internationalisation in education through telecollaboration.
- To encourage student integration through intercultural communication.
- To take advantage of the benefits of technology-based applications, virtual worlds and gamification to diversify CLIL and intercultural exchanges through games and telecollaboration.

In order to fulfil these objectives, the TeCoLa project offers a series of tools and resources which facilitate the acquisition of the necessary competences to engage in online gamified and telecollaboration teaching and learning environments. With regard to the tools, the following ones can be found:

- *Webinars*: The TeCoLa project organizes periodically a series of online training seminars addressed to instructors of the countries participating in the project on the use of the necessary strategies, techniques and tools to implement gamification and telecollaboration in the classroom.
- *TeCoLa Virtual World*: This is a tool where the different virtual scenarios are created so that telecollaboration can take place. In this virtual scenario, different small *islands* concerning different aspects of language learning can be found. These are: *Welcome Island*, where users start and find information on how to move to other islands; *Arcadia,* where users learn the basic principles to work and navigate in the virtual world; *Chatterdale*, the English language island; *Parolay*, the French language island, *Saarburg*, the German language island; *Valencia*, the Spanish language island; *3D-LAB Island*, where users learn how to build and script objects and where new tools and games can be found; and *Quest 2030 Island*, where there is a big game in which participants need to find clues to fulfil a certain objective.
- *TeCola Gamify*: It is an online gamification tool in which different gamified activities such as quiz chairs and assignment boards can be created. With regard to the first one, these are chairs which show a number of multiple choice questions as soon as an avatar sits down on the chair; as for the latter, they comprise a series of interactive boards which can be placed in the virtual world.
- *TeCola Moodle*: The project integrates the online Moodle Learning Platform in the virtual world so that both instructors and students can complete a series of assignments, tasks, cretate wikis, blogs, etc. It hosts the *Big Blue Button*, a video conferencing platform created to engage in intercultural language tasks. Through this platform, students can communicate through chat, microphone or webcam. It also allows users to share documents, pictures, videos, etc.
- *TeCola Cloud*: where materials and resources can be stored and shared by the members of the project.

As for the resources offered by the TeCoLa project, the following ones can be highlighted:

- *TeCola Tasks:* It consists of a series of tasks in different languages (English, French, German and Spanish) focused on gamified collaboration and intercultural communication which can be adapted to the instructors and students' purposes and needs. It also contains suggestions for assessment activities.

- *TeCoLa Pedagogical Guides*: They provide instructors with relevant information on how to implement and design intercultural telecollaboration and gamified activities.
- *TeCoLa Case Studies*: Although this resource is still under construction, it pursues to include results of gamified intercultural telecollaboration experiences in FLL in secondary and vocational education.
- *TeCola Teachers Talk*: Here both instructors and learners can express and share their insights and experiences from their intercultural telecollaboration exchanges.
- *TeCoLa Language Learning Report*: It shows the results of the different surveys conducted about learning diversity and communication challenges in FLL in secondary schools.

Despite the fact that this project is still ongoing and no concluding empirical results on its efficiency in language learning are still available, we consider that, due to the specificities detailed above, this is one of the most complete and comprehensive projects on gamification for FLL to date. As Clavel-Arroitia & Pennock-Speck (2015) support, the telecollaboration nature of the project allows students talking to "native speakers of a target language of a similar age" (p. 75). Additionally, its approach to diversity "allows for the acquisition of social civic and intercultural communicative competences" (Pennock-Speck & Clavel-Arroitia, 2018, p. 486). In the same vein, these authors support that the project intends to enrich content and language integrated learning (CLIL) "by complementing content knowledge development with intercultural communicative competence development" (p. 488), so one of the main challenges of the TeCoLa project is to explore the possibilities to integrate intercultural and cross-cultural teaching with foreign language instruction.

Such combination of telecollaboration in virtual gamified environments promoting authentic meaningful communication with foreign peers, grounded on a task-based approach with a focus on diversity and intercultural dimension as added values for FLL, makes this project potentially appropriate for multilingual and multicultural university contexts such as EMI settings where diversity, foreign language use and intercultural communication are the rule.

The following table summarises the main features and advantages of the gamification tools described above.

CONCLUSION

Once identified the need of incorporating innovative teaching techniques in multilingual and EMI university contexts, a revision of recent research on gamification in different educational contexts has led us to propose gamification as a possible solution to fill this gap. Most empirical studies show positive results in gamified courses, which posit gamification as an effective tool that fosters motivation and engagement in education that may eventually derive in better results (Christian, 2018; Landers & Landers, 2014; Reiners et al., 2012; Sailer et al., 2017, among others). Additionally, data points to university students as the most suitable target audience for gamified courses (Barata et al., 2013; Garland, 2015).

The analysis of the main features of Classcraft, Kahoot and the TeCoLa project as gamification tools suggests that multilingual and EMI university settings may benefit from the incorporation of these computer-based applications. Despite their differences regarding the purpose, functioning or technical considerations, all of them share a series of benefits including the promotion of classroom dynamism, collaboration, solidarity and responsibility. Additionally, besides fostering students' motivation and

Table 1. Advantages and features of ClassCraft, Kahoot and TeCoLa Project

Summary		ClassCraft	Kahoot	TeCoLa Project
FEATURES	Technical considerations	Available online (no need to download for smartphones). Free access. Requirements for classroom: computer, screen, projector and internet access.	Available online and as app (downloadable for smartphone). Free access (there is a Premium version). Requirements for classroom: computer, screen, projector and internet access.	Available online (still only to project members). Free access. Requirements for classroom: computer, screen, projector and internet access.
	Purpose and functioning	To learn and reinforce particular learning issues. Allows the creation and use of questionnaires and quizzes. Students create their own "alias" to participate (either individually or in group). Responses are time-limited. Questions can be customized. Score statistics are available for all participants.	To involve the classroom in an online a role-play. Lecturer and students create an account. Avatars created for each participant (either by the lecturer or the students themselves). Students can access the platform either inside or outside the classroom. Students work in groups (4-6 members). They can adopt different roles (e.g. Warrior, Healer, Mage).	To develop and evaluate innovative gamified telecollaboration approaches in foreign language and subject-matter educational environments (originally in secondary education). Students can engage in virtual gamified environments to communicate and collaborate through authentic task-based interactions among peers from different language, socio-cultural and educational backgrounds. It offers different tools (*webinars, virtual world, gamify, moodle and cloud*) and resources (*language tasks, pedagogical guides, case studies, teachers talks, language learning reports,* etc.)
	Rewards / points	A rewarding system can be created by the lecturer according to the number of correct answers and the answer speed.	Students get points according to behavior and positive actions (XP, AP and HP). As students get points, they can get powers (basic, intermediate and advanced powers).	A rewarding system can be created by the instructor according to the number of successfully completed tasks.
ADVANTAGES	Benefits for the classroom	Enhances students' motivation and engagement. Favors student-centered environment. Effective resource for foreign language improvement, academic achievement and formative assessment.	Promotes dynamism, collaboration, solidarity and sense of responsibility. Fosters motivation and engagement.	Fosters engagement in meaningful communication tasks. Enhances language improvement and social civic communicative competences. Enriches CLIL by combining content knowledge, language learning and intercultural competences.

engagement, as the recent literature demonstrate, both Classcraft and Kahoot enhance linguistic improvement, academic achievement and favors student-centered environments (Bicen & Kocakoyun, 2018, Papadakis & Kalogiannakis, 2018; Sanchez, Young, & Jouneau-Sion, 2017). Finally, the focus on task-based approaches, diversity and intercultural communicative competences pursued particularly by the TeCola project (Clavel-Arroitia & Pennock-Speck, 2015; Pennock-Speck & Clavel-Arroitia, 2018) constitutes an ideal scenario for both teachers and learners following courses taught in a non-native language, as suggested by the EHEA.

In order to take advantage of the implementation of gamification and its benefits in multilingual and EMI classrooms, the need to train both instructors and students on current digital and information and communication technologies (ICT) competences is essential. This can be achieved through student and teacher training – as pursued by the TeCoLa project – and the development of ICT integrated curricula. Despite the increasing amount of literature on the positive effects of using digital and ICT resources in educational contexts on teaching and learning improvement, reality shows that the development of ICT integrated curricula still remains underperformed (Arzal & Chen, 2017). Some of the reasons underlying the underdevelopment of ICT integrated curricula, especially in English as a foreign language (EFL) educational settings, range from scarce institutional support (Harendita, 2013), lack of professional development programs (Hew & Brush, 2007) or scant commitment to include training in digital resources in pre- and in-service programs (Sánchez-García et al., 2013). However, if we pursue to incorporate the positive effects of gamification in university classrooms, both practitioners and institutional authorities should become aware of the urgent need to adopt the necessary measures to integrate digital and ICT competences in higher education curricula, especially in EMI contexts, where both the learning and teaching of disciplinary content through a non-native language is doubly challenging.

We hope that this study constitutes a theoretical basis that paves the way for future empirical research and potential implementation of gamification in EMI and multilingual university contexts.

ACKNOWLEDGMENT

This study has been developed under the framework of a Teaching Innovation Project funded by the University of Almería (Spain) entitled "Classcraft como Plataforma de Gamificación en el Aula Universitaria de Lengua Extranjera".

REFERENCES

Ahmed, H. B. E. (2016). Duolingo as a bilingual learning app: A case study. *Arab World English Journal*, *7*(2), 255–267. doi:10.24093/awej/vol7no2.17

Al-Hadithy, T., & Ali, S. (2018). Gamification in learning English for academic purposes: Designing assessment for language using Kahoot with UAE undergraduate law students. In *Proceedings of 110th IASTEM International Conference*, Toronto, Canada.

Alexander, C. (2017). Student perceptions of gamification in higher education. In *Proceedings of Society for Information Technology & Teacher Education International Conference* (pp. 1428-1433). Association for the Advancement of Computing in Education (AACE).

Alvarado, N. C., Coelho, D., & Dougherty, E. (2016). Mobile apps for ELLs: Supporting language learning with engaging digital tools. *Argentinian Journal of Applied Linguistics*, *4*, 43–58.

Arzal, A. & Chen, S. (2017). Improving ICT integration in language classrooms: Voices from teachers in Indonesian Province of Gorontalo. *Proceedings of the CALL in Context Conference*, Berkeley, CA, USA. Retrieved from: http://call2017.language.berkeley.edu/wp-content/uploads/2017/07/CALL2017_proceedings.pdf

Barata, G., Gama, S., Jorge, J., & Gonçalves, D. (2013). Improving participation and learning with gamification. In L. Nacke, K. Harrigan, & N. Randall (Eds.), *Proceedings of International Conference on Gameful Design, Research, and Applications* (pp. 10– 17). Stratford, Canada: ACM.

Bicen, H., & Kocakoyun, S. (2018). Perceptions of students for gamification approach. Kahoot as a case study. *International Journal of Emerging Technologies in Learning*, *13*(2), 72–93. doi:10.3991/ijet.v13i02.7467

Brown, D. H. (2007). *Principles of Language Learning and Teaching*. White Plains, NY: Pearson Education.

Burke, B. (2014). *Gamify: How gamification motivates people to do extraordinary things*. Brookline, MA: Gartner, Inc.

Burston, J. (2014). The reality of MALL: Still on the fringes. *CALICO Journal*, *31*(1), 103–125. doi:10.11139/cj.31.1.103-125

Caton, H., & Greenhill, D. (2014). Rewards and penalties: A gamification approach for increasing attendance and engagement in an undergraduate computing module. *International Journal of Game-Based Learning*, *4*(3), 1–12. doi:10.4018/ijgbl.2014070101

Christians, G. (2018). *The Origins and Future of Gamification*. (Senior Theses). 254. Retrieved from https://scholarcommons.sc.edu/senior_theses/254

Clavel-Arroitia, B., & Pennock-Speck, B. (2015). Telecollaboration in a secondary school context: Negotiation of meaning in English as lingua Franca/Spanish tandem interactions. *Revista d'Innovació Educativa*, *15*, 74–78.

Clement, R., Dornyei, Z., & Noels, K. A. (1994). Motivation, self-confidence, and group cohesion in the foreign language classroom. *Language Learning*, *44*(3), 417–448. doi:10.1111/j.1467-1770.1994.tb01113.x

Colpaert, J. & Spruyt, E. (2018). Toward a task design model for mental acceptance and motivation: A transdisciplinary approach. In *Proceedings of the 3rd International Conference on Language Education and Testing - What a Feeling: Language Education and Emotions,* (pp. 138-145). Antwerp, Belgium. Retrieved from https://www.uantwerpen.be/en/conferences/language-education-and-emotions

De Byl, P. (2013). Factors at play in tertiary curriculum gamification. *International Journal of Game-Based Learning*, *3*(2), 1–21. doi:10.4018/ijgbl.2013040101

Dearden, J. (2015). *English as a medium of instruction–A growing global phenomenon*. London, UK: British Council.

Deterding, S., Dixon, D., Khaled, R., & Nacke, L. (2011). From game design elements to gamefulness: Defining "gamification.". In A. Lugmayr, H. Franssila, C. Safran, & I. Hammouda (Eds.), *MindTrek 2011* (pp. 9–15). New York, NY: ACM. doi:10.1145/2181037.2181040

Dicheva, D., Dichev, C., Agre, G., & Angelova, G. (2015). Gamification in education: A systematic mapping study. *Journal of Educational Technology & Society*, *18*, 75–88.

Doiz, A., Lasagabaster, D., & Sierra, J. M. (Eds.). (2012). *English-medium instruction at universities: Global challenges*. Bristol, UK: Multilingual Matters. doi:10.21832/9781847698162

Domínguez, A., Saenz-de-Navarrete, J., De-Marcos, L., Fernández-Sanz, L., Pagés, C., & Martínez Herráiz, J.-J. (2013). Gamifying learning experiences: Practical implications and outcomes. *Computers & Education*, *63*, 380–392. doi:10.1016/j.compedu.2012.12.020

Dudley-Evans, T. (1997). Five questions for LSP teacher training. In R. Howard & G. Brown (Eds.), *Teacher education for languages for specific purposes* (pp. 58–67). Clevedon, UK: Multilingual Matters.

Ellis, R. (2003). *Task-based language learning and teaching*. Oxford, UK: Oxford University Press.

Fanfarelli, J. R. & McDaniel, R. (2017). Exploring digital badges in university courses: Relationships between quantity, engagement, and performance. *Online Learning, 21*.

Figueroa, J. (2015). Using gamification to enhance second language learning. *Digital Education Review*, *27*, 32–54.

Garland, C. M. (2015). Gamification and implications for second language education: A meta analysis. *Culminating Projects in English.* Paper 40.

Göksün, O., & Gürsoy, G. (2019). Comparing success and engagement in gamified learning experiences via Kahoot and Quizizz. *Computers & Education*, *135*, 15–29. doi:10.1016/j.compedu.2019.02.015

Haaranen, L., Ihantola, P., Hakulinen, L., & Korhonen, A. (2014). How (not) to introduce badges to online exercises. In *Proceedings of the 45th ACM technical symposium on Computer science education* (pp. 33–38). New York, NY: ACM Press. 10.1145/2538862.2538921

Hanus, M. D., & Fox, J. (2015). Assessing the effects of gamification in the classroom: A longitudinal study on intrinsic motivation, social comparison, satisfaction, effort, and academic performance. *Computers & Education*, *80*, 152–161. doi:10.1016/j.compedu.2014.08.019

Harendita, M. E. (2013). Why resist? A closer look at Indonesian teachers' resistance to ICT. *International Journal of Indonesian Studies*, *1*(1), 41–57.

Hasegawa, T., Koshino, M., & Ban, H. (2015). An English vocabulary learning support system for the learner's sustainable motivation. *SpringerPlus*, *4*(99), 1–9. PMID:25830080

Hew, K. F., & Brush, T. (2007). Integrating technology into K-12 teaching and learning: Current knowledge gaps and recommendations for future research. *Educational Technology Research and Development*, *55*(3), 223–252. doi:10.100711423-006-9022-5

Huang, W., & Soman, D. (2013). *A practitioner's guide to gamification of education. Research report series: Behavioral economics in action*. University of Toronto–Rotman School of Management.

Hung, A. (2017). A critique and defense of gamification. *Journal of Interactive Online Learning*, *15*(1), 57–72.

Hung, A., Zarco, E., Yang, M., Dembicki, D., & Kase, M. (2017). Gamification in the wild: Faculty perspectives on gamifying learning in higher education. *Issues and Trends in Educational Technology*, *5*(2). University of Arizona Libraries. Retrieved from https://www.learntechlib.org/p/182049/

Hwang, S. H. (2018). *Students' perceptions about gamified methods in English as a second language acquisition*. (Theses and Dissertations). 2774.

Iaremenko, N. V. (2017). Enhancing English language learners' motivation through online games. *Information Technologies and Learning Tools*, *59*(3), 126–133. doi:10.33407/itlt.v59i3.1606

Ibáñez, M., Di-Serio, Á., & Delgado-Kloos, C. (2014). Gamification for engaging computer science students in learning activities: A case study. *IEEE Transactions on Learning Technologies*, *7*(3), 291–301. doi:10.1109/TLT.2014.2329293

Jauregi, K., & Melchor-Couto, S. (2017). The TeCoLa project: Pedagogical differentiation through telecollaboration and gaming for intercultural and content integrated language teaching. In *Proceedings of the EUROCALL 2017 Conference*, (pp. 163-169), Southampton, UK. 10.14705/rpnet.2017.eurocall2017.707

Kapp, K. M. (2012). *The gamification of learning and instruction: Game-based methods and strategies for training and education*. San Francisco, CA: Pfeiffer.

Kumar, J., & Herger, M. (2013). *Gamification at work: Designing engaging business software*. Aarhus, Denmark: The Interaction Design Foundation. doi:10.1145/2468356.2468793

Lam, S. L. (2014). Use of gamification in vocabulary learning: A case study in Macau. *Alternative Pedagogies in the English Language & Communication Classroom*, 90-97. In *4th CELL Symposium Proceedings*.

Landers, R. N., & Landers, A. K. (2014). An empirical test of the theory of gamified learning: The effect of leaderboards on time-on-task and academic performance. *Simulation & Gaming*, *45*(6), 769–785. doi:10.1177/1046878114563662

Lasagabaster, D., Doiz, A., & Sierra, J. M. (2014). *Motivation and foreign language learning: From theory to practice*. Amsterdam, The Netherlands: John Benjamins. doi:10.1075/lllt.40

Lee, J. J., & Hammer, J. (2011). Gamification in education: What, how, why bother? *Academic Exchange Quarterly*, *15*(2).

Macaro, E., Curle, S., Pun, J., An, J., & Dearden, J. (2017). A systematic review of English medium instruction in higher education. *Language Teaching*, *51*(1), 36–76. doi:10.1017/S0261444817000350

Majdoub, M. (2016). Promoting high school ESL learners' motivation and engagement through the use of gamified instructional design. Presented in partial fulfillment of the requirements for the degree of Master of Arts (Educational Technology) at Concordia University, Montreal, Canada.

Medina, E. G. L., & Hurtado, C. P. R. (2017). Kahoot! A digital tool for learning vocabulary in a language classroom. *Revista Publicando*, *4*(12), 441–449.

Melchor-Couto, S. (2019). Virtual worlds and language learning. *Journal of Gaming and Virtual Worlds, 11*(1), 29–43. doi:10.1386/jgvw.11.1.29_1

Munday, P. (2016). The case for using duolingo as part of the language classroom experience. *Revista Iberoamericana de Educación a Distancia, 19*(1), 83–109.

Murar, P. (2015, October). Edmodo as a gamification platform: Review and plans. In *Proceedings European Conference on Games Based Learning*. (p. 800). Academic Conferences International Limited.

Nelson, M. (2012). Soviet and American precursors to the gamification of work. In A. Lugmayr (Ed.), *MindTrek* (pp. 23–26). New York, NY: ACM. doi:10.1145/2393132.2393138

O'Donovan, S., Gain, J., & Marais, P. (2013). A case study in the gamification of a university-level games development course. In *Proceedings of South African Institute for Computer Scientists and Information Technologists Conference*, 245–251. 10.1145/2513456.2513469

Papadakis, S., & Kalogiannakis, M. (2018). Using gamification for supporting an introductory programming course. The case of classcraft in a secondary education classroom. In A. Brooks, E. Brooks, & N. Vidakis (Eds.), *Interactivity, Game Creation, Design, Learning, and Innovation. ArtsIT 2017, DLI 2017. Lecture Notes of the Institute for Computer Sciences, Social Informatics and Telecommunications Engineering* (Vol. 229). Cham, Switzerland: Springer. doi:10.1007/978-3-319-76908-0_35

Pede, J. (2017). *The effects of the online game Kahoot on science vocabulary acquisition.* (Doctoral dissertation), Rowan University.

Pennock-Speck, B., & Clavel-Arroitia, B. (2018). Teachers' perspectives on telecollaboration in secondary school foreign language education. *LEA. Lingue e Letterature d'Oriente e d'Occidente, 7*, 483–509.

Pérez, Á., Soto, E., Sola, M., & Serván, M. (2009). *Aprender en la Universidad. El sentido del cambio en el EEES*. Madrid, Spain: Akal.

Pérez Cañado, M. L. & Ráez Padilla, J. (2015). Introduction and Overview. In D. Marsh, M. L. Pérez Cañado, & J. Ráez Padilla (Eds.), *CLIL in Action: Voices from the Classroom* (pp. 1-12). Newcastle upon Tyne, UK: Cambridge Scholars.

Pérez Cañado, M. L. (2018). AICLE/EMI y CMC: Un tándem efectivo para el desarrollo de la interacción comunicativa escrita. In M. González-Lloret & M. Vinagre (Eds.), *Comunicación Mediada por Tecnologías: Aprendizaje y Enseñanza de la Lengua Extranjera* (pp. 35–54). Sheffield, UK: Equinox.

Perry, B. (2015). Gamifying French language learning: A case study examining a quest-based, augmented reality mobile learning-tool. *Procedia: Social and Behavioral Sciences, 174*, 2308–2315. doi:10.1016/j.sbspro.2015.01.892

Pintrich, P. L. & Schunk, D. H. (1996). Motivation in education: Theory, research and applications. Englewood.

Pixie, Y. (2010). 10 reasons why I left farmville [Weblog]. Retrieved from http://blogcritics.org/gaming/article/10-reasons-why-i-left-farmville/

Plump, C. M., & LaRosa, J. (2017). Using Kahoot! in the classroom to create engagement and active learning: A game-based technology solution for elearning novices. *Management Teaching Review*, *2*(2), 151–158. doi:10.1177/2379298116689783

Prensky, M. (2001). Digital natives, digital immigrants. From On the Horizon (MCB University Press, Vol. 9 No. 5, October 2001) © 2001 Marc Prensky.donova

Reiners, T., & Wood, L. C. (2013). Immersive virtual environments to facilitate authentic education in logistics and supply chain management. In Y. Kats (Ed.), *Learning management systems and instructional design: Best practices in online education* (pp. 323–343). Hershey, PA: IGI Global. doi:10.4018/978-1-4666-3930-0.ch017

Reiners, T., Wood, L. C., Chang, V., Guetl, C., Herrington, J., Gregory, S., & Teräs, H. (2012). Operationalising gamification in an educational authentic environment. In *IADIS 2012 International Conference on Internet Technologies and Society* (pp. 93-100). Perth, Australia: IADIS.

Rivera-Trigueros, I. (2018). *Audiovisual translation as a tool to foster the linguistic competence in the EFL classroom and the improvement of the motivation through gamification* (Unpublished Master's dissertation). University of Almería, Spain.

Rosell-Aguilar, F. (2017). State of the app: A taxonomy and framework for evaluating language learning mobile applications. *CALICO Journal*, *34*(2), 243–258.

Rosewell, J. (2012). A speculation on the possible use of badges for learning at the UK open university. In *EADTU Annual Conference: The Role Of Open And Flexible Education In European Higher Education Systems For 2020: New Models, New Markets, New Media.*

Sailer, M., Hense, J., Mandl, H., & Klevers, M. (2013). Psychological perspectives on motivation through gamification. *Interaction Design and Architecture(s) Journal - IxD&A, 19*, 28-37.

Sailer, M., Hense, J., Mayr, S., & Mandl, H. (2017). How gamification motivates: An experimental study of the effects of specific game design elements on psychological need satisfaction. *Computers in Human Behavior*, *69*, 371–380. doi:10.1016/j.chb.2016.12.033

Sánchez, E., Young, S., & Jouneau-Sion, C. (2017). Classcraft: From gamification to ludicization of classroom management. *Education and Information Technologies*, *22*(2), 497–513. doi:10.100710639-016-9489-6

Sánchez-García, A. B., & Mena Marcos, J. J. (2013). Teacher development and ICT: The effectiveness of a training program for in-service school teachers. *Procedia: Social and Behavioral Sciences*, *92*, 529–534. doi:10.1016/j.sbspro.2013.08.713

Sánchez-Pérez, M. M., & Salaberri Ramiro, M. S. (2017). Implementing plurilingualism in higher education: Teacher training needs and plan evaluation. *Porta Linguarum. Monograph*, *II*, 139–156.

Sant, C. (2014). *Leveling up: Evaluating theoretical underpinnings and applications of gamification in the classroom* (Unpublished dissertation). Peabody College of Education and Human Development, Vanderbilt University, Nashville, TN.

Savvani, S. (2019). State-of-the-art duolingo features and applications. *Advances in Intelligent Systems and Computing*, *917*, 139–148. doi:10.1007/978-3-030-11935-5_14

Schmidt, R., & Watanabe, Y. (2001). Motivation, strategy use, and pedagogical preferences in foreign language learning. In Z. Dörnyei & R. Schmidt (Eds.), *Motivation and second language acquisition* (pp. 313–359). Honolulu, HI: University of Hawai'i, Second Language Teaching and Curriculum Center.

Scott, D., & Beadle, S. (2014). *Improving the effectiveness of language learning: CLIL and computer assisted language learning*. London, UK: European Commission.

Silvers, A. (2011). On education, badges and scouting [web log message].

Sitzmann, T. (2011). A meta-analytic examination of the instructional effectiveness of computer-based simulation games. *Personnel Psychology*, *64*(2), 489–528. doi:10.1111/j.1744-6570.2011.01190.x

Smit, U. & Dafouz, E. (2012). Integrating content and language in higher education: Gaining insights into English-medium instruction at European universities (AILA vol. 25). Amsterdam, The Netherlands: John Benjamins.

TeCoLa. (2019). Pedagogical differentiation through telecollaboration and gaming for intercultural and content integrated language teaching. Retrieved from https://sites.google.com/site/tecolaproject/

Van den Branden, K. (2006). *Task-based language education: From theory to practice*. Cambridge, UK: Cambridge University Press. doi:10.1017/CBO9780511667282

Werbach, K., & Hunter, D. (2012). *For the win: How game thinking can revolutionize your business*. Wharton Digital Press.

Wood, L. C., & Reiners, T. (2015). Gamification. In M. Khosrow-Pour (Ed.), *Encyclopedia of Information Science and Technology* (3rd ed., pp. 3039–3047). Hershey, PA: IGI Global. doi:10.4018/978-1-4666-5888-2.ch297

Chapter 13
Digital Literacy and Motivation: How Students Evaluate Digital L2 Learning

Jane Vinther
https://orcid.org/0000-0002-6158-7111
University of Southern Denmark, Denmark

Jørgen T. Lauridsen
University of Southern Denmark, Denmark

ABSTRACT

The contribution of this chapter to the knowledge of motivation in relation to the learning of a foreign language lies in the extensive examination of student attitudes to a detailed field of involved factors. This chapter gives an account of affective factors in the digital classroom on the basis of the level of the digital literacy of the participating students. The digital learning of an L2 is correlated to the level of language learning anxiety, willingness to communicate in the L2, attitude to feedback, and preference for teaching methods. The results show that the level of digital literacy is so high that it should not in itself negatively impact attitude to other factors. The central finding of the investigation is that intrinsic motivation is the strongest motivational drive when compared to other motivational factors.

INTRODUCTION

In digital language learning research and in the application of technology in the classroom, it is conspicuous how affective factors rarely are taken into account. No doubt the potential of emerging technologies are enhancing learning possibilities which benefit a variety of, if not all, groups of learners, and perhaps for this reason focal points have been learning about and applying new tools while few resources have been allocated to investigating motivation and attitudes as they impact the learning processes. This chapter has taken the learner's perspective with the aim of getting a more insightful basis for planning and classroom management initiatives. For several decades now computer assisted language learning (CALL) in one form or another has had a place in language teaching, but great advances during the past decade of technological developments have speeded up the inclusion in the classroom due to

DOI: 10.4018/978-1-7998-1097-1.ch013

both increasing resources but also due to the ubiquity of ownership of mobile devices in the hands of students themselves. This has changed expectations among the student body for the inclusion of these devices as a natural merging of realities outside the classroom and inside the classroom. The current situation is thus not a transition from none-CALL to the availability of CALL; rather the process has been incremental in nature and increasing in number of options as the technological development took off provided a diversity of possibilities suitable for nearly all aspects of the diversity and complexity of language learning. The prevailing situation is no longer characterized by novelty but a steady influx of a multitude of both innovative software and hardware which makes digital learning an unavoidable and integrative necessity. This fuels the need for a sound knowledge base of the influence this has on student attitude and motivation. This chapter has as its aim to provide a contribution towards assisting teachers in making informed decisions with a view to the improvement of their students' enjoyment and benefit of outcome in learning a foreign or second language.

The language teacher has to take into account not only the nature of the language taught, the specific element of that language under consideration, the level of proficiency of the learner, but also the learning material, the level of proficiency of the learner, the age of the learner, and not least the teaching method(s) and appropriate supportive tools. The teaching professional often has to make quick choices in suddenly arisen situations, and the better the background knowledge is of the learners, the better the possibility of making the right decision is. This chapter encompasses research on motivation with social, psychological and educational perspectives while relating it to the application of digital teaching and learning in classroom instruction as embedded in actual and concrete situations.

Information on how language learners and users perceived the usefulness and the application of digital tools was collected through a detailed questionnaire available online. The questions probed into attitudes and motivation and gave learners the opportunity to point to preferred modes of learning and to voice opinions on how they perceived the goodness of fit on their own learning as well as that of the collective of the learning classroom. The subjective experience of the students form an insight into their motivation and their preferred style of learning in relation to the digitalized classroom and the complex data sets were analyzed for effect sizes of the various components. Creating an underlying general understanding of how the students view the digitalized classroom is vital for further advancement and customization of solutions that can be both technologically and pedagogically viable.

BACKGROUND

Language Learning, Technology and Motivation

The connection between language learning and technology has given occasion to coin several names of categories and attached acronyms, e.g. CALL (computer assisted language learning), TELL (technology enhanced language learning), MALL (mobile assisted language learning) and many more (Levy, 1997). Often CALL is used as a general umbrella term (Levy & Hubbard, 2005), and for the sake of indiscriminate reference to a particular mode of learning, the term CALL will be used interchangeably with 'digital' when referring to application of various devices and interfaces supporting language learning in this study. Equally, the term second language learning will be used interchangeably with foreign language learning with L2/Learning as a common mode of abbreviation.

There is a tendency for studies into digital learning to focus on aspects of learning that can be best described as technology driven in the sense that technology may shape the application of it in both theory and practice (Hubbard & Levy, 2016). These studies often measure effects and usefulness of a particular software or application of software, or there is a focus on the possibilities and affordances of a particular application, software or platform (Kukulska-Hulme, A., 2009, 2016). This has brought forth a great deal of valuable information about the connection between digital possibilities and language learning. However, few studies take the students' perspective and their attitudes to the learning situation and how motivation is affected. Since the beginning of the 2000s there has been an increased interest in language learning motivation studies generally, but only few of the relevant studies connect motivation specifically to digital interfaces embedded in the teaching and learning environments. Technology in the L2 classroom is a given fact which is no longer questioned, yet there are still significant gains to be had from investigating not just the motivation for second language learning but also how this motivation is affected by technology so it becomes possible to support language learners in their achievements and reduce their anxiety.

Our understanding of motivation and attitude has been gradually been increasing in magnitude beginning with Gardner and Lambert (1959; 1972), and the sustained research by Gardner and his associates (Gardner, 1985, 2001; Gardner, Masgoret, Tennant, & Mihic, 2004; Gardner & Tremblay, 1994; Tremblay & Gardner, 1994;Tremblay, Goldberg, & Gardner, 1995) has documented how motivation is a composite and dynamic construct encompassing integrative and instrumental kinds of motivation with a related attitude to the learning situation and material as a situated component that will deeply influence motivation. The integrative motivation finds expression in identification with the target language group while the instrumental motivation is related to the utilitarian aspects of the gains and advantages to be had from knowing the L2 in question.

Our realization of the importance of attitude and motivation was enlightened by the research Gardner and his associates and their work laid the foundation on which others have continued to build our understanding of the intricacies of motivation and the self of the learning. Dörnyei and Ushioda have explored the importance of individual differences in L2 learning. Their research has given us a systematic insight into the L2 Self (Dörnyei, 2009; Dörnyei & Ushioda, 2009; Ushioda and Dörnyei, 2009; Ushioda, 2011) and how possible and future selves integrate an Ideal Self and an Ought-to Self. This L2 Self system builds on the general psychological theory of self-determination (Deci & Ryan, 1985; Ryan & Deci, 2000) which connects the question of identity with motivation of basically two kinds: intrinsic and extrinsic motivation. The strongest motivational drive is found in the intrinsic motivation of the self-determined, self-regulating individual who acts and learns from some inner fulfilment because to the individual the activity is "inherently interesting and enjoyable" (Ryan and Deci, 2000, p.55). In L2 classroom it is therefore important to create a learning atmosphere which enhances learner autonomy and creativity. For this the digital classroom seems particularly suited in that it allows students to pursue goals in their own pace and in their personally preferred style of learning. Extrinsic motivation on the other hand is "doing something because it leads to a separable outcome" (Ryan & Deci, 2000, p.55).

Bodnar, Cucchiarini, Strik, & van Hout (2016) have evaluated a large number of recent research studies on CALL and motivation, and they have found that few studies take theories of language learning motivation into account. Ushioda (2011) addressed the issue of technology and motivation, and she sees possibilities for autonomy and new learner roles in embracing the digital options and emphasizes how it might open the learning space to go beyond the classroom. With CALL we have the tools to move towards the self-regulated learner respecting the identity of the L2 self while providing teaching and learning with a pedagogical framework observing local guidelines and regulation. Ushioda (2011, p. 207) observes that "Creating this kind of fusion between how students use technology inside and outside the language classroom will help reduce the barriers between L2 learning and life". Rosell-Aguilar (2013) investigated the effect of listening to podcasts and they had "hugely positive response" (p. 87), but simultaneously, it is pointed out that perhaps the learning aspects is not sufficiently served. Kobayashi & Little (2011) found in their study on blended learning that "the students' perceptions of the program differed according to the proficiency levels of students" (p.103). Vinther (2012) found digital possibilities enhanced cognitive processes and retention of grammatical knowledge. However, in reference to the Rosell-Aguilar study, Ushioda emphasizes the important element of motivation, "the potential value of mobile technologies may lie instead in motivating and facilitating frequent (rather than deep) engagement in language learning or language use opportunities." (Ushioda, 2013, p.3). Stockwell is one of those scholars who have devoted sustained interest to motivation of both students and teachers. Stockwell (2013) stresses the dynamic nature of motivation while noting the importance of the instructional design in a versatile learning environment, which will require motivated teachers to give students support and feedback. Stockwell (2012) concludes:

What must always be kept in mind is that CALL is about providing the optimum environment to maximize language learning opportunities for our learners. CALL is not a result of endeavoring to use the latest technologies, but rather the technological options that are available and choosing what is the most appropriate following a sound pedagogy. (pp.172-173).

The diversity of digital learning possibilities, as mentioned above, can enable and encourage learning activities to continue beyond the classroom, but as Burston (2014) found in his study of mobile-assisted learning it "has yet to realize its full potential and that achieving this aim is more a matter of pedagogy than technology" (p.344) Stockwell & Hubbard (2013) state that there may be a difference between general appreciation of digital possibilities and that of the learning application of the same, "Knowledge of how to use mobile devices for specific personal and social functions is not always a good indicator of knowledge of educational function" (p.4). The study reported below investigated a very large number of disparate devices and software applied in many different classrooms and beyond and looks into the motivational impact across the span of diversity while looking the common factor of motivation in its various manifestations and situations.

DIGITAL L2 LEARNING AND MOTIVATIONAL IMPACT

The study reported in this chapter has been interested to examine whether the students' experience of the various digital technologies has found a satisfactory expression in teaching approaches and pedagogies to such an extent that language learning and the motivation for it has been sustained or enhanced. In particular the study sought to examine motivational attitudes in respect to willingness to communicate, L2 anxiety, collaborative learning, individualized opportunities for learning, and feedback. However, it was deemed important to secure background knowledge of the learner's level of digital literacy as a lack of competence in mastering the digital learning situations would have the potential to impact the results of the inquiry into motivation and attitude resulting in the following research questions:

RQ 1: What is the level of digital literacy?
RQ 2: What is student attitude to and motivation for traditional, digital, and blended learning, respectively?
RQ 3: What is the impact of the digital environment on motivation and

a) willingness to communicate
b) level of anxiety
c) feedback

RQ 4: What are the characteristics of motivation in the digital learning environment?

THE STUDY

Participants and Data Collection

The participants of this study were upper secondary students enrolled in study lines which all comprised learning at least two foreign languages. The schools were situated in areas covering all regions of Denmark, and the students were aged 16-19. From this fact it is clear that the number of teachers and teaching approaches involved have been greatly varied. In respect to the digital learning and the platforms, devices and software harnessed in the digital classroom have been equally disparate. The results therefore constitute a statement which the researchers maintain will be generalizable in that the large number of participants and different devices and application vouch for such a variety that it grounds the findings in a general attitude to L2 learning in the digital classroom. Most studies related to motivation and attitudes have focused narrowly on for example particular software, a particular language item to be learned, or a particular approach. This chapter reports findings across these settings and across several languages.

The data were collected through an online questionnaire. The respondents accessed the questionnaire in the same week after almost a full year's instruction, i.e. just before the exam period. The number of respondents varies from item to item in the questionnaire as not all respondents chose to fill in every item, so responses vary in number from 403 to 373, depending on the item.

Statistical Method

The study investigates relationships between categories as suggested by the research questions by applying Confirmatory Factor Analysis (CFA). Confirmatory analysis is used to either confirm or reject the validity of the construct/model involving large datasets. For an introduction to the technique, see Sharma (1996, chapter 6, pp 144-184). Briefly spoken, the idea is that each of the theoretical constructs (TC) is seen as a latent (unobserved) phenomenon that partly represents itself in one or more manifest (observed) questions from the questionnaire.

The example assumes two theoretical constructs TC1 and TC2 that represents themselves in two and three questions from the questionnaire respectively, i.e. TC1 represents itself in Q1 and Q2, and TC2 represents itself in Q3, Q4 and Q5.

As an illustration, one can think of TC1 as Intrinsic Motivation, manifesting itself in the two questions "I love to learn languages" and "Languages are interesting in themselves", while TC2 could be Extrinsic Motivation, manifesting itself in the questions "I have to learn an L2 in my line of study", "It is easy for me", and "I can get a good grades".

Figure 1.

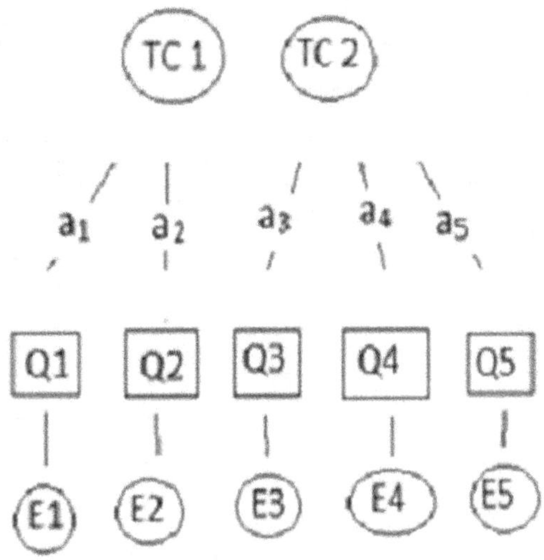

Another way to put it is that Q1 and Q2 are partly "explained" by TC1. This motivates the presence of two "unexplained" latent parts, denoted E1 and E2; similarly for Q3 to Q5 introducing E3 to E5.

The important point for the analysis is to estimate the coefficients a_1 to a_5, which expresses to which extent the questions Q1 to Q5 actually manifests TC1, and to test whether the coefficients are actually significantly different from zero. Furthermore, the statistics will also report the R^2, i.e. the percentage of variation in each of the questions Q1 to Q5 that can be ascribed to the common latent factors. Finally, a couple of Goodness of Fit measures are also applied: The Goodness of Fit (GFI) and the Adjusted GFI (AGFI); see Sharma (1996) for details (for factor analysis, see also Gardner and Lambert, 1959)

Results

Digital Literacy

The digital media offers many possibilities and varieties for language learning approaches, but successful student participation builds on a high degree of digital literacy. Stockwell & Hubbard (2013) state that there may be a difference between general appreciation of digital possibilities and that of the learning application of the same, "Knowledge of how to use mobile devices for specific personal and social functions is not always a good indicator of knowledge of educational function" (p.4). Stockwell (2008) found that some students were quite comfortable with the PC, but did not really want to learn how to use mobile devices for their language learning. The authors therefore found it paramount to investigate the actual mastery of such digital devices relative to student ability to engage with the L2 learning devices and software. The digital literacy level may play a role in the motivation that students experience in relations to instruction with CALL, and our RQ 1 was therefore: *What is the level of digital literacy*? Our data builds on six detailed questions as shown in Table 1.

The hypothesis is that learner's level of digital literacy can be represented by one latent construct, denoted Digital Literacy (DL), which manifests itself in six questions as outlined in Table 1.

Table 1. The level of digital literacy (N=398)

Construct	Question	Label	Coefficient	p-value	R^2	%Agree
DL	RQ1a	I can install apps on my device	0.83	<0.001	0.68	61
DL	RQ1b	I have many good language learning apps on my device	0.76	<0.001	0.58	35
DL	RQ1c	I know how to use many different L2 learning apps	0.90	<0.001	0.76	47
DL	RQ1d	I master many ways of using my digital device in L2 learning	0.88	<0.001	0.78	54
DL	RQ1e	I have control over how to use my digital device in L2 learning	0.75	<0.001	0.56	54
DL	RQ1f	I love to experiment with my digital device	0.70	<0.001	0.49	46
GFI	0.94					
AGFI	0.84					

It is seen that the latent construct DL is well manifested into the six questions, all with positive coefficients. Furthermore, all coefficients are statistically significant with small p-values, and the R^2 values are quite high in that an R^2 value of 0.30 is consider high, and the values for all the sub-questions are well-above that with the lowest value at 0.49. Finally, the GFI and AGFI values are really high. In other words, the effect sizes are high and the squared R values show that the questions express high levels of shared variation in the questions of the construct digital literacy.

It can thus be concluded that the subsequent insights into L2 learning aspects and digital learning environments can be founded on a basis of a satisfactory level of digital literacy. Thus, if the research questions were to reveal negative attitude or lack of participation or little enthusiasm for learning it is reasonable to believe that it is not due to lack of ability to come to terms with the technology.

The answer to RQ 1 (*What is the level of digital literacy?*) given the high levels of R^2 is that the students' level is satisfactorily high in the sense that is cannot be seen as a hindrance towards the potential benefits from digital learning. Relatively speaking, the effect size for "*I have many good language learning apps on my device*" is the second lowest. One interpretation could be that the actual apps used or that apps for L2 learning in general are considered to need improvement in quality. Stockwell (2013) has pointed to a lack of quality in CALL software. Burston (2014) has pointed to a lack of appropriate pedagogical approaches with CALL. Sydorenko, Hsieh, Ahn, & Arnold, (2017) suggest, "The groups of elearners surveyed may not have realized the potential benefits of CALL or did not fully utilize them, thereby overlooking their potential for language learning." (p.211). In the present study this is not further investigated and could be a direction for future research.

Student Attitude to Teaching Methods

In addition to the level of digital literacy among the respondents, the study's interest in finding in-depth and detailed knowledge of motivation and attitude to the learning in the digital environment made it necessary to gauge the general attitude to the teaching methods in order to provide a background on which to evaluate their responses. Respondents were therefore asked about their attitudes to the learning preference and outcomes in relation to traditional learning, digital learning and blended learning as

Table 2. Learners' attitudes to teaching methods (N=392)

Construct	Question	Label	Coefficient	p-value	R^2	%Agree
DL	RQ2a	I like digital media and methods to be involved in L2 teaching	0.41	<0.001	0.17	88
	RQ2b	Problems with hardware are seldom	0.72	<0.001	0.52	65
	RQ2c	Problems with apps and software are seldom	0.87	<0.001	0.77	59
	RQ2d	I need more instruction in apps and programs	-0.02	0.70	0.01	42
	RQ2e	The teacher gives good instruction and help	0.29	<0.001	0.08	67
TL	RQ2f	I prefer to learn without digital media	0.62	0.17	0.39	21
	RQ2g	I learn best without digital media	0.61	0.17	0.37	44
BL	RQ2h	I prefer combinations of traditional and digital teaching methods	1.00	<0.001	1.00	90
GFI	0.91					
AGFI	0.78					
		Correlation		**p-value**		
DL	TL	-0.01		0.85		
DL	BL	0.16		<0.001		
TL	BL	0.07		0.39		

reflected in responses to RQ 2: *What is student attitude to and motivation for traditional, digital, and blended learning, respectively?* Their preferences were used subsequently to correlate with aspects of L2 learning experience as reported under research question 3 and 4.

It is hypothesized that Learner's Attitude can be expressed in three latent constructs Digital Learning (ML), Traditional Learning (TL), and Blended Learning (BL), each manifesting themselves in five, two and one questions, respectively, as outlined in Table 2.

The theoretical construct Attitude towards Digital Learning seems to be quite well represented. The statement *"I need more instruction in apps and programs"* is negatively correlated indicating that the majority of students are not in agreement with the students. However, the load is not significant (0.70). A closer look at the underlying figures reveal that the student body is split almost down the middle in their need for more instruction in that 42% say they do not need more instruction which means that 58% would like more instruction in how to utilize apps and programs. Students are not homogenous with respect to need for more instructions. The question which is not answered by this result is how vital this need is. We do not know whether students feel this is causally related to a deficiency in their ability to learn certain aspects of the L2 or whether it is related to task completion. It might also mean that some students are so enthusiastic about digital learning that they want even more of the L2 learning environment to be covered by digital application. This would be a direction for further research. When exploring students' beliefs and attitudes, it appeared from previous studies that familiarity with technology is an important factor,

Learners who had higher exposure to technology in the classrooms were more motivated to use technology to learn a FL, more satisfied with the technology used in their language classrooms, or convinced of its benefits. In other words, the more technology learners are exposed to in language learning, the more they may value it and the more positive their beliefs toward CALL may be (Sydorenko, Hsieh, Ahn, & Arnold, 2017, p. 212).

The above reported high level of digital literacy of the students surveyed in the present study may in part play into the positive attitudes towards digital learning found in this this study (Table 2). It is evident from the response to statements relating to teaching methods that an L2 learning environment including digital media is given high priority by the majority of respondents in this study. The proportion of students in agreement with the statement, *I prefer combinations of traditional and digital teaching methods* is 90 percent. And the statement, *I prefer to learn without digital media* is only supported by 21%.

Furthermore, it is not surprising that Blended Learning is well manifested, given that it is only measured by one question. The result for Traditional Learning appears to be explained by the extremely high preference for blended learning. The responses for the two questions manifesting Traditional Learning indicate that few would prefer a learning environment without digital options in that only 21% agree with the statement, *I prefer to learn without digital media.* With regard to the statement, *I learn best without digital media,* the student responses show that 44% are in support (in effect saying 'yes') while 56 against, i.e. a small majority who say they do not best without digital options. This result can be interpreted at face value, which would indicate to split student preferences almost in two equal portions. However, the responses could also be open to further speculation about the attitudes of those who say they learn best without digital media (44%). Is this because they believe that they do not learn at all or learn less, and it so to what degree? It is worth noticing that a fairly high number of students do not know (7%), which would add to the interpretation that students find it hard to decide one way or the other, but the majority favor inclusion of digital learning in the L2 classroom. If compared to the vast support for digital involvement in the learning environment (88%) as expressed by the statement, *I like digital media and methods to be involved in L2 teaching*, there is reason to believe that there are aspects here that need deeper probing. The survey did not go further into this issue so it would be a fruitful area for further research. Turning to the correlations, there appears to be a positive relationship between Mobile Learning and Blended Learning, while Traditional Learning is not significantly related to the former two. Thus, it is indicated that learners with preference for mobile and blended learning are heterogeneous with respect to preferences for traditional teaching (Vinther & Lauridsen, 2018) – some like it and others do not. The above conclusions are supported by the R^2 values. Finally, the GFI and AGFI values are indicative of a good solution.

When seeking to answer to RQ 2 (*What is student attitude to and motivation for traditional, digital, and blended learning, respectively?*) it is relevant to state that all three approaches investigated (digital, traditional, and blended learning) are found acceptable by respondents. The difference between them is the preference for each of them respectively, and here it is clearly demonstrated that digital and blended learning environments are preferred by the students relative to that of traditional teaching methods.

Willingness to Communicate in the L2

One of the barriers that language teachers need to overcome, and to which many teachers devote considerable energy, is the reluctance of the students to produce sufficient amounts of language for learning and interaction to take place. The willingness to communicate is a prerequisite for all interaction in the

Table 3. Willingness to communicate in the L2 (n=377)

Construct	Question	Label	Coefficient	p-value	R^2	%Agree
WTC	RQ3a.1	DL makes me communicate in the L2 without inhibitions	0.80	<0.001	0.64	54
	RQ3a.2	DL makes me write more in the L2	0.77	<0.001	0.59	67
	RQ3a.3	DL makes me communicate more in the L2	0.80	<0.001	0.64	56
GFI	1.00					
AGFI	-					
		Correlations with learning preferences (p-values in parentheses)			WTC	
		Digital Learning (DL)			0.35 (<0.001)	
		Traditional Learning (TL)			-0.05 (0.38)	
		Blended Learning (BL)			0.26 (<0.001)	

L2. We know from SLA research that output and feedback promote proficiency (Gass & Mackay, 2006; Clément, Baker, & MacIntyre. 2003), but as Bodnar et al. (2016, p.186) state "a system can provide sufficient L2 input and opportunities for L2 output, yet fail to be pedagogically effective if learners are unwilling to participate."

Based on initial descriptive examination of our statistical method, we hypothesize that willingness to communicate can be expressed in one theoretical construct as shown in Table 3. Specifically, the theoretical construct, which we denote as Willingness to Communicate (WTC) manifests itself in three questions.

The theoretical construct 'willingness to communicate' seems to be quite well represented by the manifest questions. This conclusion is further supported by the R^2 values. Finally, the GFI is indicative of a very good construction, while the AGFI could not be calculated, given that we have only three manifest variables to one factor.

The relationship between digital learning and the willingness to communicate in the L2 as presented in the three statements and research question 3a, is reported in the correlations between the three theoretical constructs of DL (attitude to digital learning), TL (attitude to traditional learning), and BL (attitude to blended learning) and the construct of the present construct of willingness to communicate in the L2. Specifically, it is seen that Willingness to Communicate is positively related to Digital and Blended Learning, and negatively related to Traditional Learning.

The answer to RQ 3a (*What is the impact of the digital environment on motivation and willingness to communicate?*) is can be seen that the students responding to the survey indicate an increased willingness to communicate in the L2 in the digital or blended learning environment, but less so if the traditional learning environment without digital options. MacIntyre, Dörnyei, Clément, & Noels (1998) suggested that contextual factors would influence the willingness to communicate in the L2, and the findings in the present study confirm that. MacIntyre et al.'s (1998) model of components for L2 willingness to communicate had several layers and influences included intergroup relations and perceived competence in the L2. A theme for further research would be to look into whether the digital learning environment contributes towards enhancing students' perceived competence resulting in encouragement to participate in L2 communicative activities.

Anxiety

Gardner, Masgoret, Tennant, & Mihic (2004, p.3) found that "language anxiety has a debilitating effect" on student achievement. The connection to willingness to communicate is present in the reticence in classroom participation that anxiety may cause (Ellis, 1994; Horwitz, 2001). Similarly, MacIntyre, & Gardner (1994) demonstrated how language anxiety can inhibit L2 learning. Dörnyei (2009) sees a relation between student self-confidence and anxiety, which supports Gardener et al.'s (2004) finding that A-students suffer less from anxiety that B-students do. In a study on the effect of recasts Sheen (2008) found that the low-anxiety group outperformed the high-anxiety group. The affective influence on L2 learning outcome first came to prominence with Krashen's (1985) notion of the affective filter, which he theorized could severely diminish learning. Today there is ample evidence to suggest that L2 anxiety has a detrimental effect and that creating a relaxed learning atmosphere will promote learning and students' motivation to participate in activities. Tran & Moni (2015) suggest that both teachers and students are aware of language learning anxiety and that strategies to manage it might make it possible for it to be applied with a positive effect.

Based on initial descriptive examination, we hypothesize that anxiety can be expressed in two theoretical constructs as shown in Table 4. Specifically, the two theoretical constructs, which we denote as Relaxed Learning (RL) and Relaxed Communication on Language (RC) manifest themselves in three and one questions, respectively.

Table 4. Anxiety (N=377)

Construct	Question	Label	Coefficient	p-value	R^2	%Agree
RL	RQ3b.1	Digital learning relaxes the learning atmosphere in language lessons	0.78	<0.001	0.61	54
	RQ3b.2	Digital learning reduces fear of errors in language lessons	0.79	<0.001	0.63	54
RC	RQ3b.3	Digital learning makes me learn in a relaxed and natural way	0.71	<0.001	0.50	62
	RQ3b.4	Digital learning makes me communicate on the language without inhibitions	1.00	<0.001	1.00	54
GFI	0.97					
AGFI	-					
		Correlation		**p-value**		
RL	RC	0.54		<0.001		
		Correlations with learning preferences (p-values in parentheses)		RL	RC	
		Digital Learning (DL)		0.39 (<0.001)	0.31 (<0.001)	
		Traditional Learning (TL)		-0.22 (<0.001)	0.02 (0.69)	
		Blended Learning (BL)		0.36 (<0.001)	0.22 (<0.001)	

Both theoretical constructs seem to be quite well represented by the manifest questions. The relationship between the two constructs is evident and significant ($p<0.001$). The conclusion is supported by the R^2 values. Finally, the GFI is indicative of a very good construction. The AGFI could not be calculated, as there are only two manifest questions per theoretical construct.

The relationship between learning preference and L2 anxiety, we report as the correlations between the three theoretical constructs of the former and the two constructs of the present. Specifically, it is seen that Relaxed Learning is positively related to Digital and Blended Learning, and negatively to Traditional Learning. Relaxed Communication is positively correlated to all three preferences but only the correlations to Digital Learning and Blended Learning are significant. It should be noted that the correlation of Relaxed Communication to Blended Learning is weak (0.22) and so is the correlation between Relaxed Learning and Traditional Learning.

One tentative interpretation of this result in answer to RQ 3b (*What is the impact of the digital environment on motivation and level of anxiety?*) could be that in general the L2 classroom is perceived as a friendly learning environment albeit that the digital element has a significant effect in making it even less stressful and less anxiety inducing than otherwise is perceived as being the case in the traditional learning space. Lai & Gu (2011) noted that the Hong Kong students in their study were hesitant to communicate with people outside the classroom. The present study does not confirm their finding, but it needs to be noted that we did not specifically look into technology use outside the classroom although learning tasks did involve activities beyond the classroom albeit as integrated with the instructed setting.

Feedback, Collaborative Learning, and the Individual

One of the advantages of digital learning is the opportunity it lends to immediate feedback which can be both collective and personalized. Our knowledge of feedback and its various forms and effectiveness is steadily growing, and it is clear that feedback is useful, but it is less clear which forms feedback should take for which learning objectives (Dilans, 2010; Ellis & Sheen, 2006; Hawkes & Nassaji, 2015; Li, Zhu, & Ellis, 2016; Lyster & Kazuya, 2010). In a communicative approach to L2 learning the feedback type which is used most frequently appears to be recasts. Recasts can be defined as "implicit reformulations of learners' nontarget utterances" (Lyster & Izquierdo, 2009, p.454), and the implicit nature of the interaction has given rise to discussions on whether learners perceive the reformulation as correction. The preference for recasts correlates with the communicative and interactionist SLA theories and task-based learning. This study did not investigate the effectiveness nor the nature of the feedback given, but focused on the students' perceptions and experiences of feedback and its usefulness in the contexts of the digital L2 learning environment and relative to the individual and the group (Table 5).

Based on confirmatory factor analysis, it is hypothesized that feedback and collaborative learning can be expressed in two theoretical constructs as shown in Table 5. We denote the two constructs as Digital Feedback (DFB) and Digital Group Work (DGW) and they manifest themselves in seven and two questions respectively.

Both theoretical constructs seem to be quite well represented by the manifest questions. Correlations appear to indicate a positive relationship between Digital Feed-Back and Digital Group Work. The above conclusions are supported by the R^2 values. Finally, the GFI and AGFI values are indicative of a very good construction.

Table 5. Feedback (N=373)

Construct	Question	Label	Coefficient	p-value	R^2
DFB	RQ3c.1	I can help others learn with digital learning	0.38	<0.001	0.15
	RQ3c.2	I receive immediate feedback from other group members with digital learning	0.69	<0.001	0.47
	RQ3c.3	Digital learning makes it easy for me to help others in their L2 learning by offering comments on their work	0.80	<0.001	0.64
	RQ3c.4	Digital learning facilitates my L2 learning through receiving comments from others	0.81	<0.001	0.66
	RQ3c.5	Digital learning helps me improve in my L2 learning by having to comment on the work of other group members	0.78	<0.001	0.61
	RQ3c.6	With digital learning I can chose to receive comments on my work privately	1.22	<0.001	1.05
	RQ3c.7	Digital learning facilitates L2 learning through reciprocal comments and discussions with peers	0.63	<0.001	0.40
DGW	RQ3c.8	Digital learning is good for collaborative learning in the group	0.88	<0.001	0.77
	RQ3c.9	Digital learning stimulates interest and participation in group work	0.85	<0.001	0.72
GFI	0.95				
AGFI	0.90				
		Correlation		**p-value**	
DFB	DGW	0.28		<0.001	
		Correlations with learning preferences (p-values in parentheses)		DFB	DGW
		Digital Learning (DL)		0.35 (<0.001)	0.38 (<0.001)
		Traditional Learning (TL)		0.04 (56)	-0.04 (0.49)
		Blended Learning (BL)		0.26 (<0.001)	0.31 (<0.001)

To examine the relationship between learning preference (Attitude to teaching methods in RQ 2) and the theoretical construct of the present research related to feedback and collaborative learning, we report the correlations between the three theoretical constructs of the former and the two constructs of the present. Specifically, it is seen that Digital Feedback and Digital Group Work are both positively related to Mobile and Blended Learning with high significance, but unrelated to Traditional Learning. In fact, the relation between Digital Group Work and Traditional Learning is negative and not significant.

The confirmatory factory analysis indicated two latent theoretical constructs, Digital Feedback and Digital Group Work; however, a closer look at some of the statements under Digital Feedback will reveal that the feedback is also related to group activities albeit with the perspective of the individual.

This result gives rise to the assumption that there appears to be sound support for digital L2 learning possibilities related to the individual's perception and participation in group tasks and activities. In general digital learning seems to support its suitability for collaborative learning and respondents also express satisfaction with their own learning in the collaborative classroom Even though digital learning is still beset by limitations (Burston, 2014), perhaps not least pedagogically, the students in this study seem convinced of its positive influence in interaction between peers and in promoting L2 learning for the groups as well as the individual in the affordance for quick, immediate, and relevant feedback which can be given in subtle ways.

The statements presented to the respondents in Table 5 explore the attitude to feedback in a learning situation which builds on both giving and receiving feedback. When seeking to answer RQ 3c (*What is the impact of the digital environment on motivation and feedback?)* it is encouraging to note strong support for inter-peer feedback that digital learning facilitates. Peer interaction, support and scaffolding are fundamental to interactionist SLA theory, and this survey connects this in a positive way to the digital classroom. The students voice support of both individual feedback from the teacher and inter-peer feedback and is in accordance with findings of Henry &Thorsen (2018) who found that moments of engagement were important to motivation and the learning process, "In emerging relationships, moments of contact can generate immediate, conscious responses that take the form of greater engagement and increased motivational energy" (p. 230).

The findings in this study give rise to believe that students have positive attitudes to the application of digitally based teaching methods in which the individual is in interaction with the group of peers moving towards a common goal of improving their L2 proficiency while aiding and scaffolding each other through digitally facilitated communication.

The Characteristics of Motivation in the Digital Classroom

The underlying hypothesis is that the students' motivation can be expressed in four latent constructs Intrinsic Motivation (INTRIN), Extrinsic Motivation (EXTRIN), Instrumental Motivation (INSTRU) and Integrative Motivation (INTGRA), each manifesting themselves in two, three, two and three questions, respectively, as outlined in Table 6.

The three factors INTRIN, INSTRU and INTGRA seem to be very well represented. However, EXTRIN seems less well represented by the statement *"I have to learn an L2 in my line of study",* while it is well manifested in the two statements *"It is easy for me"* and *"I can get good grades"*. The fact is that *"I have to learn an L2 in my line of study"* has hardly any explanatory power with an effect size/ R^2 as low as 0.01, and it accounts for only a minute shared part of the observed factor. The differences relative to the specific languages show that L2 English and L2 German are positively related to the factor EXTRIN while L2 French and L2 Spanish are negatively related to the factor. A closer look at the effect sizes and coefficients for the individual L2 languages reveal the differentiated picture. The result probably reflects the context of the guidelines for the lines of study in which French and Spanish are options that can be selected by individual choice in addition to required English or German. The little explanatory power of the item is thus grounded in the context in which many respondents have an obligatory language (English and German) while others have an optional one of their choice (French or Spanish).

The two negative coefficients for the EXTRIN construct imply that extrinsic motivation is negatively correlated to the other three factors, which are in turn positively correlated to each other. For all languages integrative, external and intrinsic motivational factors are positively related. The strongest factor is INTRIN which means that intrinsic motivation is manifested as the strongest motivational factor for learning an L2 in the digital classroom.

Table 6. Motivation for L2 learning. 'Why are you learning a foreign language'? (N=403)

Construct	Question	Label	Co-efficient	English	German	French	Spanish	R^2
INTRIN	RQ4.a	I love learning languages	0.85	0.81	0.84	0.91	0.83	0.71
	RQ4.b	Languages are interesting in themselves	0.93	0.97	0.86	0.95	0.99	0.86
EXTRIN	RQ4.c	I have to learn an L2 in my line of study*	0.05	0.04	0.14	-0.01	-0.16	0.01
	RQ4.d	It is easy for me	-0.97	-0.94	-0.95	-0.95	-0.89	0.94
	RQ4.e	I can get good grades	-0.80	-0.79	-0.84	-0.86	-0.87	0.65
INSTRU	RQ4.f	It opens possibilities for education and jobs	0.41	0.35	0.25	0.64	0.74	0.17
	RQ4.g	I want to watch untexted TV series online	0.51	0.49	0.37	0.65	0.44	0.25
INTGRA	RQ4.h	It gives me insight in other cultures	0.68	0.64	0.56	0.74	0.87	0.46
	RQ4.i	Friends / family in the L2 country	0.41	0.32	0.44	0.39	0.39	0.17
	RQ4.j	I am interested in music /lyrics in the L2	0.58	0.56	0.65	0.77	0.42	0.34
GFI			0.91	0.89	0.91	0.83	0.90	
AGFI			0.78	0.76	0.79	0.63	0.77	
		Correlation	**p-value**	**English**	**German**	**French**	**Spanish**	
INTRIN	EXTRIN	-0.34	<0.001	-0.30	-0.27	-0.61	-0.26	
INTRIN	INSTRU	0.47	<0.001	0.42	0.42	0.70	0.65	
INTRIN	INTGRA	0.79	<0.001	0.77	0.60	0.76	0.96	
EXTRIN	INSTRU	-0.17	<0.001	-0.16	-0.11	-0.43	-0.23	
EXTRIN	INTGRA	-0.26	<0.001	-0.25	-0.18	-0.51	-0.22	
INSTRU	INTGRA	0.52	<0.001	0.47	0.43	0.75	0.57	

*The p-value for *'I have to learn an L2 in my line of study'* is non-significant at 0.32. For all the other statements the p-values are at p<0.001

The answer to RQ 4 (*What are the characteristics of motivation in the digital learning environment?)* is complex, but the overall result is that for all L2's the strongest motivation is intrinsic motivation with an effect size of 0.71 and 0.86 for the two items, which is a strong manifestation. Self-Determination Theory holds that intrinsic motivation is the types of motivation which leads to the best result as it is internal to the learner and is driven by interest and learning for its own value rather than some force external to the learner as designated by extrinsic motivation. The correlations in Table 6 show that intrinsic and extrinsic factors are negatively.

The two types of motivation from Gardner's social-educational theory are positively related to each other so the instrumental type of motivation complements the integrative type of motivation. Integrative motivation is related to the cultural aspects of the L2 group while instrumental motivation is an expression of the gains one believes to be attached to the knowledge of an L2 other than the language itself, e.g. access to education and job opportunities. Both instrumental and integrative are positively related to intrinsic motivation. The general conclusion is that there is a connection between Gardner's integrative and instrumental aspects of motivation as well as to intrinsic motivation. The one motivational aspect not supported in this survey is extrinsic motivation. Tentatively is can be said that the students found L2 learning in the digital learning environment interesting and helpful, and that they were engaged in the active learning from internal impulse rather than from coercion or from the point of a feeling of obligation to external forces.

FUTURE RESEARCH DIRECTIONS

In the examination of the issues relating to the research questions that this survey has attempted to answer it has become apparent that each answer has given rise to new aspects which has opened new venues for further research. With respect to digital literacy (Table 1) there is an unresolved aspect of the background for what lies behind some of the student attitudes to language learning apps as they do in the statement "I have many good language learning apps on my device". This needs to be investigated further, and we need more detailed probing into the issue to understand whether the applications, software or devices make student give negative expressions of digital learning, or whether it is the lack of pedagogically appropriate tasks suited to the digital learning environment that is the underlying cause.

The student attitude to different teaching approaches and their preferences for a particular learning environment to suit their L2 needs point to a dismissal of the traditional teaching methods. Students prefer learning with digital media in a blended learning environment. However, the split result from the statement "I need more instruction in apps and programs" (Table 2) leaves unanswered ensuing questions. Is the need an expression of a deficiency in quality of the currently available apps which impacts the learning outcomes, or is it an expression of a desire for even more digitally mediated learning? Further research into the underlying reason for the expression of this need should be carried out in more detail.

Another aspect connected to uncertainty in relation to RQ2 is the response to the statement I learn best without digital media, because in perspective of the overall result, it is possible to interpret the result to indicate that in general the motivation for learning among the student body is generally high but the majority see a preference for enhancing this with digital mediation. To enlighten this aspect a more detailed type of research needs to be initiated. We do not know which aspects of the mediated instruction that students are less enthusiastic about. We can see that all three teaching approaches are acceptable to the students, but we also need to be aware that the disparate types of software and tasks may give cause for a differentiated view of the digital classroom. One further aspect which needs to investigated is the role of activities beyond the classroom. Many of the participating respondents had learning tasks which involved activities beyond the classroom albeit as integrated with the instructed setting, but the role of this aspect was not directly enlightened by this survey. Grgurović, Chapelle, &, Mack completed a comprehensive meta-analysis of several aspects of CALL and its place in the L2 learning environment. One element of their analysis was centered on communication and among other things they investigated computer mediated communication in a blended learning setting but found contradictory results of negative and positive effect sizes. They could not ascertain any explanation for this and encouraged further research, "These contradicting results, for which no explanation is evident, call for more research in the area of blended learning because of the growing importance of blended learning in language classes" (2013, p.179). In our investigation we did find a positive attitude to willingness to communicate and most importantly, the students indicated an increased amount of language production. Equally, they found it easier to communicate through digital media. What our research failed to clarify was the reason for this. We cannot establish whether the increased language production was causally related to a more relaxed learning atmosphere or whether collaborative support could be the explanation. We can see, however, from the results of RQ3b, that there is a reduction in L2 anxiety, and this might indeed have had a facilitating effect on the willingness to communicate.

When it comes to the characteristics of motivation in the digital language learning classroom this study has shown that the extrinsic dimension of motivation stands out from the other dimension. This was indeed a surprise and this could be a further direction that could be render new discoveries. Likewise, it could be fruitful to investigate further into the result showing that the instrumental dimension is related to intrinsic and integrative motivation respectively. A more detailed knowledge of this relationship would perhaps bring new insights into how it is possible to create a vigorous interest in foreign language learning.

CONCLUSION

The chapter presents results which are aimed to fill in a gap in the body of research on CALL and learner beliefs about CALL. This was most explicitly pointed out by Sydorenko et al. "A significant body of research has examined language learners' attitudes toward particular technologies used foreign language classrooms. However, literature is scarce on foreign language learners' beliefs towards computer assisted language learning (CALL) in general." (2017, p.196).

The chapter presents findings indicating that L2 learners in the digital learning environment show an overall strong presence of intrinsic motivation in combination with instrumental and integrative motivation. The findings also show that the motivational forces may vary in intensity depending on the L2. It appears that the integrative and thus cultural aspect is stronger for L2s such as French and Spanish, and less so for English. The fact that the integrative motive is less strong for English could be related to the global and lingua franca aspects of English. Detailed investigations on the relationship between learner preferences for particular teaching methods indicate that the digital classroom induces less L2 anxiety and high degrees of engagement in collaborative learning and peer interaction and feedback within the group. The obvious advantage of the digital media to be able to facilitate and focus on the individual and the individual's needs is also present in the findings of this study, but most encouraging there is also support for the students' acceptance of peer support and collaborative engagement. Indeed, the peer aspect of feedback and group work seem important in the digitally mediated L2 learning.

The study found the students possessed a high level of digital competence, and this may be an important factor in the general finding that students are accepting of digital mediation of learning on levels of input, interaction as well as output. Further, it is interesting that students to such a high degree prefer blended learning environments in which the digital possibilities are integrating in the classroom and applied to tasks for which the chosen medium is suitable.

REFERENCES

Bakhoda, I., & Shabani, K. (2019). Bringing L2 learners' learning preferences in the mediating process through computerized dynamic assessment. *Computer Assisted Language Learning*, *32*(3), 210–236. doi:10.1080/09588221.2018.1483950

Bodnar, S., Cucchiarini, C., Strik, H., & van Hout, R. (2016). Evaluating the motivational impact of CALL systems: Current practices and future directions. *Computer Assisted Language Learning*, *29*(1), 186–212. doi:10.1080/09588221.2014.927365

Boulton, A. (2017). Data-Driven Learning and Language Pedagogy. In S. L. Thorne, and May, Stephen (Ed.), Language, Education and Technology (pp. 181-192). Berlin: Springer International Publishing AG. doi:10.1007/978-3-319-02237-6_15

Burston, J. (2014). MALL: The pedagogical challenges. *Computer Assisted Language Learning*, *27*(4), 344–357. doi:10.1080/09588221.2014.914539

Chapelle, C. A. (2009). The relationship between second language acquisition theory and computer-assisted language learning. *Modern Language Journal*, *93*, 741–753. doi:10.1111/j.1540-4781.2009.00970.x

Clément, R., Baker, S. C., & MacIntyre, P. D. (2003). Willingness to communicate in a language: The effects of context, norms, and vitality. *Journal of Language and Social Psychology*, *22*, 190–209. doi: 10.1177/0261927X03022002003

Deci, E. L., & Ryan, R. M. (1985). *Intrinsic motivation and self-determination in human behaviour.* New York, NY: Plenum. doi:10.1007/978-1-4899-2271-7

Dilans, G. (2010). Corrective feedback and L2 vocabulary development: Prompts and recasts in the adult ESL classroom. *Canadian Modern Language Review*, *66*(6), 787–815. doi:10.3138/cmlr.66.6.787

Dörnyei, Z. (2009). The L2 motivational self system. In Z. Dörnyei & E. Ushioda, (Eds.), Motivation, language identity and the L2 self (pp. 9-42). Bristol, UK: Multilingual Matters. doi:10.21832/9781847691293-003

Dörnyei, Z. & Ushioda, E. (Eds.) (2009). *Motivation, language identity and the L2 self.* Bristol, UK: Multilingual Matters.

Ellis, R. (1994). *The study of second language acquisition.* Oxford, UK: Oxford University Press.

Ellis, R., & Sheen, Y. (2006). Re-examining the role of recasts in L2 acquisition. *Studies in Second Language Acquisition*, *28*(04), 575–600. doi:10.1017/S027226310606027X

Gardner, R. C. (1985). *Social psychology and second language learning: The role of attitudes and motivation.* London, UK: Edward Arnold.

Gardner, R. C. (2001). Integrative motivation and second language acquisition. In Z. Dörnyei & R. Schmidt (Eds.), Motivation and Second Language Acquisition, 6, pp. 1-19. Honolulu, HI: University of Hawaii Press.

Gardner, R. C. (2001a). Language learning motivation: The student, the teacher, and the researcher. *Texas Papers in Foreign Language Education*, *6*(1), 1–18.

Gardner, R. C., & Lambert, W. E. (1959). Motivational variables in second language acquisition. *Canadian Journal of Psychology*, 13. PMID:13855818

Gardner, R. C., & Lambert, W. E. (1972). *Attitudes and motivation in second language learning*. Rowley, MA: Newbury House.

Gardner, R. C., Masgoret, A.-M., Tennant, J., & Mihic, L. (2004). Integrative motivation: Changes during a year-long intermediate-level language course. *Language Learning*, *54*(1), 1–34. doi:10.1111/j.1467-9922.2004.00247.x

Gardner, R. C., & Tremblay, P. F. (1994). On motivation: Measurement and conceptual considerations. *Modern Language Journal*, *78*(4), 524–527.

Gass, S., & Mackey, A. (2006). Input, interaction, and output: An overview. *AILA Review*, *19*, 3–17. doi:10.1075/aila.19.03gas

González-Lloret, M. (2017). Technology and Task-Based Language Teaching. In S. L. Thorne, and May, Stephen (Ed.), Language, Education and Technology (pp. 193-205). Berlin: Springer International Publishing AG.

Grgurović, M., Chapelle, C. A., & Shelley, M. C. (2013). A meta-analysis of effectiveness studies on computer-technology supported language learning. *ReCALL*, *25*(2), 165–198.

Hawkes, L., & Nassaji, H. (2015). The role of extensive recasts in error detection and correction by adult ESL students. *Studies in Second Language Learning and Teaching*, *6*(1), 19–41. doi:10.14746sllt.2016.6.1.2

Henry, A., & Thorsen, C. (2018). Teacher–student relationships and L2 motivation. *Modern Language Journal*, *102*(1), 218–241. doi:10.1111/modl.12446

Horwitz, E. K. (2001). Language Anxiety and Achievement. *Annual Review of Applied Linguistics*, *21*, 112–126. doi:10.1017/S0267190501000071

Hubackova, S. (2014). Motivation in eLearning Motivation in Language Courses. *Procedia: Social and Behavioral Sciences*, *122*, 353–356. doi:10.1016/j.sbspro.2014.01.1353

Hubbard, P. & Levy, M. (2016). Theory in computer-assisted language learning research and practice. In F. Farr & L. Murray (Eds.), *The Routledge handbook of language learning and technology*. Oxford, UK: Routledge.

Kobayashi, K., & Little, A. (2011). Learner perceptions on the usefulness of a blended learning EFL program. *JALTCALL Journal*, *7*(1), 103–117.

Krashen, S. (1985). *The input hypothesis: Issues and implications*. London, UK: Longman.

Kukulska-Hulme, A. (2009). Will mobile learning change language learning? *ReCALL*, *21*(2), 157–165. doi:10.1017/S0958344009000202

Kukulska-Hulme, A. (2016). *Personalization of language learning through mobile technologies*. Cambridge, UK: Cambridge University Press.

Lai, C., & Gu, M. Y. (2011). Self-regulated out-of-class language learning with technology. *Computer Assisted Language Learning*, *24*(4), 317–335. doi:10.1080/09588221.2011.568417

Lamb, M. (2017). The motivational dimension of language teaching. *Language Teaching*, *50*(3), 301–346. doi:10.1017/S0261444817000088

Levy, M. (1997). *Computer-assisted language learning*. Oxford, UK: Clarendon Press.

Levy, M., & Hubbard, P. (2005). Why call CALL "CALL"? *CALL Journal, 18*(3), 143–149. doi:10.1080/09588220500208884

Li, S., Zhu, Y. A. N., & Ellis, R. (2016). The effects of the timing of corrective feedback on the acquisition of a new linguistic structure. *Modern Language Journal, 100*(1), 276–295. doi:10.1111/modl.12315

Lyster, R., & Izquierdo, J. (2009). Prompts versus recasts in dyadic interaction. *Language Learning, 59*(2), 453–498. doi:10.1111/j.1467-9922.2009.00512.x

Lyster, R., & Saito, K. (2010). Oral feedback in classroom SLA: A meta-analysis. *Studies in Second Language Acquisition, 32*(2), 265–302. doi:10.1017/S0272263109990520

MacIntyre, P. D., Clément, R., Dörnyei, Z., & Noels, K. A. (1998). Conceptualizing willingness to communicate in a L2: A situational model of L2 confidence and affiliation. *Modern Language Journal, 82*(4), 545–562. doi:10.1111/j.1540-4781.1998.tb05543.x

MacIntyre, P. D., & Gardner, R. C. (1994). The subtle effects of language anxiety on cognitive processing in the second language. *Language Learning, 44*(2), 283–305. doi:10.1111/j.1467-1770.1994.tb01103.x

MacIntyre, P. D., & Legatto, J. J. (2011). A dynamic system approach to willingness to communicate: Developing an idiodynamic method to capture rapidly changing affect. *Applied Linguistics, 32*(2), 149–171. doi:10.1093/applin/amq037

Penner, N., & Schulze, M. (2010). Group work in a technology-rich environment. *Journal of Interactive Learning Research, 21*(1), 111–137.

Rosell-Aguilar, F. (2013). Podcasting for language learning through iTunes U: The learner's view. *Language Learning & Technology, 17*(3), 74–93.

Ryan, R. M., & Deci, E. L. (2000). Intrinsic and extrinsic motivations: Classic definitions and new directions. *Contemporary Educational Psychology, 25*(1), 54–67. doi:10.1006/ceps.1999.1020 PMID:10620381

Sharma, S. (1996). *Applied Multivariate Techniques*. New York, NY: John Wiley & Sons.

Sheen, Y. (2008). Recasts, Language Anxiety, Modified Output, and L2 Learning. *Language Learning, 58*(4), 835–874.

Stockwell, G. (2008). Investigating learner preparedness for and usage patterns of mobile learning. *ReCALL, 20*(3), 253–270. doi:10.1017/S0958344008000232

Stockwell, G. (Ed.). (2012). *Computer assisted language learning: Diversity in research and practice*. Cambridge, UK: Cambridge University Press. doi:10.1017/CBO9781139060981

Stockwell, G. (2013). Technology and motivation in English-language teaching and learning. In E. Ushioda (Ed.), *International perspectives on motivation: Language learning and professional challenges* (pp. 156–175). London, UK: Palgrave Macmillan UK. doi:10.1057/9781137000873_9

Stockwell, G., & Hubbard, P. (2013). *Some emerging principles for mobile-assisted language learning*. Monterey, CA: The International Research Foundation for English Language Education; Retrieved from http://www.tirfonline.org/english-in-the-workforce/mobile-assisted-language-learning

Sydorenko, T., Hsieh, C.-N., Ahn, S., & Arnold, N. (2017). Foreign language learners' beliefs about CALL: The case of a midwestern University. *CALICO, 34*(2), 196-218.

Tran, T. T. T., & Moni, K. (2015). Management of foreign language anxiety: Insiders' awareness and experiences. *Cogent Education*, *2*(1), 992593. doi:10.1080/2331186X.2014.992593

Tremblay, P. F., & Gardner, R. C. (1994). On motivation, research agendas, and theoretical framework. *Modern Language Journal*, *78*(3), 359–368.

Tremblay, P. F., Goldberg, M. P., & Gardner, R. C. (1995). Trait and state motivation and the acquisition of Hebrew vocabulary. *Canadian Journal of Behavioural Science*, *27*(3), 356–370. doi:10.1037/0008-400X.27.3.356

Tsai, C.-H., Cheng, C.-H., Yeh, D.-Y., & Lin, S.-Y. (2017). Can learning motivation predict learning achievement? A case study of a mobile game-based English learning approach. *Education and Information Technologies*, *22*(5), 2159–2173. doi:10.100710639-016-9542-5

Ushioda, E. & Dörnyei, Z. (2009). Motivation, language identities and the L2 self: A theoretical overview. In Z. Dörnyei & Ushioda, E. (Eds.), Motivation, language identity and the L2 self (pp. 1-8). Bristol, UK: Multilingual Matters.

Ushioda, E. (2011). Language learning motivation, self and identity: Current theoretical perspectives. *Computer Assisted Language Learning*, *24*(3), 199–210. doi:10.1080/09588221.2010.538701

Ushioda, E. (2013). Motivation matters in mobile language learning: A brief commentary. *Language Learning & Technology*, *17*(3), 1–5.

Vinther, J. (2011). Enhancing motivation with cultural narratives in computer-mediated communication. *Computer Assisted Language Learning*, *24*(4), 337–352. doi:10.1080/09588221.2011.572898

Vinther, J. (2012). Cognitive skills through CALL-enhanced teacher training. In F. Zhang (Ed.), *Computer-enhanced and mobile-assisted language learning: Emerging issues and trends*. Hershey, PA: IGI-Global. doi:10.4018/978-1-61350-065-1.ch008

Vinther, J., & Lauridsen, J. T. (2018). Getting to know the learner: Motivational drive, digital literacy, and preferred mode of learning. Paper presented at the XIXth International CALL Research Conference, Bruges, Belgium.

Yin, C., Song, Y., Tabata, Y., Ogata, H., & Hwang, G.-J. (2013). Developing and implementing a framework of participatory simulation for mobile learning using scaffolding. *Educ. Technol. Soc.*, *16*, 137–150.

Zarrinabadi, N., & Tanbakooei, N. (2016). Willingness to Communicate: Rise, Development, and Some Future Directions. *Language and Linguistics Compass*, *10*(1), 30–45. doi:10.1111/lnc3.12176

Chapter 14
Designing In-Service Teacher Training for Computer- and Mobile-Assisted Foreign Language Learning:
A Mixed-Methods and SWOT Analysis of the TELL-OP Training Module for Language Professionals

Alice Meurice
Université catholique de Louvain, Belgium

Fanny Meunier
https://orcid.org/0000-0003-2186-2163
Université catholique de Louvain, Belgium

ABSTRACT

The chapter discusses the importance of in-service teacher training (INSET) to promote the use of open natural language processing (NLP)-based technologies (NLPTs). The first section briefly outlines the affordances of technology for second language acquisition and emphasizes the potential of open NLPTs. The second section presents the overall INSET design used in the TELL-OP ERASMUS+ project led by a team of researchers from several European universities. Section 3 provides a quantitative and qualitative analysis of the questionnaires and feedback data from Belgian French-speaking teachers (n = 86) on the TELL-OP online training module. A SWOT analysis (strengths, weaknesses, opportunities, and threats) is used to complement the teachers' feedback. In the fourth section, the authors put the course design into perspective using several theoretical models on the use of technology and open access resources in education, and provide suggestions for improving future similar INSET initiatives.

DOI: 10.4018/978-1-7998-1097-1.ch014

INTRODUCTION

Affordances of Technology for Second Language Acquisition: Focus on NLPTs

The affordances of technology for second language acquisition have regularly been put forward in the literature (some recent references include Hinkelman 2018, Carrió-Pastor, 2016; Caws & Hamel, 2016; Vandergriff, 2016). Gee & Hayes (2011, p.1) label technologies as 'power-ups' for language learning. Whilst the use technology for educational purposes may sometimes take time to trickle down to instructed settings, it seems fair to say that (at least some) technology is currently part of many teachers' lives. As argued by Haines (2015, p.166), however, it is crucial to keep in mind that a "key aspect of affordance is that it is situated in the relationship between user and artefact, rather than being about tools that can be developed as independent components and integrated into any learning environment". She thus defines affordance "as the potential that teachers perceive in a particular technology tool that will support learning and teaching activities in their educational contexts". This chapter focusses on a type of technology, often less known to teachers, viz. open natural language processing (NLP)-based technologies (NLPTs). The rationale behind the research was twofold, i.e. assess the affordances that teachers might perceive in technologies that are often less known to them, and discuss the type of design that in-service teacher training (INSET) modules should ideally adopt for presenting, discussing and potentially using such tools in the language classroom.

Garbade (2018[1]) provides a simple definition of NLP as "the technology used to aid computers to understand the human's natural language" and whose objective is "to read, decipher, understand, and make sense of the human languages in a manner that is valuable". Meurers (2012) distinguishes two broad uses of NLP. The first one is the analysis of words, sentences, or texts produced by language learners that are analyzed by NLP systems to, for example, find mistakes. One concrete example would be the use of a programme to check the grammatical accuracy of a text. Such tools can for instance use algorithms to search part-of-speech tagged texts and look for patterns like [Plural noun + singular verb] as in *the chairs is* or [plural demonstrative determiner + singular noun] as in *those example* which would be cases of potential mistakes. The second main use of NLP tools is to analyze native language to, for instance, assess the lexical and syntactic complexity of sentences or texts and present reading material in the target language that is adapted to the level of language learners. Other examples include the generation of exercises or tests based on native language data, lemmatizers that reduce the various inflected forms of a word into a single form for easy analysis, word sense disambiguation to infer the meaning of a word based on the context. The few examples provided show the potential of NLP both for Second Language Acquisition (SLA) and for learning and teaching second/foreign languages. Regarding the link with SLA, Ziegler et al. (2017, p.209) argue that: "The strategic use of technology offers researchers novel methodological opportunities to examine how incremental changes in L2 development occur during treatment as well as how the longitudinal impacts of experimental interventions on L2 learning outcomes occur on a case-by-case basis". As for the learning/teaching perspective, Alhawiti (2014, p.72) explains that "Natural Language Processing (NLP) is an effective approach for bringing improvement in educational setting. (…) Natural Language Processing provides solution in a variety of different fields associated with the social and cultural context of language learning. It is an effective approach for teachers, students, authors and educators for providing assistance for writing, analysis, and assessment procedures" (more examples of NLP tools and further discussion on the links between NLP and Second Language Acquisition can also be found in Schulze (2007) and Ziegler et al. 2017).

Whilst some NLP tools are proprietary and not freely accessible to a larger audience, quite a few NLP tools are open (re)source(s). The *opensource.com* webpage lists some of those tools (https://opensource.com/article/19/3/natural-language-processing-tools). One of the tools listed is CogCompNLP which can be used to process text and offers processing functions such as tokenization, part-of-speech tagging, lemmatization, dependency and constituency parsing, and semantic role labeling. Many of those tools lie behind systems such as voice-to-text or text-to-speech tools, that may be familiar to teachers but that are rarely used in class. Another example of website making NLP tools freely accessible to teachers is Tom Cobb's Compleat Lexical Tutor (see https://www.lextutor.ca/). Some examples of the tools included are:

- Vocabulary Profilers that break texts down by word frequencies in a corpus;
- An N-Gram Extractor that pulls out from texts the recurring word strings (up to the maximum length of five words);
- Familizer + Lemmatizer that build headword lists from texts or full list of words.

In order to introduce such tools to language teachers and present them with the potential they can have for teaching/learning foreign languages, five European universities (i.e. Universidad de Murcia in Spain, Bath Spa University in the United Kingdom, Université Catholique de Louvain in Belgium, Justus-Liebig-Universität Gießen and Melikşah Üniversitesi in Turkey) decided to collaborate on a large-scale, two-year project with the support of the European Union via their Erasmus+ programme. The aim of the project was to promote the take-up of innovative practices in European language learning by supporting personalised learning approaches that rely on the use of information and communication technologies and open educational resources (OER) NLPTs, in particular, through mobile devices (see Pérez-Paredes et al., 2018 for an introduction to the project). As many teachers are not familiar with those NLPTs, one of the ambitions of the five partner universities was to bring those tools the fore and create connections between researchers in these fields and teaching practitioners. The group has generated several outputs on these topics and made them available on their website. One of the outputs of the project – described in the next section - is an in-service teacher training (INSET) online module on the aforementioned types of tools.

THE TELL-OP TRAINING MODULE FOR LANGUAGE PROFESSIONALS

The TELLOP project (whose acronym stands for Transforming European Learner Language into Learning Opportunities) is a KA200 Higher Education Strategic Partnership that ran from 2015 to 2017. The project website (http://www.tellop.eu/) contains numerous open education resources, among which an INSET module that was made available in four languages – German, Spanish, French and English. The module was launched on 16 January 2017 and lasted for five weeks. One researcher was assigned to handle each module in his or her mother tongue but the format was identical from one language to another. Every week, a new topic would be addressed according to the following schedule:

1. week 1: Pronunciation
2. week 2: Vocabulary acquisition
3. week 3: Interaction
4. week 4: Writing
5. week 5: Reading

Table 1. Text To Speech technologies – excerpt from TELLOP INSET online module

TEXT TO SPEECH TECHNOLOGIES*	
What is it?	Text to speech technologies (TTS) are pieces of software that convert digital text into spoken voice output. First developed for the visually impaired they are now used across education for different purposes.
How can it be used?	The user highlights a section of text or imports it into the box and the program returns the same section of text as a computerized voice or recorded human voice that reads the selection.
How does it contribute to language learning?	Language learning students benefit greatly by explicit attention in speaking skills one of the most important being pronunciation (Baker, 2015). One way to develop pronunciation is to hear how words sound by themselves and as parts of sentences (Dutoit, 1997). Text to speech technologies allow teachers and students to type in different text and hear them spoken by either a computerized voice or a recorded human voice (Meihami, 2013). Text to speech technologies have matured over the last fifteen years and have become more sophisticated in terms of pronunciation in multiple languages (Baker, 2015). Language learning programmes have taken advantage of this technology by including exercises where students can speak to an artificial intelligence and get feedback on their pronunciation (Shadiev, Hwang, Chen, & Huang, 2014). Comprehension and pronunciation are increased when language learners can hear the spoken version of a written text (Dutoit, 1997). One challenge of text to speech technology is the ability to recognize and pronounce novel words (McCully, Gillespie, and Murr, 2014).
OER examples	http://www.naturalreaders.com/index.html http://text-to-speech.imtranslator.net (available in multiple languages) http://www.fromtexttospeech.com (available in multiple languages) https://text-to-speech-demo.mybluemix.net (with expressive SSML)
OER practice	1.-Go to the natural readers site and insert a passage. You can try changing the voice output to see which is easiest to understand. 2.-Using the text-to-speech website, insert an article from CALL and look at the speech that emerges. 3.-Using the from text to speech website, you can type in different versions of sentences to see how grammar effects the voice outputs. 4.-The my blue mix site offers a wide range of services in addition to a text to speech programme. You can also use the expressive SSML utility. 5.-Search the internet for other POS taggers, test all their utilities and use the one that best covers your necessities.

A common pattern was also set for each topic, starting with a general introduction and some scientific background on the presented tools.

The following example (Table 1) is related to week 1 and to Text To Speech technologies (English version of the course).

The next step would take the teachers to a list of open NLPTs, along with an explanation of their affordances for language teaching and learning, and a link to the tools. The teachers would then access a series of instructions for exercises to be performed with each of the presented tools and finally, with some scientific references on the topic. Every weekly module would be concluded on personal reflexive feedback from the participants, to be put into perspective with their own language teaching context.

Having decided on a common framework and content, the researchers strived to leave room for a co-construction of the training sessions, offering an online collaborative space for the participants to engage in conversation. They selected the Moodle platform, which allowed the teachers, not only to post their reflection and discuss with their trainer, but also to interact with one another on each weekly topic.

The NLPTs addressed in the module include (in alphabetical order):

- Automated POS Taggers
- Corpora and corpus software
- Lemmatizers
- Online dictionaries and collocation dictionaries
- Readability Indices
- Resource-Linked Text Builders

- Social Networking Services (SNSs) for Image Sharing, Oral Discourse and Written Discourse
- Spell Checkers
- Text Summarization Tools
- Text-to-Speech Technologies
- Visual representation of word clusters
- Vocabulary profilers
- Word Lists per Frequency Counts

MIXED-METHODS AND SWOT ANALYSIS OF THE TELL-OP TRAINING MODULE

The feedback received weekly from the participants, as well as their written interactions on topical forums provided the research team with ample data to assess the TELLOP training module. On top of that, the teachers were asked to complete a questionnaire before and after taking the programme, therefore complementing the data and allowing for both quantitative and qualitative analyses. When it comes to investigating "expected changes in lecturers' (and learners') practice: looking for evidence of differences in what they do and in how they think about what they do", Masterman and Wild (2011, p.3) do indeed recommend a qualitative approach. The current chapter investigates and reports on the data collected from the 86 participants who completed the French-speaking version of the TELLOP online training programme. It also makes use of Cohen, Marnion and Morrison's *Grounded Theory* (2013, p.475) to compare the feedback from all the respondents, allowing core values and patterns to emerge.

Prior to attending the online course, the teachers were prompted to fill in a questionnaire addressing their profile as teachers and their knowledge and perceptions of NLP tools. Once the training module was completed, they were asked to answer the same questions, hence allowing for an assessment of how the TELLOP module had impacted their perception. In the following section we will outline the data, starting with the participants' profiles, then we will make use of the pre- and post-questionnaires to assess the impact of the training module on the perception of the participants regarding each of the aspects covered in the weekly sessions. A cross analysis of the teachers' answers based on their profile has provided some significant results. This article reports only on the factors that have shown to have some influence on their responses. Other items can therefore be considered as insignificant within the framework of this study.

Finally, we will complement these topical analyses with a review of the reactions of the teachers via their weekly assignments and via their forum interactions. We have focused on 4 major aspects, which emerged from the written feedback: the advantages of the tools for the teacher, the presumed benefits of the tools for language learners according to the teachers, the various issues raised by the teachers and finally, any extra input we collected.

Participants' Profiles

Figure 1 details the sample of participants, based on their teaching experience. The 86 teachers make for a quite diverse and balanced group, from novices (22%) to more seasoned teachers with 16 to 20 years of experience (23%). A rapid glance at Figure 1 attests to the balance between the five experience groups, since they range from a 12% – for 3 to 5 years – to a 25% ratio – for 11 to 15 years of experience. As depicted in Figure 2, most of the practitioners (73%) teach in secondary schools, whereas 12% of them

Figure 1. Teachers according to their teaching experience in years (respondents out of 86)

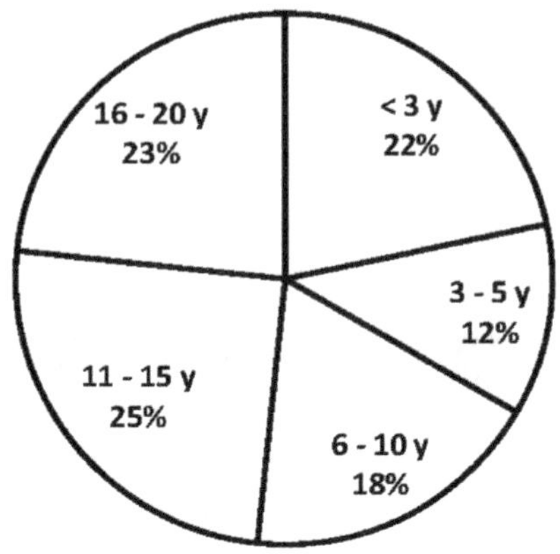

Figure 2. Teachers according to the type of institution they teach

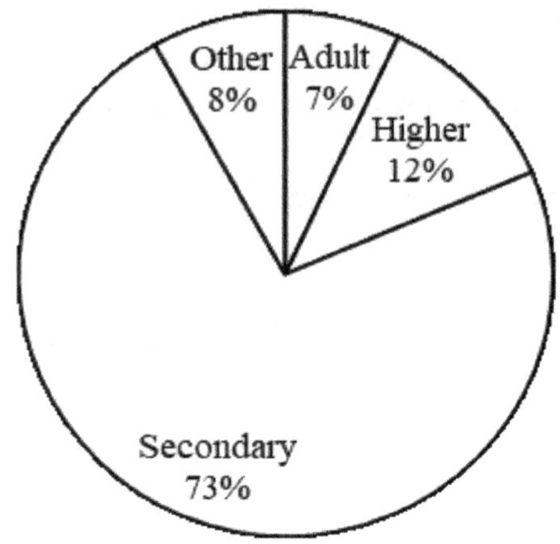

do so in a higher education context, 7% teach adults and 8% work in a type of environment that was not represented by the other three categories.

The teachers also had to express themselves on their prior exposition to OER[2] and mobile devices for education. Figure 3 shows that roughly a third (36%) of the respondents teach in an institution fostering the use of mobile devices in the classroom, leaving the majority (64%) to deal with schools which, in their perception, do not encourage them to use smartphones or tablets to teach languages. As for previous training on the matter, only 16% of the participants have been briefed on the use of mobile devices in the classroom, and a mere 10% have been trained on using OER in their practice.

Figure 4 offers more insight into the participants' perception of OER, as it details their responses on a 1 to 5 Likert scale, ranging from "totally disagree" to "totally agree". Before taking the training programme, the teachers seemed to have very little experience – 2.09 on average – and knowledge – 2.22 – about OER, therefore barely using them in their language teaching, with an average answer rate of 1.89. The teachers' view of how much their students might appreciate their use of OER or of how easy it would be to integrate them is rather neutral but the last question brings hope as to how useful – 4.05 – these could be for teaching languages.

Figure 3. Pre-questionnaire – OER- and mobile device-related questions

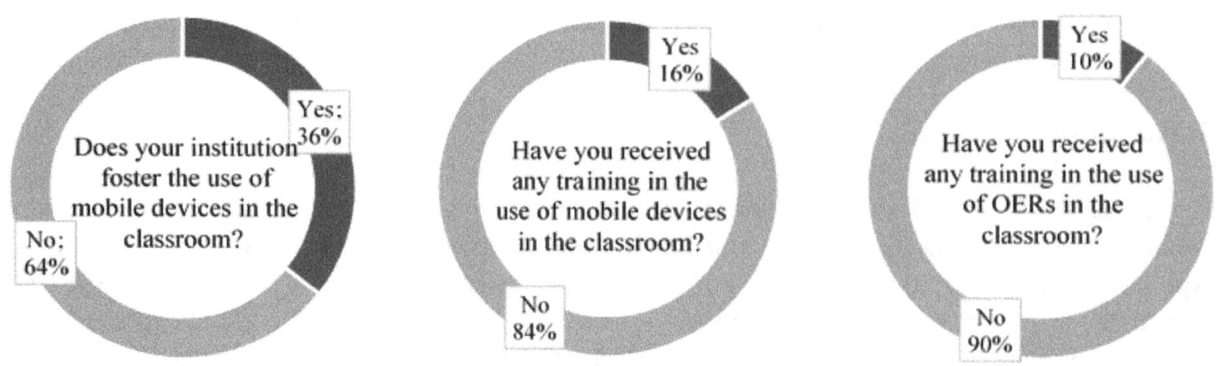

Figure 4. Pre-questionnaire – OER-related questions on a 1 to 5 Likert scale

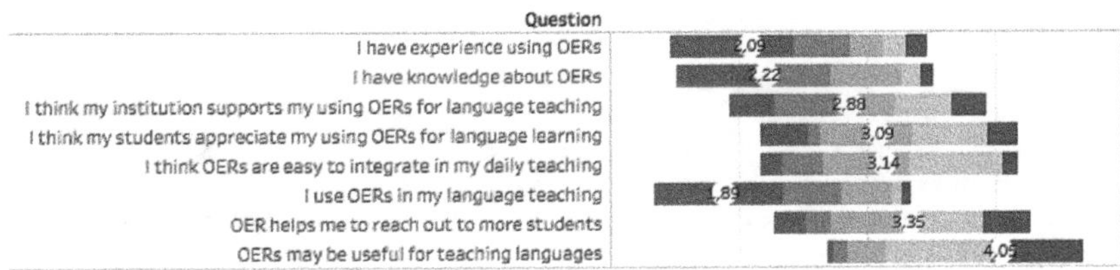

The French-speaking TELLOP training module seems to have embarked on an online adventure with 86 language teachers who had close to little knowledge of OER or of how to integrate mobile devices into their practices. However, their perceptions on those tools were rather optimistic, which can be expected, given the quite transparent title of the module. One can only postulate that teachers who had no interest whatsoever in technology were not drawn to the programme in the first place.

Impact of the Programme on the Teachers' Perception of OER in General

With a view to assessing the impact of the TELLOP training module on the participants' perception of OER, the same questions were asked to them once the programme was completed. Thirty-two of the original 86 teachers (viz. 37%) completed the survey, which represents a typical response rate for an internal survey[3]. Figure 5 details these responses based on whether the participants had received prior training on the use of OER in the classroom (see "No" and "Yes" sections of the table). In general, the teachers who had been trained on OER before taking the module seem to have a more positive take on the topic, with an average response rate of more than 4 on whether they intend to use OER in the future (4.8), on OER being useful to teach languages (4.80) and on how the online course has shown them new ways of using OER in teaching (4.6). Despite a slightly less enthusiastic response from OER novices, it seems taking the course has succeeded in showing them new ways of using OER in teaching (4.13) and convincing them that they may be useful for teaching languages (4.07). Their intention to use OER in the future is however but neutral (3.63), which can be put into perspective with their perceived ease of use – or lack thereof – of such tools (2.9). However, it may be noted that the group that had received previous training intends to make use of OER despite their similarly tepid (3.00) view of their ease of use.

Impact of the Programme on the Teachers' Perception of Text-to-Speech OER

A quick glance at Figure 6 attests to the almost total absence of prior knowledge of the participants regarding the topic for the first week of training, i.e. *pronunciation and text-to-speech technologies.* Regardless of whether they had been trained on OER or not, the teachers were vastly unfamiliar with the concept (1.89 and 2.91) and did not make use of the technology (1.34 and 1.73). Figure 7, however, demonstrates the impact of the French-speaking version of the TELLOP training programme, showing that the teachers are more familiar with text-to-speech technologies than they used to be. The OER novices went from almost complete ignorance on the matter (1.34) to feeling almost confident (3.87) about their familiarity with it. As for the practitioners who had received prior training on OER, they seem to have made less progress (from 2.91 to 3.40). The most noticeable impact of the module on both fronts is the teachers' perceived usefulness of t he technology for language teaching (4.13 and 4.20).

Figure 5. Post-questionnaire – OER questions by 'have you received any training on the use of OER in the classroom?'

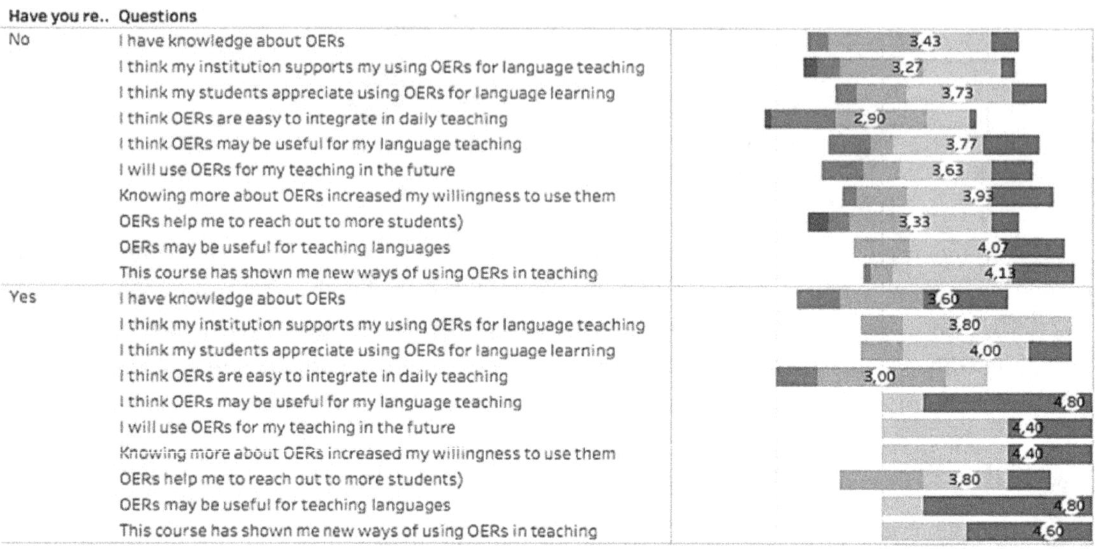

Figure 6. Pre-questionnaire – pronunciation by 'have you received any training on the use of OER in the classroom?'

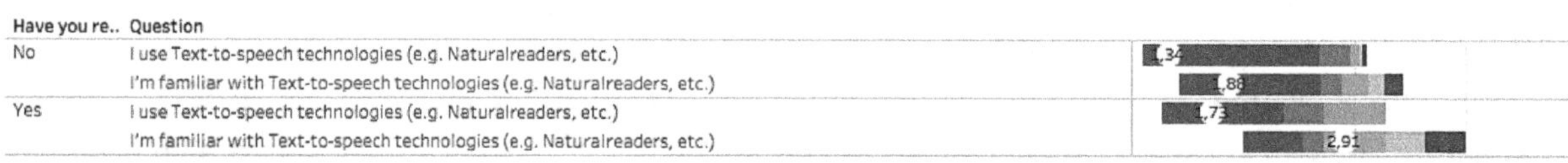

Figure 7. Post-questionnaire – pronunciation by 'have you received any training on the use of OER in the classroom?'

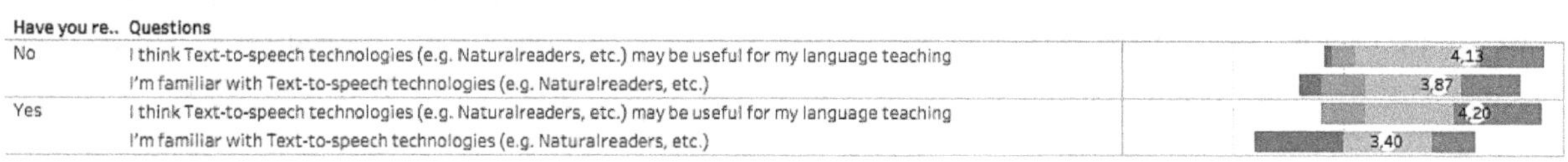

An analysis of the teachers' weekly assignments, coupled with a detailed reading of their forum interactions provided us with more information concerning their perceptions of text-to-speech tools. On the one hand, the teachers highlighted several positive aspects of the presented technology. Many of them saw an opportunity to create their own listening comprehension material in the target language, therefore appreciating the numerous accents and intonations at their disposal. They also praised text-to-speech technology in that it could allow their learners to work autonomously on pronunciation. The motivational aspects of the tools were also emphasized; an individualized ICT tool dedicated to pronunciation has the potential, according to them, to boost the self-confidence of their L2 learners, or even help pupils with learning difficulties, all the while promoting differentiated learning. On the other hand, several shortcomings were outlined, such as the fact that an internet connection was needed to work with most tools, or a lack of integrated grammar and spell check. The teachers would have also liked for the tools to feature a built-in recorder, which would play back the learner's attempt at imitating pronunciation in the target language.

Impact of the programme on the teachers' perception of vocabulary acquisition tools. Once again, Figure 8 demonstrates how little the respondents knew about and used corpora (1.36 ; 1.82 ; 1.59 ; 2.55) and visual representations of word clusters (1.53 ; 2.36 ; 2.06 ; 2.73) before taking the training programme. Online dictionaries and collocation dictionaries seemed to be better known by both parties (3.62 ; 4.55). The practitioners who had not been trained on the use of OER in the classroom did not tend to use these dictionaries (2.91), despite their familiarity with them, but the OER-trained teachers did (4.36).

After completing the online training programme – cf. Figure 9 – the perceived usefulness of Dictionaries for language teaching was even higher than before (4.13 ; 5.00). Nevertheless, the other tools encountered less enthusiasm with corpora and specialized corpora invoking a rather negative reaction (2.70 ; 2.80) on their potential usefulness for L2 teaching, which could potentially be explained by a lack of confidence from the respondents on their familiarity with this type of OER (3.4 ; 3.00). Visual representations of word clusters encountered but a tepid reaction (3.40) from those who had no previous OER knowledge, but seem to have convinced the more OER knowledgeable (4.60).

A review of the teachers' assignments and forum interactions on these tools confirms their fondness for dictionaries. They appreciate the opportunity of extending their own vocabulary database through access to collocations and authentic material. They could also envisage having their learners browse these tools to access exemplified words and expressions and reflect upon how the foreign language words autonomously. They also envisaged these tools as an opportunity to offer new vocabulary memorization techniques to their language learners. Nonetheless, they stressed the importance of training their pupils on how to navigate these complex resources. Some even expressed their concerns at how untransparent the "jargon" used in corpora tools was, fearing that the whole process of browsing these would poten-

Figure 8. Pre-questionnaire – vocabulary acquisition by 'Have you received any training on the use of OER in the classroom?'

Figure 9. Post-questionnaire – vocabulary acquisition by 'Have you received any training on the use of OER in the classroom?'

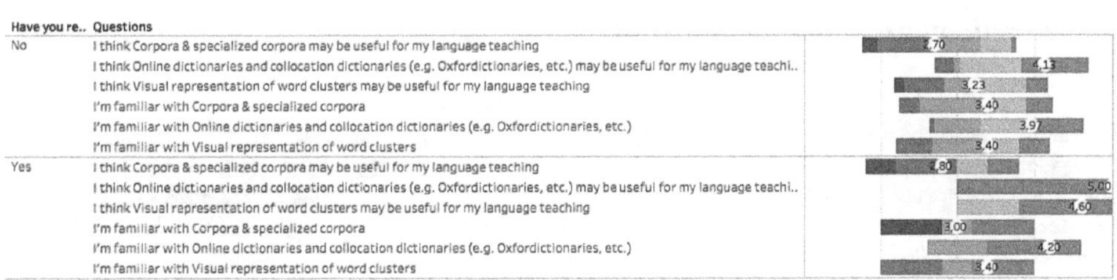

tially be time consuming and counter-productive when focusing on a communicative approach of L2 acquisition. Many were those who referred to tools, which they deemed much more user-friendly than those presented in the training module. Finally, the issue of internet access and mobile-device use at school was brought up as a major hindrance to the frequent use of online vocabulary acquisition tools.

Impact of the Programme on the Teachers' Perception of Interaction Tools

Surprisingly, the TELLOP teachers did not consider themselves knowledgeable on social networking services (2.62 ; 2.84 ; 3.49) before taking the online module – cf. Figure 10. A breakdown of the data according to teaching experience did not reveal any higher feeling of competence among the younger educators. Only the tools for written discourse were considered familiar (3.49) by the teachers, but they don't use them in class (1.91).

Post-questionnaire results – cf. Figure 11 – reveal a much higher self-confidence rate among the teachers regarding their acquaintance with social networking services, especially those who had been previously trained on OER use (4.50 ; 4.63 ; 4.63). Despite a clear improvement in their knowledge on these tools, educators still seem reluctant to implement them into their language teaching. The "previous OER training" group is barely favorable to using them for their classes (3.88 ; 3.88 ; 3.88), and the other group remains quite neutral regarding tools for oral (3.11) and written (3.37) discourse, and even negative when it comes to image sharing (2.78).

A closer look at the teachers' interactions on the module forum, as well as their assignments reveals a lot on why they remain so suspicious of social networking services. The affordances of online interaction tools were not lost on the teachers, who value their ease of use, contrary to the lack of user-friendliness attributed to vocabulary acquisition tools. The educators saw an opportunity to share information and content or give feedback to their pupils very easily, or even to have the learners collaborate online, therefore shifting the traditional "teacher-learner" paradigm within the classroom, all the while being in touch with authentic L2 content and developing their media literacy.

Nevertheless, the assignments and forums showed great reluctance from the participants to have their students use social media tools for several reasons. The issue of internet access and mobile-device use at school was brought up once more. On top of that, the secondary school teachers foresaw potential bullying or legal issues with their underage learners freely interacting on the web. They also felt responsible for any potential linguistic mistake published by their pupils. They therefore suggested integrating media literacy classes within the curriculum and providing extra guidance on web communication. Interestingly, forum discussions on the topic displayed many recommendations of online interaction tools the teachers already use, all of which allow for teacher control and private use. Finally, several participants mentioned public platforms, but only to be used for receptive purposes, i.e. having their language learners create vocabulary lists based on tweets.

Figure 10. Pre-questionnaire – Interaction Tools

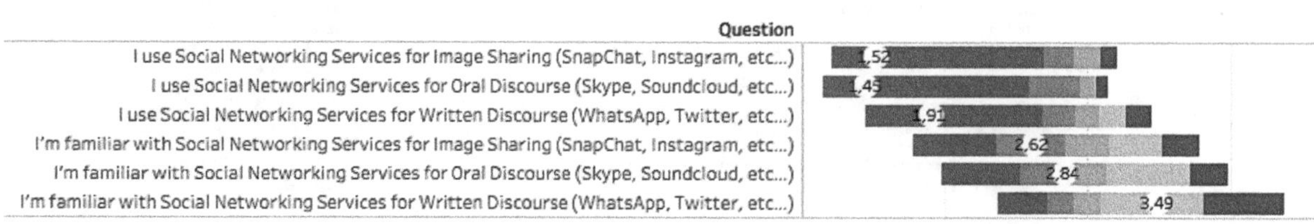

Figure 11. Post-questionnaire – interaction tools by 'Have you received any training on the use of mobile devices in the classroom?'

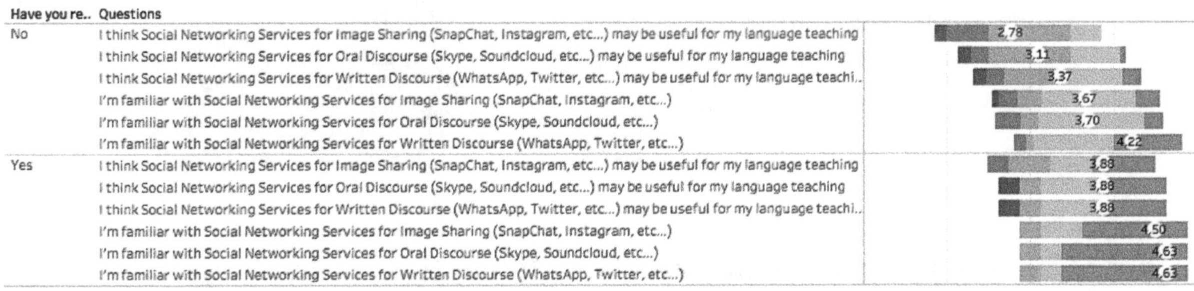

Impact of the Programme on the Teachers' Perception of Writing Tools

The results from Figure 12 show a clear lack of both knowledge and use of NLP writing tools by the participants before they took the TELLOP training programme. They only seemed to be relatively familiar with spell checkers (3.15) but not to make use of them (2.39). An analysis based on previous OER training did not reveal any significant difference between the two groups.

Post-questionnaire results – cf. Figure 13 – reveal a similar trend with a generally more positive general outlook, and a preference for spell checkers, which teachers feel familiar with (4.03) but most importantly, which they find useful for their language teaching (4.09). As for other OER writing tools, they encounter less enthusiasm with neutral reactions on resource-linked text builders (3.34 ; 3.29) and vocabulary profiling tools (3.49 ; 3.23). POS taggers are even not considered useful for language teaching (2.69).

Astonishingly, most discussions on the forums and in the assignments on the writing tools actually addressed their usefulness for reading purposes. Teachers referred to the web-based tools as a way of identifying the level of authentic texts or creating exercises. In fact, the extra information and tools suggested by the teachers were also meant for reading. The participants still acknowledged the potential of such tools to improve their learners spelling or vocabulary range in writing, but they also questioned – as they had before – the user-friendliness and relevance of complex POS taggers and resource-linked text builders for L2 learners within a communicative pedagogical approach of SLA.

Figure 12. Pre-questionnaire – writing skills

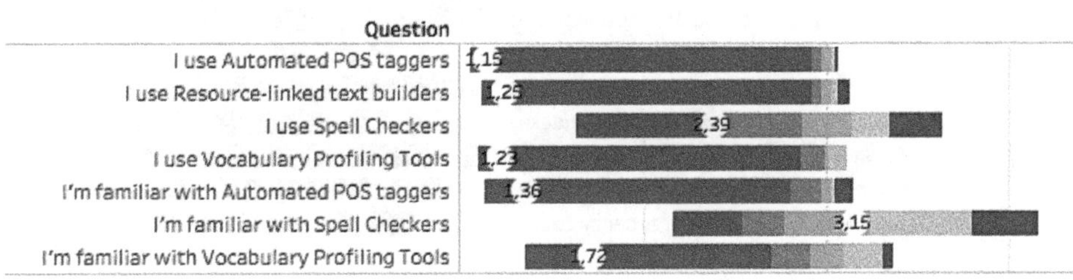

Figure 13. Post-questionnaire – writing skills

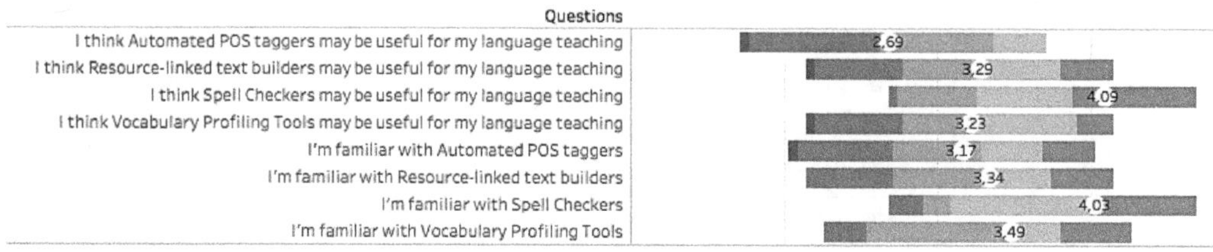

Impact of the Programme on the Teachers' Perception of Reading Tools

The pre-test – cf. Figure 14 depicts an extremely poor knowledge of reading-oriented OER by the participants, with average rankings below 2 for all the presented tools but word lists and frequency counts (2.09).

Based on their answers to the post-questionnaire, the participants seem to have become more familiar with the tools after having completed the TELLOP training programme. However, the average results on familiarity remain broadly neutral (between 3.49 and 3.63) and the perceived usefulness for the tools is, on average, barely neutral (from 3.03 to 3.40) with lemmatizers failing to convince the teachers of their usefulness for SLA.

The teachers discussed the advantages of the presented reading tools, attributing the same qualities they did for the writing tools – i.e. identifying the level of a text, creating level-appropriate reading comprehension material and exercises, etc. Advantages for the learners were also identified: immediate access to word definitions and pronunciation, leading to vocabulary extension. Finally, the lack of user-friendliness of certain tools was pointed out as a hindrance to their use by both teachers and learners. Some of the teachers would not refer their students to summarizing tools as they would prefer them to learn how to perform that particular skill themselves. They deplored the fact that no socio-cultural aspects were taken into account in readability indexes and concluded that such tools should be used at home but not in a classroom environment.

Figure 14. Pre-questionnaire – reading skills

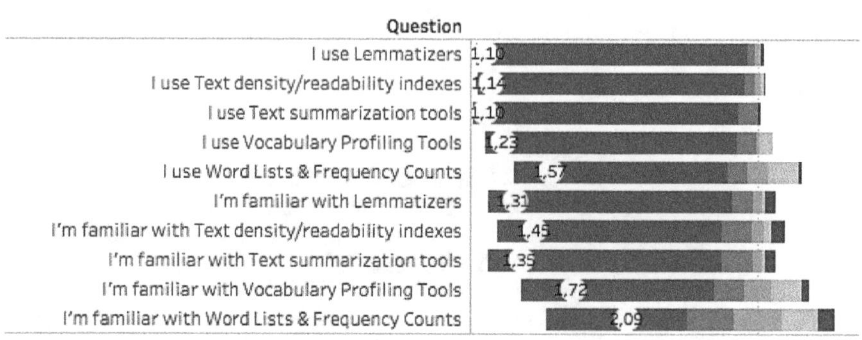

Figure 15. Post-questionnaire – reading skills

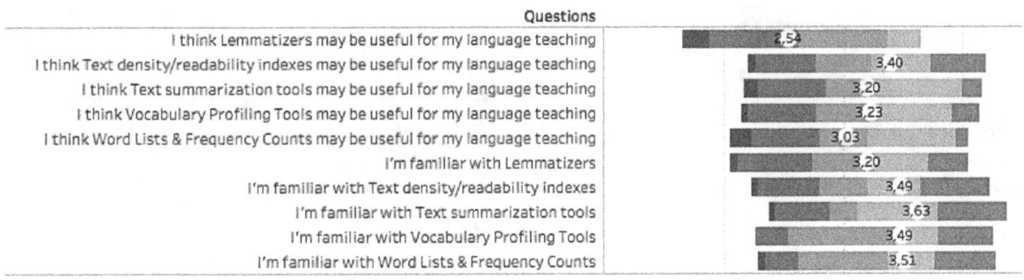

SWOT Analysis by the French-Speaking TELLOP Trainer

After having reviewed the teachers' perspective of the TELLOP module, we would like to shed a different light on the data analysis by examining the French-speaking trainer's point of view. We will therefore go through the different strengths, weaknesses, opportunities and threats (SWOT) she has identified after the training module was completed – cf. Figure 16.

First of all, the trainer highlighted the importance of an open approach to promote OER by making the training module accessible online, to any language teacher, and for free. The completion of the programme also gave each teacher an official certification, which adds to the aforementioned motivational factors for taking the TELLOP module. As previously stated, the research team had chosen an online platform to deliver the programme, allowing for interaction between the trainer and each teacher on the one hand, but also between the participants via forums. This allowed the teachers to share about their practices and to experience an online community of education professionals.

According to the trainer, several technical aspects of the module could be improved. Managing the five-week programme proved to be time-consuming, among other things because of the individual feedback prompted by each weekly assignment. She also acknowledged that some tools were complex to use, which coupled with some teachers' lack of computer literacy, probably jeopardized the sustainability of the presented resources. This could in part be explained by the fact that each type of tool had to be available in several foreign languages, whereas they do not always offer the same level of user-friendliness. English L2 teachers were also presented with many more resources to choose from than their German or Dutch-teaching counterparts. Finally, despite a focus on OER tools, no feedback was collected from the teachers on their perceptions of OER practice.

The potential for future collaboration and a follow-up module was identified as an opportunity for development by the trainer. She also saw the fact that education professionals were trained on technology, through technology, as an area to expand on. Organizing a second edition of the TELLOP training module would also pose certain potential problems, such as the ever-evolving nature of online environments, which could make some tools quickly obsolete. The whole approach used in the first edition of the programme did not cater – or not enough – for the individual needs and contexts of each language teacher, which makes their application of the new tools in their daily work less likely. On top of that, no follow-up session was planned after the five-week programme, therefore lessening the chances for application of the NLP OER by the teachers.

Figure 16. SWOT analysis of the TELLOP training module by the French-speaking trainer

Strengths	Weaknesses
Easy access to the module Free access to the module Sharing best practices Community Certification OER practice	Learners' lack of computer literacy Time-consuming (management of the course) Time-consuming (assignment feedback) Technical issues Expertise needed for the teacher More resources for English learning No feedback on OER practice Complexity of some tools
Opportunities	**Threats**
Discovering learning with technology Future collaboration Follow-up course or 2nd module session	Out-to-date tools No adaptation to the audience No application from the teachers

CONCLUSION AND RECOMMENDATIONS

Our INSET project is clearly part of general continuous professional development (CPD) which can be defined as "the holistic commitment of professionals towards the enhancement of personal skills and proficiency throughout their careers". [4]It not only includes the learning activities that professional teachers engage in, such as training workshops, conferences and events, e-learning programmes, best practice techniques or ideas sharing (Stoll & Fink, 1996, p.155), but also informal learning opportunities, management development or even preparation for retirement. INSET, as a specific type of CPD (Osamwonyi, 2016), includes activities such as professional education (widening and deepening of a teacher's theoretical perspectives by means of advanced study), professional training (development of knowledge and skills of direct applicability to daily work) or professional support (activities aimed to develop on-the-job experience and performance (Altun et al., 2007; Altun, 2011).

Approaches towards CPD and INSET have experienced several important paradigmatic shifts over the last 40 years. Roesken-Winter et al. (2015) explain that one of these shifts is the increasing significance assigned to the social dimension (from teachers working in isolation to collegial collaboration). Taking into account the significance of teachers' collaborative work within professional learning communities has also been described as central by Weißenrieder et al. (2015). Another shift that is called for by Borg (2015, p.6) is that the earlier idea of 'the teacher as a consumer of received knowledge' be transformed into that of 'the teacher as (also) a knowledge generator'. Based on literature reviews (Broad, 2006; Timperley et al., 2008), Borg (2015, p.6) explains that there is consensus that "CPD 'works' better" when it has features such as:

- relevance to the needs of teachers and their students
- teacher involvement in decisions about content and process
- teacher collaboration
- support from the school leadership
- exploration and reflection with attention to both practices and beliefs
- internal and/or external support for teachers (e.g. through mentoring)

- job-embeddedness (i.e. CPD is situated in schools and classrooms)
- contextual alignment (with reference to the institutional, educational, social and cultural milieu)
- critical engagement with received knowledge
- a valuing of teachers' experience and knowledge.

Whilst the INSET module presented and analyzed in Sections 2 and 3 has clearly helped widen and deepen teachers' perspectives and knowledge of new tools, it has mainly adopted a top-down approach (selection of tools to be presented made by researchers, presentation-practice approach). Top-down approaches do not meet all of Borg's desired features of CPD but they still seem unavoidable when new tools, hardly known to teachers, are presented. The qualitative and SWOT analysis that have been carried out have taught us a few lessons that will be implemented in future modules on NLPTs to enhance collaboration and sustainability.

Despite the integration of an online collaborative space for the participants (teachers could post their reflection, discuss with their trainer, interact with one another on each weekly topic) more attention and time should be devoted to the promotion of communities of practice (CoPs) which are defined by Wenger, McDermott & Snyder (2002, p.4) as "groups of people who share a concern, a set of problems, or a passion about a topic, and who deepen their knowledge and expertise by interacting on an ongoing basis". As put forward by MacKinnon et al. (2016) in their meta-analysis of Open Educational Communities of Practice and sustainability in higher educational policy, CoPs are typically decentralized and can improve the scalability of open projects. The authors also stress the importance of sustainability which has now become an indicator of success in educational projects, and which ranks fourth on UNESCO's list of top concerns for the international OER (Open Education Resources) community and which is well ahead of technology. This sustainability not only applies to content creators, but also to CoPs.

Taking into account the qualitative and SWOT analysis we believe that, in addition to the improvement of some technical aspects related to the platform, the following recommendations should be taken into account in the future:

1. Lengthening the duration of the module. Taking part in and managing the five-week programme proved to be time-consuming. The cognitive load related to the amount of new input and of testing phases was too heavy for some teachers. As for the trainer she did not have enough time to address the individual feedback prompted by each weekly assignment in detail.
2. Selecting and reducing the number of tools to be presented. Some tools proved too complex to use, which, combined with some teachers' lack of computer literacy, probably jeopardized the sustainability of the presented resources.
3. More focus on the individual needs and contexts of each language teacher in order to foster the use the new tools presented in their daily work.
4. Collecting more long-term feedback from teachers on their perceptions of and experience with OER tools.
5. Organizing follow-up sessions to provide what Wilson (2013, p.62) calls 'aftercare', or continuous support, to address problems that teachers may encounter when using new tools in their classroom. Such provision of aftercare is essential to prevent teachers from losing interest in the tools and from giving up using them.
6. Encouraging teachers, once they are at ease with the new tools, to prompt learners to also becoming autonomous users of some of those tools outside the classroom.

In conclusion, the quest for a smooth integration and actual use of new tools inside and outside classrooms requires time, patience and long-term repeated interactions between teachers and teacher trainers. It also often requires a shift both in INSET and in classroom practices, where beyond the technical aspects, cultural and functional habits need to be progressively changed. As suggested by Haines (2015, p.166) "identifying the affordances that teachers might see and the way that such perceptions develop over time is therefore an important part of extending our knowledge about new tools and how they can be usefully integrated into language learning classrooms". We therefore plead for the adoption of what Snoder & [author] (accepted) call 'a circular bottom-up approach' that requires close teacher-researcher collaboration as it often appears that cutting-edge tools and research findings in applied language acquisition studies fail to reach L2 teachers and their learners (see for instance Erlam, 2008; Ellis, 2010; Tavakoli & Howard, 2012; Tavakoli, 2015; Sato & Loewen, 2018). New professional development schemes are in order to make sure that L2 research addresses teachers' and learners' needs in their specific contexts. It is of utmost importance not only to single out what L2 research can offer to address these needs but also to provide teachers with supporting pedagogical materials and suitable teaching methodologies and practices that concretely integrate the use of new tools and techniques. We hope that the lessons learnt from our TELLOP experience will help inform future sustainable and open INSET schemes.

ACKNOWLEDGMENT

We would like to acknowledge the collaborative work done with our ERASMUS+ and TELLOP project partners. TELLOP is a KA200 Higher Education Strategic Partnership (ref. 2014-1-ES01-KA203-004782).

Special thanks go to Ms Julie Van de Vyver, French-speaking TELLOP trainer for her insights on the training module.

REFERENCES

Alhawiti, K. (2014). Natural language processing and its use in education. *International Journal of Advanced Computer Science and Applications*, *5*(10), 72–75.

Baker, F. S. (2015). Emerging realities of text-to-speech software for nonnative-English-speaking community college students in the freshman year. *Community College Journal of Research and Practice*, *39*(5), 423–441. doi:10.1080/10668926.2013.835290

Carrió-Pastor, M. L. (2016). *Technology implementation in second language teaching and translation studies: New tools, new approaches*. Singapore: Springer. doi:10.1007/978-981-10-0572-5

Caws, C., & Hamel, M. (2016). *Language-learner computer interactions: Theory, methodology and CALL applications*. Amsterdam, The Netherlands: John Benjamins. doi: 10.1075/lsse.2

Cohen, L., Manion, L., & Morrison, K. (2013). *Research methods in education*. Retrieved from http://books.google.com/books?hl=en&lr=&id=mLh0Oza3V1IC&oi=fnd&pg=PR3&dq=info:YWZJAiLelagJ:scholar.google.com&ots=SOBLHpp9pm&sig=5PN7dflZLpa6oikITWQLcXn-Q5w

Dutoit, T. (1997). *An introduction to text-to-speech synthesis* (Vol. 3). Springer Science & Business Media. doi:10.1007/978-94-011-5730-8

Ellis, R. (2010). Second language acquisition, teacher education and language pedagogy. *Language Teaching*, *43*(2), 182–201. doi:10.1017/S0261444809990139

Erlam, R. (2008). What do you researchers know about language teaching? Bridging the gap between SLA research and language pedagogy. *International Journal of Innovation in Language Learning and Teaching*, *2*(3), 253–267. doi:10.1080/17501220802158958

Gee, J. P., & Hayes, E. R. (2011). *Language and learning in the digital age.* Taylor and Francis; doi:10.4324/9780203830918

Haines, K. J. (2015). Learning to identify and actualize affordances in a new tool. *Language Learning & Technology*, *19*(1), 165–180.

Hinkelman, D. (2018). *Blending technologies in second language classrooms.* doi: 10.1057/978-1-137-53686-0

MacKinnon, T., Pasfield-Neofitou, S., Manns, H., & Grant, S. (2016). A meta-analysis of open educational communities of practice and sustainability in higher educational policy. [Social dynamics in open educational language practice.]. *ALSIC*, *19*, 1.

Masterman, L. & Wild, J. (2011). JISC-funded OER impact study. Retrieved from http://www.jisc.ac.uk/media/documents/programmes/elearning/oer/JISCOERImpactStudyResearchReportv1-0.pdf

McCulley, Y., Gillespie, C., & Murr, A. H. (2014). Assessing the effectiveness of text-to-speech software in incarcerated adult literacy education. *Journal of Correctional Education*, *65*(2), 2.

Meihami, H. (2013). Text-to-speech software: a new perspective in learning and teaching word stress, word intonation, pitch contour, and fluency of English reading. *International Letters of Social and Humanistic Sciences*, *8*, 24–33. doi:10.18052/www.scipress.com/ILSHS.8.24

Meurers, D. (2012). Natural language processing and language learning. In C. A. Chapelle (Ed.), *Encyclopedia of Applied Linguistics* (pp. 4193–4205). Oxford, UK: Blackwell. doi:10.1002/9781405198431.wbeal0858

Pérez-Paredes, P., Ordoñana Guillamón, C., & Aguado Jiménez, P. (2018). Language teachers' perceptions on the use of OER language processing technologies in MALL. *Computer Assisted Language Learning*, *31*(5-6), 5–6, 522–545. doi:10.1080/09588221.2017.1418754

Sato, M., & Loewen, S. (2018). Do teachers care about research? The research-pedagogy dialogue. *ELT Journal.*

Schulze, M. (2007). Modeling SLA processes using NLP. In C. Chapelle, Y.-R. Chung, & J. Xu (Eds.), *Towards adaptive CALL: Natural language processing for diagnostic assessment* (pp. 149–166). Iowa State University.

Shadiev, R., Hwang, W. Y., Chen, N. S., & Huang, Y. M. (2014). Review of speech-to-text recognition technology for enhancing learning. *Journal of Educational Technology & Society*, *17*(4), 65–84.

Snoder, P. & [author] (accepted). *Word Up! Boosting L2 learners' vocabulary through ecological communities of practice.* Presentation at the VOCAB@Leuven Conference. Katholieke Universiteit Leuven, 1-3 July 2019.

Tavakoli, P. (2015). Connecting research and practice in TESOL: A community of practice perspective. *RELC Journal, 46*(1), 37–52. doi:10.1177/0033688215572005

Tavakoli, P., & Howard, M. (2012). Teaching English to speakers of other languages teachers' views on the relationship between research and practice. *European Journal of Teacher Education, 35*(2), 229–242. doi:10.1080/02619768.2011.643398

Vandergriff, I. (2016). *Second-language discourse in the digital world: Linguistic and social practices in and beyond the networked classroom.* Amsterdam, The Netherlands: John Benjamins. doi:10.1075/lllt.46

Wenger, E., McDermott, R., & Snyder, W. M. (2002). *Cultivating communities of practice: A guide to managing knowledge. Harvard.* Harvard Business School Press.

Wilson, J. (2013). Technology, pedagogy and promotion. How can we make the most of corpora and data-driven learning (DDL) in language learning and teaching? *The Higher Education Academy* [online]. Retrieved from https://www.heacademy.ac.uk/system/files/corpus_technology_pedagogy_promotion2.pdf

Ziegler, N., Meurers, D., Rebuschat, P., Ruiz, S., Moreno-Vega, J.-L., Chinkina, M., Li, W., & Grey, S. (2017). Interdisciplinary research at the intersection of CALL, NLP, and SLA: Methodological implications from an input enhancement project. Special Issue: *Currents in language learning series: Experimental, computational, and corpus-based approaches to language learning: Evidence and interpretation*, 209-231.

KEY TERMS AND DEFINITIONS

CPD: Continuous Professional Development.
INSET: In-Service Teacher Training.
NLP: Natural Language Processing.
NLPTs: Natural Language Processing-Based Technologies.
OER: Open Educational Resources.

ENDNOTES

1 https://becominghuman.ai/a-simple-introduction-to-natural-language-processing-ea66a1747b32
2 Note that we used the general OER term (Open Educational Resource) in the questionnaires, as a couple of the tools presented were not NLPTs per se.
3 Internal surveys generally receive a 30-40% response rate (compared to an average 10-15% response rate for external surveys). See https://www.surveygizmo.com/survey-blog/survey-response-rates/
4 (see https://cpduk.co.uk/explained)

Chapter 15
Analysing WhatsApp and Instagram as Blended Learning Tools

Alberto Andujar
https://orcid.org/0000-0002-8865-9509
University of Almeria, Spain

ABSTRACT

This chapter explores, from a theoretical point of view, the literature investigating the use of WhatsApp and Instagram applications to foster language learning as well as the outcomes of the studies carried out in these two fields. The possibilities of these two apps to develop blended learning environments are also regarded. The theoretical frameworks in the existing literature for the possible implementation of these social networking tools in classroom contexts are considered. Findings indicate a growing number of studies making use of the WhatsApp application to develop either blended learning models or online tasks in foreign language courses while the research exploring the use of Instagram for language development remains very scarce.

INTRODUCTION

The constant growth of a globalized education model in which information and communication technology (ICT) becomes a fundamental element to carry out ubiquitous teaching and learning processes (Altbach, 2004) has involved changes at all levels. Specifically, at a university level, the need to catch up with the latest educational tools has transformed the learning environment to a context in which technology and in-class tuition coexist. This situation, which is relatively new, has led to the exploration of blended learning processes in which in-class tuition is combined with an online environment in which between 20% to 79% of course contents are delivered online (Arbaugh, 2014). Blended learning becomes a flexible and autonomous environment (Im & Kim, 2015) which need to be investigated from different points of view – technological, educational, and practical among others – in order to develop the most appropriate learning environments to fulfill students' needs. In this context, most of the present literature exploring Computer-

DOI: 10.4018/978-1-7998-1097-1.ch015

assisted Language Learning (CALL) or Mobile-assisted Language Learning (MALL) involves a certain degree of blended learning. The new online environments, in which students normally face a high cognitive load due to their exposure to discourse in a foreign language, have become a support during language learning, in many cases facilitating the learning process (Ball and Lindsay, 2013). Positive effects of this learning model have been found in previous literature (Miyazoe & Anderson, 2010; López-Pérez, Pérez-López & Rodríguez-Ariza, 2011; Law, Geng, & Li, 2019) and different factors such as higher participation or motivation levels and an increase in students' autonomy have been emphasized.

By contrast, it also seems necessary to understand the impact of certain tools in the development of blended learning models and how students' previous knowledge of that tool, the level of engagement generated or its accessibility may entail different outcomes during the development of blended learning processes. In this sense, this chapter will explore from a theoretical perspective, the potential of two of the most common social networking platforms among students, Instagram and WhatsApp. The reason behind this choice has to do with several theoretical and practical considerations. First, the outstanding number of users in each of the applications as of August 2019, WhatsApp with approximately 1.5 billion monthly active users and 1 billion daily active users, and Instagram with 1 billion monthly active users and 600 million daily active users (Stout, 2019). Second, theoretical considerations acknowledged by experts in the field such as Kukulska-Hulme and Shield (2007) who stated that student do not use technology they may consider intrusive or those technological tools and software which may require a long time to master. Thus, from a teacher perspective, it seems necessary to explore how the most used social networking platforms can be used to develop blended learning models and how the existing literature into the two applications can shape teacher practices and course design.

The third reason behind this choice is the growing amount of literature dealing with language learning and social networking platforms. These online environments, normally referred as 2.0 platforms and used for socializing, sharing information or meeting (Boyd, 2014), are visited by a great number of young people (Counts & Fisher, 2010; Dunn, 2013). Interestingly, although the majority of social media platforms have a web version, most of them are accessed from mobile devices (Global Digital Statistics, 2019). This fact needs to be taken into consideration by teachers and institutions willing to implement blended learning models through social networks. Apart from the fact that studies into language learning highlighted differences in students' behavior in computer and mobile environments (e.g. Andujar & Salaberri-Ramiro, 2019), the accessibility and ubiquity of mobile devices becomes a fundamental aspect to guarantee students' engagement during the learning process.

In this context, the chapter aims to shed light on the possibilities of WhatsApp and Instagram to develop blended learning models in which these two social networking platforms can foster language learning development. In order to tackle this aim, the existing literature in each of the two platforms together with researchers' insights for an appropriate implementation of these applications in language learning contexts will be explored.

SOCIAL NETWORKS AND LANGUAGE LEARNING

Social networks have been used in classroom environments since their appearance and for different learning purposes (Mao, 2014). From an early use of Wikis and Blogs (Seaman & Tinti-Kane, 2013) to the use of more widespread social networks such as Facebook, Linkedin, Wechat or WhatsApp among others, social networks have allowed students to interact through new different ways of communication (Mazer, Murphy & Simonds, 2007). This context seems particularly interesting from the language learning perspective as different modes of communication may foster different language abilities such as reading, speaking, listening or writing among others. Moreover, students' social network dependence becomes another factor which needs to be taken into consideration when selecting the best tools for developing blended learning models.

Before mentioning the most relevant studies in the field making use of the social networking platforms investigated in this chapter, it is worth mentioning the most widespread social networks in use as well as a short description so language teachers and practitioners know what to expect from each of them. Table 1 presents the most common ones and their current use.

As observed in the aforementioned descriptions, some of the social networks may cover a wide spectrum of communication modes that could be exploited from a language learning perspective. By contrast, others such as Twitter or Pinterest may only be used to develop students' writing skills and collaborative work. Thus, it seems necessary to understand the potential of each social networking tool to appropriately develop effective and successful learning environments. These virtual environments may give students the opportunity of interacting using the foreign language, working in groups and improving their linguistic competence (Blattner & Fiori, 2011; Mills, 2011; Blattner & Lomicka, 2012; Farr & Murray, 2016).

Table 1. Description of the most common social networking platforms

Nombre	Descripción
Edmodo	Edmodo is an educational webpage that is sometimes used as language management system (LMS) and that shares design with other popular social networking platforms. It provides a safe and easy way of connecting peers and collaborating. Students can also share content and notes as well as access homework and messages from the school or university.
Facebook	An online social network service that was originally designed for secondary students but later extended to the general public. Among the different functionalities found in this platform, it is worth mentioning the possibility of posting pictures, using individual or group chats, commenting on the photos and sharing information.
Google Hangout	This platform takes conversations to real life through pictures, emoticons and even free group calls. Similar to Skype, you can connect with friends through computers, or Android or Apple devices.
GroupMe	GroupMe is a mobile instant messaging application which belongs to Microsoft and is similar to already existing instant messaging applications such as WhatsApp, Wechat, Line, Telegram or KaoTalk.
Instagram	This social network is normally accessed from mobile devices and provides a platform where users can share pictures and videos, giving users the possibility of applying different filters, sharing video stories with other users as well as sharing these contents in other social networks such as Facebook, Twitter, Tumbler and Flickr.
LinkedIN	This social network has similar functionalities to Facebook but it is oriented to the business world.
Pinterest	This tool gathers pictures that people use to compile ideas for their projects and interests. People create and share collections of images called "Pins" that are used to do different things such as planning a trip, organizing events or keeping food recipes.
PodOmatic	This webpage is specialized into the creation of tools and services that help find, create, distribute and foster the listening of audio and video podcasts.
Snapchat	This mobile application allows users to take pictures and create short videos. Its use is very similar to the Instagram "Stories". Users control the visibility of the video and decide when they want the video to disappear.
Twitter	This social and "microblogging" network gives users the possibility of sending short texts with a limit of 140 characters, also called "tweets". Registered users can read and publish tweets from their personal accounts.
Voice Thread	Voice Thread is an interactive tool to share content and collaborate that allows users to add images, documents and videos. Likewise, users can add voice, text, audio files and comments.

Blyth (2008) distinguished between 4 categories to be considered in language learning research: technological, psycholinguistic, sociocultural and ecological. These 4 categories have been part of studies exploring social networks in language learning contexts. In this sense, those studies addressing the technological aspect such as the present book chapter have emphasized the possible benefits of a certain tool that could be useful for other teachers willing to put into practice a particular social network in a language learning environment. Those studies addressing the psycholinguistic aspect are focused on social interaction and can specifically explore the role of social networks as learning tools, emphasizing the connection between peers as a stepping stone towards language development (Lomicka & Lord, 2016). With regard to the sociocultural approach, which lies on Vygotsky's (1978) theory of mediation and zone of proximal development, the role of communication and interaction becomes indispensable for language acquisition and learning. Social networks become a very suitable environment for this type of approach to language learning where interactions with native speakers and non-native speakers favor an environment where language-related episodes (Swain & Lapkin, 1998) can occur. These language related-episodes have been found to be conducive to language development as students reflect on their language productions as well as the language used by their peers (Long, 1996). Likewise, factors such as students' observation and collaborative work are direct consequences of this sociocultural approach as well (Ryberg & Christiansen, 2008). As for those studies using an ecological approach, social networks have not specifically been explored. However, this type of approach which focuses primarily on the quality of learning opportunities, of classroom interaction and of educational experience (Van Lier, 2010), can be applied to any virtual space (Lomicka & Lord, 2016).

Although different studies yielded positive outcomes as result of using social networks for language learning purposes (e.g. Alvarez Valencia, 2016; Andujar & Salaberri-Ramiro, 2019; Donmus, 2010; Stevenson & Liu, 2013), negative aspects have been also found in the literature. Shih (2011) highlighted that the use of social networks for language learning purposes involves an increase in teachers' workload as they have to control the language and contents used. Further aspects pointed out in the literature such as the type of language production which often has an informal character and takes place out of the classroom environment may lead to abbreviations, misspellings or a high number of interventions (Kabilan, Ahmad & Abidin, 2010). However, as stated by Andujar (2016) and Andújar-Vaca and Cruz-Martínez (2017), students who are using a foreign language to interact with their peers through social networks normally aim for accuracy, whether writing or speaking, in order to achieve an understanding. Furthermore, they are not aware of the abbreviations in the target language and thus are unable to put them into practice. Finally, other negative aspects such as privacy issues have been found when implementing social networks for language learning purposes due to the necessary information exchange between peers and between students and teachers (Gettman & Cortijo, 2015; Teclehaimanot & Hickman, 2011). All these issues together with the distractive nature of social networks which are normally associated with informal learning have led to a lack of curricular integration into the course contents at all levels of the educational system. Thus, from a research point of view, it seems necessary to shed light on the possibilities of certain social networks for language development and for developing blended learning processes that can contribute to incorporating technological tools into the classroom environment as well as facilitating the learning process.

TOWARDS A FRAMEWORK FOR INTEGRATING BLENDED LEARNING TOOLS

When making use of technological tools to develop blended learning processes, it seems necessary to understand first the most widespread frameworks for blended learning models in the existing literature. One of these theoretical frameworks is the Revised Community of Inquiry (RCOI) (Shea et al., 2012) that although was not particularly designed for blended models, most of its features such as cognitive presence, learner presence, social presence and teaching presence are relevant to this type of learning environments. Figure 1 presents an example of such model.

This framework emphasizes communication and collaborative interaction between peers as well as between teachers and students. This type of interaction, from a language learning perspective may lead to foster knowledge building without students' awareness that such learning is occurring. Nevertheless, although this type of learning may be unconscious and unintentional, learners may generate different LREs (Swain & Lapkin, 1998) as well as notice the gap between their current language and the target language (Gass, 2003) which may foster L2 development.

Other blended learning frameworks such as Shea's (2007) pyramid framework or Khan's (2011) octagonal model have explored blended learning from different angles. However, more recent frameworks such as the Complex Adaptative Blended Learning Systems (Wang, Han & Yang, 2015) have attempted to incorporate not only all the elements present in blended learning models but how these elements work individually and with one another in order to offer a more complete picture of the blended learning environment. This system measures the changes brought about by the interaction with new elements in the system as could be the inclusion of any given technological tool. The authors of this framework identified six systems in their framework: the learner, the teacher, the technology, the content, the learning support, and the institution (Wang, Han, & Yang 2015). Within each system there are different subsystems with their own internal characteristics that interact with the rest of the systems to create a blended learning system. Figure 2 shows the different systems in the framework.

Figure 1. The revised community of inquiry framework (RCOI)
Source: Kim, Kim, Kera, and Getman (2014).

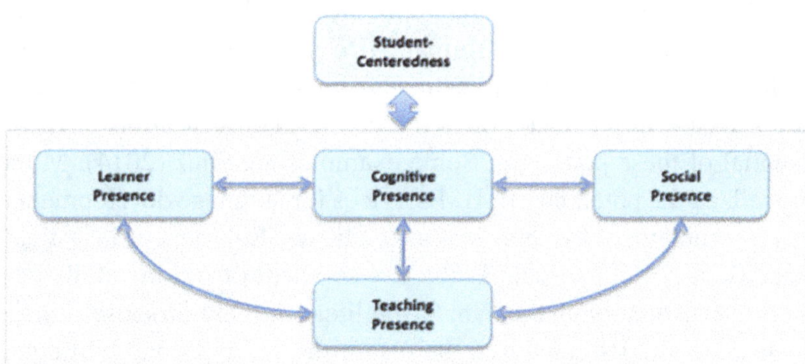

Figure 2. The framework of complex adaptive blended learning systems (CABLS)
Source: Wang, Han and Yang (2015)

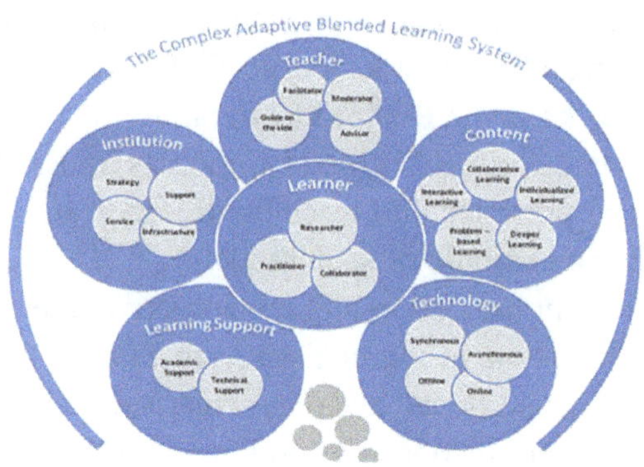

The systems presented in this framework, as opposed to other frameworks that present a more static nature, are open and dynamic and can self-organize, adapt to and evolve with the different changes happening in the blended model. This type of open theoretical framework seems a solid base to incorporate any potential blended learning tools such as WhatsApp or Instagram to the abovementioned "technology" and "learning support" systems.

WHATSAPP

The number of studies investigating and putting into practice this mobile instant messaging (MIM) application has grown steadily since 2013. Early studies into MIM such as Cavus & Ibrahim's (2009), Kennedy & Levy 's (2008) or Levy and Kennedy's (2007) explored the use of short test messages for language development using SMS to support language learning. However, with the appearance of mobile instant messaging platforms which allow users to maintain individual and group conversations online as well as share videos, voice messages and images, the landscape of MIM changed radically. This change also involved new possibilities for language development due to the various modes of communication available in these platforms which led to a more specific exploitation of MIM functionalities. In this sense, early studies into WhatsApp or other similar MIM platforms such as WeChat, Kaotalk, Line or Telegram explored the potential of these applications from a general point of view, analysing students' perception and reactions to the implementation of these tools in the classroom environment and focusing on the learning potential of these platforms. Some examples are Díaz (2014), Morato-Payá (2014) or Padron (2013) who realized the potential of MIM services for language development and focused on the use of these tools as a complement for their language classes. Subsequent investigations into the field focused on more specific aspects, particularly the development of writing skills which have been one of the main areas under investigation in the WhatsApp literature. In this sense studies such as Castrillo, Martín-Monje and Bárcena's (2014) analysed the responses, attitudes and participation of students of German as well as the meaning negotiation strategies taking place in the application. The researchers in this investigation also highlighted the adequacy of WhatsApp technology to develop students' interactive skills and to promote language learning at early stages of acquisition. Likewise, Andujar's (2016) investigation into the potential of WhatsApp to develop students writing skills explored the interaction in the application of 80 Spanish students taking an intermediate English course. Students' grammatical, lexical and mechanical accuracy as well as syntactic complexity was analysed over six months. An experimental design yielded significant differences between the control and experimental group in favour of the latter in terms of accuracy, but no syntactic differences were found between the two groups. Other studies exploring students' writing development through WhatsApp such as Fattah's (2015) investigation analysed sentence structure, punctuation and generating ideas between two groups, one in which writing was taught through the prescribed book and another in which the application was used. Results yielded significant effects on students' writing as a result of the use of the application.

Although further literature into the use of WhatsApp to foster writing development is sought in order to achieve a better understanding of how to exploit the application for writing purposes, Andujar-Vaca & Cruz's (2017) investigation explored the voice messaging functionality of the app to foster speaking skills. Students' chat conversation took place through voice messages and the samples collected in the application were subsequently analysed to measure the type and triggers of the language-related episodes (LREs) originated. A mixed analysis approach and a temporal axis were used to measure differences between a control and an experimental group. Findings highlighted negotiations of meaning as the most common LREs and phonetic triggers as the most abundant during the interaction. Likewise, factors such as ubiquitous meaning negotiation, students' reflection and evaluation of their own performance and authentic interaction and feedback were emphasized. Students were able to listen several times to their language productions perceiving gaps between their language and the target language. The use of the voice service was also considered by Hsieh, Wu and Marek (2017) who, using a very similar application in this case "Line", investigated oral and written communication in the app to learn English idioms. The application was used to develop a flipped learning model in which 48 students participated during their English training sessions. The questionnaires and interviews carried out revealed that the application had a better perception than the traditional method of tuition and aspects such as motivation, increase knowledge and an improvement of the oral abilities were emphasized.

More recent studies into WhatsApp such as Samaie, Mansouri Nejad, & Qaracholloo's (2018) explored the use of the application for self- and peer-assessments. The attitude of 30 Iranian students towards the two types of assessment were explored through the use of 4 questionnaires before and after the implementation of each assessment type. Results showed that participants adopted negative attitudes towards mobile-assisted assessments through the use of the app. Other recent studies such as Alqahtani, Bhaskar, Vadakalur Elumalai, Abumelha's (2018) explored students' perceptions towards the use of the application as a learning platform which could be used to develop and support their learning skills. 300 students replied to a survey exploring factors related with WhatsApp and language learning. Results yielded positive opinions from participants with regard to the use of WhatsApp to assist students during the learning process. Moreover, students highlighted the necessity of creating courses in which the access to class materials as well as the objectives and outcomes of the course were associated with the use of the app to develop a blended learning model. Finally, Ahmed (2019) investigated the use of the application to foster motivational levels of Yemeni EFL students in order to develop reading and writing abilities. Through the creation of chat group, participants commented on news articles during a period of two months. Pre and post-tests as well as a questionnaire were used to evaluate participants taking part in the chat conversation. Although the findings of this study found the application to be effective to improve students' motivation during the language classes, the difficulty to measure the motivation construct only through the use of a questionnaire poses certain threats with regard to the validity of the results.

Overall, as for the use of WhatsApp in language learning, it seems necessary to develop further studies exploring specific aspects of language development. Many times, the use of the application in the existing literature has been investigated vaguely without carrying out in-depth analysis of language development or exploring factors that are conducive to second language acquisition. Thus, more detailed cross-cultural studies involving higher number of participants are sought in order to obtain relevant and conclusive results that could have an impact on the educational community.

INSTAGRAM

Due to the fact that the social network "Instagram" is a relatively new application which was created in 2010, the number of studies making use of this platform for language learning is very scarce. However, and as aforementioned in the introduction of this chapter, the outstanding popularity of the application has led to the investigation of the possibilities of this application for language development. In this sense, early studies into Instagram and language learning such as such as Al-Ali's (2014) explored the use of the application combining the different functionalities offered by the app. 40 female students had to upload a picture of their vacations that was used as an starting point to develop in-class speaking activities. The second activity involved the exploration of the hashtags in order to find an inspiring picture form which they had to write a story. As for the third activity, students had to take a picture during their school time and publish an Instagram story related to this picture. Although the design of the activities seems to exploit the potential of the application, the research instruments - in this case a questionnaire – and the small number of participants cannot provide conclusive results of the overall experience. Students were satisfied with the use of the app for language learning purposes but several challenges were also found such as privacy issues or students feeling reluctant to upload their homework into their personal profiles.

Other researchers such as Ayuni, Kasuma and Mydin (2017) explored the use of Instagram to develop students' writing abilities. 101 students had to participate in a writing task through the use of the application in which they had to comment on pictures taken by themselves. Different surveys were given to participants in order to explore students' perception of the task carried out. Most of the students indicated that they had improved their language competence and described their Instagram experience as positive. Participants also highlighted that the application could be used for educational purposes. However, challenges were also found such as the distractive nature of the application which led students to different publication from the ones intended and also a certain degree of fear of feeling judged or criticized when writing in a foreign language.

As opposed to previous studies in which the use of Instagram for language learning was focused on the use of images to develop different tasks, Mansor and Rahim (2017) made use of the video service in the app to make students upload presentations that were later commented by their peers. Although the number of students was small (n =20), results indicated an increase in students' motivation towards learning and a higher degree of interaction between peers. Results regarding participation were also significant going from a 50% on the first day to a 95% at the end of the project. More recent investigations into this field are also presented in this book such as Çakmak (2019) who explored the use of Instagram to assess students' oral communication skills. The performance scores of participants on a speaking task delivered through Instagram and in class, and students' scores on the Big Five personality traits were investigated from a statistical point of view. Findings highlighted that the application facilitated students' performance in the oral communication skill and that personality traits are not predictors of students' performance on Instagram.

As aforementioned, the literature regarding the use of Instagram for language learning is still very scarce and although the application provides different options to develop blended learning models and to foster language development, research into this field is almost non-existent.

DISCUSSION

The distinctive nature of mobile devices that allow students to maintain conversations "anytime" and "anywhere" (Traxler, 2009) becomes a fertile environment for language interaction and the development of blended learning models. However, most of the literature which explores the use of technological devices to foster language development is mainly focused on the use of computer devices rather than mobile phones. In this sense, Burston (2014) pointed that the reality of MALL is "still on the fringes" as there is a lack of curricular integration as well as a strong theoretical deficit in studies making use of mobile devices to foster language development. As for the applications tackled in this chapter, the situation is very similar. Studies addressing mobile instant messaging applications are still scarce and there is only a small number of in-depth studies exploring which factors can make a difference when making use of these applications.

In this vein, the possibilities offered by MIM services such as WhatsApp are different from previous chat services and several factors need to be considered: the synchronous and asynchronous nature of mobile chat-based interaction; an easy access to contents; a wider range of chat features that allow teachers to exploit different modes of communication; greater ease when sharing voice, image or text messages; and the availability of using Internet resources anytime and anywhere. These differential characteristics should mark as well the design of blended learning environments in which the affordances of the applications should guide the course design and tasks implemented. MIM apps like WhatsApp or Line can be used to develop blended environments in which the role of the technological devices should shift from a complementary role to a more central one during the language learning process. In other words, MIM services could not only be used as classroom activities but also integrated within the course design to foster autonomous learning outside the class. This has already been carried out by some researchers in the field (e.g. Hew, Tang, Lo, & Zhu, 2018; Hsieh, Wu and Marek, 2017) with very positive results but further empirical studies are sought to better understand and develop successful blended learning environments.

With regard to Instagram, the enormous popularity of the platform has led developers to include many add-ons that allow teachers and language practitioners to use different types of communication. As aforementioned, the literature in this field is very scarce and, as in WhatsApp, the existing studies make use of the application to develop certain tasks instead of taking a more central role during language learning process. An example of this central role could be blended learning processes such as flipped learning (Bergmann and Sams, 2012), which involves a video explanation of the class contents before the start of the lecture while the class time is devoted to collaborative tasks and group work (Hwang, Lai & Wang, 2015). The video contents could be delivered through Instagram instead of using any other platform or LMS students are not familiar with. This may help students' engagement within the blended learning model which has been found to be one of the main challenges in this type of learning environments (e.g. Ancliff and Kang, 2017; Kim, Kim, Khera, and Getman, 2014) as well as decrease truancy (e.g. Yao, 2018).

Furthermore, it also seems necessary to overcome the challenges for the integration of blended learning models using mobile phones, which are not only presented at a classroom level but also at an institutional one as mobile restrictions are still present in many secondary and higher education institutions (Cortesi, Haduong, Gasser, Aricak, Saldaña, & Lerner, 2014). These challenges do not exclusively affect the education system but also the technological development that is still limited in certain parts of the world in which the Internet is not a reality yet. Similarly, in the developing world, the arrival of 5G connectivity may contribute to accelerating technological integration in the classroom as well as increase the possibilities of using mobile phones for language learning.

CONCLUSION

This chapter has attempted to shed light on the current situation of these two social networking tools in the language learning field as well as their possibilities to develop blended learning models. As shown in the literature review carried out for each of the apps, the amount of research is still very scarce and mainly focused on small case studies using experimental designs. Moreover, there is a lack of cross-cultural studies exploring the use of these application and thus, the information presented is usually very limited. In this sense, it seems necessary to further investigate the possibilities of these two apps for language development, particularly Instagram in which the literature seems to be underdeveloped. Likewise, there is a lack of studies using Instagram to carry out blended learning models and only a few using MIM (e.g. Hew, Tang, Lo, & Zhu, 2018; Hsieh, Wu and Marek, 2017). This indicates that the implementation of these technological tools, although interesting from a language development and motivational perspective, has not reached an integration into the course contents and course design. Thus, researchers into CALL and MALL should not exclusively investigate the use of these tools in language learning tasks as a complement to in-class tuition but incorporate them as a relevant element of the blended learning process.

Moreover, due to the fact that these tools are relatively new, there is also a lack of models to follow when making use of these social networking tools as well as certain misunderstanding regarding their use. In this sense, the complex adaptative blended learning systems (CABLS) (Wang, Han, & Yang 2015) framework explained in this chapter becomes a stepping stone towards the development of an open and dynamic blended learning framework that is able to change and evolve, and in which the different systems making up the framework can work individually and with one another. From a theoretical perspective, it is also worth mentioning the importance of Vygotsky's (1978) sociocultural theory that plays a fundamental role when putting into practice social networking tools in language learning environments. Vygotsky's social development theory, which stresses the indispensable role of social interaction in the development of cognition, is very present in social networking platforms. In this sense, as stated by Lantolf and Thorne (2006), individual development cannot be understood without making reference to the social and cultural context of that individual. This theory which supports the idea that higher mental processes have their origin in social processes is directly related with the use of social networking platforms for language development.

With these ideas in mind, it seems necessary to implement technological tools and applications that address the social nature of language and that rely on the interaction to construct knowledge. Social networking tools are a clear example of such tools in which the social element becomes an indispensable part of language development. Thus, together with the development of new technological advances than can contribute to improving the language learning process, researchers into CALL and MALL should also turn their focus towards accessible existing applications and tools that can help foster students' competence and that are still to be explored.

REFERENCES

Ahmed, S. T. S. (2019). Chat and learn: Effectiveness of using whatsapp as a pedagogical tool to enhance EFL learners' reading and writing skills. *International Journal of English Language and Literature Studies*, *8*(2), 61–68.

Akhiar, A., Mydin, A. A., & Kasuma, S. A. A. (2017). Students' perceptions and attitudes towards the use of Instagram in English language writing. *Education*, *47*, 72.

Al-Ali, S. (2014). Embracing the selfie craze: Exploring the possible use of Instagram as a language mlearning tool. *Issues and Trends in Educational Technology, 2*(2).

Alqahtani, S. M., Bhaskar, C. V., Vadakalur Elumalai, K., & Abumelha, M. (2018). *WhatsApp: An online platform for university-level English language education. Arab World English Journal*, 9. AWEJ.

Altbach, P. G. (2004). Globalisation and the university: Myths and realities in an unequal world. *Tertiary Education and Management*, *10*(1), 3–25. doi:10.1080/13583883.2004.9967114

Álvarez Valencia, J. A. (2016). Language views on social networking sites for language learning: The case of Busuu. *Computer Assisted Language Learning*, *29*(5), 853–867. doi:10.1080/09588221.2015.1069361

Ancliff, M., & Kang, A. (2017). Flipping an EMI physics class: Implications of student motivation and learning strategies for the design of course contents. *International Journal of Contents*, *13*(4).

Andujar, A. (2016). Benefits of mobile instant messaging to develop ESL writing. *System*, *62*, 63–76. doi:10.1016/j.system.2016.07.004

Andujar, A., & Salaberri-Ramiro, M. S. (2019). Exploring chat-based communication in the EFL class: Computer and mobile environments. *Computer Assisted Language Learning*, 1–28. doi:10.1080/09588221.2019.1614632

Andújar-Vaca, A., & Cruz-Martínez, M. S. (2017). Mobile instant messaging: WhatsApp and its potential to develop oral skills. *Comunicar*, *25*(50), 43–52. doi:10.3916/C50-2017-04

Arbaugh, J. B. (2014). What might online delivery teach us about blended management education? Prior perspectives and future directions. *Journal of Management Education*, *38*(6), 784–817. doi:10.1177/1052562914534244

Ball, P. & Lindsay, D. (2013). Language demands and support for English-medium instruction in tertiary education. Learning from a specific context. *English-medium instruction at universities: Global challenges, 4466.*

Bergmann, J. & Sams, A. (2012). Flip your classroom: Reach every student in every class every day. *International Society for Technology In Education.*

Blattner, G., & Fiori, M. (2011). Virtual social network communities: An investigation of language learners' development of sociopragmatic awareness and multiliteracy skills. *CALICO Journal, 29*(1), 24–43. doi:10.11139/cj.29.1.24-43

Blattner, G. & Lomicka, L. (2012). Facebooking and the social generation: A new era of language learning. *Alsic. Apprentissage des Langues et Systèmes d'Information et de Communication, 15*(1).

Blyth, C. (2008). Research perspectives on online discourse and foreign language learning. *Mediating discourse online*, 47-70.

Boyd, D. (2014). *It's complicated: The social lives of networked teens*. London, UK: Yale University Press.

Burston, J. (2014). The reality of MALL: Still on the fringes. *CALICO Journal, 31*(1), 103–125. doi:10.11139/cj.31.1.103-125

Çakmak, F. (2019). Social networking and language learning: Use of Instagram (IG) for evaluating oral communication skill. In Andujar (Ed.), Recent tools for computer – and mobile-assisted foreign language learning. Hershey, PA: IGI Global.

Castrillo, M. D., Martín-Monje, E., & Bárcena, E. (2014). *Mobile-based chatting for meaning negotiation in foreign language learning*. International Association for the Development of the Information Society.

Cavus, N., & Ibrahim, D. (2009). m-Learning: An experiment in using SMS to support learning new English language words. *British Journal of Educational Technology, 40*(1), 78–91. doi:10.1111/j.1467-8535.2007.00801.x

Chen Hsieh, J. S., Wu, W. C. V., & Marek, M. W. (2017). Using the flipped classroom to enhance EFL learning. *Computer Assisted Language Learning, 30*(1-2), 1–21. doi:10.1080/09588221.2015.1111910

Cortesi, S., Haduong, P., Gasser, U., Aricak, O., Saldaña, M., & Lerner, Z. (2014, Jan. 15). Youth perspectives on tech in schools: From mobile devices to restrictions and monitoring. *Berkman Center Research Publication*, (2014-3).

Counts, S., & Fisher, K. E. (2010). Mobile social networking as information ground: A case study. *Library & Information Science Research, 32*(2), 98–115. doi:10.1016/j.lisr.2009.10.003

Diaz, J. (2014). *El WhatsApp como herramienta de intervencion didáctica para fomentar el aprendizaje cooperativo* [WhatsApp as a tool of educational intervention to foster cooperative learning]. (Unpublished Paper presented at X Jornadas De Material Didáctica y Experiencias Innovadoras En Educación Superior), Buenos Aires, Argentina. Retrieved from http://eprints.rclis.org/23597/

Donmus, V. (2010). The use of social networks in educational computer-game based foreign language learning. *Procedia: Social and Behavioral Sciences, 9*, 1497–1503. doi:10.1016/j.sbspro.2010.12.355

Dunn, L. (2013). Teaching in higher education: can social media enhance the learning experience? *6th Annual conference of teaching and learning*, University of Glasgow, UK.

Farr, F., & Murray, L. (Eds.). (2016). *The Routledge handbook of language learning and technology.* Routledge. doi:10.4324/9781315657899

Fattah, S. (2015). The effectiveness of using WhatsApp messenger as one of mobile learning techniques to develop students' writing skills. *Journal of Education and Practice*, *6*(32), 115–127.

Gass, S. (2003). Input and interaction. In C. Doughty & M. Long (Eds.), *The handbook of second language acquisition* (pp. 224–255). Oxford, UK: Blackwell. doi:10.1002/9780470756492.ch9

Gettman, H. J., & Cortijo, V. (2015). "Leave me and my Facebook alone!" Understanding college students' relationship with Facebook and its use for academic purposes. *International Journal for the Scholarship of Teaching and Learning*, *9*(1), 8.

Global Digital Statistics. (2019). Retrieved from https://www.smartinsights.com/social-media-marketing/social-media-strategy/new-global-social-media-research/

Hew, K. F., Tang, Y., Lo, C. K., & Zhu, Y. (2018). Examining a WeChat-supported 5E-flipped classroom pedagogical approach. *International Journal of Services and Standards*, *12*(3-4), 224–242. doi:10.1504/IJSS.2018.100217

Hwang, G. J., Lai, C. L., & Wang, S. Y. (2015). Seamless flipped learning: A mobile technology-enhanced flipped classroom with effective learning strategies. *Journal of Computers in Education*, *2*(4), 449-473.

Im, J. H., & Kim, J. (2015). Use of blended learning for effective implementation of English-medium instruction in a non-English higher education context. *International Education Studies*, *8*(11), 1–15. doi:10.5539/ies.v8n11p1

Kabilan, M. K., Ahmad, N., & Abidin, M. J. Z. (2010). Facebook: An online environment for learning of English in institutions of higher education? *The Internet and higher education*, *13*(4), 179–187. doi:10.1016/j.iheduc.2010.07.003

Kennedy, C., & Levy, M. (2008). L'italiano al telefonino: Using SMS to support beginners' language learning. *ReCALL*, *20*(3), 315–330. doi:10.1017/S0958344008000530

Khan, B. H. (2001). A framework for web-based learning. In B. H. Khan (Ed.), *Web-based training* (pp. 75–98). Englewood Cliffs, NJ: Educational Technology Publications.

Kim, M. K., Kim, S. M., Khera, O., & Getman, J. (2014). The experience of three flipped classrooms in an urban university: An exploration of design principles. *The Internet and Higher Education*, *22*, 37–50. doi:10.1016/j.iheduc.2014.04.003

Kukulska-Hulme, A. & Shield. (2007). An overview of mobile assisted language learning: Can mobile devices support collaborative practice in speaking and listening. In *EuroCALL*, 2007.

Lantolf, J., & Thorne, S. (2006). *Sociocultural theory and the genesis of second language development*. Oxford, UK: Oxford University Press.

Law, K. M., Geng, S., & Li, T. (2019). Student enrollment, motivation and learning performance in a blended learning environment: The mediating effects of social, teaching, and cognitive presence. *Computers & Education*, *136*, 1–12. doi:10.1016/j.compedu.2019.02.021

Levy, M., & Kennedy, C. (2007). Learning Italian via mobile SMS. In *Mobile learning* (pp. 92–99). Routledge.

Lomicka, L. & Lord, G. (2016). Social networking and language learning. The Routledge handbook of language learning and technology, 255-268.

Long, M. H. (1996). The role of the linguistic environment in second language acquisition. In W. Ritchie & T. Bathia (Eds.), *Handbook of research on second language acquisition* (pp. 413–469). New York, NY: Academic Press. doi:10.1016/B978-012589042-7/50015-3

López-Pérez, M. V., Pérez-López, M. C., & Rodríguez-Ariza, L. (2011). Blended learning in higher education: Students' perceptions and their relation to outcomes. *Computers & Education*, *56*(3), 818–826. doi:10.1016/j.compedu.2010.10.023

Mansor, N., & Rahim, N. A. (2017). Instagram in ESL classroom. *Man in India*, *97*(20), 107–114.

Mao, J. (2014). Social media for learning: A mixed methods study on high school students' technology affordances and perspectives. *Computers in Human Behavior*, *33*, 213–223. doi:10.1016/j.chb.2014.01.002

Mazer, J. P., Murphy, R. E., & Simonds, C. J. (2007). I'll see you on "Facebook": The effects of computer-mediated teacher self-disclosure on student motivation, affective learning, and classroom climate. *Communication Education*, *56*(1), 1–17. doi:10.1080/03634520601009710

Mills, N. (2011). Situated learning through social networking communities: The development of joint enterprise, mutual engagement, and a shared repertoire. *CALICO Journal*, *28*(2), 345–368. doi:10.11139/cj.28.2.345-368

Miyazoe, T., & Anderson, T. (2010). Learning outcomes and students' perceptions of online writing: Simultaneous implementation of a forum, blog, and wiki in an EFL blended learning setting. *System*, *38*(2), 185–199. doi:10.1016/j.system.2010.03.006

Morato-Paya, A. (2014). El WhatsApp como complemento del aprendizaje en la clase E/LE [WhatsApp as a learning complement in the classroom of E/LE]. Revista Foro De Profesores De E/LE, 10(1), 165-173.

Padron, C. J. (2013). Teaching strategies based on WhatsApp instant messaging application only for phones (mobile learning) and its use to promote collaborative learning. *Revista De Tecnología De Información y Comunicación En Educación*, *7*(2), 123–134.

Ryberg, T., & Christiansen, E. (2008). Community and social network sites as technology enhanced learning environments. *Technology, Pedagogy and Education*, *17*(3), 207–219. doi:10.1080/14759390802383801

Samaie, M., Mansouri Nejad, A., & Qaracholloo, M. (2018). An inquiry into the efficiency of WhatsApp for self-and peer-assessments of oral language proficiency. *British Journal of Educational Technology*, *49*(1), 111–126. doi:10.1111/bjet.12519

Seaman, J., & Tinti-Kane, H. (2013). *Social media for teaching and learning*. London, UK: Pearson Learning Systems.

Shea, P. (2007). Towards a conceptual framework for learning in blended environments. In A. G. Picciano & C. D. Dziuban (Eds.), *Blended learning: Research perspectives* (pp. 19–35). Retrieved from http://elab.learningandteaching.dal.ca/dalblend2013-files/blended-learning-research-perspectives-book.pdf#page=30

Shea, P., Hayes, S., Smith, S. U., Vickers, J., Bidjerano, T., Pickett, A., ... Jian, S. (2012). Learning presence: Additional research on a new conceptual element within the Community of Inquiry (CoI) framework. *The Internet and Higher Education*, *15*(2), 89–95. doi:10.1016/j.iheduc.2011.08.002

Shih, R. C. (2011). Can Web 2.0 technology assist college students in learning English writing? Integrating Facebook and peer assessment with blended learning. *Australasian Journal of Educational Technology*, *27*(5), 829–845. doi:10.14742/ajet.934

Stevenson, M. P., & Liu, M. (2013). Learning a language with Web 2.0: Exploring the use of social networking features of foreign language learning websites. *CALICO Journal*, *27*(2), 233–259. doi:10.11139/cj.27.2.233-259

StoutD. (2019). Retrieved from https://dustinstout.com/social-media-statistics/#whatsapp-stats

Swain, M., & Lapkin, S. (1998). Interaction and second language learning: Two adolescent French immersion students working together. *Modern Language Journal*, *82*(3), 320–337. doi:10.1111/j.1540-4781.1998.tb01209.x

Teclehaimanot, B., & Hickman, T. (2011). Student-teacher interaction on Facebook: What students find appropriate. *TechTrends*, *55*(3), 19–30. doi:10.100711528-011-0494-8

Traxler, J. (2009). Learning in a mobile age. [IJMBL]. *International Journal of Mobile and Blended Learning*, *1*(1), 1–12. doi:10.4018/jmbl.2009010101

Van Lier, L. (2010). The ecology of language learning: Practice to theory, theory to practice. *Procedia: Social and Behavioral Sciences*, *3*, 2–6. doi:10.1016/j.sbspro.2010.07.005

Vygotsky, L. (1978). *Mind in society: The development of higher psychological processes*. Cambridge, MA: Harvard University Press.

Wang, Y., Han, X., & Yang, J. (2015). Revisiting the blended learning literature: Using a complex adaptive systems framework. *Journal of Educational Technology & Society*, *18*(2), 380–393.

Yao, Z. (2018). Research on flipped classroom teaching mode based on MOODLE. In *Proceedings of the 3rd International Conference on Education & Education Research* (EDUER 2018), 656-660.

Compilation of References

(Forthcoming). Rialtas na hÉireann (2019). Plean Digiteach don Ghaeilge 2019-2029. *Digital Plan for the Irish Language.*

ABAIR. (2019). *ABAIR: An Sintéiseoir Gaeilge – The Irish Language Synthesiser.* Retrieved from www.abair.ie

Abbott, W., Donaghey, J., Hare, J., & Hopkins, P. (2013). An Instagram is worth a thousand words: An industry panel and audience Q&A. *Library Hi Tech News, 30*(7), 1–6. doi:10.1108/LHTN-08-2013-0047

Abdalhamid, F. (2012). *Listening comprehension strategies of Arabic-speaking ESL learners* (Doctoral dissertation), Colorado State University, Colorado.

Abernathy Marsh-Soloway, M., & Michel Delers, O. (2019). New directions in eTandem: An expanded vision of capabilities and practices. *Revista do GEL, 15*(3), 176–193. doi:10.21165/gel.v15i3.2395

AbuSa'aleek, A. (2014). A review of emerging technologies: Mobile Assisted Language Learning (MALL). Asian Journal of Education and e-Learning, 2(6), 469-475. Retrieved from https://www.academia.edu/9844553/A_Review_of_Emerging_Technologies_Mobile_Assisted_Language_Learning_MALL_

Afflerbach, P., & Cho, B. (2009). Determining and describing reading strategies: Internet and traditional forms of reading. In H. S. Waters & W. Schneider (Eds.), *Metacognition, strategy use, and instruction* (pp. 201–225). New York, NY: Guilford.

Afzali, P., Shabani, S., Basir, Z., & Ramazani, M. (2017). Mobile-assisted vocabulary learning: A review study. *Advances in Language and Literary Studies, 8*(2), 190-195. doi:10.7575/aiac.alls.v.8n.2p.190

Ahmed, H. B. E. (2016). Duolingo as a bilingual learning app: A case study. *Arab World English Journal, 7*(2), 255–267. doi:10.24093/awej/vol7no2.17

Ahmed, S. T. S. (2019). Chat and learn: Effectiveness of using whatsapp as a pedagogical tool to enhance EFL learners' reading and writing skills. *International Journal of English Language and Literature Studies, 8*(2), 61–68.

Akbari, E., Naderi, A., Simons, R. J., & Pilot, A. (2016). Student engagement and foreign language learning through online social networks. *Asian-Pacific Journal of Second and Foreign Language Education, 1*(1), 1–22. doi:10.118640862-016-0006-7

Akçayır, M., & Akçayır, G. (2017). Advantages and challenges associated with augmented reality for education: A systematic review of the literature. *Educational Research Review, 20*, 1–11. doi:10.1016/j.edurev.2016.11.002

Akhiar, A., Mydin, A. A., & Kasuma, S. A. A. (2017). Students' perceptions and attitudes towards the use of Instagram in English language writing. *Malaysian Journal of Learning and Instruction* [Special Issue], 47–72.

Akhiar, A., Mydin, A. A., & Kasuma, S. A. A. (2017). Students' perceptions and attitudes towards the use of Instagram in English language writing. *Education, 47*, 72.

Akiyama, Y. (2015). Task-based investigation of learner perceptions: Affordances of video-based eTandem learning. In E. Dixon & M. Thomas (Eds.), *Researching language learner interaction online: From social media to MOOCs, CALICO Monograph Series* (Vol. 13, pp. 149–170). San Marcos, TX: CALICO.

Akiyama, Y., & Cunningham, J. (2018). Synthesizing the practice of SCMC-based telecollaboration: A scoping review. *CALICO Journal*, *35*(1), 49–76. doi:10.1558/cj.33156

Al-Adwan, A. S., Al-Madadha, A., & Zvirzdinaite, Z. (2018). Modeling students' readiness to adopt mobile learning in higher education: An empirical study. *International Review of Research in Open and Distributed Learning*, *19*(1). doi:10.19173/irrodl.v19i1.3256

Al-Ali, S. (2014). Embracing the selfie craze: Exploring the possible use of Instagram as a language mlearning tool. *Issues and Trends in Educational Technology, 2*(2).

Al-Ali, S. (2014). Embracing the selfie craze: Exploring the possible use of Instagram as a language m-learning tool. *Issues and Trends in Educational Technology*, *2*(2), 1–16. doi:10.2458/azu_itet_v2i2_ai-ali

Al-Emran, M., Elsherif, H. M., & Shaalan, K. (2016). Investigating attitudes towards the use of mobile learning in higher education. *Computers in Human Behavior*, *56*, 93–102. doi:10.1016/j.chb.2015.11.033

Alexander, C. (2017). Student perceptions of gamification in higher education. In *Proceedings of Society for Information Technology & Teacher Education International Conference* (pp. 1428-1433). Association for the Advancement of Computing in Education (AACE).

AlGhamdi, M. A. (2018). Arabic learners' preferences for Instagram English lessons. *English Language Teaching*, *11*(8), 103–110. doi:10.5539/elt.v11n8p103

Al-Hadithy, T., & Ali, S. (2018). Gamification in learning English for academic purposes: Designing assessment for language using Kahoot with UAE undergraduate law students. In *Proceedings of 110th IASTEM International Conference*, Toronto, Canada.

Alhawiti, K. (2014). Natural language processing and its use in education. *International Journal of Advanced Computer Science and Applications*, *5*(10), 72–75.

Allen, D. (2004). *Oxford placement test*. Oxford, UK: Oxford University Press.

Allik, J., Hrebícková, M., & Realo, A. (2018). Unusual configurations of personality traits indicate multiple patterns of their coalescence. *Frontiers in Psychology*, 1–14. PMID:29515499

Aloraini, N. (2018). Investigating Instagram as an EFL learning tool. *Arab World English Journal*, *4*(4), 174–184. doi:10.24093/awej/call4.13

Alqahtani, S. M., Bhaskar, C. V., Vadakalur Elumalai, K., & Abumelha, M. (2018). *WhatsApp: An online platform for university-level English language education. Arab World English Journal*, 9. AWEJ.

Al-Qudaimi, K. (2013). Mobile-Assisted Language Learning: A literature review. 1-15. Unpublished. Retrieved from https://www.academia.edu/6369026/Mobile-Assisted_Language_Learning_A_Literature_Review

Altbach, P. G. (2004). Globalisation and the university: Myths and realities in an unequal world. *Tertiary Education and Management*, *10*(1), 3–25. doi:10.1080/13583883.2004.9967114

Alvarado, N. C., Coelho, D., & Dougherty, E. (2016). Mobile apps for ELLs: Supporting language learning with engaging digital tools. *Argentinian Journal of Applied Linguistics*, *4*, 43–58.

Álvarez Valencia, J. A. (2016). Language views on social networking sites for language learning: The case of Busuu. *Computer Assisted Language Learning*, *29*(5), 853–867. doi:10.1080/09588221.2015.1069361

Amaral, L., & Meurers, D. (2007). Conceptualizing student models for ICALL. In C. Conati, K. McCoy, & G. Paliouras (Eds.), *User modeling 2007* (pp. 340–344). Heidelberg, Germany: SpringerVerlag. doi:10.1007/978-3-540-73078-1_41

An Caighdeán Oifigiúil. (2017). Gramadach na Gaeilge – An Caighdeán Oifigiúil. (The official standard for writing in the Irish language). Tithe an Oireachtais. Retrieved from https://data.oireachtas.ie/ie/oireachtas/caighdeanOifigiul/2017/2017-08-03_an-caighdean-oifigiuil-2017_en.pdf

Ancliff, M., & Kang, A. (2017). Flipping an EMI physics class: Implications of student motivation and learning strategies for the design of course contents. *International Journal of Contents*, *13*(4).

Anderson, T., & Elloumi, F. (2004). *Theory and practice of online learning* (2nd ed.). Edmonton, Canada: AU Press.

Andujar, A. & Buchner, J. (2019). The potential of 3D virtual reality (VR) for language learning: an overview. In I. A. Sanchez, P. Isaias, P. Ravesteijn, & G. Ongena (Hrsg.), *Proceedings of the 15th International Conference Mobile Learning 2019* (S. 153–156). Utrecht, The Netherlands: iadis. 10.33965/ml2019_201903R002

Andujar, A. (2019). Shaping the Future of Telecollaboration: Web RTC. In Educational Technology and Resources for Synchronous Learning in Higher Education (pp. 151-172). Hershey, PA: IGI Global. doi:10.4018/978-1-5225-7567-2.ch008

Andujar, A. (2016). Benefits of mobile instant messaging to develop ESL writing. *System*, *62*, 63–76. doi:10.1016/j.system.2016.07.004

Andujar, A., & Medina-López, C. (2019). Exploring new ways of eTandem and telecollaboration through the WebRTC protocol: Students' engagement and perceptions. *International Journal of Emerging Technologies in Learning*, *14*(5), 200–217. doi:10.3991/ijet.v14i05.9612

Andujar, A., & Salaberri-Ramiro, M. S. (2019). Exploring chat-based communication in the EFL class: Computer and mobile environments. *Computer Assisted Language Learning*, 1–28. doi:10.1080/09588221.2019.1614632

Andújar-Vaca, A., & Cruz-Martínez, M. S. (2017). Mobile instant messaging: WhatsApp and its potential to develop oral skills. *Comunicar*, *25*(50), 43–52. doi:10.3916/C50-2017-04

Angelova, M., & Zhao, Y. (2016). Using an online collaborative project between American and Chinese students to develop ESL teaching skills, cross-cultural awareness and language skills. *Computer Assisted Language Learning*, *29*(1), 167–185. doi:10.1080/09588221.2014.907320

Antoniadou, V. (2011). Using activity theory to understand the contradictions in an online transatlantic collaboration between student-teachers of English as a foreign language. *ReCALL*, *23*(3), 233–251. doi:10.1017/S0958344011000164

Appel, C. & Pujolà, J. T. (2015). Pedagogical and technological issues in the instructional design of a tandem MOOC. In *Proceedings of E-Learn: World Conference on E-Learning in Corporate, Government, Healthcare, and Higher Education* (pp. 1696-1705). Kona, Hawaii: Association for the Advancement of Computing in Education (AACE).

Arbaugh, J. B. (2014). What might online delivery teach us about blended management education? Prior perspectives and future directions. *Journal of Management Education*, *38*(6), 784–817. doi:10.1177/1052562914534244

Arslan, R., & Şahin-Kızıl, A. (2010). How can the use of blog software facilitate the writing process of English language learners? *Computer Assisted Language Learning*, *23*(3), 183–197. doi:10.1080/09588221.2010.486575

Arzal, A. & Chen, S. (2017). Improving ICT integration in language classrooms: Voices from teachers in Indonesian Province of Gorontalo. *Proceedings of the CALL in Context Conference*, Berkeley, CA, USA. Retrieved from: http://call2017.language.berkeley.edu/wp-content/uploads/2017/07/CALL2017_proceedings.pdf

Asmalı, M. (2014). The relationship between the Big Five personality traits and language learning strategies. Balıkesir University Journal of Social Sciences Institute, 17(32), 1–18.

Azlan, N. A. B., Zakaria, S. B., & Yunus, M. M. (2019). Integrative task-based learning: Developing speaking skill and increase motivation via Instagram. *International Journal of Academic Research in Business and Social Sciences*, *9*(1), 620–636. doi:10.6007/IJARBSS/v9-i1/5463

Azuma, R. (1997). A survey of augmented reality. *Presence (Cambridge, Mass.)*, *6*(6), 355–385. doi:10.1162/pres.1997.6.4.355

Bajner, M. (2018). Digital competence through authentic, self-made texts. *R&E-SOURCE - Open Online Journal for Research and Education*, (12). Retrieved from https://journal.ph-noe.ac.at/index.php/resource/article/view/583

Baker, F. S. (2015). Emerging realities of text-to-speech software for nonnative-English-speaking community college students in the freshman year. *Community College Journal of Research and Practice*, *39*(5), 423–441. doi:10.1080/10668926.2013.835290

Bakhoda, I., & Shabani, K. (2019). Bringing L2 learners' learning preferences in the mediating process through computerized dynamic assessment. *Computer Assisted Language Learning*, *32*(3), 210–236. doi:10.1080/09588221.2018.1483950

Baleghizadeh, S., & Oladrostam, E. (2010). The effect of mobile assisted language learning (MALL) on grammatical accuracy of EFL students. *Mextesol Journal*, *34*(2), 1–10.

Ball, P. & Lindsay, D. (2013). Language demands and support for English-medium instruction in tertiary education. Learning from a specific context. *English-medium instruction at universities: Global challenges, 4466*.

Bandura, A. (1986). *Social foundations of thought and action: A social cognitive theory*. Englewood Cliffs, NJ: Prentice-Hall.

Bangert-Drowns, R. L., Kulik, J. A., & Kulik, C.-L. C. (1985). Effectiveness of computer-based education in secondary schools. *Journal of Computer-Based Instruction*, *12*, 59–68.

Barata, G., Gama, S., Jorge, J., & Gonçalves, D. (2013). Improving participation and learning with gamification. In L. Nacke, K. Harrigan, & N. Randall (Eds.), *Proceedings of International Conference on Gameful Design, Research, and Applications* (pp. 10– 17). Stratford, Canada: ACM.

Barnes, E., Ní Chiaráin, N., & Ní Chasaide, A. (2017). Departures from the "norm": how the phonology, morphology and orthography of the Irish language impact on literacy instruction and acquisition. Exploring the literacy landscape: Celebrating 40 years of research and practice, *Literacy Association of Ireland 40th Annual Conference, Marino Institute of Education,* Dublin, Sept. 29-30, 2016, Literacy Association of Ireland, pp. 22–32.

Baron, N. (2015). *Words onscreen: The fate of reading in a digital world.* Oxford, UK: Oxford University Press.

Barzillai, M., & Thomson, J. M. (2018). Children learning to read in a digital world. *First Monday*, *23*(10). doi:10.5210/fm.v23i10.9437

Barzillai, M., Thomson, J. M., Schroeder, S., & Van den Broek, P. W. (2018). Learning to read in a digital world. In M. Barzillai, J. Thomson, S. Schroeder, & P. W. Van den Broek (Eds.), *Learning to Read in a Digital World* (pp. 225–238). Amsterdam, The Netherlands: John Benjamins. doi:10.1075wll.17

Batardière, M.-T. & Helm, F. (2016). Fostering students' engagement with topical issues through different modes of online exchange. In S. Jager, M. Kurek, & B. O'Rourke (Eds.), *New directions in telecollaborative research and practice: Selected papers from the second conference on telecollaboration in higher education*, 313-319. Research-publishing.net. doi:10.14705/rpnet.2016.telecollab2016.52

Bates, T. (2012, Aug. 5). What's right and what's wrong about Coursera-style MOOCs? Retrieved from https://www.tonybates.ca/2012/08/05/whats-right-and-whats-wrong-about-coursera-style-moocs/

Bax, S. (2003). CALL—Past, present and future. *System*, *31*(1), 13–28. doi:10.1016/S0346-251X(02)00071-4

Bayne, S., & Ross, R. (2015). MOOC pedagogy. In P. Kim (Ed.), *Massive open online courses* (pp. 23–45). New York, NY: Routledge.

Beaven, A. (2013). Using MOOCs in an academic English course at university level. In A. Beaven, A. Comas-Quinn, & B. Sawhill (Eds.), Case studies of openness in the language classroom (pp. 217-227). © Research-publishing.net. doi:10.14705/rpnet.2013.000122

Bebell, D., & O'Dwyer, L. M. (2010). Educational outcomes and research from 1:1 computing settings. *The Journal of Technology, Learning, and Assessment*, *9*(1), 5–15.

Beqiri, M. S., Chase, N. M., & Bishka, A. (2010). Online course delivery: An empirical investigation of factors affecting student satisfaction. *Journal of Education for Business*, *85*(2), 95–100. doi:10.1080/08832320903258527

Bergmann, J. & Sams, A. (2012). Flip your classroom: Reach every student in every class every day. *International Society for Technology In Education.*

Bicen, H., & Kocakoyun, S. (2018). Perceptions of students for gamification approach. Kahoot as a case study. *International Journal of Emerging Technologies in Learning*, *13*(2), 72–93. doi:10.3991/ijet.v13i02.7467

Biedroń, A. (2016). Personality factors as predictors of foreign language aptitude. *Studies in Second Language Learning and Teaching*, *1*(4), 467–489. doi:10.14746sllt.2011.1.4.2

Billinghurst, M., Kato, H., & Poupyrev, I. (2001). The MagicBook-Moving seamlessly between reality and virtuality. *IEEE Computer Graphics and Applications*, *21*(1), 6–9. doi:10.1109/38.920621

Blattner, G. & Lomicka, L. (2012). Facebooking and the social generation: A new era of language learning. *Alsic. Apprentissage des Langues et Systèmes d'Information et de Communication, 15*(1).

Blattner, G., & Fiori, M. (2011). Virtual social network communities: An investigation of language learners' development of sociopragmatic awareness and multiliteracy skills. *CALICO Journal*, *29*(1), 24–43. doi:10.11139/cj.29.1.24-43

Blattner, G., & Lomicka, L. (2012). Facebooking and the social generation: A new era of language learning. *Apprentissage Des Langues et Systèmes d'Information et de Communication*, *15*, 1–32.

Blommaert, J. (2005). *Discourse: A critical introduction.* Cambridge, UK: Cambridge University Press. doi:10.1017/CBO9780511610295

Blyth, C. (2008). Research perspectives on online discourse and foreign language learning. *Mediating discourse online*, 47-70.

BMBWF (Bundesministerium für Bildung, Wissenschaft und Forschung). (2018). *Verordnung des Bundesministers für Unterricht und Kunst vom 14. November 1984 über die Lehrpläne der allgemeinbildenden höheren Schulen* [Regulation by the Austrian Federal ministry of Education, Science and Research from Nov. 14, 1984 about the curricula in secondary forms].

Bocchi, J., Eastman, J. K., & Swift, C. O. (2004). Retaining the online learner: Profile of students in an online MBA program and implications for teaching them. *Journal of Education for Business*, *79*(4), 245–253. doi:10.3200/JOEB.79.4.245-253

Bodnar, S., Cucchiarini, C., Strik, H., & van Hout, R. (2016). Evaluating the motivational impact of CALL systems: Current practices and future directions. *Computer Assisted Language Learning, 29*(1), 186–212. doi:10.1080/09588221.2014.927365

Bonk, C. J., & Graham, C. R. (2005). *The handbook of blended learning: Global perspectives, local designs.* Hoboken, NJ: John Wiley & Sons.

Bosker, H. R., Pinget, A. F., Quene, H., Sanders, T., & de Jong, N. H. (2012). What makes speech sound fluent? The contribution of pauses, speed and repairs. *Language Testing, 30*(2), 159–175. doi:10.1177/0265532212455394

Bosker, H. R., Quené, H., Sanders, T., & de Jong, N. H. (2014). The perception of fluency in native and nonnative speech. *Language Learning, 64*(3), 579–614. doi:10.1111/lang.12067

Botero, G. G., Questier, F., Cincinnato, S., He, T., & Zhu, C. (2018). Acceptance and usage of mobile assisted language learning by higher education students. *Journal of Computing in Higher Education, 30*(3), 426–451. doi:10.100712528-018-9177-1

Boulton, A. (2017). Data-Driven Learning and Language Pedagogy. In S. L. Thorne, and May, Stephen (Ed.), Language, Education and Technology (pp. 181-192). Berlin: Springer International Publishing AG. doi:10.1007/978-3-319-02237-6_15

Bowden, P. (2018, Jan. 7). Five years of MOOC trends: Pat Bowden recounts her experiences of taking 89 MOOCs, and analyzes the trends she has noticed in that time. [Web log post]. Retrieved from https://www.class-central.com/report/five-years-mooc-trends/

Bower, M., Howe, C., McCredie, N., Robinson, A., & Grover, D. (2014). Augmented reality in education - Cases, places and potentials. *Educational Media International, 15*(2), 1–15. doi:10.1080/09523987.2014.889400

Boyd, D. (2014). *It's complicated: The social lives of networked teens.* London, UK: Yale University Press.

Bozdoğan, D. (2015). MALL revisited: Current trends and pedagogical implications. *Procedia: Social and Behavioral Sciences, 195*, 932-939. doi:10.1016/j.sbspro.2015.06.373

Bradley, L., Lindström, N. B., & Hashemi, S. S. (2017). Integration and language learning of newly arrived migrants using mobile technology. *Journal of Interactive Media in Education, 2017*(1), 3. doi:10.5334/jime.434

Brandhofer, G., ... (2016). *digi.kompP - Kompetenzmodell* [dig.kompP - Competence Model]. Retrieved from http://www.virtuelle-ph.at/wp-content/uploads/2016/09/digi.kompP-Grafik-und-Deskriptoren-1.pdf

Breslow, L., Pritchard, D., DeBoer, J., Stump, G., Ho, A.,D., & Seaton, D. (2013). Studying learning in the worldwide classroom: Research into edX's first MOOC.

Brick, B. (2012). How effective are web 2.0 language learning sites in facilitating language learning? *Compass: The Journal of Learning and Teaching, 3*, 57–63.

Brinthaupt, T. M., & Shin, C. M. (2001). The relationship of academic cramming to flow experience. *College Student Journal, 35*(3), 457–471.

Broin, Ó., D., Barnes, E., Ní Chiaráin, N., & Ní Chasaide, A. (2018). A story based game for teaching phonics for the Irish language. In *Proceedings of the 12th European Conference on Games Based Learning*, Sophia Antipolis, France, pp. 495-503.

Brown, D. H. (2007). *Principles of Language Learning and Teaching.* White Plains, NY: Pearson Education.

Brown, S. (2006). *Teaching Listening.* New York, NY: Cambridge University Press.

Bruegge, B., & Dutoit, A. H. (2002). *Object-Oriented software engineering. Conquering complex and changing systems.* DF, Mexico: Prentice Hall.

Bruff, D. O., Fisher, D. H., McEwen, K. E., & Smith, B. E. (2013). Wrapping a MOOC: Student perceptions of an experiment in blended learning. *Journal of Online Learning and Teaching / MERLOT*, *9*(2), 187–199.

Buchner, J. & Zumbach, J. (2018). Promoting intrinsic motivation with a mobile augmented reality learning environment. In I. A. Sanchez & P. Isaias (Hrsg.), *Proceedings of the 14th International Conference Mobile Learning 2018* (S. 55–61). Lisbon, Portugal: iadis.

Buchner, J., & Höfler, E. (2018). Flipped learning mit augmented reality. In J. Buchner, C. Freisleben-Teutscher, J. Haag, & E. Rauscher (Eds.), *Inverted Classroom: Vielfältiges Lernen* (pp. 61–66). Retrieved from http://skill.fhstp.ac.at/wp-content/uploads/2017/09/23489_TdL_sh_270218_final.pdf#page=63

Buck, G. (2001). *Assessing Listening*. New York, NY: Cambridge University Press. doi:10.1017/CBO9780511732959

Bull, S. (2000, June). Individualized recommendations for learning strategy use. In G. Gauthier, C. Frasson, & K. VanLehn (Eds.), *Intelligent tutoring systems.* In *Proceedings 5th International Conference, ITS 2000,* Montréal, Canada, (pp. 594–603). Berlin, Germany: Springer Verlag 10.1007/3-540-45108-0_63

Burke, B. (2014). *Gamify: How gamification motivates people to do extraordinary things*. Brookline, MA: Gartner, Inc.

Burston, J. (2014). MALL: The pedagogical challenges. *Computer Assisted Language Learning*, *27*(4), 344–357. doi:10.1080/09588221.2014.914539

Burston, J. (2014). The reality of MALL: Still on the fringes. *CALICO Journal*, *31*(1), 103–125. doi:10.11139/cj.31.1.103-125

Burston, J. (2015). Twenty years of MALL project implementation: A meta-analysis of learning outcomes. *ReCALL*, *27*(01), 4–20. doi:10.1017/S0958344014000159

Busselle, R., & Bilandzic, H. (2008). Fictionality and perceived realism in experiencing stories: A model of narrative comprehension and engagement. *Communication Theory*, *18*(2), 255–280. doi:10.1111/j.1468-2885.2008.00322.x

Butler, Y. & Zeng, W. (2015). Young learners' interactional development in task-based paired-assessment in their first and foreign languages: A case of English learners in China. *Education 3-13, 43*(3), 292–321. doi:10.1080/03004279.2013.813955

Butler, Y., Someya, Y., & Fukuhara, E. (2014). Online games for young learners' foreign language learning. *ELT Journal*, *68*(3), 265–275. doi:10.1093/elt/ccu008

Byrne, J., & Diem, R. (2014). Profiling mobile English language learners. *JALT CALL Journal*, *10*(1), 3–19.

Çakmak, F. (2019). Social networking and language learning: Use of Instagram (IG) for evaluating oral communication skill. In Andujar (Ed.), Recent tools for computer – and mobile-assisted foreign language learning. Hershey, PA: IGI Global.

Calderon, D. (2015). Introducción: El derecho a aprender inglés. *Sorry, el aprendizaje del inglés en México*, Mexicanos Primero, 28. Recuperado desde http://www.mexicanosprimero.org/

Çam Aktaş, B., & Can, Y. (2019). The effect of Whatsapp usage on the attitudes of students toward English self-efficacy and English courses in foreign language Education outside the school. *International Electronic Journal of Elementary Education*, *11*(3), 247–256. doi:10.26822/iejee.2019349249

Carrell, P. L., Prince, M. S., & Astika, G. G. (1996). Personality types and language learning in an EFL context. *Language Learning*, *46*(1), 75–99. doi:10.1111/j.1467-1770.1996.tb00641.x

Carrió-Pastor, M. L. (2016). *Technology implementation in second language teaching and translation studies: New tools, new approaches*. Singapore: Springer. doi:10.1007/978-981-10-0572-5

Carr, N. (2011). *The shallows: What the internet is doing to our brains*. New York, NY: W.W. Norton.

Castrillo, M. C., Martín-Monje, E., & Bárcena, E. (2014). Mobile-based chatting for meaning negotiation in foreign language learning. In *Proceedings of the 10th International Conference on Mobile Learning*. (pp. 49-59). International Association for Development of the Information Society.

Castrillo, M. D., Martín-Monje, E., & Bárcena, E. (2014). *Mobile-based chatting for meaning negotiation in foreign language learning*. International Association for the Development of the Information Society.

Caton, H., & Greenhill, D. (2014). Rewards and penalties: A gamification approach for increasing attendance and engagement in an undergraduate computing module. *International Journal of Game-Based Learning*, *4*(3), 1–12. doi:10.4018/ijgbl.2014070101

Caulfield, M., Collier, A., & Halawa, S. (2013, Oct. 6). Rethinking online community in MOOCs used for blended learning [Web log post]. Retrieved from http://www.educause.edu/ero/article/rethinking-online-community-moocs-used-blended-learning

Cavus, N., & Ibrahim, D. (2009). m-Learning: An experiment in using SMS to support learning new English language words. *British Journal of Educational Technology*, *40*(1), 78–91. doi:10.1111/j.1467-8535.2007.00801.x

Caws, C., & Hamel, M. (2016). *Language-learner computer interactions: Theory, methodology and CALL applications*. Amsterdam, The Netherlands: John Benjamins. doi:10.1075/lsse.2

Çelik, Ö., & Yavuz, F. (2018). An extensive review of literature on teaching vocabulary through mobile applications. *Bilecik Şeyh Edebali University Journal of Social Sciences Institute*, 3(1), 56-91. Retrieved from https://www.researchgate.net/publication/325540454_An_Extensive_Review_Of_Literature_On_Teaching_Vocabulary_Through_Mobile_Applications

Center for Self-Determination Theory. (2019). Intrinsic Motivation Inventory (IMI). Retrieved from https://selfdeterminationtheory.org/intrinsic-motivation-inventory/

Çetinkaya, L., & Sütçü, S. S. (2018). The effects of Facebook and WhatsApp on success in English vocabulary instruction. *Journal of Computer Assisted Learning*, *34*(5), 504–514. doi:10.1111/jcal.12255

Ch'ng, L. K., & Samsudin, Z. (2013). Integration of mobile devices into ubiquitous learning by the 21st century teenagers. *Education*, *3*(6), 362–374. doi:10.5923/j.edu.20130306.12

Chacón-Beltrán, R. (2014). Massive online open courses and language learning: The case for a beginners' English course. *Procedia: Social and Behavioral Sciences*, *141*, 242–246. doi:10.1016/j.sbspro.2014.05.042

Chapelle, C. (2001). *Computer applications in second language acquisition: Foundations for teaching, testing, and research*. Cambridge, UK: Cambridge University Press. doi:10.1017/CBO9781139524681

Chapelle, C. A. (2003). *English language learning and technology*. Amsterdam, The Netherlands: John Benjamins. doi:10.1075/lllt.7

Chapelle, C. A. (2009). The relationship between second language acquisition theory and computer-assisted language learning. *Modern Language Journal*, *93*, 741–753. doi:10.1111/j.1540-4781.2009.00970.x

Chen Hsieh, J. S., Wu, W. C. V., & Marek, M. W. (2017). Using the flipped classroom to enhance EFL learning. *Computer Assisted Language Learning*, *30*(1-2), 1–21. doi:10.1080/09588221.2015.1111910

Cheng, Y. (2015). Towards an understanding of the factors affecting m-learning acceptance: Roles of technological characteristics and compatibility. *Asia Pacific Management Review*, *20*(3), 109–119. doi:10.1016/j.apmrv.2014.12.011

Chen, I. J., & Chang, C. C. (2011). Content presentation modes in mobile language listening tasks: English proficiency as a moderator. *Computer Assisted Language Learning*, *24*(5), 451–470. doi:10.1080/09588221.2011.577749

Chen, P. S. D., Lambert, A. D., & Guidry, K. R. (2010). Engaging online learners: The impact of web-based learning technology on college student engagement. *Computers & Education*, *54*(5), 1222–1232. doi:10.1016/j.compedu.2009.11.008

Chen, X.-B. (2013). Tablets for informal language learning: Student usage and attitudes. *Language Learning & Technology*, *17*(1), 20–36.

Chen, Y. H., & Chen, P. J. (2015). MOOC study group: Facilitation strategies, influential factors, and student perceived gains. *Computers & Education*, *86*, 55–70. doi:10.1016/j.compedu.2015.03.008

Chinnery, G. (2006). Going to the MALL: Mobile Assisted Language Learning. *Language Learning & Technology*, *10*(1), 9-16. Retrieved from http://llt.msu.edu/vol10num1/emerging/default.html

Cho, K., Lee, S., Joo, M.-H., & Becker, B. (2018). The effects of using mobile devices on student achievement in language learning: A meta-analysis. *Education in Science*, *8*(3), 1-16. doi:10.3390/educsci8030105

Christians, G. (2018). *The Origins and Future of Gamification.* (Senior Theses). 254. Retrieved from https://scholarcommons.sc.edu/senior_theses/254

Chun, D. M. (2011). Developing intercultural communicative competence through online exchanges. *CALICO Journal*, *28*(2), 392–419. doi:10.11139/cj.28.2.392-419

Çiftçi, E. Y., & Savaş, P. (2018). [2017]). The role of telecollaboration in language and intercultural learning: A synthesis of studies published between 2010 and 2015. *ReCALL*, *30*(3), 278–298. doi:10.1017/S0958344017000313

Clavel-Arroitia, B., & Pennock-Speck, B. (2015). Telecollaboration in a secondary school context: Negotiation of meaning in English as lingua Franca/Spanish tandem interactions. *Revista d'Innovació Educativa*, *15*, 74–78.

Clément, R., Baker, S. C., & MacIntyre, P. D. (2003). Willingness to communicate in a language: The effects of context, norms, and vitality. *Journal of Language and Social Psychology*, *22*, 190–209. doi:10.1177/0261927X03022002003

Clement, R., Dornyei, Z., & Noels, K. A. (1994). Motivation, self-confidence, and group cohesion in the foreign language classroom. *Language Learning*, *44*(3), 417–448. doi:10.1111/j.1467-1770.1994.tb01113.x

Cohen, L., Manion, L., & Morrison, K. (2013). *Research methods in education.* Retrieved from http://books.google.com/books?hl=en&lr=&id=mLh0Oza3V1IC&oi=fnd&pg=PR3&dq=info:YWZJAiLelagJ:scholar.google.com&ots=SOBLHpp9pm&sig=5PN7dflZLpa6oikITWQLcXn-Q5w

Cohen, L., Manion, L., & Morrison, K. (2013). *Research Methods in Education*. London, UK: Routledge. doi:10.4324/9780203720967

Colpaert, J. & Spruyt, E. (2018). Toward a task design model for mental acceptance and motivation: A transdisciplinary approach. In *Proceedings of the 3rd International Conference on Language Education and Testing - What a Feeling: Language Education and Emotions,* (pp. 138-145). Antwerp, Belgium. Retrieved from https://www.uantwerpen.be/en/conferences/language-education-and-emotions

Comer, D. R., Lenaghan, J. A., & Sengupta, K. (2015). Factors that affect students' capacity to fulfill the role of online learner. *Journal of Education for Business*, *90*(3), 145–155. doi:10.1080/08832323.2015.1007906

Conde Gafaro, B. (2019). Repurposing MOOCs for self-regulated language learning in an English for academic purposes course. In A. Beaven, A. Comas-Quinn, & B. Sawhill (Eds.), *Case studies of openness in and beyond the language classroom. © Research-publishing.net.* doi:10.14705/rpnet.2019.37.970

Cong-Lem, N. (2018). A review of research in mobile-assisted collaborative language learning. *VNU Journal of Foreign Studies*, 34(4), 71–82. Retrieved from https://www.researchgate.net/publication/327315618_A_REVIEW_OF_RESEARCH_IN_MOBILE_ASSISTED_COLLABORATIVE_LANGUAGE_LEARNING

Coniam, D. (2013). Computerized dictation for assessing listening proficiency. *CALICO Journal, 13*(2&3).

Conijn, R., Van, B., & Cuijpers, P. (2018). Predicting student performance in a blended MOOC. *Journal of Computer Assisted Learning, 34*(5), 615–628. doi:10.1111/jcal.12270

Cormier, D. (2008, Oct. 2). *The CCK08 MOOC – Connectivism course, 1/4 way.* [Web log post]. Retrieved from http://davecormier.com/edblog/2008/10/02/the-cck08-mooc-connectivism-course-14-way/

Cormier, D. (2010, Dec. 8). *What is a MOOC?* Retrieved from https://www.youtube.com/watch?v=eW3gMGqcZQc

Correa, T., Hinsley, A. W., & de Zúñiga, H. G. (2010). Who interacts on the web? The intersection of users' personality and social media use. *Computers in Human Behavior, 26*(2), 247–253. doi:10.1016/j.chb.2009.09.003

Cortesi, S., Haduong, P., Gasser, U., Aricak, O., Saldaña, M., & Lerner, Z. (2014, Jan. 15). Youth perspectives on tech in schools: From mobile devices to restrictions and monitoring. *Berkman Center Research Publication*, (2014-3).

Costa, C., Tyner, K., Henriques, S., & Sousa, C. (2017). *Games for media and information literacy - developing media and information literacy skills in children through digital games creation.* 2997–3004. doi:10.21125/edulearn.2017.1627

Council of Europe. (2018). *Common European framework of reference for languages: Learning, teaching, assessment: Companion volume with new descriptors.* Retrieved from https://rm.coe.int/cefr-companion-volume-with-new-descriptors-2018/1680787989

Council of Europe. (2018). *Global scale - Table 1 (CEFR 3.3): Common reference levels.* Retrieved from https://www.coe.int/en/web/common-european-framework-reference-languages/table-1-cefr-3.3-common-reference-levels-global-scale

Council of Europe. (2018). Qualitative aspects of spoken language use - Table 3 (CEFR 3.3): Common reference levels. Retrieved from https://www.coe.int/en/web/common-european-framework-reference-languages/table-3-cefr-3.3-common-reference-levels-qualitative-aspects-of-spoken-language-use

Counts, S., & Fisher, K. E. (2010). Mobile social networking as information ground: A case study. *Library & Information Science Research, 32*(2), 98–115. doi:10.1016/j.lisr.2009.10.003

CSO. (2016) Census of Population 2016 – Profile 10 Education, Skills and the Irish Language. Retrieved from https://www.cso.ie/en/releasesandpublications/ep/p-cp10esil/p10esil/ilg/

Cucchiarini, C., Nejjari, W., & Strik, H. (2012). My pronunciation coach: Improving English pronunciation with an automatic coach that listens. *Language Learning in Higher Education, 1*(2), 365–376. doi:10.1515/cercles-2011-0024

Dallimore, E. J., Hertenstein, J. H., & Platt, M. B. (2017, March 27). How do students learn from participation in class discussion? *Faculty Focus: Higher Ed Teaching & Learning.*

Damron, J. & Dobberfuhl-Quinlan, J. (2018, July 4-6). An exploratory study of student learning. In *CALL Proceedings.* Universiteit Antwerpen.

Daniel, J. (2012). Making sense of MOOCs: Musings in a maze of myth, paradox and possibility. *Journal of Interactive Media in Education, 18*, 1–20. Retrieved from http://wwwjime.open.ac.uk/jime/article/view/2012-18

Darmi, R., & Albion, P. (2014). A review of integrating mobile phones for language learning. In *Proceedings of 10th International Conference Mobile Learning* (pp. 93–100). International Association for Development of the Information Society; Retrieved from http://files.eric.ed.gov/fulltext/ED557201.pdf

Davis, A. (2003). *The native speaker: Myth and reality* (2nd ed.). Tonawanda, NY: Multilingual Matters. doi:10.21832/9781853596247

Davis, F. D. (1989). Perceived usefulness, perceived ease of use and user acceptance of information technology. *Management Information Systems Quarterly*, *13*(3), 319–339. doi:10.2307/249008

De Byl, P. (2013). Factors at play in tertiary curriculum gamification. *International Journal of Game-Based Learning*, *3*(2), 1–21. doi:10.4018/ijgbl.2013040101

de Jong, N. H. & Wempe, T. (2008). Praat script syllable nuclei [Programming script]. Retrieved from https://sites.google.com/site/speechrate/Home/praat-script-syllable-nuclei

de Jong, N. H. (2016). Predicting pauses in L1 and L2 speech: The effects of utterance boundaries and word frequency. *International Review of Applied Linguistics in Language Teaching*, *54*(2), 113–132. doi:10.1515/iral-2016-9993

de Jong, N. H., & Mora, J. C. (2017). Does having good articulatory skills lead to more fluent speech in first and second language? *Studies in Second Language Acquisition*, 1–13. doi:10.1017/S0272263117000389

de Waard, I. & Demeulenaere, K. (2017). The MOOC-CLIL project: Using MOOCs to increase language, and social and online learning skills for 5th grade K-12 students. In Q. Kan & S. Bax (Eds.), Beyond the language classroom: Researching MOOCs and other innovations (pp. 29-42). Research-publishing.net. doi:10.14705/ rpnet.2017.mooc2016.669

de Waard, I. (2015). MOOC factors influencing teachers in formal education. *Revista Mexicana de Bachillerato a distancia, 7*(13).

de Waard, I., Kukulska-Hulme, A., & Sharples, M. (2015). Self-directed learning in trial FutureLearn courses. [University Catholique de Louvain, Belgium.]. In *Proceedings of the European Stakeholder Summit EMOOCS, 2015*, 234–243.

De Zúñiga, H. G., Diehl, T., Huber, B., & Liu, J. (2017). Personality traits and social media use in 20 Countries: How personality relates to frequency of social media use, social media news use, and social media use for social interaction. *Cyberpsychology, Behavior, and Social Networking*, *20*(9), 540–552. doi:10.1089/cyber.2017.0295 PMID:28922034

Dearden, J. (2015). *English as a medium of instruction–A growing global phenomenon*. London, UK: British Council.

Deci, E. L., & Ryan, R. M. (1985). *Intrinsic motivation and self-determination in human behaviour*. New York, NY: Plenum. doi:10.1007/978-1-4899-2271-7

Deci, E. L., & Ryan, R. M. (2000). The “what” and “why” of goal pursuits: Human needs and the self-determination of behavior. *Psychological Inquiry*, *11*(4), 227–268. doi:10.1207/S15327965PLI1104_01

Decker, G. L. (2014). MOOCology 1.0. In S. D. Krause & C. Lowe (Eds.), *Invasion of the MOOCs: The promises and perils of massive open online courses* (pp. 3–13). Anderson, SC: Parlor Press.

Derakhshan, A., & Khodabakhshzadeh, H. (2011). Why CALL not MALL: An in-depth review of text-message vocabulary learning. *Theory and Practice in Language Studies*, *1*(9), 1150-1159. doi:10.4304/tpls.1.9.1150-1159

Derwing, T. M. (2017). L2 Fluency development. In S. Loewen & M. Masatoshi (Eds.), *The Routledge handbook of instructed second language acquisition* (pp. 246–259). New York, NY: Routledge. doi:10.4324/9781315676968-14

Deterding, S., Dixon, D., Khaled, R., & Nacke, L. (2011). From game design elements to gamefulness: Defining "gamification.". In A. Lugmayr, H. Franssila, C. Safran, & I. Hammouda (Eds.), *MindTrek 2011* (pp. 9–15). New York, NY: ACM. doi:10.1145/2181037.2181040

Dewey, J. (1997). *Experience and education.* New York, NY: Simon & Schuster.

Di Serio, Á., Ibáñez, M. B., & Kloos, C. D. (2013). Impact of an augmented reality system on students' motivation for a visual art course. *Computers & Education, 68*, 586–596. doi:10.1016/j.compedu.2012.03.002

Diaz, J. (2014). *El WhatsApp como herramienta de intervencion didáctica para fomentar el aprendizaje cooperativo* [WhatsApp as a tool of educational intervention to foster cooperative learning]. (Unpublished Paper presented at X Jornadas De Material Didáctica y Experiencias Innovadoras En Educación Superior), Buenos Aires, Argentina. Retrieved from http://eprints.rclis.org/23597/

Dicheva, D., Dichev, C., Agre, G., & Angelova, G. (2015). Gamification in education: A systematic mapping study. *Journal of Educational Technology & Society, 18*, 75–88.

Dickey, M. D. (2011). Murder on Grimm Isle: The impact of game narrative design in an educational game-based learning environment. *British Journal of Educational Technology, 42*(3), 456–469. doi:10.1111/j.1467-8535.2009.01032.x

digi.komp: Informatik 5. Klasse [digi.komp: ICT 5th form]. (n.d.). Retrieved from https://digikomp.at/?id=585

Dilans, G. (2010). Corrective feedback and L2 vocabulary development: Prompts and recasts in the adult ESL classroom. *Canadian Modern Language Review, 66*(6), 787–815. doi:10.3138/cmlr.66.6.787

Dixon, E. & Fuchs, C. (2015). Face to face, online, or MOOC–How the format impacts content, objectives, assignments, and assessments. In E. Dixon & M. Thomas (Eds.), Researching language student interaction online: From social media to MOOCs. (pp. 89–105). San Marcos, TX: CALICO monograph.

Doiz, A., Lasagabaster, D., & Sierra, J. M. (Eds.). (2012). *English-medium instruction at universities: Global challenges.* Bristol, UK: Multilingual Matters. doi:10.21832/9781847698162

Doleck, T., Bazelais, P., & Lemay, D. J. (2017). Need for self-expression on Instagram: A technology acceptance perspective. In *3rd International Conference on Computational Intelligence & Communication Technology (CICT)*, 1-3. IEEE. Retrieved from https://ieeexplore.ieee.org/stamp/stamp.jsp?arnumber=7977305

Domínguez, A., Saenz-de-Navarrete, J., De-Marcos, L., Fernández-Sanz, L., Pagés, C., & Martínez Herráiz, J.-J. (2013). Gamifying learning experiences: Practical implications and outcomes. *Computers & Education, 63*, 380–392. doi:10.1016/j.compedu.2012.12.020

Donmus, V. (2010). The use of social networks in educational computer-game based foreign language learning. *Procedia: Social and Behavioral Sciences, 9*, 1497–1503. doi:10.1016/j.sbspro.2010.12.355

Dooly, M. (Ed.). (2008). *Telecollaborative language learning: A guidebook to moderating intercultural collaboration online.* Peter Lang.

Dooly, M. A., & O'Dowd, R. (2018). Telecollaboration in the foreign language classroom: A review of its origins and its application to language teaching Practices. In M. A. Dooly & R. O'Dowd (Eds.), *In this together: Teachers' experiences with transnational, telecollaborative language learning projects* (pp. 11–34). New York, NY: Peter Lang. doi:10.3726/b14311

Dörnyei, Z. & Ushioda, E. (Eds.) (2009). *Motivation, language identity and the L2 self.* Bristol, UK: Multilingual Matters.

Dörnyei, Z. (2009). The L2 motivational self system. In Z. Dörnyei & E. Ushioda, (Eds.), Motivation, language identity and the L2 self (pp. 9-42). Bristol, UK: Multilingual Matters. doi:10.21832/9781847691293-003

Dörnyei, Z. (2005). *The psychology of the language learner individual differences in second language acquisition.* Mahwar, NJ: Lawrence Erlbaum.

Douglas, Y., & Hargadon, A. (2001). The pleasure of immersion and engagement schemas, scripts and the fifth business. *Digital Creativity*, *12*(3), 153–166. doi:10.1076/digc.12.3.153.3231

Downes, S. (2009, April 25). New technology supporting informal learning [Web log post]. Retrieved from https://halfanhour.blogspot.co.uk/2009/04/new-technology-supporting-informal.html

Dressman, M. (2010). *Let's poem: The essential guide to teaching poetry in a high-stakes, multimodal world.* New York, NY: Teachers College Press.

Ducate, L., & Arnold, N. (2006). *Calling on CALL: From theory and research to new directions in foreign language teaching*. TX: CALICO.

Dudeney, G., Hockly, N., & Pegrum, M. (2013). *Digital literacies: Research and resources in language teaching*. Harlow, UK: Pearson.

Dudley-Evans, T. (1997). Five questions for LSP teacher training. In R. Howard & G. Brown (Eds.), *Teacher education for languages for specific purposes* (pp. 58–67). Clevedon, UK: Multilingual Matters.

Dugas, C. A. (2005). *Adopter characteristics and teaching styles of faculty adopters and nonadopters of a course management system.* (Unpublished doctoral dissertation), Indiana State University, Terre Haute, IN.

Duman, D., Orhon, G., & Gedik, N. (2015). Research trends in mobile assisted language learning from 2000 to 2012. *ReCALL, 27*(2), 197-216. doi:10.1017/S0958344014000287

Dunleavy, M., Dede, C., & Mitchell, R. (2009). Affordances and limitations of immersive participatory augmented reality simulations for teaching and learning. *Journal of Science Education and Technology*, *18*(1), 7–22. doi:10.100710956-008-9119-1

Dunn, L. (2013). Teaching in higher education: can social media enhance the learning experience? *6th Annual conference of teaching and learning*, University of Glasgow, UK.

Dutoit, T. (1997). *An introduction to text-to-speech synthesis* (Vol. 3). Springer Science & Business Media. doi:10.1007/978-94-011-5730-8

E. U. Science Hub. (2018). *Digital competence framework for educators (DigCompEdu)*. Retrieved from https://ec.europa.eu/jrc/en/digcompedu

Eccles, J., & Wang, M. T. (2012). Part I commentary: So what is student engagement anyway? In *Handbook of research on student engagement* (pp. 133–145). Boston, MA: Springer; doi:10.1007/978-1-4614-2018-7_6

Edmunds, R., Thorpe, M., & Conole, G. (2012). Student attitudes towards and use of ICT in course study, work and social activity: A technology acceptance model approach. *British Journal of Educational Technology*, *43*(1), 71–84. doi:10.1111/j.1467-8535.2010.01142.x

edX Help Center. (2018a). What is self-paced or instructor-paced? Which is my course? Retrieved from https://support.edx.org/hc/en-us/articles/115011202847-What-is-self-paced-or-instructor-paced-Which-is-my-course-

edX Help Center. (2018b). What time is the class? Retrieved from https://support.edx.org/hc/en-us/articles/206503538-What-time-is-the-class-

Egbert, J., Paulus, T., & Nakamichi, Y. (2002). The impact of CALL instruction on classroom computer use: A foundation for rethinking technology in teacher education. Language Learning & Technology, *6*(3), 108-126. Retrieved from https://eric.ed.gov/?id=EJ653846

Elgort, I. (2018). Technology-mediated second language vocabulary development: A review of trends in research methodology. *CALICO Journal*, *35*(1), 1-29. Retrieved from https://journals.equinoxpub.com/index.php/CALICO/article/view/34554/pdf

El-Hariri, Y. (2016). Learner perspectives on task design for oral–visual eTandem language learning. *Innovation in Language Learning and Teaching*, *10*(1), 49–72. doi:10.1080/17501229.2016.1138578

Ellison, N., & Boyd, D. M. (2008). Social network sites: Definition, history, and scholarship. *Journal of Computer-Mediated Communication*, *13*, 210–230.

Ellis, R. (1994). *The study of second language acquisition.* Oxford, UK: Oxford University Press.

Ellis, R. (2003). *Task-based language learning and teaching.* Oxford, U.K.: Oxford University Press.

Ellis, R. (2010). Second language acquisition, teacher education and language pedagogy. *Language Teaching*, *43*(2), 182–201. doi:10.1017/S0261444809990139

Ellis, R., & Sheen, Y. (2006). Re-examining the role of recasts in L2 acquisition. *Studies in Second Language Acquisition*, *28*(04), 575–600. doi:10.1017/S027226310606027X

Ellis, R., & Shintani, N. (2014). *Exploring language pedagogy through second language acquisition research.* London, UK: Routledge.

Erben, T., Ban, R., & Castañeda, M. (2009). *Teaching English language learners through technology.* New York, NY: Routledge.

Erçetin, G., Lee, S., Kadel, R., Madden, A., & Gazi, Y. (2018, July 4). How can learner analytics data inform language MOOC design? In *Proceedings of the 19th International CALL Conference*, Bruges, Belgium. Academic Press.

Erlam, R. (2008). What do you researchers know about language teaching? Bridging the gap between SLA research and language pedagogy. *International Journal of Innovation in Language Learning and Teaching*, *2*(3), 253–267. doi:10.1080/17501220802158958

European Union. (2017). *Being young in Europe today - digital world.* Retrieved from http://ec.europa.eu/eurostat/statistics-explained/index.php?title=Being_young_in_Europe_today_-_digital_world

European Union. (2018). *DigiComp: The European Digital Competence Framework.*

Eysenck, H. J., & Eysenck, M. W. (1985). *Personality and individual differences.* New York, NY: Plenum. doi:10.1007/978-1-4613-2413-3

Fanfarelli, J. R. & McDaniel, R. (2017). Exploring digital badges in university courses: Relationships between quantity, engagement, and performance. *Online Learning, 21.*

Farr, F., & Murray, L. (Eds.). (2016). *The Routledge handbook of language learning and technology.* Routledge. doi:10.4324/9781315657899

Fattah, S. (2015). The effectiveness of using WhatsApp messenger as one of mobile learning techniques to develop students' writing skills. *Journal of Education and Practice*, *6*(32), 115–127.

Fazal, M. B. (1996). *Effectiveness and costs of computer-based instruction in higher education and defense training: A meta-analysis* (Doctoral dissertation). University of Southern California, CA.

Fendler, R. J., Ruff, C., & Shrikhande, M. (2016). Evaluating characteristics of top and bottom performance: Online versus in-class. *American Journal of Distance Education, 30*(2), 109–120. doi:10.1080/08923647.2016.1153350

Ferguson, R., & Sharples, M. (2014). Innovative pedagogy at massive scale: Teaching and learning in MOOCs. In Proceedings of 9th European Conference on Technology Enhanced Learning - EC-TEL 2014 (pp. 98–111). Graz, Austria: Springer.

Ferguson, R., & Shum, S. B. (2012). Towards a social learning space for open educational resources. In A. Okada, T. Connolly, & P. Scott (Eds.), Collaborative Learning 2.0: Open educational resources (pp. 309–327). Hershey, PA: IGI Global. doi:10.4018/978-1-4666-0300-4.ch017

Ferguson, R., Clow, D., Beale, R., Cooper, A. J., Morris, N., Bayne, S., & Woodgate, A. (2015). Moving through MOOCS: Pedagogy, learning design and patterns of engagement. In G. Conole, T. Klobucar, C. Rensing, J. Konert, & E. Lavoué (Eds.), *Design for teaching and learning in a networked world* (pp. 70–84). Lecture Notes in Computer Science Springer International. doi:10.1007/978-3-319-24258-3_6

Figueroa, J. (2015). Using gamification to enhance second language learning. *Digital Education Review, 27*, 32–54.

Fitzsimmons, G. (2016). *The influence of hyperlinks on reading on the web: An empirical approach* (Doctoral thesis). University of Southampton, UK.

Fleischer, H. (2012). What is our current understanding of one-to-one computer projects: A systematic narrative research review. *Educational Research Review, 7*(2), 107–122. doi:10.1016/j.edurev.2011.11.004

Flowerdew, J., & Miller, L. (2005). *Second language listening theory and practice*. New York, NY: Cambridge University Press. doi:10.1017/CBO9780511667244

Foster, P., & Snyder Ohta, A. (2005). Negotiation for meaning and peer assistance in second language classrooms. *Applied Linguistics, 26*(3), 402–430. doi:10.1093/applin/ami014

Fraillon, J., Ainley, J., Schulz, W., Friedman, T., & Gebhardt, E. (2014). *Preparing for life in a digital age*. Cham, Switzerland: Springer International Publishing. doi:10.1007/978-3-319-14222-7

Fredricks, J. A., Blumenfeld, P. C., & Paris, A. H. (2004). School engagement: Potential of the concept, state of the evidence. *Review of Educational Research, 74*(1), 59–109. doi:10.3102/00346543074001059

Freed, B. F. (1995). What makes us think that students who study abroad become fluent? In B. Freed (Ed.), *Second language acquisition in a study abroad context* (pp. 123–148). Amsterdam, The Netherlands: John Benjamins. doi:10.1075ibil.9.09fre

Freed, B. F., Segalowitz, N., & Dewey, D. P. (2004). Context of learning and second language fluency in French: Comparing regular classroom, study abroad, and intensive domestic immersion program. *Studies in Second Language Acquisition, 26*(02), 275–301. doi:10.1017/S0272263104262064

Frydenberg, M., & Andone, D. (2016). Creating micro-videos to demonstrate technology learning and digital literacy. *Interactive Technology and Smart Education, 13*(4), 261–273. doi:10.1108/ITSE-09-2016-0030

Fu, D., & Matoush, M. M. (2015). *Focus on literacy*. Oxford, UK: Oxford University Press.

Fukada, A., Hirotani, M., & Matsumoto, K. (2019). *Tools for generating objective measures for oral proficiency research* [Computer software]. Retrieved from http://tell.cla.purdue.edu/CAF-calculator/

Fukada, A., Hirotani, M., Matsumoto, K., & Huston, S. M. (2015a). *Tools for generating objective measures for oral proficiency research.* Retrieved from http://tell.cla.purdue.edu/fluency-calculator/

Fukada, A., Hirotani, M., Matsumoto, K., & Huston, S. M. (2015b). *Fluency Calculator* [Computer software]. Retrieved from http://tell.cla.purdue.edu/fluency-calculator/

Gamper, J., & Knapp, J. (2002). A review of intelligent CALL systems. *Computer Assisted Language Learning, 15*(4), 329–342. doi:10.1076/call.15.4.329.8270

Gardner, R. C. (2001). Integrative motivation and second language acquisition. In Z. Dörnyei & R. Schmidt (Eds.), Motivation and Second Language Acquisition, 6, pp. 1-19. Honolulu, HI: University of Hawaii Press.

Gardner, R. C. (1985). *Social psychology and second language learning: The role of attitudes and motivation.* London, UK: Edward Arnold.

Gardner, R. C. (2001a). Language learning motivation: The student, the teacher, and the researcher. *Texas Papers in Foreign Language Education, 6*(1), 1–18.

Gardner, R. C., & Lambert, W. E. (1959). Motivational variables in second language acquisition. *Canadian Journal of Psychology*, 13. PMID:13855818

Gardner, R. C., & Lambert, W. E. (1972). *Attitudes and motivation in second language learning.* Rowley, MA: Newbury House.

Gardner, R. C., Masgoret, A.-M., Tennant, J., & Mihic, L. (2004). Integrative motivation: Changes during a year-long intermediate-level language course. *Language Learning, 54*(1), 1–34. doi:10.1111/j.1467-9922.2004.00247.x

Gardner, R. C., & Tremblay, P. F. (1994). On motivation: Measurement and conceptual considerations. *Modern Language Journal, 78*(4), 524–527.

Garland, C. M. (2015). Gamification and implications for second language education: A meta analysis. *Culminating Projects in English.* Paper 40.

Gass, S. (2003). Input and interaction. In C. Doughty & M. Long (Eds.), *The handbook of second language acquisition* (pp. 224–255). Oxford, UK: Blackwell. doi:10.1002/9780470756492.ch9

Gass, S. M. (2017). *Input, interaction, and the second language learner* (2nd ed.). Routledge Linguistics Classics Edition. doi:10.4324/9781315173252

Gass, S., & Mackey, A. (2006). Input, interaction, and output: An overview. *AILA Review, 19*, 3–17. doi:10.1075/aila.19.03gas

Gee, J. P., & Hayes, E. R. (2011). *Language and learning in the digital age.* Taylor and Francis; doi:10.4324/9780203830918

Gerrig, R. J. (1993). *Experiencing narrative worlds: On the psychological activities of reading.* New Haven, CT: Yale University Press.

Gershenfeld, A. (2011). From player to designer: Engaging and empowering youth through making video games. *Knowledge Quest, 40*(1), 7.

Gesellschaft für Informatik. (2016). *Dagstuhl-Erklärung: Bildung in der digitalen vernetzten Welt* [Dagstuhl Declaration: Education in a digital and interconnected society]. Retrieved from https://www.gi.de/aktuelles/meldungen/detailansicht/article/dagstuhl-erklaerung-bildung-in-der-digitalen-vernetzten-welt.html

Gettman, H. J., & Cortijo, V. (2015). "Leave me and my Facebook alone!" Understanding college students' relationship with Facebook and its use for academic purposes. *International Journal for the Scholarship of Teaching and Learning*, *9*(1), 8.

Gikas, J., & Grant, M. M. (2013). Mobile computing devices in higher education: Student perspectives on learning with cellphones, smartphones & social media. *The Internet and Higher Education*, *19*, 18–26. doi:10.1016/j.iheduc.2013.06.002

Gillen, J. (2014). *Digital literacies*. London, UK: Routledge. doi:10.4324/9781315813530

Gimeno-Sanz, A. (2018). Learner expectations and satisfaction in a US-Spain intercultural telecollaboration project. *Bellaterra Journal of Teaching & Learning Language & Literature*, *11*(3), 5–38. doi:10.5565/rev/jtl3.776

Ginther, A., Dimova, S., & Yang, R. (2010). Conceptual and empirical relationships between temporal measures of fluency and oral English proficiency with implications for automated scoring. *Language Testing*, *27*(3), 379–399. doi:10.1177/0265532210364407

Glaser, B. G., & Strauss, A. L. (1967). *The discovery of grounded theory: Strategies for qualitative research*. New Brunswick, NJ: Transaction Publishers. [2006 reprint]

Glenn, D. (2007). Why cramming doesn't work [1]. *The Chronicle of Higher Education*, *53*(40).

Global Digital Statistics. (2019). Retrieved from https://www.smartinsights.com/social-media-marketing/social-media-strategy/new-global-social-media-research/

Goertler, S., & Gacs, A. (2018). Assessment in online German: Assessment methods and results, linguistics and language behavior abstracts (LLBA). *Die Unterrichtspraxis*, *51*(2), 1–21. doi:10.1111/tger.12071

Göksün, O., & Gürsoy, G. (2019). Comparing success and engagement in gamified learning experiences via Kahoot and Quizizz. *Computers & Education*, *135*, 15–29. doi:10.1016/j.compedu.2019.02.015

Goldberg, L. R. (1992). The development of markers for the Big-Five factor structure. *Psychological Assessment*, *4*(1), 26–42. doi:10.1037/1040-3590.4.1.26

González-Lloret, M. (2017). Technology and Task-Based Language Teaching. In S. L. Thorne, and May, Stephen (Ed.), Language, Education and Technology (pp. 193-205). Berlin: Springer International Publishing AG.

González-Lloret, M. (2016). Technology and task-based language teaching. In S. Thorne & S. May (Eds.), *Language, education and technology. Encyclopedia of language and education* (3rd ed.). Cham, Switzerland: Springer. doi:10.1007/978-3-319-02328-1_16-1

Greenhow, C., Robelia, E., & Hughes, J. (2009). Web 2.0 and classroom research: What path should we take now? *Educational Researcher*, *38*, 246–259. doi:10.3102/0013189X09336671

Green, M. C., & Brock, T. C. (2000). The role of transportation in the persuasiveness of public narratives. *Journal of Personality and Social Psychology*, *79*(5), 701–721. doi:10.1037/0022-3514.79.5.701 PMID:11079236

Grgurović, M., Chapelle, C. A., & Shelley, M. C. (2013). A meta-analysis of effectiveness studies on computer-technology supported language learning. *ReCALL*, *25*(2), 165–198.

Griffith, R. (1991). Personality and second language learning: theory, research and practice. In E. Sadtonom (Ed.), *Language acquisition and the second/foreign language classroom*. Singapore: SEAMEO.

Griffith, S., & Liyanage, L. (2008). An introduction to the potential of social networking sites in education. In I. Olney, G. Lefoe, J. Mantei, & J. Herrington (Eds.), *Proceedings of the Second Emerging Technologies Conference 2008* (pp. 76–81). Australia: University of Wollongong. Retrieved from http://ro.uow.edu.au/etc08/9

Gupta, P. & Schulze, M. (2012). Human language technologies (HLT). Module 3.5. In G. Davies (Ed.), *Information and Communications Technology for Language Teachers (ICT4LT),* Slough, Thames Valley University [Online]. Retrieved from http://www.ict4lt.org/en/en_mod3-5.htm

Gutiérrez, A. (2014). La ficción progresiva de Leonardo Valencia, o el autor somos todos. *Bulletin of Spanish Studies, 91*(4), 575–594. doi:10.1080/14753820.2014.886900

Gutiérrez-Colón Plana, M., Gibert-Escofet, M. I., Triana Figueras, I., Gimeno, A., Appel, C., & Hopkins, J. (2013). *Improving learners' reading skills through instant short messages: A sample study using WhatsApp. Global perspectives on Computer-Assisted Language Learning* (pp. 80–84). Glasgow, UK: WorldCALL.

Gutiérrez-Colon Plana, M., Gimeno-Sanz, A., Appel, C., & Hopkins, J. (2015). Improving learners' reading skills through instant short messages: A sample study using WhatsApp. In A. Gimeno-Sanz, M. Levy, D. Barr, & F. Blin (Eds.), *WorldCALL: Sustainability and computer-assisted language learning* (pp. 266–281). London, UK: Bloomsbury.

Gutiérrez-Rojas, I., Alario-Hoyos, C., Pérez-Sanagustín, M., Leony, D., & Delgado-Kloos, C. (2014). Scaffolding self-learning in MOOCs. [Lausanne, Switzerland: Ecole Polytechnique Federale de Lausanne.]. *Proceedings of the European MOOC Stakeholder Summit, 2014*, 43–49.

Haaranen, L., Ihantola, P., Hakulinen, L., & Korhonen, A. (2014). How (not) to introduce badges to online exercises. In *Proceedings of the 45th ACM technical symposium on Computer science education* (pp. 33–38). New York, NY: ACM Press. 10.1145/2538862.2538921

Haines, K. J. (2015). Learning to identify and actualize affordances in a new tool. *Language Learning & Technology, 19*(1), 165–180.

Hamid, S., Waycott, J., Kurnia, S., & Chang, S. (2015). Understanding students' perceptions of the benefits of online social networking use for teaching and learning. *Internet and Higher Education, 26*, 1–9. doi:10.1016/j.iheduc.2015.02.004

Handayani, F. (2016). Instagram as a teaching tool? Really? In *Proceedings of the Fourth International Seminar on English Language and Teaching (ISELT-4)*, 320-327. Atlantis Press.

Handayani, F. (2017). Students' attitudes towards using Instagram in teaching writing. *Journal of Educative Studies, 2*(1), 23–29.

Handley, Z. (2009). Is text-to-speech synthesis ready for use in computer-assisted language learning? *Speech Communication, 51*(10), 906–919. doi:10.1016/j.specom.2008.12.004

Hanus, M. D., & Fox, J. (2015). Assessing the effects of gamification in the classroom: A longitudinal study on intrinsic motivation, social comparison, satisfaction, effort, and academic performance. *Computers & Education, 80*, 152–161. doi:10.1016/j.compedu.2014.08.019

Harendita, M. E. (2013). Why resist? A closer look at Indonesian teachers' resistance to ICT. *International Journal of Indonesian Studies, 1*(1), 41–57.

Harlow, L. L., & Muyskens, J. A. (1994). Priorities for intermediate-level language instruction. *Modern Language Journal, 78*(2), 141–154. doi:10.1111/j.1540-4781.1994.tb02025.x

Hasegawa, T., Koshino, M., & Ban, H. (2015). An English vocabulary learning support system for the learner's sustainable motivation. *SpringerPlus, 4*(99), 1–9. PMID:25830080

Hassan Taj, I., Sulan, N., Sipra, M., & Ahmad, W. (2016). Impact of Mobile-Assisted Language Learning (MALL) on EFL: A meta-analysis. *Advances in Language and Literary Studies, 7*, 2. Retrieved from http://www.journals.aiac.org.au/index.php/alls/article/view/2164/1914

Hattie, J. (2008). *Visible learning: A synthesis over 800 meta-analyses relating to achievement.* London, UK: Routledge. doi:10.4324/9780203887332

Hawkes, L., & Nassaji, H. (2015). The role of extensive recasts in error detection and correction by adult ESL students. *Studies in Second Language Learning and Teaching, 6*(1), 19–41. doi:10.14746sllt.2016.6.1.2

Haythornthwaite, C. (2019). Learning, connectivity and networks. *Information and Learning Science, 120*(1-2), 19–38. doi:10.1108/ILS-06-2018-0052

Hazaea, A. N., & Alzubi, A. A. (2016). The effectiveness of using mobile on EFL learners' reading practices in Najran University. *English Language Teaching, 9*(5), 8–21. doi:10.5539/elt.v9n5p8

Heift, T., & Schulze, M. (2007). *Errors and intelligence in CALL: Parsers and pedagogues.* New York, NY: Routledge.

Helm, F. & Guth, S. (2016). Telecollaboration and language learning. The Routledge Handbook of Language Learning and Technology, 241-254.

Helm, F. (2018). The long and winding road…. In C. Fuchs, C. Kennedy, Satar, & H. Müge (Eds.), Journal of Virtual Exchange, 1, 41-63. Research-publishing.net. doi:10.14705/rpnet.2018.jve.3

Henderson, M., Auld, G., Holkner, B., Russell, G., Seah, W. T., Fernando, A., & Romeo, G. (2010). *Students creating digital video in the primary classroom: Student autonomy, learning outcomes, and professional learning communities. 24*(2), 9.

Henrie, C. R., Bodily, R., Manwaring, K. C., & Graham, C. R. (2015). Exploring intensive longitudinal measures of student engagement in blended learning. *The International Review of Research in Open and Distributed Learning, 16*(3). doi:10.19173/irrodl.v16i3.2015

Henrie, C. R., Larsen, R., Manwaring, K., Halverson, L. R., & Graham, C. R. (2016). Validation of a longitudinal activity-level measure of student engagement. *Contemporary Educational Psychology.*

Henry, A., & Thorsen, C. (2018). Teacher–student relationships and L2 motivation. *Modern Language Journal, 102*(1), 218–241. doi:10.1111/modl.12446

Hew, K. F., & Brush, T. (2007). Integrating technology into K-12 teaching and learning: Current knowledge gaps and recommendations for future research. *Educational Technology Research and Development, 55*(3), 223–252. doi:10.100711423-006-9022-5

Hew, K. F., Tang, Y., Lo, C. K., & Zhu, Y. (2018). Examining a WeChat-supported 5E-flipped classroom pedagogical approach. *International Journal of Services and Standards, 12*(3-4), 224–242. doi:10.1504/IJSS.2018.100217

Hinkelman, D. (2018). *Blending technologies in second language classrooms.* doi:10.1057/978-1-137-53686-0

Hitosugi, C. I. (2011). Using a social networking site in Japanese class. In E. Forsythe, T. Gorham, M. Grogan, D. Jarrell, R. Chartrand, & P. Lewis (Eds.), *CALL: What's your motivation?* (pp. 72–83). Nagoya, Japan: JALT CALL SIG.

Ho, C. M. L., Nelson, M. E., & Müeller-Wittig, W. (2011). Design and implementation of a student-generated virtual museum in a language curriculum to enhance collaborative multimodal meaning-making. *Computers & Education, 57*(1), 1083–1097. doi:10.1016/j.compedu.2010.12.003

Holotescu, C., Grosseck, G., Cretu, V., & Naaji, A. (2014). Integrating MOOCs in blended courses. In *Proceedings of the International Scientific Conference of eLearning and Software for Education*, 1, pp. 243-250. Bucharest, Romania: eLSE. doi:10.12753/2066-026X-14-034

Hood, N., Littlejohn, A., & Milligan, C. (2015). Context counts: How students' contexts influence learning in a MOOC. *Computers & Education*, *91*, 83–91. doi:10.1016/j.compedu.2015.10.019

Hopkyns, S. (2017). The Instagram identity project: Improving paragraph writing through social media. In *10th International Conference on ICT for Language Learning Conference Proceedings*, 224–228. Libreria Universitaria.

Horwitz, E. K. (2001). Language Anxiety and Achievement. *Annual Review of Applied Linguistics*, *21*, 112–126. doi:10.1017/S0267190501000071

Houston, T. (2005). Outcomes assessment for beginning and intermediate Spanish: One program's process and results. *Foreign Language Annals*, *38*(3), 366–374. doi:10.1111/j.1944-9720.2005.tb02223.x

Hoven, D., & Palalas, A. (2013). The design of effective mobile-enabled tasks for ESP students: A longitudinal study. *CALICO Journal*, *30*, 137–165.

Hsu, C. K., Hwang, G. J., & Chang, C. K. (2013). A personalized recommendation-based mobile learning approach to improving the reading performance of EFL students. *Computers & Education*, *63*, 327–336. doi:10.1016/j.compedu.2012.12.004

Hsu, L. (2013). English as a foreign language learners' perception of mobile assisted language learning: A cross-national study. *Computer Assisted Language Learning*, *26*(3), 197–213. doi:10.1080/09588221.2011.649485

Huang, S., Liu, C., & Gao, Z. (2005). Computer-assisted item generation for listening cloze tests and dictation practice in English. In *Proceedings of ICWL*, 197–208. 10.1007/11528043_19

Huang, W., & Soman, D. (2013). *A practitioner's guide to gamification of education. Research report series: Behavioral economics in action.* University of Toronto–Rotman School of Management.

Huang, Y.-T., & Su, S.-F. (2018). Motives for Instagram use and topics of interest among young adults. *Future Internet*, *10*(77), 1–12.

Hubackova, S. (2014). Motivation in eLearning Motivation in Language Courses. *Procedia: Social and Behavioral Sciences*, *122*, 353–356. doi:10.1016/j.sbspro.2014.01.1353

Hubbard, P. & Levy, M. (2016). Theory in computer-assisted language learning research and practice. In F. Farr & L. Murray (Eds.), *The Routledge handbook of language learning and technology*. Oxford, UK: Routledge.

Hubbard, P. (2008). Twenty-five years of theory in the CALICO Journal. *CALICO Journal*, *25*(3), 387–399. Retrieved from https://www.learntechlib.org/p/74286/

Hubbard, P. (2009a). *Computer assisted language learning. Critical Concepts in Linguistics series*. New York, NY: Routledge.

Hubbard, P. (2009b). Developing CALL theory: A new frontier. In *Proceedings of the JALTCALL Conference*. Nagoya, Japón.

Hubbard, P., & Levy, M. (2016). Theory in computer-assisted language learning research and practice. In F. Farr & L. Murray (Eds.), *The Routledge handbook of language learning and technology* (pp. 24–38). Routledge Handbooks Online.

Hult, F. M. (2014). Covert bilingualism and symbolic competence: Analytical reflections on negotiating insider/outsider positionality in Swedish speech situations. *Applied Linguistics*, *35*(1), 63–81. doi:10.1093/applin/amt003

Hung, A., Zarco, E., Yang, M., Dembicki, D., & Kase, M. (2017). Gamification in the wild: Faculty perspectives on gamifying learning in higher education. *Issues and Trends in Educational Technology*, *5*(2). University of Arizona Libraries. Retrieved from https://www.learntechlib.org/p/182049/

Hung, A. (2017). A critique and defense of gamification. *Journal of Interactive Online Learning, 15*(1), 57–72.

Hwang, G. J., Lai, C. L., & Wang, S. Y. (2015). Seamless flipped learning: A mobile technology-enhanced flipped classroom with effective learning strategies. *Journal of Computers in Education, 2*(4), 449-473.

Hwang, S. H. (2018). *Students' perceptions about gamified methods in English as a second language acquisition*. (Theses and Dissertations). 2774.

Hwang, H. S., & Cho, J. (2018). Why Instagram? Intention to continue using Instagram among Korean college students. *Social Behavior and Personality, 46*(8), 1305–1315. doi:10.2224bp.6961

Hwang, W.-Y., Chen, H. S. L., Shadiev, R., Huang, R. Y.-M., & Chen, C.-Y. (2014). Improving English as a foreign language writing in elementary schools using mobile devices in familiar situational contexts. *Computer Assisted Language Learning, 27*(5), 359–378. doi:10.1080/09588221.2012.733711

Hyman, J. A., Moser, M. T., & Segala, L. N. (2014). Electronic reading and digital library technologies: Understanding learner expectation and usage intent for mobile learning. *Educational Technology Research and Development, 62*(1), 35–52. doi:10.100711423-013-9330-5

Iaremenko, N. V. (2017). Enhancing English language learners' motivation through online games. *Information Technologies and Learning Tools, 59*(3), 126–133. doi:10.33407/itlt.v59i3.1606

Ibáñez Moreno, A., & Vermeulen, A. (2015). Profiling a MALL app for English oral practice: A case study. *International Journal of Computers and Technology, 21*(10), 1339–1136. doi:10.3217/jucs-021-10-1339

Ibáñez, M., Di-Serio, Á., & Delgado-Kloos, C. (2014). Gamification for engaging computer science students in learning activities: A case study. *IEEE Transactions on Learning Technologies, 7*(3), 291–301. doi:10.1109/TLT.2014.2329293

Ikrath, P. & Speckmayr, A. (2016). Digitale Kompetenzen für eine digitalisierte Lebenswelt. Eine Jugendstudie der AK Wien, durchgeführt vom Institut für Jugendkulturforschung [Digital skills for a digital world. A youth survey by the AK Vienna, conducted by the Institute for youth culture research]. Vienna, Austria.

Im, J. H., & Kim, J. (2015). Use of blended learning for effective implementation of English-medium instruction in a non-English higher education context. *International Education Studies, 8*(11), 1–15. doi:10.5539/ies.v8n11p1

Ishizaki, A. (2005). How does a learner leave a pause when reading Japanese aloud? A comparison of English, French, Chinese and Korean learners of Japanese and native Japanese speakers. *Japanese-Language Education around the Globe 15*, 75–89.

Ishizaki, A. (2004). Sakubun ondoku ni okeru shokyū gakushūsha no pōzu no tokuchō: eigo bogo washa yon-mei no ōdan-teki shiryō o moto ni (Characteristics of pauses produced by novice learners in an oral reading task: based on the longitudinal study of four native English speakers. *Acquisition of Japanese as a Second Language, 7*, 26–44.

Israel, M. (2015). Effectiveness of integrating MOOCs in traditional classrooms for undergraduate students. *International Review of Research in Open and Distance Learning, 16*(5), 102–118. doi:10.19173/irrodl.v16i5.2222

Iwasaki, N. (2002). Nihongo nōryoku kani shiken (SPOT) no tokuten to ACTFL kōtō nōryoku (OPI) no reberu no kankē ni tsuite [Simple performance-oriented test (SPOT) vs. oral proficiency interview (OPI)]. *Journal of Japanese Language Teaching, 114*, 100–105.

Iwashita, N., Brown, A., McNamara, T., & O'Hagan, S. (2008). Assessed levels of second language speaking proficiency: How distinct? *Applied Linguistics, 29*(1), 24–49. doi:10.1093/applin/amm017

Jabeen, S. & Jesse Thomas, A. (2015). Effectiveness of online language learning [2]. In *Proceedings of the World Congress on Engineering and Computer Science*, 1. Academic Press.

Jauregi, K., De Graaff, R., van den Bergh, H., & Kriz, M. (2012). Native/non-native speaker interactions through video-web communication: A clue for enhancing motivation? *Computer Assisted Language Learning*, *25*(1), 1–19. doi:10.1080/09588221.2011.582587

Jauregi, K., & Melchor-Couto, S. (2017). The role of coaching in teacher competence development for telecollaboration. *Alsic*, *20*(2). doi:10.4000/alsic.3218

Jauregi, K., & Melchor-Couto, S. (2017). The TeCoLa project: Pedagogical differentiation through telecollaboration and gaming for intercultural and content integrated language teaching. In *Proceedings of the EUROCALL 2017 Conference*, (pp. 163-169), Southampton, UK. 10.14705/rpnet.2017.eurocall2017.707

J-CAT project team, & International Student Center at University of Tsukuba (n.d.) J-CAT (Japanese computerized adaptive test) Retrieved from http://www.j-cat.org/en/

Jilg, T., & Southgate, M. (2017). Students helping students: A learning partnership initiative for distance language learners. *Language Learning Journal*, *45*(2), 245–262. doi:10.1080/09571736.2013.877067

Johnson, R. (1997). Examining the validity structure of qualitative research. *Education*, *118*(2), 282–292.

Johnson, W., Krueger, R. F., Bouchard, T. J. Jr, & McGue, M. (2002). The personalities of twins: Just ordinary folks. *Twin Research*, *5*(2), 125–131. doi:10.1375/twin.5.2.125 PMID:11931690

Joosen, V., & Vloeberghs, K. (2012). *Uitgelezen jeugdliteratuur. Een ontmoeting met traditie en vernieuwing*. Leuven, Belgium: Lannoo Campus.

Ju, I. (2010). Second language proficiency and pausing: Syllable duration and pausing in the Kahng, J. (2014). Exploring utterance and cognitive fluency of L1 and L2 English speakers: Temporal measures and stimulated recall. *Language Learning*, *64*(4), 809–854.

Ju, I. (2010). Second language proficiency and pausing: Syllable duration and pausing in the oral reading of Japanese learners. *Bulletin of the Graduate School of Education. Hiroshima University*, *59*(Part II), 229–238.

Kaakinen, J. K., Papp-Zipernovsky, O., Werlen, E., Castells, N., Bergamin, P., Baccino, T., & Jacobs, A. M. (2018). Emotional and motivational aspects of digital reading. In M. Barzillai, J. Thomson, S. Schroeder, & P. W. Van den Broek (Eds.), *Learning to Read in a Digital World* (pp. 141–164). Amsterdam, The Netherlands: John Benjamins. doi:10.1075wll.17.06kaa

Kabilan, M. K., Ahmad, N., & Abidin, M. J. Z. (2010). Facebook: An online environment for learning of English in institutions of higher education? *The Internet and higher education*, *13*(4), 179–187. doi:10.1016/j.iheduc.2010.07.003

Kahng, J. (2018). The effect of pause location on perceived fluency. *Applied Psycholinguistics*, *39*(3), 569–591. doi:10.1017/S0142716417000534

Kaiser, H. F. (1960). The application of electronic computers to factor analysis. *Educational and Psychological Measurement*, *20*(1), 141–151. doi:10.1177/001316446002000116

Kamarainen, A., Metcalf, S., Grotzer, T., Mazzzuca, D., Tutwiler, M., & Dede, C. (2013). EcoMOBILE: Integrating augmented reality and porbeware with environmental education field trips. *Computer Education*, *68*, 545–556. doi:10.1016/j.compedu.2013.02.018

Kapp, K. M. (2012). *The gamification of learning and instruction: Game-based methods and strategies for training and education*. San Francisco, CA: Pfeiffer.

Karadağ, Ş., & Kaya, Ş. D. (2019). The effects of personality traits on willingness to communicate: A study on university students. *Manas Sosyal Araştırmalar Dergisi*, *8*(1), 397–410.

Kearney, M. (2011). A learning design for student-generated digital storytelling. *Learning, Media and Technology*, *36*(2), 169–188. doi:10.1080/17439884.2011.553623

Kearney, M., & Schuck, S. (2006). Spotlight on authentic learning: Student developed digital video projects. *Australasian Journal of Educational Technology*, *22*(2). doi:10.14742/ajet.1298

Kennedy, C., & Levy, M. (2008). L'italiano al telefonino: Using SMS to support beginners' language learning. *ReCALL*, *20*(3), 315–330. doi:10.1017/S0958344008000530

Kesim, M., & Ozarslan, Y. (2012). Augmented reality in education: Current technologies and the potential for education. *Procedia: Social and Behavioral Sciences*, *47*, 297–302. doi:10.1016/j.sbspro.2012.06.654

Kessler, G. (2013). Teaching ESL/EFL in a world of social media, mash-ups, and hyper-collaboration. *TESOL Journal*, *4*(4), 615–632. doi:10.1002/tesj.106

Khalitova, L. & Gimaletdinova, G. (2016). Mobile technologies in teaching English as a foreign language in higher education: A case study of using mobile application Instagram. In *ICERI2016 Proceedings, 1*, 6155-6161. IATED.

Khan, B. H. (2001). A framework for web-based learning. In B. H. Khan (Ed.), *Web-based training* (pp. 75–98). Englewood Cliffs, NJ: Educational Technology Publications.

Khany, R., & Ghoreyshi, M. (2013). The nexus between Iranian EFL students' Big Five personality traits and foreign language speaking confidence. *European Online Journal of Natural and Social Sciences*, *2*(2), 601–611.

Khlaisang, J., & Mingsiritha, K. (2016). Engaging virtual learning environment system to enhance communication and collaboration skills among ASEAN higher education learners. *International Journal of Emerging Technologies in Learning*, *11*(4), 103–113. doi:10.3991/ijet.v11i04.5503

Kim, D., Rueckert, D., Kim, D.-J., & Seo, D. (2013). Students' perceptions and experiences of mobile learning. *Language Learning & Technology*, *17*(3), 52–73.

Kim, M. K., Kim, S. M., Khera, O., & Getman, J. (2014). The experience of three flipped classrooms in an urban university: An exploration of design principles. *The Internet and Higher Education*, *22*, 37–50. doi:10.1016/j.iheduc.2014.04.003

King, A. (2016). *Blended language learning: Part of the Cambridge Papers in ELT series*. Cambridge, UK: Cambridge University Press.

Kırkağaç, Ş. & Öz, H. (2017). The role of Big Five personality traits in predicting prospective EFL teachers' academic achievement. *International Online Journal of Education and Teaching (IOJET), 4*(4), 317-328. Retrieved from http://iojet.org/index.php/IOJET/article/view/243/174

Kizilcec, R. F., Pérez-Sanagustín, M., & Maldonado, J. J. (2017). Self-regulated learning strategies predict student behavior and goal attainment in massive open online courses. *Computers & Education*, *104*, 18–33. doi:10.1016/j.compedu.2016.10.001

Kline, P. (1994). *An easy guide to factor analysis*. Newbury Park, CA: Sage.

Kloos, C. D., Muñoz-Merino, P. J., Alario-Hoyos, C., Ayres, I. E., & Fernández-Panadero, C. (2015). Mixing and blending MOOC technologies with face-to-face pedagogies. In *Proceedings of Global Engineering Education Conference (EDUCON*), 2015 IEEE (pp. 967-971). New York, NY: Institute of Electrical and Electronics Engineers (IEEE).

Klopfer, E., & Sheldon, J. (2010). Augmenting your own reality: Student authoring of science-based augmented reality games. *New Directions for Youth Development*, *128*(128), 85–94. doi:10.1002/yd.378 PMID:21240956

Knox, J. (2014). Digital culture clash: "Massive" education in the E-learning and digital cultures MOOC. *Distance Education, 35*(2), 164–177. doi:10.1080/01587919.2014.917704

Kobayashi, K., & Little, A. (2011). Learner perceptions on the usefulness of a blended learning EFL program. *JALTCALL Journal, 7*(1), 103–117.

Kohn, K., & Hoffstaedter, P. (2017). Learner agency and non-native speaker identity in pedagogical lingua franca conversations: Insights from intercultural telecollaboration in foreign language education. *Computer Assisted Language Learning, 30*(5), 351–367. doi:10.1080/09588221.2017.1304966

Kolodner, J. L., Gray, J., & Fasse, B. B. (2003). Promoting transfer through case-based reasoning: Rituals and practices in learning by DesignTM classrooms. *Cognitive Science Quarterly*, 119–170.

Kolodner, J. L., Camp, P. J., Crismond, D., Fasse, B., Gray, J., Holbrook, J., ... Ryan, M. (2003). Problem-based learning meets case-based reasoning in the middle-school science classroom: Putting learning by design into practice. *Journal of the Learning Sciences, 12*(4), 495–547. doi:10.1207/S15327809JLS1204_2

Kormos, J. (2006). *Speech production and second language acquisition*. London, UK: Routledge.

Kormos, J., & Csizér, K. (2008). Age-related differences in the motivation of learning English as a foreign language: Attitudes, selves, and motivated learning behavior. *Language Learning, 58*(2), 327–355. doi:10.1111/j.1467-9922.2008.00443.x

Kormos, J., & Denes, M. (2004). Exploring measures and perceptions of fluency in the speech of second language learners. *System, 32*(2), 145–164. doi:10.1016/j.system.2004.01.001

Kothgassner, O. D., Felnhofer, A., Hauk, N., Kastenhofer, E., Gomm, J., & Kryspin-Exner, I. (2013). *TUI: Technology Usage Inventory*. Retrieved from https://www.ffg.at/sites/default/files/allgemeine_downloads/thematische%20programme/programmdokumente/tui_manual.pdf

Kramsch, C. (2002). *Language acquisition and socialization. Ecological perspectives*. London, UK: Continuum.

Kramsch, C. (2006). From communicative competence to symbolic competence. *Modern Language Journal, 90*(2), 249–252. doi:10.1111/j.1540-4781.2006.00395_3.x

Kramsch, C. (2009). Discourse, the symbolic dimension of intercultural competence. In A. Hu & M. Byram (Eds.), *Interkulturelle kompetenz und fremdsprachliches lerneen. Modelle, empirie, evaluation*. Tübingen: Gunter Narr.

Kramsch, C. (2011). The symbolic dimensions of the intercultural. *Language Teaching, 44*(3), 354–367. doi:10.1017/S0261444810000431

Kramsch, C., & Vork Steffensen, S. (2008). Ecological perspectives on second language acquisition and socialization. In *Encyclopedia of Language and Education* (pp. 2595–2606). Boston, MA: Springer US; doi:10.1007/978-0-387-30424-3_194

Kramsch, C., & Whiteside, A. (2008). Language ecology in multilingual settings. Towards a theory of symbolic competence. *Applied Linguistics, 29*(4), 645–671. doi:10.1093/applin/amn022

Krashen, S. (1985). *The input hypothesis: Issues and implications*. London, UK: Longman.

Kukulska-Hulme, A. & Shield. (2007). An overview of mobile assisted language learning: Can mobile devices support collaborative practice in speaking and listening. In *EuroCALL*, 2007.

Kukulska-Hulme, A., & Shield, L. (2007). An overview of Mobile Assisted Language Learning: Can mobile devices support collaborative practice in speaking and listening? EuroCALL Virtual Strand. Retrieved from http://vsportal2007.googlepages.com/collaborativepractice

Kukulska-Hulme, A. (2009). Will mobile learning change language learning? *ReCALL, 21*(2), 157–165. doi:10.1017/S0958344009000202

Kukulska-Hulme, A. (2012). Chapter One: Language learning defined by time and place: A framework for next generation designs. In J. Díaz-Vera (Ed.), *Left to my own devices: Learner autonomy and mobile-assisted language learning* (pp. 1–20). BRILL.

Kukulska-Hulme, A. (2016). *Personalization of language learning through mobile technologies.* Cambridge, UK: Cambridge University Press.

Kukulska-Hulme, A., Lee, H., & Norris, L. (2017). Mobile learning revolution: Implications for language pedagogy. In C. A. Chapelle & S. Sauro (Eds.), *The handbook of technology and second language teaching and learning* (pp. 217–233). Oxford, UK: Wiley & Sons. doi:10.1002/9781118914069.ch15

Kukulska-Hulme, A., & Shield, L. (2008). An overview of Mobile Assisted Language Learning: From content delivery to supported collaboration and interaction. *ReCALL, 20*(3), 271-289. doi:10.1017/S0958344008000335

Kukulska-Hulme, A., & Viberg, O. (2018). Mobile collaborative language learning: State of the art. *British Journal of Educational Technology, 49*(2), 207–218. doi:10.1111/bjet.12580

Kulik, C.-L. C., & Kulik, J. A. (1991). Effectiveness of computer-based instruction: An updated analysis. *Computers in Human Behavior, 7*(1-2), 75–94. doi:10.1016/0747-5632(91)90030-5

Kumar, J., & Herger, M. (2013). *Gamification at work: Designing engaging business software.* Aarhus, Denmark: The Interaction Design Foundation. doi:10.1145/2468356.2468793

Kurniawan, A. & Kastuhandani, L. A. (2016). Utilizing Instagram for engaging students in their creative writing. In Proceedings of Indonesia Technology Enhanced Language Learning, 4–7. Salatiga, Indonesia: Satya Wacana University Press Universitas Kristen Satya Wacana

Laghos, A., & Nisiforou, E. (2018). Computer assisted language learning social networks: What are they talking about? *Social Networking, 7*(3), 170–180. doi:10.4236n.2018.73014

Lai, C., & Gu, M. Y. (2011). Self-regulated out-of-class language learning with technology. *Computer Assisted Language Learning, 24*(4), 317–335. doi:10.1080/09588221.2011.568417

Lai, C., & Zheng, D. (2018). Self-directed use of mobile devices for language learning beyond the classroom. *ReCALL, 30*(3), 299–318. doi:10.1017/S0958344017000258

Lam, S. L. (2014). Use of gamification in vocabulary learning: A case study in Macau. *Alternative Pedagogies in the English Language & Communication Classroom*, 90-97. In *4th CELL Symposium Proceedings.*

Lambert, W. & Wolfgang, A. (1973). Culture and language as factors in learning and education. En Annual Simp on Cultural Factors in Learning, Washington.

Lambert, C., & Kormos, J. (2014). Complexity, accuracy, and fluency in task-based L2 research: Toward more developmentally based measures of second language acquisition. *Applied Linguistics, 53*(5), 607–614. doi:10.1093/applin/amu047

Lamb, M. (2017). The motivational dimension of language teaching. *Language Teaching, 50*(3), 301–346. doi:10.1017/S0261444817000088

Lam, Y. (2000). Technophilia vs. technophobia: A preliminary look at why second-language teachers do or do not use technology in their classrooms. *Canadian Modern Language Review, 56*(3), 389–420. doi:10.3138/cmlr.56.3.389

Landers, R. N., & Landers, A. K. (2014). An empirical test of the theory of gamified learning: The effect of leaderboards on time-on-task and academic performance. *Simulation & Gaming*, *45*(6), 769–785. doi:10.1177/1046878114563662

Landow, G. (1997). *Hypertext 2.0: The convergence of contemporary critical theory and technology*. Baltimore, MD: Johns Hopkins University Press.

Lantolf, J., & Thorne, S. (2006). *Sociocultural theory and the genesis of second language development*. Oxford, UK: Oxford University Press.

Larson, L. C. (2015). E-books and audiobooks: Extending the digital reading experience. *The Reading Teacher*, *69*(2), 169–177. doi:10.1002/trtr.1371

Lasagabaster, D., Doiz, A., & Sierra, J. M. (2014). *Motivation and foreign language learning: From theory to practice*. Amsterdam, The Netherlands: John Benjamins. doi:10.1075/lllt.40

Law, K. M., Geng, S., & Li, T. (2019). Student enrollment, motivation and learning performance in a blended learning environment: The mediating effects of social, teaching, and cognitive presence. *Computers & Education*, *136*, 1–12. doi:10.1016/j.compedu.2019.02.021

Lawson, M. A., & Lawson, H. A. (2013). New conceptual frameworks for student engagement research, policy, and practice. *Review of Educational Research*, *83*(3), 432–479. doi:10.3102/0034654313480891

Lederman, D. (2018). Who is studying online (and where). Inside Higher Ed.

Lee, Y.-S., Sung, Y.-T., Chang, K.-E., Liu, T.-C., & Chen, W.-C. (2014). A meta-analysis of the effects of learning languages with mobile devices. Lecture Notes in Computer Science, 8699, 106-113. Retrieved from https://link.springer.com/chapter/10.1007%2F978-3-319-13296-9_12

Lee, J. J., & Hammer, J. (2011). Gamification in education: What, how, why bother? *Academic Exchange Quarterly*, *15*(2).

Lee, L., & Markey, A. (2014). A study of learners' perceptions of online intercultural exchange through Web 2.0 technologies. *ReCALL*, *26*(3), 281–297. doi:10.1017/S0958344014000111

Lee, Y., & Choi, J. (2011). A review of online course dropout research: Implications for practice and future research. *Educational Technology Research and Development*, *59*(5), 593–618. doi:10.100711423-010-9177-y

Leier, V. (2018). Using Instagram in a tertiary German language course: Students' perception and approaches. *The New Zealand Language Teacher*, *44*, 77–90.

Lennon, P. (1990). Investigating fluency in EFL: A quantitative approach. *Language Learning*, *3*(3), 387–417. doi:10.1111/j.1467-1770.1990.tb00669.x

Levelt, W. (1989). *Speaking: From intention to articulation*. Cambridge, MA: MIT Press.

Levy, M. (1997). *Computer-assisted language learning*. Oxford, UK: Clarendon Press.

Levy, M. (1997). *Computer-assisted language learning: Context and conceptualization*. Oxford, UK: Oxford University Press.

Levy, M. Y., & Stockwell, G. (2006). *CALL dimensions: Options and issues in computer assisted language learning*. Mahwah, NJ: Lawrence Erlbaum Associates.

Levy, M., & Hubbard, P. (2005). Why call CALL "CALL"? *CALL Journal*, *18*(3), 143–149. doi:10.1080/09588220500208884

Levy, M., & Kennedy, C. (2007). Learning Italian via mobile SMS. In *Mobile learning* (pp. 92–99). Routledge.

Levy, Y. (2007). Comparing dropouts and persistence in e-learning courses. *Computers & Education, 48*(2), 185–204. doi:10.1016/j.compedu.2004.12.004

Lewis, T., & O'Dowd, R. (2016). Online intercultural exchange and foreign language learning: A systematic review. In R. O'Dowd & T. Lewis (Eds.), *Online intercultural exchange: Policy, pedagogy, and practice* (pp. 21–66). New York, NY: Routledge.

Liakin, D., Cardoso, W., & Liakina, N. (2017). The pedagogical use of mobile speech synthesis (TTS): Focus on French liaison. *Computer Assisted Language Learning, 30*(3-4), 3–4, 325–342. doi:10.1080/09588221.2017.1312463

Liamputtong, P. (2011). *Focus group methodology: Principles and practice*. London, UK: Sage. doi:10.4135/9781473957657

Lightbrown, M. P., & Spada, N. (2006). *How languages are learned.* Oxford, UK: Oxford University Press.

Li, J., & Cummins, J. (2019). Effect of using texting on vocabulary instruction for English learners. *Language Learning & Technology, 23*(2), 43–64.

Li, J., & Qin, X. (2006). Language learning styles and learning strategies of tertiary level English learners in China. *RECL Journal, 37*, 67–89.

Li, K. (2019). MOOC learners' demographics, self-regulated learning strategy, perceived learning and satisfaction: A structural equation modeling approach. *Computers & Education, 132*, 16–30. doi:10.1016/j.compedu.2019.01.003

Lin, C.-C., Lin, V., Liu, G.-Z., Kou, X., Kulikova, A., & Lin, W. (2019). Mobile-assisted reading development: A review from the Activity Theory perspective. Computer Assisted Language Learning. Retrieved from https://www.tandfonline.com/doi/pdf/10.1080/09588221.2019.1594919?needAccess=true

Lin, J.-J., & Lin, H. (2019). Mobile-assisted ESL/EFL vocabulary learning: A systematic review and meta-analysis. *Computer Assisted Language Learning*, 1-42. Retrieved from https://www.tandfonline.com/doi/full/10.1080/09588221.2018.1541359?scroll=top&needAccess=true

Li, S., Zhu, Y. A. N., & Ellis, R. (2016). The effects of the timing of corrective feedback on the acquisition of a new linguistic structure. *Modern Language Journal, 100*(1), 276–295. doi:10.1111/modl.12315

Listiani, G. (2016). The effectiveness of Instagram writing compared to teacher centered writing to teach recount text to students with high and low motivation. *Journal of English Language Teaching, 5*(1), 1–8.

Little, D., Dam, L., & Legenhausen, L. (2017). *Language Learner Autonomy: Theory, Practice and Research.* UK: Multilingual Matters.

Littlejohn, A. & Hood, N. (2018). Reconceptualising learning in the digital age: The [un]democratizing potential of MOOCs. Gateway East, Singapore: Springer.

Littlejohn, A. & Milligan, C. (2015). Designing MOOCs for professional students: Tools and patterns to encourage self-regulated learning. *eLearning Papers, 42*, 38–45.

Littlejohn, A. (2013). Understanding massive open online courses. *CEMCA EdTech Notes*, 1–12.

Littlejohn, A., Hood, N., Milligan, C., & Mustain, P. (2016). Learning in MOOCs: Motivations and self-regulated learning in MOOCs. *The Internet and Higher Education, 29*, 40–48. doi:10.1016/j.iheduc.2015.12.003

Litzler, M. F., Huguet-Jérez, M., & Bakieva, M. (2018). Prior experience and student satisfaction with e-tandem language learning of Spanish and English. *International Journal of Interactive Mobile Technologies, 12*(4), 4–20. doi:10.3991/ijim.v12i4.9196

Liu, G. Z., Kuo, F. R., Shi, Y. R., & Chen, Y. W. (2015). Dedicated design and usability of a context-aware ubiquitous learning environment for developing receptive language skills: A case study. *International Journal of Mobile Learning and Organisation*, *9*(1), 49–65. doi:10.1504/IJMLO.2015.069717

Liu, P.-H., & Tsai, M. (2013). Using augmented-reality-based mobile learning material in EFL English composition: An exploratory case study. *British Journal of Educational Technology*, *44*(1), E1–E4. doi:10.1111/j.1467-8535.2012.01302.x

Liu, T., & Chu, Y. (2010). Using ubiquitous games in an English listening and speaking course: Impact on learning outcomes and motivation. *Computers & Education*, *55*(2), 630–643. doi:10.1016/j.compedu.2010.02.023

Liu, T., Shadiev, R., & Hwang, W.-Y. (2017). Review of research on MALL in familiar authentic environments. *Computer Assisted Language Learning*, *30*(3-4), 284-303. doi:10.1080/09588221.2017.1308383

Lomicka, L. & Lord, G. (2016). Social networking and language learning. The Routledge handbook of language learning and technology, 255-268.

Lomicka, L., & Lord, G. (2009). Introduction to social networking, collaboration, and Web 2.0 tools. In L. Lomicka & G. Lord (Eds.), *The next generation: Social networks and online collaboration in foreign language learning*. San Marcos, TX: CALICO.

Lomicka, L., & Lord, G. (2016). Social networking and language learning. In F. Farr & M. Murray (Eds.), *The Routledge handbook of language learning and technology* (pp. 255–268). New York, NY: Routledge.

Long, M. H. (1991). Focus on form: A design feature in language teaching methodology. In K. de Bot, D. Coste, R. Ginsberg, & C. Kramsch (Eds.), *Foreign language research in crosscultural perspectives* (pp. 39–52). Amsterdam, The Netherlands: John Benjamins. doi:10.1075ibil.2.07lon

Long, M. H. (1996). The role of the linguistic environment in second language acquisition. In W. Ritchie & T. Bathia (Eds.), *Handbook of research on second language acquisition* (pp. 413–469). New York, NY: Academic Press. doi:10.1016/B978-012589042-7/50015-3

Lopes, J., Trancoso, I., Correia, R., Pellegrini, T., Meinedo, H., Mamede, N., & Eskenazi, M. (2010). Multimedia Learning Materials. In *Proceedings IEEE Workshop on Spoken Language Technology SLT*, Berkeley, CA. pp. 109–114.

López-Pérez, M. V., Pérez-López, M. C., & Rodríguez-Ariza, L. (2011). Blended learning in higher education: Students' perceptions and their relation to outcomes. *Computers & Education*, *56*(3), 818–826. doi:10.1016/j.compedu.2010.10.023

Luckin, R., Holmes, W., Griffiths, M., & Forcier, L. B. (2016). *Intelligence Unleashed. An argument for AI in Education*. London, UK: Pearson.

Lys, F. (2013). The development of advanced learner oral proficiency using iPads. *Language Learning & Technology*, *17*(3), 94–116.

Lyster, R., & Izquierdo, J. (2009). Prompts versus recasts in dyadic interaction. *Language Learning*, *59*(2), 453–498. doi:10.1111/j.1467-9922.2009.00512.x

Lyster, R., & Saito, K. (2010). Oral feedback in classroom SLA: A meta-analysis. *Studies in Second Language Acquisition*, *32*(2), 265–302. doi:10.1017/S0272263109990520

Macaro, E., Curle, S., Pun, J., An, J., & Dearden, J. (2017). A systematic review of English medium instruction in higher education. *Language Teaching*, *51*(1), 36–76. doi:10.1017/S0261444817000350

MacIntyre, P. D., Clément, R., Dörnyei, Z., & Noels, K. A. (1998). Conceptualizing willingness to communicate in a L2: A situational model of L2 confidence and affiliation. *Modern Language Journal, 82*(4), 545–562. doi:10.1111/j.1540-4781.1998.tb05543.x

MacIntyre, P. D., & Gardner, R. C. (1994). The subtle effects of language anxiety on cognitive processing in the second language. *Language Learning, 44*(2), 283–305. doi:10.1111/j.1467-1770.1994.tb01103.x

MacIntyre, P. D., & Legatto, J. J. (2011). A dynamic system approach to willingness to communicate: Developing an idiodynamic method to capture rapidly changing affect. *Applied Linguistics, 32*(2), 149–171. doi:10.1093/applin/amq037

MacKinnon, T., Pasfield-Neofitou, S., Manns, H., & Grant, S. (2016). A meta-analysis of open educational communities of practice and sustainability in higher educational policy. [Social dynamics in open educational language practice.]. *ALSIC, 19*, 1.

Mahdi, H. (2017). Effectiveness of mobile devices on vocabulary learning: A meta-analysis. *Journal of Educational Computing Research, 55*(3), 1-21. Retrieved from https://kku-sa.academia.edu/HassanMahdi

Majdoub, M. (2016). Promoting high school ESL learners' motivation and engagement through the use of gamified instructional design. Presented in partial fulfillment of the requirements for the degree of Master of Arts (Educational Technology) at Concordia University, Montreal, Canada.

Mangen, A. (2008). Hypertext fiction reading: Haptics and immersion. *Journal of Research in Reading, 31*(4), 404–419. doi:10.1111/j.1467-9817.2008.00380.x

Mangen, A., & Kuiken, D. (2014). Lost in an iPad: Narrative engagement on paper and tablet. *Scientific Study of Literature, 4*(2), 150–178. doi:10.1075sol.4.2.02man

Mansor, N., & Rahim, N. A. (2017). Instagram in ESL classroom. *Man in India, 97*(20), 107–114.

Mantle, M. W., & Lichty, R. (2012). *Managing the unmanageable: Rules, tools, and insights for managing software people and teams.* Addison-Wesley.

Mao, J. (2014). Social media for learning: A mixed methods study on high school students' technology affordances and perspectives. *Computers in Human Behavior, 33*, 213–223. doi:10.1016/j.chb.2014.01.002

Margaryan, A., Bianco, M., & Littlejohn, A. (2015). Instructional quality of massive open online courses (MOOCs). *Computers & Education, 80*, 77–83. doi:10.1016/j.compedu.2014.08.005

Martin-Monje, E. & Castrillo, M. D. (2016). A redefinition of the teacher and student roles in language MOOCs: The example of 'how to succeed in the English B1 level exam'. In *LLAS 11th Annual e-learning Symposium,* Southampton, UK.

Martin-Monje, E. (2017). *Mind the L in LMOOCs: The importance of language learning in massive courses.* Venice, Italy: Conference: IV Congresso della Societa di Didattica delle Lingue e Linguistica Educativa (DILLE) 2017.

Martín-Monje, E., & Bárcena, E. (Eds.). (2014). *Language MOOCs: Providing learning, transcending boundaries.* Berlin, Germany: De Gruyter Open.

Martín-Monje, E., Castrillo, M. D., & Mañana-Rodríguez, J. (2018). Understanding online interaction in language MOOCs through learning analytics. *Computer Assisted Language Learning, 31*(3), 251–272. doi:10.1080/09588221.2017.1378237

Maske, S. S., Kamble, P. H., Kataria, S. K., Raichandani, L., & Dhankar, R. (2018). Feasibility, effectiveness, and students' attitude toward using WhatsApp in histology teaching and learning. *Journal of Education and Health Promotion, 7*, 158. doi:10.4103/jehp.jehp_30_18 PMID:30693295

Masterman, L. & Wild, J. (2011). JISC-funded OER impact study. Retrieved from http://www.jisc.ac.uk/media/documents/programmes/elearning/oer/JISCOERImpactStudyResearchReportv1-0.pdf

Mathews, J. M. (2010). Using a studio-based pedagogy to engage students in the design of mobile-based media. *English Teaching*, *9*(1), 87–102.

Maynard, S. (2010). The impact of e-books on young children's reading habits. *Publishing Research Quarterly*, *26*(4), 236–248. doi:10.100712109-010-9180-5

Mazer, J. P., Murphy, R. E., & Simonds, C. J. (2007). I'll see you on "Facebook": The effects of computer-mediated teacher self-disclosure on student motivation, affective learning, and classroom climate. *Communication Education*, *56*(1), 1–17. doi:10.1080/03634520601009710

McAndrew, P., & Scanlon, E. (2013). Open learning at a distance: Lessons for struggling MOOCs. *Science*, *342*(6165), 1450–1451. doi:10.1126cience.1239686 PMID:24357302

McAuley, A., Stewart, B., Siemens, G., & Cormier, D. (2010). *The MOOC model for digital practice.* Retrieved from https://oerknowledgecloud.org/sites/oerknowledgecloud.org/files/MOOC_Final.pdf

McBride, K. (2009). Social networking sites in foreign language classes: Opportunities for re-creation. In L. Lomicka & G. Lord (Eds.), *The next generation: Social networking and online collaboration in foreign language learning (CALICO Monograph Series)* (pp. 35–58). San Marcos, TX: CALICO.

McCrae, R. R., & Costa, P. T. Jr. (2003). *Personality in adulthood: A Five-Factor Theory perspective* (2nd ed.). New York, NY: Guilford. doi:10.4324/9780203428412

McCulley, Y., Gillespie, C., & Murr, A. H. (2014). Assessing the effectiveness of text-to-speech software in incarcerated adult literacy education. *Journal of Correctional Education*, *65*(2), 2.

McGeown, S. P., Osborne, C., Warhurst, A., Norgate, R., & Duncan, L. G. (2016). Understanding children's reading activities: Reading motivation, skill and child characteristics as predictors. *Journal of Research in Reading*, *39*(1), 109–125. doi:10.1111/1467-9817.12060

McGuirk, R. (2015). *Exploration of the use of Irish language synthesis with a screen reader in the teaching of Irish to pupils with vision impairment.* (Unpublished MPhil dissertation). Trinity College, Ireland.

McLoughlin, L. & Magnoni, F. (2017). The Move-Me project: Reflecting on xMOOC and cMOOC structure and pedagogical implementation. In Q. Kan & S. Bax (Eds.), Beyond the language classroom: researching MOOCs and other innovations (pp. 59-69). Research-publishing.net. .mooc2016.671 doi:10.14705/rpnet.2017.mooc2016.671

McQuiggan, S., Kosturko, L., McQuiggan, J., & Sabourin, J. (2015). *Mobile learning: A handbook for developers, educators, and learners*. Hoboken, NJ: John Wiley & Sons.

Medienpädagogischer Forschungsverbund Südwest (mpfs). (2017). *JIM-Studie 2017. Jugend, Information, (Multi-) Media. Basisstudie zum Medienumgang 12- bis 19-Jähriger* [JIM Survey 2017. Youth, information (multi-)media. Basic study on media usage]. Retrieved from https://www.mpfs.de/fileadmin/files/Studien/JIM/2016/JIM_Studie_2016.pdf

Medina, E. G. L., & Hurtado, C. P. R. (2017). Kahoot! A digital tool for learning vocabulary in a language classroom. *Revista Publicando*, *4*(12), 441–449.

Mehdizadeh, S. (2010). Self-presentation 2.0: Narcissism and self-esteem on Facebook. *Cyberpsychology, Behavior, and Social Networking*, *13*(4), 357–364. doi:10.1089/cyber.2009.0257 PMID:20712493

Meihami, H. (2013). Text-to-speech software: a new perspective in learning and teaching word stress, word intonation, pitch contour, and fluency of English reading. *International Letters of Social and Humanistic Sciences*, *8*, 24–33. doi:10.18052/www.scipress.com/ILSHS.8.24

Melchor-Couto, S. (2019). Virtual worlds and language learning. *Journal of Gaming & Virtual Worlds*, *11*(1), 29–43. doi:10.1386/jgvw.11.1.29_1

Merchant, G. (2011). Unravelling the social network: Theory and research. *Learning, Media and Technology*, *37*(1), 4–19. doi:10.1080/17439884.2011.567992

Merchant, Z., Goetz, E., Cifuentes, L., Keeney-Kennicutt, W., & Davis, T. J. (2014). Effectiveness of virtual reality-based instruction on students' learning outcomes in K-12 and higher education: A meta-analysis. *Computers & Education, 70*, 29–40. doi:10.1016/j.compedu.2013.07.033

Meskill, C. & Quah, J. (2013). Researching language learning in the age of social media. En T. Michael, H. Reinders, & W. Mark (Eds.), Contemporary computer-assisted language learning (pp. 39–54). London, UK: Bloomsbury.

Messick, S. (1994). The matter of style: Manifestations of personality in cognitive, learning and teaching. *Educational Psychologist*, *29*(3), 121–136. doi:10.120715326985ep2903_2

Metruk, R. (2019). The call of the MALL: The use of smartphones in higher education. A literature review. Revista Dilemas Contemporáneos: Educación, Política y Valores, 6(3), np. Retrieved from https://www.academia.edu/38990631/The_Call_of_the_MALL_the_Use_of_Smartphones_in_Higher_Education._A_Literature_Review?email_work_card=view-paper

Meurers, D. (2012). Natural language processing and language learning. In C. A. Chapelle (Ed.), *Encyclopedia of Applied Linguistics* (pp. 4193–4205). Oxford, UK: Blackwell. doi:10.1002/9781405198431.wbeal0858

Mewald, C., Wallner, S., & Buchner, J. (2018). *How-to: Lexical notebooks 1 & 2*. Video. Retrieved from https://www.youtube.com/watch?v=etjnHfKWzN4

Mewald, C., & Wallner, S. (2018). Digitalised materials for young foreign language learners on the interactive website PALM. In M. Bouckaert, M. Konings, & M. van Winkelhof (Eds.), *Meaning-focused materials for language learning* (pp. 189–204). Newcastle-upon-Tyne, UK: Cambridge Scholars.

Miall, D. S. (2004). Reading hypertext – theoretical ambitions and empirical studies. Retrieved from http://computerphilologie.uni-muenchen.de/jg03/miall.html

Miall, D. S., & Dobson, T. (2001). Reading hypertext and the experience of literature. *Journal of Digital Information*, *2*. Retrieved from https://journals.tdl.org/jodi/index.php/jodi/ article/view/35/37

Miall, D. S., & Kuiken, D. (2002). A feeling for fiction: Becoming what we behold. *Poetics*, *30*(4), 221–241. doi:10.1016/S0304-422X(02)00011-6

Michel, M. C. (2011). Effects of task complexity and interaction in L2 performance. In P. Robinson (Ed.), *Second language task complexity: Researching the cognition hypothesis of language learning and performance* (pp. 141–174). Amsterdam, The Netherlands: John Benjamins. doi:10.1075/tblt.2.12ch6

Mifsud, C. L., & Petrová, Z. (2018). Literacy education in the digital age. In M. Barzillai, J. Thomson, S. Schroeder, & P. W. Van den Broek (Eds.), *Learning to Read in a Digital World* (pp. 165–183). Amsterdam, The Netherlands: John Benjamins. doi:10.1075wll.17.07mif

Milligan, C., & Littlejohn, A. (2016). How health professionals regulate their learning in massive open online courses. *The Internet and higher education*, *31*, 113–121. doi:10.1016/j.iheduc.2016.07.005 PMID:27928198

Milligan, C., Littlejohn, A., & Hood, N. (2016). Learning in MOOCs: A comparison study. In *Proceedings of the 4th European MOOCs Stakeholder Summit on experiences and best practices in and around MOOCs (EMOOCS 2016).* (pp. 15-26). University of Graz, Austria.

Mills, N. (2011). Situated learning through social networking communities: The development of joint enterprise, mutual engagement, and a shared repertoire. *CALICO Journal*, *28*(2), 345–368. doi:10.11139/cj.28.2.345-368

Miranda, T., Williams-Rossi, D., Johnson, K., & McKenzie, N. (2011). Reluctant readers in middle school: Successful engagement with text using the e-reader. *International Journal of Applied Science and Technology*, *1*(6), 81–91.

Mitchell, R., Myles, F., & Marsden, E. (2013). *Second language learning theories.* New York, NY: Routledge.

Miyazoe, T., & Anderson, T. (2010). Learning outcomes and students' perceptions of online writing: Simultaneous implementation of a forum, blog, and wiki in an EFL blended learning setting. *System*, *38*(2), 185–199. doi:10.1016/j.system.2010.03.006

MLA. (2007). Foreign languages and higher education: New structures for a changed world. *Profession*, (1), 234–245. doi:10.1632/prof.2007.2007.1.234

Mora, J. C., & Valls-Ferrer, M. (2012). Oral fluency, accuracy and complexity in formal instruction and study abroad learning contexts. *TESOL Quarterly*, *46*(4), 610–641.

Morato-Paya, A. (2014). El WhatsApp como complemento del aprendizaje en la clase E/LE [WhatsApp as a learning complement in the classroom of E/LE]. Revista Foro De Profesores De E/LE, 10(1), 165-173.

Morgan, D. L. (1998). *The focus group guidebook.* London, UK: Sage. doi:10.4135/9781483328164

Morris, N., Hotchkiss, S., & Swinnerton, B. (2015a). Can demographic information predict MOOC learner outcomes? In Proceedings of the European MOOC Stakeholder Summit, 199-207, Mons, Belgium.

Morshidi, A., Embi, M. A., & Hashim, H. (2019). Instagram application: An active tool in cultivating reading behavior. *Journal of Information Systems and Technology Management*, *4*(11), 95–106.

Morsünbül, U. (2014). The validity and reliability study of the Turkish version of Quick Big Five personality test. *Düşünen Adam*, *27*(4), 316–322. doi:10.5350/DAJPN2014270405

Motzo, A. & Proudfoot, A. (2017). MOOCs for language learning–opportunities and challenges: The case of the open university Italian beginners' MOOCs. In Q. Kan & S. Bax (Eds.), Beyond the language classroom: Researching MOOCs and other innovations (pp. 85-97). Research-publishing.net. doi:.mooc2016.673 doi:10.14705/rpnet.2017

Munday, P., Delaney, Y., & Bosque, A. (2016). #InstagramELE: Learning Spanish through a Social Network. *Languages Faculty Publications.* Retrieved from http://digitalcommons.sacredheart.edu/lang_fac/57

Munday, P. (2016). The case for using duolingo as part of the language classroom experience. *Revista Iberoamericana de Educación a Distancia*, *19*(1), 83–109.

Muñoz-Merino, P. J., Ruipérez-Valiente, J. A., Delgado Kloos, C., Auger, M. A., Briz, S., de Castro, V., & Santalla, S. N. (2016). Flipping the classroom to improve learning with MOOCs technology. *Computer Applications in Engineering Education*, *25*(1), 15–25. doi:10.1002/cae.21774

Murar, P. (2015, October). Edmodo as a gamification platform: Review and plans. In *Proceedings European Conference on Games Based Learning.* (p. 800). Academic Conferences International Limited.

Myers-Briggs, I., McCaulley, M. H., Quenk, N. L., & Hammer, A. L. (1998). *MBTI manual: A guide to the development and use of the Myers Briggs type indicator.* Palo Alto, CA: Consulting Psychologists Press.

Nation, I., & Newton, J. (2009). *Teaching ESL/EFL Listening and Speaking*. New York, NY: Routledge Taylor and Francis Groups.

Nell, V. (1988). *Lost in a book: The psychology of reading for pleasure*. New Haven, CT: Yale University Press. doi:10.2307/j.ctt1ww3vk3

Nelson, M. (2012). Soviet and American precursors to the gamification of work. In A. Lugmayr (Ed.), *MindTrek* (pp. 23–26). New York, NY: ACM. doi:10.1145/2393132.2393138

Nerbonne, J. A. (2003). Natural language processing in computer-assisted language learning. In R. Mitkov (Ed.), *The Oxford handbook of computational linguistics* (pp. 670–698). Oxford.

Ní Chasaide, A., Ní Chiaráin, N., Wendler, C., Berthelsen, H., Murphy, A., Barnes, E., & Gobl, C. (2019). Leveraging Phonetic and Speech Research for Irish Language Revitalisation and Maintenance. Ifn S. Calhoun, P. Escudero, M. Tabain & P. Warren (Eds.) Proceedings of the International Congress of Phonetic Sciences, Melbourne, Australia 2019 (pp. 994-998). Canberra, Australia: Australasian Speech Science and Technology Association Inc.

Ní Chasaide, A., Ní Chiaráin, N., Berthelsen, H., Wendler, C., & Murphy, A. (2015). Speech technology as documentation for endangered language preservation: The case of Irish, *Proceedings of the 18th International Congress of Phonetic Sciences*, Glasgow, UK. pp. 1-5.

Ní Chasaide, A., Ní Chiaráin, N., Wendler, C., Berthelsen, H., Murphy, A., & Gobl, C. (2017). *The ABAIR initiative: Bringing spoken Irish into the digital gpace* (pp. 2113–2117). Stockholm, Sweden: INTERSPEECH.

Ní Chiaráin, N. & Ní Chasaide, A. (2015). Evaluating synthetic speech in an Irish CALL application: Influences of predisposition and of the holistic environment. In *Proceedings of SLaTE 2015: 6th Workshop on Speech and Language Technologies in Education*, pp. 149–154.

Ní Chiaráin, N. & Ní Chasaide, A. (2016a). The Digichaint interactive game as a virtual learning environment for Irish. In *Proceedings of CALL communities and culture - short papers from EUROCALL 2016,* Limassol, Cyprus, Research-publishing.net, pp. 330–336. 10.14705/rpnet.2016.eurocall2016.584

Ní Chiaráin, N. & Ní Chasaide, A. (2016b, November) Faking intelligent CALL: the Irish context and the road ahead. *Proceedings of the Joint Workshop on NLP for Computer Assisted Language Learning and NLP for Language Acquisition at SLTC*, Umeå, Sweden, (130), Linköping University Electronic Press, Linköpings universitet, pp. 60–65.

Ní Chiaráin, N. & Ní Chasaide, A. (2016c). Chatbot technology with synthetic voices in the acquisition of an endangered language: Motivation, development and evaluation of a platform for Irish. In *Proceedings of the Tenth International Conference on Language Resources and Evaluation (LREC 2016)*, Portoro, Slovenia, May 23-28, 2016, European Language Resources Association (ELRA), pp. 3429-3435.

Ní Chiaráin, N. & Ní Chasaide, A. (2019). Integrating dialogue within an iCALL platform for Irish. *Proceedings of DiGo 2019: Dialog for Good - Workshop on Speech and Language Technology Serving Society*. Stockholm, Sweden (forthcoming).

Ní Chiaráin, N. (2014). Text-to-speech synthesis in computer-assisted language learning for Irish: Development and evaluation. [Doctoral thesis] CLCS, Trinity College, Ireland.

Nohara, Y., & Takamura, M. (2010). A study on nonfluent utterances that are difficult to understand by Japanese learners: Focusing on corrections and pauses. *Journal of the Graduate School of Humanities and Sciences. Ochanomizu University*, *13*, 117–125.

Norman, D. (2013). *The design of everyday things: Revised and expanded edition*. AZ: Basic Books.

Nunnally, J. (1978). *Psychometric theory* (2nd ed.). New York, NY: McGraw Hill.

O'Donovan, S., Gain, J., & Marais, P. (2013). A case study in the gamification of a university-level games development course. In *Proceedings of South African Institute for Computer Scientists and Information Technologists Conference*, 245–251. 10.1145/2513456.2513469

O'Dowd, R. (2005). Negotiating sociocultural and institutional contexts: The case of Spanish–American telecollaboration. *Language and Intercultural Communication*, *5*(1), 40–56. doi:10.1080/14708470508668882

O'Dowd, R. (2018). From telecollaboration to virtual exchange: State-of-the-art and the role of UNICollaboration in moving forward. *Journal of Virtual Exchange*, *1*, 1–23. doi:10.14705/rpnet.2018.jve.1

O'Dowd, R. (2019). A transnational model of virtual exchange for global citizenship education. *Language Teaching*, 1–14. doi:10.1017/S0261444819000077

O'Rourke, B. & Stickler, U. (2017). Synchronous communication technologies for language learning: Promise and challenges in research and pedagogy. *Language Learning in Higher Education, 7*(1), 1-20. http://dx.doi.org.uncfsu.idm.oclc.org/10.1515/cercles-2017-0009

OECD. (2015). *Students, computers and learning: Making the connection. PISA*. OECD Publishing.

Ogunduyile, A. O. (2013). Towards the integration of mobile phones in the teaching of English language in secondary schools in Akure, Nigeria. *Theory and Practice in Language Studies*, *3*, 1149–1153. doi:10.4304/tpls.3.7.1149-1153

Ok, M., & Ratliffe, K. (2017). Use of mobile devices for English language learner students in the United States: A research synthesis. *Journal of Educational Computing Research*, *56*(4), 538-562. doi:10.1177/0735633117715748

Oliveira, L., & Schleppegrell, J. (2015). *Focus on grammar and meaning*. Oxford, UK: Oxford University Press.

Oliver, R., Philp, J., & Duchesne, S. (2017). Children working it out together: A comparison of younger and older learners collaborating in task-based interaction. *System*, *69*, 1–14. doi:10.1016/j.system.2017.08.001

Oriji, A., & Fanny, A. (2019). Social media in teaching-learning process: Investigation of the seu of Whatsapp in teaching and learning in University of Port Harcourt. *European Scientific Journal*, *15*(4), 15–39. doi:10.19044/esj.2019.v15n4p15

Orsini-Jones, M. (2015). *Innovative pedagogies series: Integrating a MOOC into the MA in English language teaching at Coventry University-Innovation in blended learning practice*. Retrieved from https://www.heacademy.ac.uk/system/files/marina_orsini_jones_final_1.pdf

Orsini-Jones, M., Conde Gafaro, B., & Altamimi, S. (2017a). Integrating a MOOC into the postgraduate ELT curriculum: reflecting on students' beliefs with a MOOC blend. In Q. Kan & S. Bax (Eds.), Beyond the language classroom: Researching MOOCs and other innovations (pp. 71-83). Research-publishing.net. doi:10.14705/rpnet.2017.mooc2016.672

Orsini-Jones, M., Conde, B., Borthwick, K., Zou, B., & Ma, W. (2018). B-MELTT blending MOOCS for English language teacher training. *ELT Research Papers 18.02*. 1-41. Retrieved from https://www.teachingenglish.org.uk/sites/teacheng/files/Pub_J121_Blending%20MOOCs%20for%20English%20language%20teacher%20training_FINAL_Web.pdf

Orsini-Jones, M., Pibworth-Dolinski, L., Cribb, M., Brick, B., Gazeley-Eke, Z., Leinster, H., & Lloyd, E. (2015). Learning about language learning on a MOOC: How massive, open, online and "course"? In *Proceedings of the 2015 EUROCALL Conference* (pp. 450-457). Research-publishing.net. doi:10.14705/rpnet.2015.000374

Orsini-Jones, M., Zou, B., Borthwick, K., & Conde, B. (2017b). B-MELTT (blending MOOCs for English language teacher training): A distributed MOOC flip to explore local and global ELT contexts and beliefs. In *Proceedings of 18th International CALL Conference: CALL in Context* (pp. 608–625). Berkeley, CA: University of California. Retrieved from http://call2017.language.berkeley.edu/wp-content/uploads/2017/07/CALL2017_proceedings.pdf

Orús, C., Barlés, M. J., Belanche, D., Casaló, L., Fraj, E., & Gurrea, R. (2016). The effects of learner-generated videos for YouTube on learning outcomes and satisfaction. *Computers & Education*, *95*, 254–269. doi:10.1016/j.compedu.2016.01.007

Oskoz, A., Gimeno-Sanz, A., & Sevilla-Pavón, A. (2018). Examining L2 learners' use of engagement strategies in telecollaborative written interactions. In B. Mousten, S. Vandepitte, E. Arnó, & B. Maylath (Eds.), *Multilingual writing and pedagogical cooperation in virtual learning environments* (pp. 200–220). Hershey, PA: IGI-Global; doi:10.4018/978-1-5225-4154-7.ch008

Öz, H. (2016). The importance of personality traits in students' perceptions of metacognitive awareness. *Procedia: Social and Behavioral Sciences*, *232*, 655–667. doi:10.1016/j.sbspro.2016.10.090

Padron, C. J. (2013). Teaching strategies based on WhatsApp instant messaging application only for phones (mobile learning) and its use to promote collaborative learning. *Revista De Tecnología De Información y Comunicación En Educación*, *7*(2), 123–134.

Papadakis, S., & Kalogiannakis, M. (2018). Using gamification for supporting an introductory programming course. The case of classcraft in a secondary education classroom. In A. Brooks, E. Brooks, & N. Vidakis (Eds.), *Interactivity, Game Creation, Design, Learning, and Innovation. ArtsIT 2017, DLI 2017. Lecture Notes of the Institute for Computer Sciences, Social Informatics and Telecommunications Engineering* (Vol. 229). Cham, Switzerland: Springer. doi:10.1007/978-3-319-76908-0_35

Papathoma, T. (2019). *MOOC educators: who they are and how they learn.* (Doctoral dissertation). Retrieved from http://oro.open.ac.uk/59302/7/Thesis%20Final%20Papathoma_03.03.2019.pdf

Papert, S., & Harel, I. (1991). *Constructionism.* Westport, CT: Ablex Publishing.

Pappano, L. (2012, Nov. 2). The year of the MOOC. Education life. *The New York Times.* Retrieved from https://www.nytimes.com/2012/11/04/education/edlife/massive-open-online-courses-are-multiplying-at-a-rapid-pace.html

Park, S. Y., Nam, M., & Cha, S. (2012). University students' behavioral intention to use mobile learning: Evaluating the technology acceptance model. *British Journal of Educational Technology*, *43*(4), 592–605. doi:10.1111/j.1467-8535.2011.01229.x

Partnership for 21. Century Learning. (2015). Retrieved from https://www.p21.org//our-work/p21-framework

Pede, J. (2017). *The effects of the online game Kahoot on science vocabulary acquisition.* (Doctoral dissertation), Rowan University.

Pegrum, M. (2016). Three agendas for MALL. In A. Palalas & M. Ally (Eds.), The international handbook of mobile-assisted language learning (pp. 86-112). China Central Radio & TV University Press Co.

Pegrum, M. (2009). *From blogs to bombs: The future of digital technologies in education.* Crawley, Australia: UWA Publishing.

Pegrum, M. (2014). *Mobile learning: Languages, literacies and cultures.* Basingstoke, UK: Palgrave Macmillan. doi:10.1057/9781137309815

Pegrum, M. (2015). Language and literacies for digital lives. In E. Martín-Monje, I. Elorza, & B. García Riaza (Eds.), *Technological advances in specialized linguistic domains: Learning on the move* (pp. 16–36). Londres, UK: Routledge.

Pellas, N., & Kazanidis, I. (2013). On the value of Second Life for students' engagement in blended and online courses: A comparative study from the Higher Education in Greece. *Education and Information Technologies*, *20*(3), 445–466. doi:10.100710639-013-9294-4

Pellegrini, T., Costa, Â., & Trancoso, I. (2012). Less errors with TTS? A dictation experiment with foreign language learners. In *Proceedings Thirteenth Annual Conference of the International Speech Communication Association.*

Penner, N., & Schulze, M. (2010). Group work in a technology-rich environment. *Journal of Interactive Learning Research, 21*(1), 111–137.

Pennock-Speck, B. & Clavel-Arroitia, B. (2018). Teachers' perspectives on telecollaboration in secondary school foreign language education. *LEA - Lingue e letterature d'Oriente e d'Occidente, 7*, 483-509. doi:10.13128/LEA-1824-484x-24405

Pennock-Speck, B., & Clavel-Arroitia, B. (2018). Teachers' perspectives on telecollaboration in secondary school foreign language education. *LEA. Lingue e Letterature d'Oriente e d'Occidente, 7*, 483–509.

Pérez Cañado, M. L. & Ráez Padilla, J. (2015). Introduction and Overview. In D. Marsh, M. L. Pérez Cañado, & J. Ráez Padilla (Eds.), *CLIL in Action: Voices from the Classroom* (pp. 1-12). Newcastle upon Tyne, UK: Cambridge Scholars.

Pérez Cañado, M. L. (2018). AICLE/EMI y CMC: Un tándem efectivo para el desarrollo de la interacción comunicativa escrita. In M. González-Lloret & M. Vinagre (Eds.), *Comunicación Mediada por Tecnologías: Aprendizaje y Enseñanza de la Lengua Extranjera* (pp. 35–54). Sheffield, UK: Equinox.

Pérez, Á., Soto, E., Sola, M., & Serván, M. (2009). *Aprender en la Universidad. El sentido del cambio en el EEES.* Madrid, Spain: Akal.

Pérez-Paredes, P., Ordoñana Guillamón, C., & Aguado Jiménez, P. (2018). Language teachers' perceptions on the use of OER language processing technologies in MALL. *Computer Assisted Language Learning, 31*(5-6), 5–6, 522–545. doi:10.1080/09588221.2017.1418754

Pérez-Sanagustín, R. & Maldonado, M. J. J. (2016). How to design tools for supporting self-regulated learning in MOOCs? Lessons learned from a literature review from 2008 to 2016. In *Proceedings of Computing Conference (CLEI), 2016 XLII Latin American* (pp. 1-12). New York, NY: Institute of Electrical and Electronics Engineers (IEEE).

Pérez-Sanagustín, M., Hilliger, I., Alario-Hoyos, C., Kloos, C. D., & Rayyan, S. (2017). H-MOOC framework: Reusing MOOCs for hybrid education. *Journal of Computing in Higher Education, 29*(1), 47–64. doi:10.100712528-017-9133-5

Perry, B. (2015). Gamifying French language learning: A case study examining a quest-based, augmented reality mobile learning-tool. *Procedia: Social and Behavioral Sciences, 174*, 2308–2315. doi:10.1016/j.sbspro.2015.01.892

Persson, V., & Nouri, J. (2018). A systematic review of second language learning with mobile technologies. *International Journal of Emerging Technologies in Learning, 13*(2), 188-210. doi:10.3991/ijet.v13i02.8094

Petrangeli, S., Pauwels, D., van der Hooft, J., Ziak, M., Slowack, J., Wauters, T., & De Turck, F. (2019). A scalable WebRTC-based framework for remote video collaboration applications. *Multimedia Tools and Applications, 78*(6), 7419–7452. doi:10.100711042-018-6460-0

Pica, T. (1994). Research on negotiation: What does it reveal about second-language learning conditions, processes, and outcomes? *Language Learning, 44*(3), 493–527. doi:10.1111/j.1467-1770.1994.tb01115.x

Picton, I. (2014). *The impact of eBooks on the reading motivation and reading skills of children and young people: A rapid literature review.* London, UK: National Literacy Trust.

Pillay, H. K. (1994). Cognitive load and mental rotation: Structuring orthographic projection for learning and problem solving. *Instructional Science, 22*(2), 91–113. doi:10.1007/BF00892159

Pintrich, P. L. & Schunk, D. H. (1996). Motivation in education: Theory, research and applications. Englewood.

Pintrich, P. R., Smith, D. A. F., Garcia, T., & Mckeachie, W. J. (1991). A manual for the use of the motivated strategies for learning questionnaire (MSLQ). Retrieved from https://files.eric.ed.gov/fulltext/ED338122.pdf

Pixie, Y. (2010). 10 reasons why I left farmville [Weblog]. Retrieved from http://blogcritics.org/gaming/article/10-reasons-why-i-left-farmville/

Plan Ceibal. (2017). *10 años Plan Ceibal: Hicimos historia haciendo futuro.*

Plump, C. M., & LaRosa, J. (2017). Using Kahoot! in the classroom to create engagement and active learning: A game-based technology solution for elearning novices. *Management Teaching Review*, *2*(2), 151–158. doi:10.1177/2379298116689783

Pourfeiz, J. (2015). Exploring the relationship between global personality traits and attitudes toward foreign language learning. *Procedia: Social and Behavioral Sciences*, *186*, 467–473. doi:10.1016/j.sbspro.2015.04.119

Prefontaine, Y., Kormos, J., & Johnson, D. E. (2015). How do utterance measures predict raters' perceptions of fluency in French as a second language? *Language Testing*, *33*(1), 53–73. doi:10.1177/0265532215579530

Prensky, M. (2001). Digital natives, digital immigrants. From On the Horizon (MCB University Press, Vol. 9 No. 5, October 2001) © 2001 Marc Prensky.donova

Prensky, M. (2001). Digital natives, digital immigrants part 1. *On the Horizon*, *9*(5), 1–6. doi:10.1108/10748120110424816

Quirk, M., Unrau, N., Ragusa, G., Rueda, R., Lim, H., Velasco, A., ... Loera, G. (2010). Teacher beliefs about reading motivation and their enactment in classrooms: The development of a survey questionnaire. *Reading Psychology*, *31*(2), 93–120. doi:10.1080/02702710902754051

Raad, B., & Mlačić, B. (2015). Big Five factor model, theory and structure. In J. D. Wright (Ed.), *International encyclopedia of the social & behavioral sciences* (pp. 559–566). Oxford, UK: Elsevier. doi:10.1016/B978-0-08-097086-8.25066-6

Radu, I. (2014). Augmented reality in education: A meta-review and cross-media analysis. *Personal and Ubiquitous Computing*, *18*(6), 1533–1543. doi:10.100700779-013-0747-y

Rahmah, R. E. (2018). Sharing photograms on Instagram boosts students' self-confidence in speaking English. *Pedagogy Journal of English Language Teaching*, *6*(2), 148–156. doi:10.32332/pedagogy.v6i2.1335

Rambe, P., & Chipunza, C. (2013). Using mobile devices to leverage student access to collaboratively-generated resources: A case of WhatsApp instant messaging at a South African University. *British Journal of Educational Technology*, *44*(4), 544–561. doi:10.1111/bjet.12057

Rat für Kulturelle Bildung. (2019). *Jugend/Youtube/Kulturelle Bildung. Horizont 2019* [Youth/Youtube/Cultural Education. Horizont 2019]. Retrieved from https://www.rat-kulturelle-bildung.de/fileadmin/user_upload/pdf/Studie_YouTube_Webversion_final.pdf

Read, T., Bárcena, E., & Kukulska-Hulme, A. (2016). Exploring the application of a conceptual framework in a social MALL app. In A. Pareja-Lora, C. Calle-Martínez, & P. Rodríguez-Arancón (Eds.), New perspectives on teaching and working with languages in the digital era (pp. 223-232). Research-Publishing.net. doi:10.14705/rpnet.2016.tislid2014.436

Read, T., Bárcena, E., & Kukulska-Hulme, A. (2016). Mobile and massive language learning. In E. Martín-Monje, I. Elorza, & B. G. Riaza (Eds.), *Technology-enhanced language learning for specialized domains: Practical applications and mobility* (pp. 151–161). New York, NY: Routledge.

Read, T., Bárcena, E., & Rodrigo, C. (2010). Modelling ubiquity for second language learning. *International Journal of Mobile Learning and Organisation*, *4*(2), 130–149. doi:10.1504/IJMLO.2010.032633

Reiners, T., Wood, L. C., Chang, V., Guetl, C., Herrington, J., Gregory, S., & Teräs, H. (2012). Operationalising gamification in an educational authentic environment. In *IADIS 2012 International Conference on Internet Technologies and Society* (pp. 93-100). Perth, Australia: IADIS.

Reiners, T., & Wood, L. C. (2013). Immersive virtual environments to facilitate authentic education in logistics and supply chain management. In Y. Kats (Ed.), *Learning management systems and instructional design: Best practices in online education* (pp. 323–343). Hershey, PA: IGI Global. doi:10.4018/978-1-4666-3930-0.ch017

Reinhardt, J. (2019). Social media in second and foreign language teaching and learning: Blogs, wikis, and social networking. *Language Teaching*, *52*(1), 1–39. doi:10.1017/S0261444818000356

Reinking, D. (2005). Multimedia learning of reading. In R. E. Mayer (Ed.), Cambridge handbook of multimedia learning (pp. 355-74). Cambridge, UK: Cambridge University Press. doi:10.1017/CBO9780511816819.024

Reiser, R., & Dempsey, J. (2002). *Trends and issues in instructional design and technology.* Boston, MA: Pearson.

Riazantseva, A. (2001). Second language proficiency and pausing: A study of Russian speakers of English. *Studies in Second Language Acquisition*, *23*(4), 497–526. doi:10.1017/S027226310100403X

Richards, J. (2008). *Teaching listening and speaking.* Cambridge, UK: Cambridge University Press.

Richards, J., & Renandya, W. (2002). *Methodology in language teaching: An anthology of current practice.* Cambridge University Press. doi:10.1017/CBO9780511667190

Riggenbach, H. (1991). Toward an understanding of fluency: A microanalysis of nonnative speaker conversations. *Discourse Processes*, *14*(4), 423–441. doi:10.1080/01638539109544795

Rivera, G. M., & Matsuzawa, C. (2007). Multiple-language program assessment: Learners' perspectives on first- and second-year college foreign language programs and their implications for program improvement. *Foreign Language Annals*, *40*(4), 569–583. doi:10.1111/j.1944-9720.2007.tb02881.x

Rivera-Trigueros, I. (2018). *Audiovisual translation as a tool to foster the linguistic competence in the EFL classroom and the improvement of the motivation through gamification* (Unpublished Master's dissertation). University of Almería, Spain.

Ronaghi, F., Saberi, A., & Trumbore, A. (2015). NovoEd, a social learning environment. In P. Kim (Ed.), *Massive Open Online Courses* (pp. 96–105). New York, NY: Routledge.

Roschelle, J., Rafanan, K., Estrella, G., Nussbaum, M., & Claro, S. (2010). From handheld collaborative tool to effective classroom module: Embedding CSCL in a broader design framework. *Computers & Education*, *55*(3), 1018–1026. doi:10.1016/j.compedu.2010.04.012

Rosell-Aguilar, F. (2013). Podcasting for language learning through iTunes U: The learner's view. *Language Learning & Technology*, *17*(3), 74–93.

Rosell-Aguilar, F. (2017). State of the app: A taxonomy and framework for evaluating language learning mobile applications. *CALICO Journal*, *34*(2), 243–258.

Rosewell, J. (2012). A speculation on the possible use of badges for learning at the UK open university. In *EADTU Annual Conference: The Role Of Open And Flexible Education In European Higher Education Systems For 2020: New Models, New Markets, New Media.*

Rubio, F. (2015). The role of interaction in MOOCs and traditional technology-enhanced language courses. In E. Dixon & M. Thomas (Eds.), Researching language student interaction online: From social media to MOOCs. (pp. 63–88). San Marcos, TX: CALICO monograph.

Rubio, F., Fuchs, C., & Dixon, E. (2016). Language MOOCs: Better by design. In E. Martín-Monje, I. Elorza, & B. G. Riaza (Eds.), *Technology-enhanced language learning for specialized domains: Practical applications and mobility* (pp. 177–188). New York, NY: Routledge.

Ryan, R. M. (1982). Control and information in the intrapersonal sphere: An extension of cognitive evaluation theory. *Journal of Personality and Social Psychology*, *43*(3), 450–461. doi:10.1037/0022-3514.43.3.450

Ryan, R. M., & Deci, E. L. (2000). Intrinsic and extrinsic motivations: Classic definitions and new directions. *Contemporary Educational Psychology*, *25*(1), 54–67. doi:10.1006/ceps.1999.1020 PMID:10620381

Ryan, S., Kaufman, J., Greenhouse, J., She, R., & Shi, J. (2016). The effectiveness of blended online learning courses at the community college level. *Community College Journal of Research and Practice*, *40*(4), 285–298. doi:10.1080/10668926.2015.1044584

Ryberg, T., & Christiansen, E. (2008). Community and social network sites as technology enhanced learning environments. *Technology, Pedagogy and Education*, *17*(3), 207–219. doi:10.1080/14759390802383801

Sabah, N. M. (2016). Exploring students' awareness and perceptions: Influencing factors and individual differences driving m-learning adoption. *Computers in Human Behavior*, *65*, 522–533. doi:10.1016/j.chb.2016.09.009

Saidouni, K. & Bahloul, A. (2016) Teachers and students' attitudes towards using mobile-assisted language learning in higher education, Arab World English Journal Special Issue on CALL No. 3 July 2016 (pp.123-140).

Sailer, M., Hense, J., Mandl, H., & Klevers, M. (2013). Psychological perspectives on motivation through gamification. *Interaction Design and Architecture(s) Journal - IxD&A, 19*, 28-37.

Sailer, M., Hense, J., Mayr, S., & Mandl, H. (2017). How gamification motivates: An experimental study of the effects of specific game design elements on psychological need satisfaction. *Computers in Human Behavior*, *69*, 371–380. doi:10.1016/j.chb.2016.12.033

Sakoda, K., Konishi, M., Sasaki, A., Suga, W., & Hosoi, Y. (2016). International corpus of Japanese as a second language. *NINJAL Project Review*, *6*(3), 93–110.

Sakuragi, T. (2011). The construct validity of the measures of complexity, accuracy, and fluency: Analyzing the speaking performance of learners of Japanese. *Japanese Association for Language Teaching Journal*, *33*, 157–173.

Salmerón, L., Strømsø, H. I., Kammerer, Y., Stadtler, M., & Van den Broek, P. W. (2018). Comprehension processes in digital reading. In M. Barzillai, J. Thomson, S. Schroeder, & P. W. Van den Broek (Eds.), *Learning to Read in a Digital World* (pp. 91–120). Amsterdam, The Netherlands: John Benjamins. doi:10.1075wll.17.04sal

Samaie, M., Nejad Mansouri, A., & Qaracholloo, M. (2018). An inquiry into the efficiency of WhatsApp for self- and peer-assessments of oral language proficiency. *British Journal of Educational Technology*, *49*(1), 111–112. doi:10.1111/bjet.12519

Sánchez, E., Young, S., & Jouneau-Sion, C. (2017). Classcraft: From gamification to ludicization of classroom management. *Education and Information Technologies*, *22*(2), 497–513. doi:10.100710639-016-9489-6

Sánchez-García, A. B., & Mena Marcos, J. J. (2013). Teacher development and ICT: The effectiveness of a training program for in-service school teachers. *Procedia: Social and Behavioral Sciences*, *92*, 529–534. doi:10.1016/j.sbspro.2013.08.713

Sánchez-Pérez, M. M., & Salaberri Ramiro, M. S. (2017). Implementing plurilingualism in higher education: Teacher training needs and plan evaluation. *Porta Linguarum. Monograph, II*, 139–156.

Sandeen, C. (2013). Integrating MOOCS into traditional higher education: The emerging "MOOC 3.0" era. *Change: The Magazine of Higher Learning*, *45*(6), 34–39. doi:10.1080/00091383.2013.842103

Sang, G., Valcke, M., van Braak, J., & Tondeur, J. (2010). Student teachers' thinking processes and ICT integration: Predictors of prospective teaching behaviors with educational technology. *Computers & Education*, *54*(1), 103–112. doi:10.1016/j.compedu.2009.07.010

Sant, C. (2014). *Leveling up: Evaluating theoretical underpinnings and applications of gamification in the classroom* (Unpublished dissertation). Peabody College of Education and Human Development, Vanderbilt University, Nashville, TN.

Sato, M., & Loewen, S. (2018). Do teachers care about research? The research-pedagogy dialogue. *ELT Journal*.

Saumell, V. (2016). Principle for meaningful technology integration. *Modern English Teacher*, *25*(2), 15–17.

Savvani, S. (2019). State-of-the-art duolingo features and applications. *Advances in Intelligent Systems and Computing*, *917*, 139–148. doi:10.1007/978-3-030-11935-5_14

Scannell, K. (2013). An Gramadóir: An open source grammar checking engine, Version 0.70. Retrieved from https://cadhan.com/gramadoir/

Schlochtermeier, L. H., Pehrs, C., Kappelhoff, H., Kuchinke, L., & Jacobs, A. M. (2015). Emotion processing in different media types: Realism, complexity, and immersion. *Journal of Systems and Integrative Neuroscience*, *1*(2), 41–47. doi:10.15761/JSIN.1000109

Schmidt, R. (1990). The role of consciousness in second language learning. *Applied Linguistics*, *11*(2), 129–158. doi:10.1093/applin/11.2.129

Schmidt, R. (2001). Attention. In P. Robinson (Ed.), *Cognition and second language instruction* (pp. 3–32). Cambridge, UK: Cambridge University Press. doi:10.1017/CBO9781139524780.003

Schmidt, R., & Watanabe, Y. (2001). Motivation, strategy use, and pedagogical preferences in foreign language learning. In Z. Dörnyei & R. Schmidt (Eds.), *Motivation and second language acquisition* (pp. 313–359). Honolulu, HI: University of Hawai'i, Second Language Teaching and Curriculum Center.

Schmitt, N., & McCarthy, M. (Eds.). (1997). *Vocabulary: Description, acquisition and pedagogy*. Cambridge, UK: Cambridge University Press.

Schuck, S., & Kearny, M. (2006). Capturing learning through student-generated digital video. *Australian Educational Computing*, *21*(1), 15–20.

Schulze, M. (2007). Modeling SLA processes using NLP. In C. Chapelle, Y.-R. Chung, & J. Xu (Eds.), *Towards adaptive CALL: Natural language processing for diagnostic assessment* (pp. 149–166). Iowa State University.

Schulze, M. (2008). AI in CALL – Artificially inflated or almost imminent? *CALICO Journal*, *25*(3), 510–527. doi:10.1558/cj.v25i3.510-527

Scott, D., & Beadle, S. (2014). *Improving the effectiveness of language learning: CLIL and computer assisted language learning*. London, UK: European Commission.

Seaman, J. E., Allen, I. E., Seaman, J., & Babson Survey Research Group. (2018). *Grade increase: Tracking distance education in the United States*. Place of publication not identified: distributed by ERIC Clearinghouse.

Seaman, J., & Tinti-Kane, H. (2013). *Social media for teaching and learning*. London, UK: Pearson Learning Systems.

Segalowitz, N. (2010). *Cognitive bases of second language fluency*. New York, NY: Routledge. doi:10.4324/9780203851357

Segalowitz, N., & Freed, B. F. (2004). Context, contact, and cognition in oral fluency acquisition: Learning Spanish in at home and study abroad contexts. *Studies in Second Language Acquisition*, *26*(02), 173–200. doi:10.1017/S0272263104262027

Segalowitz, N., French, L., & Guay, J. D. (2018). What features best characterize adult second language utterance fluency and what do they reveal about fluency gains in short-term immersion? *Canadian Journal of Applied Linguistics*, *20*(2), 90–116. doi:10.7202/1050813ar

Sevilla-Pavón, A., & Haba Osca, J. (2016). Te das cuenta de que el mundo puede ser tan distinto y similar al mismo tiempo": Telecolaboración y desarrollo de la competencia intercultural en la educación superior. *Didáctica. Lengua y Literatura*, *28*, 263–284. doi:10.5209/DIDA.54088

Shadiev, R., Hwang, W. Y., Chen, N. S., & Huang, Y. M. (2014). Review of speech-to-text recognition technology for enhancing learning. *Journal of Educational Technology & Society*, *17*(4), 65–84.

Shadiev, R., Liu, T., & Hwang, W.-Y. (2019). Review of research on mobile-assisted language learning in familiar, authentic environments: Review of research on MALL in familiar contexts. *British Journal of Educational Technology*, 1-12. Retrieved from https://onlinelibrary.wiley.com/doi/pdf/10.1111/bjet.12839

Sha, G. (2010). Using TTS voices to develop audio materials for listening comprehension: A digital approach. *British Journal of Educational Technology*, *41*(4), 632–641. doi:10.1111/j.1467-8535.2009.01025.x

Sharma, S. (1996). *Applied Multivariate Techniques*. New York, NY: John Wiley & Sons.

Sharp, A. (2008). Personality and second language learning. *Asian Social Science*, *4*(11), 17–25.

Sharples, M. (2019). *Practical pedagogy: 40 new ways to teach and learn*. London, UK: Routledge.

Shazali, S. S., Shamsudin, Z. H., & Yunus, M. M. (2019). Instagram: A platform to develop student's writing ability. *International Journal of Academic Research in Business and Social Sciences*, *9*(1), 88–98. doi:10.6007/IJARBSS/v9-i1/5365

Shea, P. (2007). Towards a conceptual framework for learning in blended environments. In A. G. Picciano & C. D. Dziuban (Eds.), *Blended learning: Research perspectives* (pp. 19–35). Retrieved from http://elab.learningandteaching.dal.ca/dalblend2013- files/blended-learning-research-perspectives-book.pdf#page=30

Shea, P., Hayes, S., Smith, S. U., Vickers, J., Bidjerano, T., Pickett, A., ... Jian, S. (2012). Learning presence: Additional research on a new conceptual element within the Community of Inquiry (CoI) framework. *The Internet and Higher Education*, *15*(2), 89–95. doi:10.1016/j.iheduc.2011.08.002

Sheen, Y. (2008). Recasts, Language Anxiety, Modified Output, and L2 Learning. *Language Learning*, *58*(4), 835–874.

Shih, R. C. (2011). Can Web 2.0 technology assist college students in learning English writing? Integrating Facebook and peer assessment with blended learning. *Australasian Journal of Educational Technology*, *27*(5), 829–845. doi:10.14742/ajet.934

Shrader, S., Wu, M., Owens, D., & Santa Ana, K. (2016). Massive open online courses: Participant activity, demographics, and satisfaction. *Online Learning Journal*, *20*(2), 199.

Siemens, G. (2005). Connectivism: A learning theory for the digital age. *International journal of instructional technology and distance learning, 2*(1), 3-10.

Silvers, A. (2011). On education, badges and scouting [web log message].

Simons, M., T'Sas, J., & Mommaerts, M. (2014). *Leesplezier en e-hype: good practices in(ge)beeld en ingezet*. Universiteit Antwerpen.

Sinha, T., Jermann, P., Li, N., & Dillenbourg, P. (2014). Your click decides your fate: Inferring information processing and attrition behavior from MOOC video clickstream interactions. In *Proceedings of the EMNLP 2014 Workshop on Analysis of Large Scale Social Interaction in MOOCS.* 10.3115/v1/W14-4102

Sirait, J. B., & Marlina, L. (2018). Using Instagram as a tool for online peer-review activity in writing descriptive text for senior high school students. *Journal of English Language Teaching, 7*(1), 291–302. Retrieved from http://ejournal.unp.ac.id/index.php/jelt/article/view/9771/pdf

Sitzmann, T. (2011). A meta-analytic examination of the instructional effectiveness of computer-based simulation games. *Personnel Psychology, 64*(2), 489–528. doi:10.1111/j.1744-6570.2011.01190.x

Skehan, P. (1998). *A cognitive approach to language learning.* Oxford, UK: Oxford University Press.

Skehan, P. (2003). Task-based instruction. *Language Teaching, 36*(1), 1–14. doi:10.1017/S026144480200188X

Skehan, P. (2009). Modelling second language performance: Integrating complexity, accuracy, fluency, and lexis. *Applied Linguistics, 30*(4), 510–532. doi:10.1093/applin/amp047

Skehan, P. (2014). Limited attentional capacity, second language performance, and task-based pedagogy. In P. Skehan (Ed.), *Processing perspectives on task performance* (pp. 211–260). Amsterdam, The Netherlands: John Benjamins. doi:10.1075/tblt.5.08ske

Skehan, P., & Foster, P. (2005). Strategic and on-line planning: The influence of surprise information and task time on second language performance. In E. Rod (Ed.), *Planning and Task Performance in Second Language* (pp. 193–216). Amsterdam, The Netherlands: John Benjamins. doi:10.1075/lllt.11.12ske

Slattery, M., & Willis, J. (2001). *English for primary teachers: A handbook of activities and classroom language* (Vol. 1). Oxford, UK: Oxford University Press.

Smit, U. & Dafouz, E. (2012). Integrating content and language in higher education: Gaining insights into English-medium instruction at European universities (AILA vol. 25). Amsterdam, The Netherlands: John Benjamins.

Smith, A. (2015). *U.S. Smartphone use in 2015.* Retrieved from http://www.pewinternet.org/files/2015/03/PI_Smartphones_ 0401151.pdf

Snoder, P. & [author] (accepted). *Word Up! Boosting L2 learners' vocabulary through ecological communities of practice.* Presentation at the VOCAB@Leuven Conference. Katholieke Universiteit Leuven, 1-3 July 2019.

Sokolik, M. (2016). Academic writing in MOOC environments. In E. Martín-Monje, I. Elorza, & B. G. Riaza (Eds.), *Technology-enhanced language learning for specialized domains: Practical applications and mobility* (pp. 165–176). New York, NY: Routledge.

So, S. (2016). Mobile instant messaging support for teaching and learning in higher education. *The Internet and Higher Education, 31*, 32–42. doi:10.1016/j.iheduc.2016.06.001

Soviyah, S., & Etikaningsih, D. R. (2018). Instagram use to enhance ability in writing descriptive texts. *Indonesian EFL Journal, 4*(2), 32–38. doi:10.25134/ieflj.v4i2.1373

SPOT. (n.d.) *About the SPOT (simple performance-oriented test).* Retrieved from http://ttbj-tsukuba.org/mt/p2_e.html

Statista. (2017, September). *Number of daily active Instagram users.* Retrieved from https://www.statista.com/statistics/657823/number-of-daily-active-instagram-users/

Steel, C. (2015). Students' perspectives on the affordances and constraints of using mobile devices and applications for learning languages. In A. Gimeno-Sanz, M. Levy, D. Barr, & F. Blin (Eds.), *WorldCALL: Sustainability and computer-assisted language learning* (pp. 230–243). London, UK: Bloomsbury.

Stevenson, B., Länsitie, J., Kogler, C., & Bauer, P. (2015). *Exploring co-creation of educational videos in an international collaborative context, 11*(2), 10.

Stevenson, M. P., & Liu, M. (2013). Learning a language with Web 2.0: Exploring the use of social networking features of foreign language learning websites. *CALICO Journal, 27*(2), 233–259. doi:10.11139/cj.27.2.233-259

Stockwell, G. & Hubbard, P. (2013). Some emerging principles for mobile-assisted language learning. *The International Research Foundation for English Language Education,* 1-15.

Stockwell, G. (2008). Investigating learner preparedness for and usage patterns of mobile learning. *ReCALL, 20*(3), 253–270. doi:10.1017/S0958344008000232

Stockwell, G. (2010). Using mobile phones for vocabulary activities: Examining the effect of platform. *Language Learning & Technology, 14*(2), 95–110.

Stockwell, G. (2013). Technology and motivation in English-language teaching and learning. In E. Ushioda (Ed.), *International perspectives on motivation: Language learning and professional challenges* (pp. 156–175). London, UK: Palgrave Macmillan UK. doi:10.1057/9781137000873_9

Stockwell, G. (Ed.). (2012). *Computer assisted language learning: Diversity in research and practice*. Cambridge, UK: Cambridge University Press. doi:10.1017/CBO9781139060981

Stockwell, G., & Hubbard, P. (2013). *Some emerging principles for mobile-assisted language learning*. Monterey, CA: The International Research Foundation for English Language Education; Retrieved from http://www.tirfonline.org/english-in-the-workforce/mobile-assisted-language-learning

Stockwell, G., & Liu, Y. C. (2015). Engaging in mobile phone-based activities for learning vocabulary: An investigation in Japan and Taiwan. *CALICO Journal, 32*(2), 299–322. doi:10.1558/cj.v32i2.25000

StoutD. (2019). Retrieved from https://dustinstout.com/social-media-statistics/#whatsapp-stats

STRATAPP. Gamifying academic English skills in higher education: Reading academic English app (2016-2019). Erasmus + Project. Available at http://stratapp.eu

Sukri, H. I. M., Mustapha, L., Othman, M., Aralas, D., & Ismail, L. (2018). Social media: Engaging language learning. *International Journal of Academic Research in Business and Social Sciences, 8*(12), 287–294. doi:10.6007/IJARBSS/v8-i12/5013

Sung, Y. T., Chang, K. E., & Liu, T. C. (2016). The effects of integrating mobile devices with teaching and learning on students' learning performance: A meta-analysis and research synthesis. *Computers & Education, 94*, 252–275. doi:10.1016/j.compedu.2015.11.008

Sung, Y.-T., Chang, K.-E., & Yang, J.-M. (2015). How effective are mobile devices for language learning? A meta-analysis. *Educational Research Review, 16*, 68-84. doi:10.1016/j.edurev.2015.09.001

Sun, Z., Lin, C. H., You, J., Shen, H. J., Qi, S., & Luo, L. (2017). Improving the English-speaking skills of young learners through mobile social networking. *Computer Assisted Language Learning, 30*(3–4), 304–324. doi:10.1080/09588221.2017.1308384

Swain, M., & Lapkin, S. (1998). Interaction and second language learning: Two adolescent French immersion students working together. *Modern Language Journal*, *82*(3), 320–337. doi:10.1111/j.1540-4781.1998.tb01209.x

Sweet, A. P., Guthrie, J. T., & Ng, M. M. (1998). Teacher perceptions and student reading motivation. *Journal of Educational Psychology*, *90*(2), 210–223. doi:10.1037/0022-0663.90.2.210

Sydorenko, T., Hsieh, C.-N., Ahn, S., & Arnold, N. (2017). Foreign language learners' beliefs about CALL: The case of a midwestern University. *CALICO, 34*(2), 196-218.

Sykes, J. M., Oskoz, A., & Thorne, S. L. (2008). Web 2.0, synthetic immersive environments, and mobile resources for language education. *CALICO Journal*, *25*(3), 528–246. doi:10.1558/cj.v25i3.528-546

Tasgin, A., & Cosku, G. (2018). The relationship between academic motivations and university students' attitudes towards learning. *International Journal of Instruction*, *11*(4), 935–950. doi:10.12973/iji.2018.11459a

Taşkıran, A., Koral-Gümüşoğlu, E., & Aydın, B. (2018). Fostering foreign language learning with Twitter: Reflections from English learners. *Turkish Online Journal of Distance Education-TOJDE*, *19*(1), 100–116. doi:10.17718/tojde.382734

Tavakoli, P., Nakatsuhara, F., & Hunter, A. M. (2017). Scoring validity of the aptis speaking test: Investigating fluency across tasks and levels of proficiency. *ARAGs Research Reports Online*. Retrieved from http://centaur.reading.ac.uk/73379/

Tavakoli, P. (2011). Pausing patterns: Differences between L2 learners and native speakers. *ELT Journal*, *65*(1), 71–79. doi:10.1093/elt/ccq020

Tavakoli, P. (2015). Connecting research and practice in TESOL: A community of practice perspective. *RELC Journal*, *46*(1), 37–52. doi:10.1177/0033688215572005

Tavakoli, P. (2016). Fluency in monologic and dialogic task performance: Challenges in defining and measuring L2 fluency. *International Review of Applied Linguistics in Language Teaching*, *54*(2), 133–150. doi:10.1515/iral-2016-9994

Tavakoli, P. (2018). L2 development in an intensive study abroad EAP context. *System*, *72*, 62–74. doi:10.1016/j.system.2017.10.009

Tavakoli, P., Campbell, C., & McCormack, J. (2016). Development of speech fluency over a short period of time: Effects of pedagogic intervention. *TESOL Quarterly*, *50*(2), 447–471. doi:10.1002/tesq.244

Tavakoli, P., & Foster, P. (2008). Task design and second language performance: The effect of narrative type on learner output. *Language Learning*, *58*(2), 439–473. doi:10.1111/j.1467-9922.2008.00446.x

Tavakoli, P., & Howard, M. (2012). Teaching English to speakers of other languages teachers' views on the relationship between research and practice. *European Journal of Teacher Education*, *35*(2), 229–242. doi:10.1080/02619768.2011.643398

Tavakoli, P., & Skehan, P. (2005). Planning task structure, and performance testing. In R. Ellis (Ed.), *Planning and task performance in a second language* (pp. 239–273). Amsterdam, The Netherlands: John Benjamins. doi:10.1075/lllt.11.15tav

Teclehaimanot, B., & Hickman, T. (2011). Student-teacher interaction on Facebook: What students find appropriate. *TechTrends*, *55*(3), 19–30. doi:10.100711528-011-0494-8

TeCoLa. (2019). Pedagogical differentiation through telecollaboration and gaming for intercultural and content integrated language teaching. Retrieved from https://sites.google.com/site/tecolaproject/

Tejada Molina, G., Pérez Cañado, M. L., & Luque Agulló, G. (2005). Current approaches and teaching methods. *Bilingual programmes. TEFL in Secondary Education.* Granada, Spain: Editorial Universidad de Granada. Available at http://www4.ujaen.es/~gluque/Chapter4HANDBOOKDEFINITIVO.pdf

Telles, J. A. (2015). Teletandem and performativity. *Revista Brasileira de Lingüística Aplicada, 15*(1), 1–30. doi:10.1590/1984-639820155536

TELL-OP. Transforming European learner language into learning opportunities. (2015-2017). OER survey: Language processing technologies. Erasmus + Project. Available at http://www.tellop.eu/wp-content/uploads/2015/09/TELL-OP-Survey-ENG.pdf

Thomas, M., Reinders, H., & Warschauer, M. (Eds.). (2013). *Contemporary Computer-Assisted Language Learning.* New York, NY: Continuum.

Thorne, S. L. (2010). The 'intercultural turn' and language learning in the crucible of new media. In F. Helm & S. Guth (Eds.), *Telecollaboration 2.0 for Language and Intercultural Learning* (pp. 139–164). Bern, Switzerland: Peter Lang.

Towell, R., Hawkins, R., & Bazergui, N. (1996). The development of fluency in advanced learners of French. *Applied Linguistics, 17*(1), 84–119. doi:10.1093/applin/17.1.84

Tran, T. T. T., & Moni, K. (2015). Management of foreign language anxiety: Insiders' awareness and experiences. *Cogent Education, 2*(1), 992593. doi:10.1080/2331186X.2014.992593

Traxler, J. (2009). Learning in a mobile age. [IJMBL]. *International Journal of Mobile and Blended Learning, 1*(1), 1–12. doi:10.4018/jmbl.2009010101

Tremblay, P. F., & Gardner, R. C. (1994). On motivation, research agendas, and theoretical framework. *Modern Language Journal, 78*(3), 359–368.

Tremblay, P. F., Goldberg, M. P., & Gardner, R. C. (1995). Trait and state motivation and the acquisition of Hebrew vocabulary. *Canadian Journal of Behavioural Science, 27*(3), 356–370. doi:10.1037/0008-400X.27.3.356

Tsai, C.-H., Cheng, C.-H., Yeh, D.-Y., & Lin, S.-Y. (2017). Can learning motivation predict learning achievement? A case study of a mobile game-based English learning approach. *Education and Information Technologies, 22*(5), 2159–2173. doi:10.100710639-016-9542-5

Tse, L. (2000). Student perceptions of foreign language study: A qualitative analysis of foreign language autobiographies. *Modern Language Journal, 84*(1), 69–84. doi:10.1111/0026-7902.00053

TTBJ. (Tsukuba Test-Battery of Japanese) (n.d.). Interpretation of TTBJ scores. Retrieved from http://ttbj-tsukuba.org/mt/p2_e.html

Tuttle, H. G. (2013). Improving students' modern language speaking skills through mobile learning. In Z. L. Berge & L. Y. Muilenburg (Eds.), *Handbook of mobile learning*. New York, NY: Routledge.

Tveist, A. K., & Mangen, A. (2014). A joker in the class: Teenage readers' attitudes and preferences to reading on different devices. *Library & Information Science Research, 36*(3-4), 179–184. doi:10.1016/j.lisr.2014.08.001

United Nations Educational, Scientific and Cultural Organization. (2012). Mobile learning and policies. Key issues to consider. Paris, France: UNESCO. Available at https://unesdoc.unesco.org/ark:/48223/pf0000217638

United Nations Educational, Scientific and Cultural Organization. (2013). The future of mobile learning. Implications for policy makers and planners. Paris, France: UNESCO. Available at https://unesdoc.unesco.org/ark:/48223/pf0000219637

Unrau, N., & Schlackman, J. (2006). Motivation and its relationship with reading achievement in an urban middle school. *The Journal of Educational Research, 100*(2), 81–101. doi:10.3200/JOER.100.2.81-101

Ushioda, E. & Dörnyei, Z. (2009). Motivation, language identities and the L2 self: A theoretical overview. In Z. Dörnyei & Ushioda, E. (Eds.), Motivation, language identity and the L2 self (pp. 1-8). Bristol, UK: Multilingual Matters.

Ushioda, E. (2011). Language learning motivation, self and identity: Current theoretical perspectives. *Computer Assisted Language Learning, 24*(3), 199–210. doi:10.1080/09588221.2010.538701

Ushioda, E. (2013). Motivation matters in mobile language learning: A brief commentary. *Language Learning & Technology, 17*(3), 1–5.

Van de Vyver, J. (2016). Mobile-assisted language learning community and culture in French-speaking Belgium: the teachers' perspective. In S. Papadima-Sophocleous, L. Bradley, & S. Thouësny (Eds.), CALL communities and culture – short papers from EUROCALL 2016, (pp. 450-455). Research-Publishing.net.

Van den Branden, K. (2006). *Task-based language education: From theory to practice.* Cambridge, UK: Cambridge University Press. doi:10.1017/CBO9780511667282

van Dijk, J. (2006). *The network society: social aspects of new media.* Thousand Oaks, CA: Sage.

Van Lier, L. (2004). *The ecology and semiotics of language learning. A sociocultural perspective.* Dordrecht, The Netherlands: Kluwer Academic. doi:10.1007/1-4020-7912-5

Van Lier, L. (2010). The ecology of language learning: Practice to theory, theory to practice. *Procedia: Social and Behavioral Sciences, 3,* 2–6. doi:10.1016/j.sbspro.2010.07.005

van't Hooft, M. (2016, März 30). Taking it to the street: Using QR codes to tell student-created (hi)stories on location. Retrieved from https://www.socialstudies.org/publications/socialeducation/march-april2013/taking_it_to_street

Vandergriff, I. (2016). *Second-language discourse in the digital world: Linguistic and social practices in and beyond the networked classroom.* Amsterdam, The Netherlands: John Benjamins. doi:10.1075/lllt.46

Venkatesh, V., & Bala, H. (2008). Technology acceptance model 3 and a research agenda on interventions. *Decision Sciences, 39*(2), 273–315. doi:10.1111/j.1540-5915.2008.00192.x

Venkatesh, V., & Davis, F. D. (2000). A theoretical extension of the technology acceptance model: Four longitudinal field studies. *Management Science, 46*(2), 186–204. doi:10.1287/mnsc.46.2.186.11926

Venkatesh, V., Morris, M. G., Davis, G. B., & Davis, F. D. (2003). User acceptance of information technology: Toward a unified. *Management Information Systems Quarterly, 27*(3), 425–478. doi:10.2307/30036540

Verhoeven, L., & Vermeer, A. (2002). Communicative competence and personality dimensions in first and second language learners. *Applied Psycholinguistics, 23*(3), 361–374. doi:10.1017/S014271640200303X

Vermulst, A. A., & Gerris, J. R. M. (2005). *QBF: Quick Big Five personality test manual.* Leeuwarden, The Netherlands: LDC Publications.

Viberg, O. & Grönlund, Å. (2012). Mobile Assisted Language Learning: A literature review. *Proceedings of 11th World Conference on Mobile and Contextual Learning, mLearn 2012* (pp. 1-8). Retrieved from http://du.diva-portal.org/smash/get/diva2:549644/REFERENCES01

Vinall, K. (2012). ¿Un legado histórico?: Symbolic competence and the construction of multiple histories. *Journal of Linguistics and Language Teaching, 4*(1), 102–123.

Vinall, K. (2016). "Got Llorona?": Teaching for the development of symbolic competence. *Journal of Linguistics and Language Teaching, 8*(1), 1–16.

Vinther, J., & Lauridsen, J. T. (2018). Getting to know the learner: Motivational drive, digital literacy, and preferred mode of learning. Paper presented at the XIXth International CALL Research Conference, Bruges, Belgium.

Vinther, J. (2011). Enhancing motivation with cultural narratives in computer-mediated communication. *Computer Assisted Language Learning*, *24*(4), 337–352. doi:10.1080/09588221.2011.572898

Vinther, J. (2012). Cognitive skills through CALL-enhanced teacher training. In F. Zhang (Ed.), *Computer-enhanced and mobile-assisted language learning: Emerging issues and trends*. Hershey, PA: IGI-Global. doi:10.4018/978-1-61350-065-1.ch008

Vlaams Ministerie Onderwijs en Vorming. (2018). *Statistisch Jaarboek 2017-2018*. Brussel. Retrieved from https://onderwijs.vlaanderen.be/nl/statistisch-jaarboek-van-het-vlaams-onderwijs-2017-2018

Vygotsky, L. (1978). *Mind in society: The development of higher psychological processes*. Cambridge, MA: Harvard University Press.

Wakefield, R., & Whitten, D. (2006). Mobile computing: A user study on hedonic/utilitarian mobile device usage. *European Journal of Information Systems*, *15*(3), 292–300. doi:10.1057/palgrave.ejis.3000619

Wang, H.-Y., Liu, G.-Z., & Hwang, G.-J. (2017). Integrating socio-cultural contexts and location-based systems for ubiquitous language learning in museums: A state of the art review of 2009-2014. *British Journal of Educational Technology*, *48*(2), 653-671. doi:10.1111/bjet.12424

Wang, Y., Han, X., & Yang, J. (2015). Revisiting the blended learning literature: Using a complex adaptive systems framework. *Journal of Educational Technology & Society*, *18*(2), 380–393.

Ware, P., & Hellmich, E. (2014). CALL in the K-12 context: Language learning outcomes and opportunities. *CALICO Journal*, *31*(2), 140–157. doi:10.11139/cj.31.2.140-157

Warschauer, M. (1996). Computer-assisted language learning: An introduction. In S. Fotos (Ed.), *Multimedia language teaching* (pp. 3–20). Tokyo, Japan: Logos International.

Warschauer, M. (2007). A teacher's place in the digital divide. *Yearbook of the National Society for the Study of Education*, *106*(2), 147–166. doi:10.1111/j.1744-7984.2007.00118.x

Warschauer, M., & Healey, D. (1998). Computers and language learning: An overview. *Language Teaching*, *31*(2), 57–71. doi:10.1017/S0261444800012970

Warschauer, M., & Liaw, M. (2011). Emerging technologies for autonomous language learning. *Studies in Self-Access Learning Journal*, *2*(3), 107–118.

Watson, J., & Murin, A. (2010). *Keeping pace with K-12 online learning. An Annual Review of Policy and Practice*. Evergreen Education Group.

Wehrli, S. (2008). Personality on social network sites: An application of the Five Factor Model. *ETH Zürich Sociology Working Paper 7*, 1–16.

Weldy, T. G. (2018). Traditional, blended, or online: Business student preferences and experience with different course formats. *E-Journal of Business Education & Scholarship of Teaching*, *12*(2), 55–62.

Wenger, E., McDermott, R., & Snyder, W. M. (2002). *Cultivating communities of practice: A guide to managing knowledge. Harvard.* Harvard Business School Press.

Werbach, K., & Hunter, D. (2012). *For the win: How game thinking can revolutionize your business*. Wharton Digital Press.

Wigfield, A., Guthrie, J., Tonks, S., & Perencevich, K. (2004). Children's motivation for reading: Domain specificity and instructional influences. *The Journal of Educational Research*, *97*(6), 299–309. doi:10.3200/JOER.97.6.299-310

Wik, P. (2011). The virtual language teacher: Models and applications for language learning using embodied conversational agents. [PhD Thesis]. KTH Royal Institute of Technology.

Wik, P., & Hjalmarsson, A. (2009). Embodied conversational agents in computer assisted language learning. *Speech Communication*, *51*(10), 1024–1037. doi:10.1016/j.specom.2009.05.006

Wilde, M., Bätz, K., Kovaleva, A., & Urhahne, D. (2009). Überprüfung einer Kurzskala intrinsischer Motivation (KIM) [Testing a short scale of intrinsic motivation]. *Zeitschrift für Didaktik der Naturwissenschaften*, *15*, 31–45.

Wiley, D. (2015). The MOOC misstep and the open education infrastructure. In C. J. Bonk, M. M. Lee, T. C. Reeves, & T. H. Reynolds (Eds.), MOOCs and Open Education around the world (pp. 43–51). New York, NY: Routledge. doi:10.4324/9781315751108-2

Wilson, J. (2013). Technology, pedagogy and promotion. How can we make the most of corpora and data-driven learning (DDL) in language learning and teaching? *The Higher Education Academy* [online]. Retrieved from https://www.heacademy.ac.uk/system/files/corpus_technology_pedagogy_promotion2.pdf

Witton-Davies, G. (2014). *The Study of Fluency and its Development in Monologue and Dialogue*. (Unpublished doctoral dissertation), Lancaster University, UK.

Wolf, M. (2016). *Tales of literacy for the 21st century*. Oxford, UK: Oxford University Press.

Wolf, M. (2018). *Reader, come home: The reading brain in a digital world*. New York, NY: Harper Collins.

Wolf, M., & Barzillai, M. (2009). The importance of deep reading. *Educational Leadership*, *66*(6), 32–37.

Wong, C.-H., Tan, G. W.-H., Loke, S.-P., & Ooi, K.-B. (2015). Adoption of mobile social networking sites for learning? *Online Information Review*, *39*(6), 762–778. doi:10.1108/OIR-05-2015-0152

Wood, L. C., & Reiners, T. (2015). Gamification. In M. Khosrow-Pour (Ed.), *Encyclopedia of Information Science and Technology* (3rd ed., pp. 3039–3047). Hershey, PA: IGI Global. doi:10.4018/978-1-4666-5888-2.ch297

Wössner, S. (2019). Immersives Fremdsprachenlernen [Immersive language learning]. *Computer+Unterricht*, *114*, 28–31.

Wrigglesworth, J. (2019). Using smartphones to extend interaction beyond the EFL classroom. *Computer Assisted Language Learning*, 1–22. doi:10.1080/09588221.2019.1569067

Wu, H.-K., Wen-Yu Lee, S., Chang, H.-Y., & Liang, J.-C. (2013). Current status, opportunities and challenges of augmented reality in education. *Computers & Education*, *62*, 41–49. doi:10.1016/j.compedu.2012.10.024

Wulandari, M. (2019). Improving EFL learners' speaking proficiency through IG Vlog. *LLT Journal: A Journal on Language and Language Teaching*, *22*(1), 111–125.

Wu, S. L., & Ortega, L. (2013). Measuring global oral proficiency in SLA research: A new elicited imitation test of L2 Chinese. *Foreign Language Annals*, *46*(4), 680–704. doi:10.1111/flan.12063

Yadegarfar, H., & Simin, S. (2016). Effects of using Instagram on learning grammatical accuracy of word classes among Iranian undergraduate TEFL students. *International Journal of Research Studies in Educational Technology*, *5*(2), 49–60. doi:10.5861/ijrset.2016.1572

Yang, J. (2013). Mobile-Assisted Language Learning: Review of the recent applications of emerging mobile technologies. *English Language Teaching*, *6*(7), 19-25. doi:10.5539/elt.v6n7p19

Yang, S. (2013). Understanding undergraduate students' adoption of mobile learning model: A perspective of the extended UTAUT2. *Journal of Convergence Information Technology*, *8*(10), 969–979. doi:10.4156/jcit.vol8.issue10.118

Yao, Z. (2018). Research on flipped classroom teaching mode based on MOODLE. In *Proceedings of the 3rd International Conference on Education & Education Research* (EDUER 2018), 656-660.

Ybarra, R. & Green T. (2003). Using technology to help ESL/EFL students develop language skills. The Internet TESL Journal, *9*(3).

Yin, C., Song, Y., Tabata, Y., Ogata, H., & Hwang, G.-J. (2013). Developing and implementing a framework of participatory simulation for mobile learning using scaffolding. *Educ. Technol. Soc., 16*, 137–150.

Yuan, L. & Powell, S. (2013). MOOCs and open education: Implications for higher education- A white paper. *JISC CETIS* 1-21. Retrieved from https://publications.cetis.org.uk/wp-content/uploads/2013/03/MOOCs-and-Open-Education.pdf

Yükselir, C. (2017). A Meta-synthesis of qualitative research about Mobile-Assisted Language Learning (MALL) in foreign language teaching. *Arab World English Journal, 8*(3), 302-318. doi:10.24093/awej/vol8no3.20

Zahn, C., Schaeffeler, N., Giel, K. E., Wessel, D., Thiel, A., Zipfel, S., & Hesse, F. W. (2014). Video clips for YouTube: Collaborative video creation as an educational concept for knowledge acquisition and attitude change related to obesity stigmatization. *Education and Information Technologies, 19*(3), 603–621. doi:10.100710639-013-9277-5

Zarrinabadi, N., & Tanbakooei, N. (2016). Willingness to Communicate: Rise, Development, and Some Future Directions. *Language and Linguistics Compass, 10*(1), 30–45. doi:10.1111/lnc3.12176

Zeng, S. (2018). *English learning in the digital age-Agency, technology, and context.* Singapore: Springer. doi:10.1007/978-981-13-2499-4

Zheng, B., & Warschauer, M. (2015). Participation, interaction, and academic achievement in an online discussion environment. *Computers & Education, 84*, 78–89. doi:10.1016/j.compedu.2015.01.008

Ziegler, N., Meurers, D., Rebuschat, P., Ruiz, S., Moreno-Vega, J.-L., Chinkina, M., Li, W., & Grey, S. (2017). Interdisciplinary research at the intersection of CALL, NLP, and SLA: Methodological implications from an input enhancement project. Special Issue: *Currents in language learning series: Experimental, computational, and corpus-based approaches to language learning: Evidence and interpretation*, 209-231.

Zimmerman, B. J. (1989). A social cognitive view of self-regulated academic learning. *J. Educ. Psychol., 81*(3), 329–339. doi:10.1037/0022-0663.81.3.329

Zimmerman, B. J. (2000). Attaining self-regulation: A social cognitive perspective. In M. Boekaerts, M. Zeidner, & P. Pintrich (Eds.), *Handbook of self-regulation* (pp. 13–39). San Diego, CA: Academic Press. doi:10.1016/B978-012109890-2/50031-7

Zourou, K., Potolia, A., & Zourou, F. (2017). Informal social networking sites for language learning: Insights into autonomy stances. In T. Lewis, A. Rivens Mompean, & T. Cappellini (Eds.), CALICO monograph "Learner autonomy and Web 2.0", 141–167.

Zucker, A. A., & Light, D. (2009). Laptop programs for students. *Science, 323*(5910), 82–85. doi:10.1126cience.1167705 PMID:19119224

About the Contributors

Alberto Andujar is lecturer at the University of Almería (Spain). He specializes into second language learning and technology, computer-assisted language and mobile-assisted language learning. He is also author of different international research articles in these fields and is currently developing projects into the use of virtual reality for second language learning.

* * *

Pablo Alcaraz-Valencia is a professor of the Universidad de Colima, who specializes in technologies for supporting second language learning. He has more than 5 years of experience as a developer.

Rocio Maciel Arellano is a Full Time Professor at the Universidad de Guadalajara. She is appointed to the System Information Department. She also collaborates with the Smart Cities Lab.

Androulla Athanasiou is an English Language Instructor at the Language Centre of the Cyprus University of Technology (CUT). Androulla holds a BA in English Literature and Classical Studies from Keele University (1999), an MA in English and Language Studies and Methods from Warwick University (2000), an MA in Educational Leadership from the European University Cyprus (2012) and a PhD in English Language Teaching from Warwick University (2005). During her teaching experience (12 years), she has engaged in course development and material design, both at an undergraduate and a postgraduate level. Apart from the ESAP (English for Specific Academic Purposes) courses, she also teaches on the MA in CALL. Dr Athanasiou's research interest lie in material design, use of technology in language teaching/learning, Mobile-Assisted Language Learning, learner autonomy, collaborative learning, the CEFR.

Josef Buchner Scientific assistant and lecturer for media didactics, PhD student at the University of Salzburg. Founder of Flipped Classroom Austria (flipped-classroom-austria.at).

Jack Burston holds the position of Honorary Research Fellow in the Language Centre of the Cyprus University of Technology. He is a language-teaching specialist with a formal background in theoretical and applied linguistics, second language acquisition and testing. His current research is focused on Mobile-Assisted Language Learning (MALL) and advanced-level foreign language instruction. Jack is a current member of the Editorial Board of the ReCALL Journal and Language Learning & Technology Journal. He also served for many years on the Editorial Board of the CALICO Journal and was the Software Review Editor of the CALICO Journal for 13 years.

Fidel Çakmak is assistant professor in the Department of Foreign Language Education at Alanya Alaaddin Keykubat University, Antalya, Turkey. She is interested in research topics such as learning and teaching EFL in the digital age, mobile assisted language learning, social CALL, learner analytics, and 21st-century teacher education.

Julie Damron received her B.A. from Brigham Young University in English, her M.A.T. in TESOL, from The School for International Teaching in Brattleboro, Vermont, and her Ph.D. in Linguistics from Purdue University, in 2000. She is an Associate Professor and Associate Section Head of Korean in the Department of Asian and Near Eastern Languages. She oversees all TAs in the Korean section, directs internships and direct enrollment and oversees the foreign language house. Dr. Damron developed and teaches all of BYU's independent study Korean courses and BYU's new "BYU Online" hybrid Korean classes. She is active in the profession with yearly or semi-yearly presentations and publications including her recent book, Korean Stories for Language Learners (2018). Dr. Damron serves as a reviewer for highly respected journals and organizations such as Foreign Language Annals, Fulbright-Hays, Routledge Publishers.

Atsushi Fukada is Professor of Japanese and Linguistics in the School of Languages and Cultures and the Director of the Center for Technology-Enhanced Language Learning at Purdue University. His research interests include pragmatics, computational linguistics, language pedagogy, and CALL.

Barbara Conde Gafaro is a PhD candidate studying MOOCs and foreign language learning at The Open University. She graduated from the MA in English Language Teaching and Applied Linguistics at Coventry University, with distinction in 2017. She is also a lecturer in Spanish as a foreign language at Coventry University. She has published articles based on her work and presented at national and international conferences.

Sara Hernández Gallardo is a Full Time Professor at the Universidad de Guadalajara. She is appointed to the System Information Department.

Laura Gaytán-Lugo is an Assistant Professor at the Universidad de Colima in Mexico, appointed at the School of Mechanical and Electrical Engineering. She focuses on Human-Computer Interaction. Her research interests include serious games for education, children-computer interaction, user experience design, social computing and evaluation methods. She holds a Ph.D. in Information Technologies from the Universidad de Guadalajara and is a member of Mexico's National System of Researchers. She serves as research vice chair of the ACM SIGCHI Latin America HCI Community and as a SIGCHI Communication Ambassador. She is a liaison for the ACM SIGCHI Across Borders Initiative (SABI). Finally, she is part of the HCI Across Borders (HCIxB) community.

Ana Gimeno is a Full Professor of English Language in the Department of Applied Linguistics at the Universitat Politècnica de València, Spain. She is Head of the CAMILLE Research Group and is President of WorldCALL. She is also editor-in-chief of EuroCALL's online scientific journal, The EuroCALL Review and Associate Editor of ReCALL (CUP).

Maki Hirotani is Professor of Japanese in the Department of Humanities, Social Sciences, and the Arts at Rose-Hulman Institute of Technology. Her research interests include CALL, second language learning in speaking and writing, fluency, computer-mediated-communication, and telecollaboration.

Vanessa Joosen specializes in English literature and children's literature studies at the Faculty of Arts (University of Antwerp). Her research interests include the construction of childhood, adulthood and old age in literature, and digital approaches to children's literature studies.

Jørgen Lauridsen is a professor in health economics with an expertise in statistical and quantitative analyses, and he contributed to the project with confirmatory factor analyses. He holds a PhD in Econometrics and has published a large number of articles in different disciplines with emphasis on quantitative methods and their applications. For further information, see www.sdu.dk/ansat/jtl.aspx.

Alicia Galera Masegosa is a doctor in English Linguistics at the University of Almería. Her primary research interests include cognitively-oriented constructionist approaches to language, as well as conceptual theory of metaphor, metonymy and other cognitive operations. Other interests include teaching English as a Second Language and English for Specific Purposes.

Kazumi Matsumoto is Associate Professor of Japanese in the Department of Modern Languages and Classics at Ball State University. Her research interests include second language writing, fluency and Kanji recognition. She is actively involved in community engagement to promote Japanese teaching and learning.

Fanny Meunier is Professor of English language, linguistics and didactics at UCLouvain, Belgium. She has been involved in learner corpus research for over 25 years and her main research interest is the link between second language acquisition (SLA) studies and pedagogical applications. She is also actively involved in pre- and in-service teacher training and is collaborating to several international research projects on, among other aspects, bi- and multi-literacies and digital literacies.

Alice Meurice is a research assistant in modern language didactics and English for specific purposes teacher at UCLouvain (Belgium). Her main research area is foreign language teaching and, more specifically, the integration of information and communication technologies in the foreign language learning classroom and curriculum. Other research interests include the use of OERs in language learning in primary and secondary education.

Valentina Morgana is Adjunct Professor of English Linguistics in the Faculty of Language Sciences and Foreign Literature at the Università Cattolica del Sacro Cuore in Milan, Italy. Her research includes mobile language teaching and learning in secondary and higher education, technology-mediated task-based language teaching and the use of virtual exchanges in higher education internationalization.

Ailbhe Ní Chasaide is Professor of Phonetics and Director of the Phonetics and Speech Lab. at the School of Linguistic, Speech and Communication Sciences, Trinity College, Dublin. She has directed over 20 funded research projects, on voice analysis and the role of voice in communication, on the prosody of Irish, and on the development of speech technology and linguistic resources for Irish.

Neasa Ní Chiaráin is Ussher Assistant Professor in Irish Speech and Language Technology at the School of Linguistic, Speech and Communication Sciences, Trinity College, Dublin. She is working on the development of Irish language synthetic voices as part of the ABAIR initiative (see www.abair.ie). As well as working on the development of the underpinning linguistic resources needed for synthesis, her research focuses on exploring the application of speech technology in a CALL context, particularly for the teaching/learning of Irish.

Maria Del Mar Sanchez Perez is a Lecturer at the Department of Philology of University of Almería (Spain). Graduated in Translation and Interpreting at the University of Granada (Spain), she completed her PhD in 2016 at the University of Almeria. From 2011-2018, she held the position of Responsible for the Campus of International Excellence at this university and cooperates actively in the management of the Plurilingualism Promotion Plan. She has participated in different research projects on multilingualism English Language Teaching and has attended plenty of conferences on the matter both in Europe and the rest of the world.

Jennifer Quinlan is the manager of language curriculum for the Missionary Training Center, a partner organization of Brigham Young University. She received her M.A in French Studies from University of Wisconsin-Madison and her PhD in Instructional Psychology and Technology with an emphasis in Second Language Acquisition from Brigham Young University. Her expertise is in second language acquisition in blended and online forums and the role of technology in education.

Jose M. Franco Rodríguez is Professor of Spanish at Fayetteville State University, North Carolina. His research spans from Spanish Golden-Age dialogues, to ethnolinguistic vitality in linguistic landscapes, to virtual exchanges in second language acquisition.

Mathea Simons is associate professor and teacher educator at the Antwerp School of Education (University of Antwerp), more specifically in the field of language teaching. Her research interests are in teacher education and foreign language education.

Julie Van de Vyver is a PhD student and teaching assistant in English linguistics and didactics at UCLouvain, Belgium. Her main research area is foreign language teaching and, more specifically, the integration of mobile technologies in the foreign language learning classroom and curriculum. Other research interests include the use of OERs on language learning in secondary education.

Claudio Vanhees is a language teacher and a doctoral candidate at the Antwerp School of Education (University of Antwerp). His research focuses specifically on the role of technology in the reading process of adolescent learners and its impact on reading motivation and immersion.

Jane Vinther holds a PhD in computer-assisted language learning and second language acquisition. She has extensive teaching and research experience in CALL, second-language acquisition, and cognitive processes of learning. Her special interests lie in motivation, grammar, language and culture. For further information, see www.sdu.dk/ansat/jvinther.aspx.

Julia Weissenboeck, MSc, BA, works as English teacher for lower- and upper Secondary at a school in Salzburg, Austria. She also works at the University of Salzburg where she teaches EFL classes. Her research focuses on digital tools in EFL teaching with a special focus on mobile learning, flipped learning and the 21st Century Skills. Currently she is writing her phD thesis about speaking vlogs.

Index

H

I

K

L

M

N

O

P

R

S

T

V

W

www.ingramcontent.com/pod-product-compliance
Lightning Source LLC
LaVergne TN
LVHW060301011026
849161LV00011B/2755